T0211579

Lecture Notes
in Business Information Processing   **281**

More information about this series at http://www.springer.com/series/7911

Marlon Dumas · Marcelo Fantinato (Eds.)

Business Process Management Workshops

BPM 2016 International Workshops
Rio de Janeiro, Brazil, September 19, 2016
Revised Papers

 Springer

Editors
Marlon Dumas
Institute of Computer Science
University of Tartu
Tartu
Estonia

Marcelo Fantinato
School of Arts, Sciences and Humanities
University of São Paulo
Sao Paulo, São Paulo
Brazil

ISSN 1865-1348 ISSN 1865-1356 (electronic)
Lecture Notes in Business Information Processing
ISBN 978-3-319-58456-0 ISBN 978-3-319-58457-7 (eBook)
DOI 10.1007/978-3-319-58457-7

Library of Congress Control Number: 2017939546

Printed on acid-free paper

This Springer imprint is published by Springer Nature
The registered company is Springer International Publishing AG
The registered company address is: Gewerbestrasse 11, 6330 Cham, Switzerland

Preface

After decades of development, business process management (BPM) is currently a highly established discipline with a rich body of techniques, methods, and tools. It is also a mature yet highly dynamic field of research, which brings together scholars and industrial researchers from the computer science, information systems, and management fields.

The International Conference Series on Business Process Management had its 14th edition in Rio de Janeiro, Brazil, during September 18–22, 2016. As per its traditions, the three-day main conference was preceded by a full day packed with workshops on a range of BPM-related topics. The collective goal of these workshops is to promote work-in-progress research that has not yet reached its full maturity but has a clear goal and approach, and promising insights or preliminary results.

In 2016, the following ten workshops were hosted:

- BPI 2016 – 12th International Workshop on Business Process Intelligence
- BPMO 2016 – First Workshop on Workshop on Business Process Management and Ontologies
- BPMS2 2016 – 9th Workshop on Social and Human Aspects of Business Process Management
- DeMiMoP 2016 – 4th International Workshop on Decision Mining and Modeling for Business Processes
- IWPE 2016 – Second International Workshop on Process Engineering
- PQ 2016 – First International Workshop on Process Querying
- ReMa 2016 – First Workshop on Resource Management in Business Processes
- PRAISE 2016 – First International Workshop on Runtime Analysis of Process-Aware Information Systems
- SABPM 2016 – First International Workshop on Sustainability-Aware Business Process Management
- TAProViz 2016 – 5th International Workshop on Theory and Application of Visualizations and Human-Centric Aspects in Processes

Some of these workshops are well established, having already been held for several years. They address long-standing challenges such as how to seamlessly incorporate social and human aspects into process improvement and innovation methods, or how to leverage data analytics techniques for process and decision management. More than half of the workshops, however, were held for the first time this year. These latter workshops address emerging concerns such as sustainability, or tap into emerging opportunities raised, for example, by the ongoing convergence of runtime analysis and predictive business process monitoring.

We would like to express our gratitude to the individual workshop chairs for their tremendous effort to promote, coordinate, and animate their respective workshops. We are grateful to the service of the countless reviewers, who supported the workshop chairs and provided valuable feedback to the authors. Several workshops had invited keynote presentations that framed the presented research papers. We would like to thank the keynote speakers for sharing their insights. Finally, we would like to thank Ralf Gerstner and the team at Springer for their support in the publication of this LNBIP volume.

December 2016

Marcelo Fantinato
Marlon Dumas

Contents

2nd International Workshop on Process Engineering (IWPE 2016)

First International Workshop on Process Querying (PQ 2016)

**5th International Workshop on Theory and Application of Visualizations
and Human-centric Aspects in Processes**

12th International Workshop on Business Process Intelligence (BPI 2016)

Introduction to the 12th International Workshop on Business Process Intelligence (BPI 2016)

Boudewijn F. van Dongen[1]([⊠]), Jochen De Weerdt[2], Andrea Burattin[3], and Jan Claes[4]

[1] Eindhoven University of Technology, Eindhoven, The Netherlands
b.f.v.dongen@tue.nl
[2] KU Leuven, Leuven, Belgium
jochen.deweerdt@kuleuven.be
[3] University of Innsbruck, Innsbruck, Austria
andrea.burattin@uibk.ac.at
[4] Ghent University, Ghent, Belgium
Jan.Claes@UGent.be

1 Aims and Scope

Business Process Intelligence (BPI) is a growing area both in industry and academia. BPI refers to the application of data- and process-mining techniques to the field of Business Process Management. In practice, BPI is embodied in tools for managing process execution by offering several features such as analysis, prediction, monitoring, control, and optimization.

The main goal of this workshop is to promote the use and development of new techniques to support the analysis of business processes based on runtime data about the past executions of such processes. We aim at bringing together practitioners and researchers from different communities, e.g. Business Process Management, Information Systems, Database Systems, Business Administration, Software Engineering, Artificial Intelligence, and Data Mining, who share an interest in the analysis and optimization of business processes and process-aware information systems. The workshop aims at discussing the current state of research and sharing practical experiences, exchanging ideas and setting up future research directions that better respond to real needs. In a nutshell, it serves as a forum for shaping the BPI area.

The 12th edition of this workshop attracted 12 international submissions. Each paper was reviewed by at least three members of the Program Committee. From these submissions, the top six were accepted as full papers for presentation at the workshop. The papers presented at the workshop provide a mix of novel research ideas, evaluations of existing process mining techniques, as well as new tool support. *Ackermann, Schönig* and *Jablonski* present an approach capable of converting DPIL multi-perspective declarative constraints into a logic language, called Alloy, and to simulate them. *De Smedt, Di Ciccio, Mendling* and *Vanthienen* focus on model checking of mixed-paradigm process models (Declare and Petri Net) in the context of

FusionMinerFul. *Rehse* and *Fettke* introduce an approach to learn a hierarchy of reference models from broad but general to detailed but narrow. *Vogelgesang, Rinderle-Ma* and *Appelrath* devise a framework for multidimensional process mining, which provides more interactivity to the analysis. *De Koninck* and *De Weerdt* investigate a novel multi-objective trace clustering approach. Finally, *Ferreira* and *Santos* discuss how often-used abstraction of event logs (i.e., the calculation of direct-follows relation) can be distributed on multiple cores or GPUs.

For the first time this year, all authors were allotted 20 minutes to present their work, followed by 20 minutes of discussion. The discussions this year were very fruitful and many points for future collaboration within the community were identified.

As has become tradition, this year's BPI workshop was accompanied by the BPI Challenge. Due to the sponsorship of GradientECM and their tool Minit, the winners of the BPI challenge were present at the workshop and presented their analysis results in detail.

For the first time, the process discovery contest was co-organized at BPI. This contest was sponsored by Celonis and the winner of this contest was also given time to present their results.

As with previous editions of the workshop, we hope that the reader will find this selection of papers useful to keep track of the latest advances in the BPI area, and we are looking forward to keep bringing new advances in future editions of the BPI workshop. As organizers, we look back at a very succesful workshop and we are looking forward to next year's edition.

Acknowledgement. The organizers thank the PC members for their valuable comments and active participation!

2 Program Committee

Wil van der Aalst	Eindhoven University of Technology, The Netherlands
Andrea Burattin	University of Innsbruck, Austria
Artur Caetano	IST, University of Lisbon, Portugal
Jan Claes	Ghent University, Belgium
Jochen De Weerdt	KU Leuven, Belgium
Boudewijn van Dongen	Eindhoven University of Technology, The Netherlands
Diogo R. Ferreira	IST, University of Lisbon, Portugal
Walid Gaaloul	Computer Science Department Télécom SudParis, France
Gianluigi Greco	University of Calabria, Italy
Daniela Grigori	Laboratoire LAMSADE, University Paris-Dauphine, France
Antonella Guzzo	University of Calabria, Italy
Anna Kalenkova	Higher School of Economics, Russia
Michael Leyer	University of Rostock, Germany
Ana Karla de Medeiros	Achmea, The Netherlands
Jan Mendling	Wirtschaftsuniversität Wien, Austria
Jorge Munoz-Gama	Pontificia Universidad Católica de Chile, Chile

Viara Popova	University of Tartu, Estonia
Manfred Reichert	University of Ulm, Germany
Andreas Rogge-Solti	Vienna University of Economics and Business, Austria
Michael Rosemann	Queensland University of Technology, Australia
Anne Rozinat	Fluxicon, The Netherlands
Domenico Sacca	University of Calabria, Italy
Pnina Soffer	University of Haifa, Israel
Alessandro Sperduti	University of Padua, Italy
Suriadi Suriadi	Massey University, Australia
Seppe Vanden Broucke	KU Leuven, Belgium
Eric Verbeek	Eindhoven University of Technology, The Netherlands
Hans Weigand	Tilburg University, The Netherlands

Summary of the Business Process Intelligence Challenge (BPI Challenge 2016)

Marcus Dees[1,2] and Boudewijn F. van Dongen[1]

[1] Eindhoven University of Technology, Eindhoven, The Netherlands
{b.f.v.dongen,m.dees}@tue.nl
[2] UWV, Amsterdam, Netherlands
marcus.dees@uwv.nl

1 Background

Since 2011, the IEEE Task Force on process mining organizes a yearly Business Process Intelligence Challenge, or BPI Challenge. The goal of this challenge is to bring together practitioners and researchers in the field to show the direct impact of academic work when facing the challenges real-life cases bring. For the BPI challenge, we provide participants with a real-life event log, and we ask them to analyze these data using whatever techniques available, focusing on one or more of the process owner's questions or proving other unique insights into the process captured in the event log.

For 2016, the data was provided by the UWV (Employee Insurance Agency) and the challenge was hosted at the BPI Workshop in Rio de Janeiro, on September 18, 2016. The challenge was sponsored by Gradient ACM, and their tool Minit v.2.

2 Case

The UWV (Employee Insurance Agency) is a Dutch autonomous administrative authority (ZBO) and is commissioned by the Ministry of Social Affairs and Employment (SZW) to implement employee insurances and provide labour market and data services in the Netherlands.

The Dutch employee insurances are provided for via laws such as the Unemployment Insurance Act (WW). The data in this year's data collection pertains to customer contacts over a period of 8 months and UWV is looking for insights into their customers' journeys. The data is focused on customers in the unemployment benefits (WW) process.

Data has been collected from several different sources, namely:

1. Clickdata from the site www.werk.nl of visitors that were not logged in,
2. Clickdata from the customer specific part of the site www.werk.nl,
3. Contact data, showing when customers sent messages to UWV through a digital channel called "werkmap",
4. Call data from the call center, showing when customers contacted the call center by phone, and
5. Complaint data showing when customers complained.

UWV was interested in insights on how their channels are being used, when customers move from one contact channel to the next and why and if there are clear customer profiles to be identified in the behavioural data. Furthermore, recommendations are sought on how to serve customers without the need to change the contact channel. The full dataset is available from https://data.4tu.nl/repository/uuid:360795c8-1dd6-4a5b-a443-185001076eab [1] and further information is available at http://www.win.tue.nl/bpi/doku.php?id=2016:challenge including various documents detailing the information in the data.

3 Sponsors

This year's challenge was sponsored by GRADIENT ECM (http://www.gradientecm.com/). They not only provided free Minit (http://www.minitlabs.com/) licenses for participants, but they also allowed for two selected winners to come to Rio de Janeiro to present their work at the 12th international BPI Workshop held there and to receive the award during the dinner of the International Conference on Business Process Management (BPM2016).

4 The Results

We received several submissions from all over the world, both from academia and from industry. The jury was pleased with the high quality of the contributions in general, but in the end, two submissions were selected as the best one to present their work in Rio:

- Ube van der Ham, with his submission entitled "Marking up the right tree: understanding the customer process at UWV", showing that by manually inspecting the data and using relatively standard data analysis tools many insights can be obtained, and
- Sharam Dadashnia, Tim Niesen, Philip Hake, Peter Fettke, Nijat Mehdiyev and Joerg Evermann, with their submission entitled: "dentification of Distinct Usage Patterns and Prediction of Customer Behavior" showing an innovative technique to predict the next action undertaken by users on the basis of the preceding ten tasks.

Representatives of both submissions came to Rio and presented their analysis to the BPI audience. Their presentations were well-received and showed true professional value and direct applicability of academic research in the field of business process intelligence.

Acknowledgement. The organizers thank the jury members for their valuable comments and active participation!

Reference

1. Dees, M., van Dongen, B.: Bpi challenge 2016 (2016). https://doi.org/10.4121/uuid:360795c8-1dd6-4a5b-a443-185001076eab

Summary of the Process Discovery Contest 2016

Josep Carmona[1], Massimiliano de Leoni[2], Benoît Depaire[3], and Toon Jouck[3]

[1] Universitat Politècnica de Catalunya (UPC), Barcelona, Spain
jcarmona@cs.upc.edu
[2] Eindhoven University of Technology, Eindhoven, The Netherlands
m.d.leoni@tue.nl
[3] Hasselt University, Hasselt, Belgium
{benoit.depaire,toon.jouck}@uhasselt.be

1 Background

Process Mining is a relatively young research discipline that aims to discover, monitor and improve processes based on real facts (and not assumptions) by extracting knowledge from event logs readily available in today's (information) systems [1]. The lion's share of attention of Process Mining has been devoted to Process Discovery, namely extracting process models - mainly business process models - from an event log.

In the last decade, several new techniques for process discovery have been put forward. Each technique has been evaluated on separate event data, thus making it difficult to perform a comparative evaluation. However, in light of a continuously growing of strength and interest in Process Mining as a discipline, it becomes crucial to finally foster a comparison of existing discovery techniques. With this need at hand, we organized the first edition of the Process-Discovery contest, which was co-located with the BPM-2016 Conference in Rio de Janeiro (Brazil).

2 Objectives and Context

The Process Discovery Contest aims to compare the efficiency of techniques for what concerns discovering process models that provide a proper balance between *overfitting* and *underfitting*. A process model is overfitting (the event log) if it is too restrictive, disallowing behavior which is part of the underlying process. This typically occurs when the model only allows for the behavior recorded in the event log. Conversely, it is underfitting (the reality) if it is not restrictive enough, allowing behavior which is not part of the underlying process. This typically occurs if it overgeneralizes the example behavior in the event log. Interested readers are referred to [2] (e.g., Sect. 6.4.3).

The starting point was 10 different *reference* process models that were randomly generated and contained the most typical process-model constructs. For each process model, in February 2016, a *training* event log with 1000 compliant traces was generated and made available to the contestants to be used as input for their techniques to discover the underlying process model. Clearly, the ideal situation was that the reference process model was rediscovered. In fact, in many situations, the reference model was different from the models discovered by contestants.

But, how can one measure how far are the discovered models from the reference models? Since we do not want to give preference to any modeling notation, we could not leverage on existing measures of overfitting and underfitting, which are notation dependent. Therefore, we used a classification perspective to evaluate the quality of a discovered model. For each reference model, we generated a *test* event log containing 20 traces, out of which 10 were compliant and 10 were not with respect to the reference model. A model is good in balancing overfitting and underfitting if it is able to correctly classify the traces in the test event log: Given a trace representing real process behavior, the model should classify it as allowed; Given a trace representing a behavior not related to the process, the model should classify it as disallowed. *With a classification view, the winner is the group that can correctly classify the largest number of traces in all the test event logs. All event logs will have the same weight.*

It is also worth mentioning that two *calibration* event logs were shared on 15 April and 15 May 2016. The *calibration* logs had the same structure as the test logs, namely 10 compliant and 10 non-compliant traces. However, we did not disclose which traces were (not) compliant. The contestants could submit their classification attempt and we replied stating how many traces were correctly classified. The feedback loops were intended to support participants with assessing the algorithm effectiveness and, consequently, with adjusting their techniques. In fact, as discussed below, the best performing groups profited from the calibration event log. Further information is available at http://www.win.tue.nl/ieeetfpm/doku.php?id=shared:edition_2016.

3   Report on the Results

The result and the winner group were announced on September, 18th, 2016 during the BPI workshop, co-located with the BPM-2016 conference. The winner group was also given a chance to present the work during the workshop. The contest was successful and attracted 14 submissions from Europe, Australia and Asia. The remainder of this section reports on the three groups that scored the best.

The winner was the group composed by **H.M.W. Verbeek** and **F. Mannhardt**, from Eindhoven University of Technology (The Netherlands), with the so-called *DrFurby Classifier* [3]. For each model, it takes the training log and a test log and classifies every trace in the test log whether it matches the training log (positive trace) or not (negative trace). To reduce the number of misclassifications, the DrFurby Classifier uses a combination of two orthogonal approaches. To reduce the number of false negatives (i.e. compliant traces classified as non-compliant), the DrFurby Classifier only uses process-discovery techniques that generates models that classify all training-log traces as compliant. Also, multiple techniques matching this criterion are combined to reduce the number of false positives (i.e. non-compliant traces classified as compliant). This means that, for each reference model, multiple models are discovered: A trace is classified as compliant if and only if the trace is compliant with all discovered models. In particular, Verbeek and Mannhardt employ two techniques that guarantee perfect fitness: the Inductive Miner with maximal decomposition and the Hybrid ILP

Miner with no decomposition. The choices fall on these two techniques because they provide the best classification on the calibration event logs. Their approach was able to correctly classify 193 out of 200 test-log traces (i.e., 20 traces for each of the 10 processes).

We want to give special mention to two runner-ups: Their approaches could correctly classify 192 traces, namely just one trace less than the winner. The first runner-up is **Raji Ghawi**, from American University of Beirut (Lebanon) [4]. For five models, dr. Ghawi employed the Inductive Miner for five processes and the ILP Miner with maximum decomposition for the other five. Similarly to the winner, the choice whether to opt for Inductive Miner or ILP Miner with maximum decomposition for a specific process was driven by the outcomes obtained on the calibration event logs in April and May. Interesting enough, the winner and one runner-up have obtained very good results by employing decomposition. However, the additional trace correctly classified by the winner group, which made the difference, was due to the employment of two discovery techniques to reduce the false positives. The second runner-up was the group composed by **Moshe Steiner** and **Liat Bodaker** under the supervision of **Arik Senderovich**, from Technion–Israel Institute of Technology (Israel) [5]. The approach is based on the Alpha+ algorithm and is used by all 10 models. To overcome the limitations of Alpha+, the mined models are improved/repaired, based on the log footprints. Last but not least, some models were subsequently improved in an ad-hoc fashion. This submission is worth of interest because it tries to overcome the limitation of Alpha+; but, on the other hand, many adjustments are rather ad-hoc and, hence, they are not generally applicable.

It is worth concluding by mentioning that two groups submitted approaches based on Recurrent Neural Networks (by N. Tax and N. Sidorova, Eindhoven University of Technology, The Netherlands) and on Bayesian Networks (by B. Blaskovic, University of Zagreb, Croatia). Although their approaches do not provide traditional process models, they are perfectly legitimate in consideration of the classification nature of the contest. Looking at these submissions, pure data-mining techniques seem to be outperformed by process-mining techniques. This is yet another case that illustrates the importance of process mining how it differs from data mining: Process mining promotes time- and sequence-related information as first-class citizens.

Given the large success, we plan to repeat the experience at BPM 2017 in Barcelona. We plan to improve the contest in the light of several valuable comments received during the BPM-2016 conference.

Acknowledgement. The organizers want to thank all contestants that made an invaluable effort to participate. Special mention goes to the winner and the runner-up groups to be willing to prepare detailed technical reports, which are cited in this summary.

References

1. van der Aalst, W.M.P., et al.: Process mining manifesto. In: Daniel, F., et al. (eds.) BPM 2011 Workshops, Part I. LNBIP, vol. 99, pp. 169–194. Springer, Berlin (2012)

2. van der Aalst, W.M.P.: Data science in action. In: Process Mining, 2nd edn. Springer, Heidelberg (2016)
3. Verbeek, H., Mannhardt, F.: The DrFurby classifier submission to the process discovery contest @ BPM 2016. Technical report BPM center report BPM-16-08, bpmcenter.org (2016)
4. Ghawi, R.: Submission to the process discovery contest @ BPM2016. Technical report (2016). arXiv:1610.07989
5. Shteiner, M., Bodaker, L., Senderovich, A.: Process discovery contest 2016: Heuristic Alpha+ Miner (HAM). Faculty of Industrial Engineering and Management Report Number IE/IS-2016-02, Technion–Israel Institute of Technology (Israel) (2016)

Mining Reference Process Models from Large Instance Data

Jana-Rebecca Rehse[(✉)] and Peter Fettke

Institute for Information Systems (IWi),
German Research Center for Artificial Intelligence (DFKI GmbH)
and Saarland University, Campus D3 2, 66123 Saarbruecken, Germany
{Jana-Rebecca.Rehse,Peter.Fettke}@iwi.dfki.de

Abstract. Reference models provide generic blueprints of process models that are common in a certain industry. When designing a reference model, stakeholders have to cope with the so-called 'dilemma of reference modeling', viz., balancing generality against market specificity. In principle, the more details a reference model contains, the fewer situations it applies to. To overcome this dilemma, the contribution at hand presents a novel approach to mining a reference model hierarchy from large instance-level data such as execution logs. It combines an execution-semantic technique for reference model development with a hierarchical-agglomerative cluster analysis and ideas from Process Mining. The result is a reference model hierarchy, where the lower a model is located, the smaller its scope, and the higher its level of detail. The approach is implemented as proof-of-concept and applied in an extensive case study, using the data from the 2015 BPI Challenge.

Keywords: Reference Model Mining · Dilemma of reference modeling · Reference model hierarchy · Inductive reference model development · Trace clustering

1 Introduction

Reference models are templates for process models that are common in a certain industry. By providing a generic blueprint of a process model, they facilitate a resource-efficient implementation of the respective process and its adaption to the individual needs of an organization. This way, companies may benefit from best practices and industry-specific experience. The use of reference models is associated with a higher quality of processes and process models, as it simplifies internal communications by introducing a common terminology and considerably reduces the resources required for business process management [1].

When designing a reference model, stakeholders have to cope with the so-called 'dilemma of reference modeling', i.e. balancing a reference model's generality against its market specificity. In general, the more details a reference model contains, the smaller its scope of application. A very specific reference model may provide a good fit to the needs of an individual customer, however,

© Springer International Publishing AG 2017
M. Dumas and M. Fantinato (Eds.): BPM 2016 Workshops, LNBIP 281, pp. 11–22, 2017.
DOI: 10.1007/978-3-319-58457-7_1

its application potential and thus its sales market are rather small [2]. Finding the right degree of generality versus specificity depends on the intended scope, use, and market of a reference model and cannot be universally decided, however, its appropriate determination will influence its acceptance, assuming a reuse-oriented conceptualization of reference models [3].

In order to provide stakeholders with an appropriate basis for decision-making, the contribution at hand presents a novel approach to mining reference models from instance-level data. In order to determine an appropriate degree of reference model specificity, a given input log is hierarchically clustered and a separate reference model is mined for each cluster separately. This results in a hierarchically structured set of reference models, where the lower a model is located, the smaller its scope and the higher its level of detail.

The paper follows the design-science research paradigm [4]. The design of the approach is outlined in the following Sect. 2. Section 3 describes its demonstration as a proof-of-concept as well as an experimental evaluation. We report on Related Work in Sect. 4, before concluding the article with a discussion in Sect. 5.

2   Conceptual Design of the Approach

2.1   Main Idea and Outline

The objective of this contribution is to overcome the dilemma of reference modeling, i.e. the challenge that a very detailed reference model may only have a small sales market, but a rather generic model requires lots of adaptations to be of practical use. Hence, we introduce the concept of a reference model hierarchy.

Definition 1 (Reference Model Scope). *Given a reference model RM_i, its scope is defined as the set of activities it contains, i.e. $scope(RM_i) = \bigcup_{a \in RM_i} a$.*

Definition 2 (Reference Model Hierarchy). *A reference model hierarchy H is a connected, directed, acyclic graph $H = (RM, E)$ (i.e. a tree), where $RM = RM_1, ..., RM_n$ is a set of reference models and $E \subset RM \times RM$ is a set of edges connecting them, such that:*

- *$(RM_i, RM_j) \in E \Rightarrow scope(RM_j) \subseteq scope(RM_i)$, i.e. the scope of a reference model is fully contained in the scope of its parent model*
- *$\forall i : \bigcup_{(RM_i, RM_k) \in E} scope(RM_k) = scope(RM_i)$, i.e. the scope of a reference model is represented by the unified scopes of its children*

While it is possible to mine either a *type-based* or *instance-based* reference model hierarchy, we focus on using instance-level data in the following. Figure 1 shows the main idea of our approach for the automatic derivation of a reference model hierarchy. Note that while it looks similar to Fig. 3 in [5], it illustrates a bottom-up instead of a top-down approach. Also, our main objective is not to increase precision in Process Discovery, but to determine an appropriate scope for a reference model.

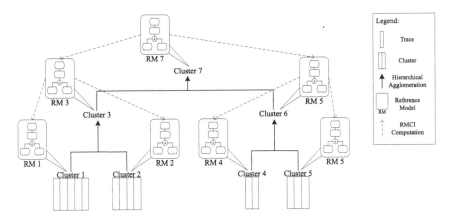

Fig. 1. Main idea of our approach for mining reference models

The provided data is hierarchically clustered, resulting in a tree structure, where the higher a cluster is located, the more and the more diverse traces it contains. For each cluster, a reference model is mined, constructing a model hierarchy in analogy to the cluster structure. The higher a reference model is located, the more traces it represents and thus, the more generic it is. Our approach consists of four major steps, as outlined in Fig. 2. Each step is described in detail in the following subsections.

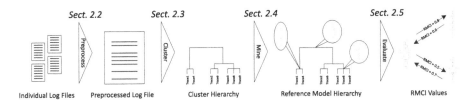

Fig. 2. Conceptual outline of our approach for mining reference models

2.2 Preprocessing Log Files

The primary input data for our approach is a set of individual log files, each containing execution data from one process implementation. We assume that the logs only contain execution data of one process and that the same process (in sense of process goals and intentions, not necessarily activities and structure) is represented in each log. In order to be mined effectively, the logs have to be preprocessed first. As this contribution is focused on mining the activities and the control flow of a business process, all additional data is considered to be irrelevant. From here on, we consider a log file as a set of event traces, i.e. ordered sequences of activity identifiers.

Definition 3 (Trace). *An* event *e* *denotes the execution of an activity. A* trace *t* = $(e_1, ..., e_n)$ *is a finite sequence of events. Its length len(t)* = *n is defined as the number of events it consists of.*

Definition 4 (Log). *A* log *L* = $\{(t_1)^{m_1}, ..., (t_n)^{m_n}\}$ *is a multiset of traces. The multiplicity* m_i *of trace* t_i *denotes the frequency of* t_i *in l.* $\mathcal{A}(L) = \{e \mid \exists t \in L : e \in t\}$ *denotes the set of events that are contained in a log.*

In the following, we also make two important assumptions:

– Activities are recorded on the same level of granularity across the different log files, i.e. a process step that is represented by one activity in the first log may not be split into several activities in the second log. If this is not the case, we are not able to discover similarities among process traces.
– Activities are uniquely identifiable by their label, i.e. there are no duplicate labels within one process and equal activities from different processes always have the same label. This is not necessarily given, especially if the logs are recorded in different organizations, which may use a different vocabulary.

If these assumptions do not hold, activities in the log files may be aligned by means of aggregating and renaming, given an appropriate matching of the labels. Such a matching can be constructed manually or by means of an automated approach to Business Process Model Matching [6]. The log files are merged into one log file, containing all traces from the individual logs.

In order to derive a meaningful reference model, the log should contain correct and typical process behavior. Since a reference model is supposed to be a template for the design of new process models, it should abstract from exceptional or overly specific process executions. Therefore, we have to remove all process traces that are incomplete, atypical, or just infrequent. This is assured by removing all traces that do not obtain a certain minimum frequency. The *result* of the preprocessing stage is a unified log file, which contains all traces that should be considered for mining the reference model.

2.3 Clustering Execution Traces

In order to determine the structure of the reference model hierarchy, the traces from the execution log are clustered in the next step. Therefore, we need to determine a suitable similarity measure between the traces as well as an applicable clustering approach. Both aspects are extensively documented in the respective literature, such that we do not need to cover any details here [7,8]. Efficient implementations of many clustering techniques are publicly available and may be employed for this purpose, as the user sees fit. As we intend to examine a possible division of the given reference model scope into several sub-scopes, we employ a hierarchical-agglomerative clustering approach. This approach yields a strict hierarchy, which later allows for a distinct determination of the reference model feasibility. As two child clusters constitute a parent cluster, we also ensure that the scope of each mined reference model consists of the unified scopes of its children. As each clustering object is contained in multiple clusters, the number of clusters exceeds the number of initial objects.

One aspect worth mentioning is that trace clustering, while still in polynomial complexity, is a fairly computationally intensive problem, especially when considering a large number of traces. Depending on the clustering approach used, clustering a log of a few thousand traces (which is a realistic size) may take up to several hours. The runtime complexity is further increased, when the number of clusters is not predetermined, but has to be defined by the algorithm itself. Hence, the factor runtime should be taken into account when determining and parametrizing a suitable clustering approach. However, as the determination of a reference model is not a repetitive task, the computational cost should be feasible. The *result* of the clustering stage is a hierarchical cluster of execution traces, represented by a dendrogram or an according textual representation.

2.4   Mining Reference Process Models from Log Data

In the next step, we mine a reference process model for each of the identified clusters. Therefore, we adapt our own approach to inductive reference model development, based on the execution semantics of a given set of process models [9]. As process traces are a representation of the intended semantics of a reference model, the RMM-2 approach is suitable for our use case. It analyzes the semantic behavior of process models in terms of behavioral profiles and computes a reference model subsuming the specified behavior. In order to adapt the RMM-2 approach to consider process traces instead of process models, we have to adapt the ordering relations used to represent process behavior. As traces—other than process models—are strictly linear, we can reduce the number of possible relations between activities as follows.

Definition 5 (Trace-based ordering relations). *Let L be a log, $\mathcal{A}(L)$ its set of identifiers, and $a, b \in \mathcal{A}(L)$.*

- $a *_L a$ *if and only if there is a trace $t = (e_1, ..., e_n)$ and $1 \leq i < n$, such that $t \in L$ and $e_i = a$ and $e_{i+1} = a$ (one-node loop)*
- $a \bullet_L b$ *if and only if there is a trace $t = (e_1, ..., e_n)$ and $1 \leq i < n - 1$, such that $t \in L$ and $e_i = a$ and $e_{i+1} = b$ and $e_{i+2} = a$ (two-node loop)*
- $a \to_L b$ *if and only if there is a trace $t = (e_1, ..., e_n)$ and $1 \leq i < n$, such that $t \in L$ and $e_i = a$ and $e_{i+1} = b$ (following)*
- $a \parallel_L b$ *if and only if $a \to_L b$ and $b \to_L a$ (parallel)*
- $a \#_L b$ *if and only if there is a trace t, such that $t \in L$ and $a, b \in t$, but $a \not\to_L b$ and $b \not\to_L a$ (none)*
- $a -_L b$ *otherwise*

Given these ordering relations, a separate behavioral profile is computed for each execution trace. These profiles are then used to derive the reference model. The reader is referred to the original publication for details on its definition and implementation [9]. The *result* of this step is a collection of reference models, which are ordered in a hierarchical way, analogous to the cluster structure from the previous step.

2.5 Determining the Feasibility of a Reference Model

The reference model hierarchy provides us with a collection of reference models of differing sizes, scopes, and abstraction levels. We now evaluate the feasibility of each model with regard to the specified scope, i.e. give a concrete solution to the dilemma of reference modeling. Given a pragmatic notion of reference model quality, we measure the usefulness in certain situations by means of the necessary adaptations. Numerically, this is expressed by the Reference Model Change Indicator (RMCI), a basic metric for determining the fitness of a reference model with regard to a certain company-specific model [10]. The RMCI is computed by dividing the number of change operations required for deriving a company-specific model RM^* from the reference model RM by the overall number of elements in the specific model.

$$RMCI(RM, RM^*) = \frac{|RM \cup RM^*| - |RM \cap RM^*|}{|RM^*|} \tag{1}$$

The numerator of Eq. 1 accounts for the number of change operations that is necessary for transforming RM into RM^*. Every element in RM that is not contained in R^* has to be removed, while every element in R^* that is not contained in R has to be inserted. This number is set in relation to the overall size of the adapted model to be able to compare the RMCI across different adapted models. A RMCI value of 0 indicates equality between the two models, i.e. no change operations are necessary to derive R^*. It is important to note that the RMCI is not normalized, as it can reach values greater than 1. If this is the case, the number of change operations surpasses the size of the adapted model, indicating a low degree of similarity between the two models. Also, the RMCI is not symmetric, as only the size of the target model is considered in the denominator. The above formula can be applied to all model elements, but an individual consideration for each element (i.e. activities, connectors, edges) is more meaningful, as it its computation it in both directions and across hierarchy levels. The higher the value of the RMCI, the more adaptations are necessary to transform the starting model into the target model. The *result* of this stage is a collection of RMCI values for the reference model hierarchy.

3 Proof-of-Concept and Experimental Evaluation

Our approach is prototypically implemented as a proof-of-concept in the RefMod-Miner research prototype[1].

1. Preprocess Log: The implementation assumes that the provided MXML log file contains only instance data, i.e. ordered sequences of activities and that all activity names have been harmonized. The log is then preprocessed and transformed into feature vectors, based on the respective activity profiles. These vectors indicate which activities are contained in which trace.

---

[1] http://refmod-miner.dfki.de/.

2. Cluster Traces: These feature vectors are used as input for a hierarchical-agglomerative clustering approach. Therefore, the RefMod-Miner prototype uses the available implementation *hclust()* in the statistical computing language *R*. The Euclidean vector distance is chosen as similarity measure.
3. Mine Reference Model Hierarchy: For each cluster contained in the trace cluster hierarchy, a separate reference model is mined as an Event-Driven Process Chain (EPC), using the implementation of the RMM-2 approach already present in the RefMod-Miner.
4. Compute RMCI: For each pair of reference models in the cluster hierarchy that is connected by a directed path (i.e. transitive parent and child models), the RMCI is computed in both directions. This information is added to the reference model hierarchy.

We evaluate our approach by applying it to the data set used in the BPI Challenge 2015 [11]. It consists of five log files, recording the execution of a building permit applications process in Dutch municipalities. As the processes underlying the logs should be identical, we assume that events are identical if and only if they are identically labeled. The merged log consists of 5649 traces, with an average length of 46.49. The entire log contains 356 distinct activities, with an average frequency of 737.72.

Table 1 summarizes the evaluation results. Computing the reference model hierarchy for the merged log yields 11297 different reference models, one for every cluster contained in the hierarchy. On average, the models contain almost 57 nodes, 45 activities, and 12 connectors. As is typical for process models, they are quite sparse, containing only 65.05 edges on average. Most of the models are connected, since the average number of connected components is less than 2. It can be noted that most of the unconnected models contain one large and a few rather small connected components.

Table 1. Characteristics of the computed reference model hierarchy

Reference Model Hierarchy		Contained Reference Models	
Size	11297	Number of Models	11297
Number of Leaves	5649	Average Size	56.87
Number of Internal Nodes	5648	Average Number of Activities	45.31
Maximum Depth	134	Average Number of Connectors	11.56
Minimum Depth	3	Average Number of Edges	65.05
Average Depth	30.15	Average Number of Components	1.73

Figure 3 shows the top three layers of the reference model hierarchy with RMCI values. These numbers indicate the general RMCI, allowing for a first assessment of the reference model feasibility. RM 1 and RM 3 have a RMCI of 0.0, indicating equality, while the RMCI between RM 1 and its other child, RM 4, is fairly high. One explanation might be that Cluster 3 is considerably larger than Cluster 4, so the activities of cluster 4 are not considered in the reference model. We can conclude that RM 1 should be a suitable reference model for Cluster 3, but not necessarily for Cluster 4. The same can be said for RM 0, which has a fairly low RMCI with regard to RM 2, but a rather high one with regard to RM 1. We can also observe differences in the RMCI values for RM 2 and RM 6. Transforming the former into the latter is almost twice as costly as vice versa. These differences are caused by the differing sizes of the two models. As RM 6 is about half as big as RM 2, the number of adaptations has a higher impact on the RMCI value.

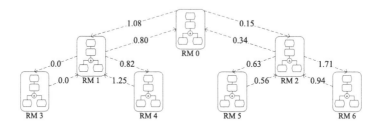

Fig. 3. Excerpt of the computed reference model hierarchy with RMCI values

Table 2 breaks down the RMCI values of Fig. 3 into three separate numbers, i.e. the RMCI with regard to activities, connectors, and edges. Examining them separately helps to analyze differences between the reference models. Considering only the activities measures the model similarity in terms of its scope, but disregards its structure. This may be helpful for high-level reference models, where the specific structure is less relevant. Both scope and structure are taken into account by measuring the RMCI for connectors and edges, as they are only

Table 2. Detailed RMCI computation

R	R^*	$RMCI_{Activities}$	$RMCI_{Connectors}$	$RMCI_{Edges}$	$RMCI_{Overall}$
RM 0	RM 1	0.51	2.0	1.13	1.08
RM 1	RM 0	0.39	1.14	0.86	0.80
RM 0	RM 2	0.10	0.69	0.04	0.15
RM 2	RM 0	0.10	0.74	0.32	0.34
RM 1	RM 3	0.0	0.0	0.0	0.0
RM 3	RM 1	0.0	0.0	0.0	0.0
RM 1	RM 4	0.44	1.26	0.80	0.82
RM 4	RM 1	0.66	3.74	0.84	1.25
RM 2	RM 5	0.22	1.5	0.64	0.63
RM 5	RM 2	0.19	1.09	0.6	0.56
RM 2	RM 6	0.66	4.14	2.15	1.71
RM 6	RM 2	0.49	1.21	1.09	0.94

identical if they have identical adjacent nodes. In our evaluation, the reference model structure differs more than the scope. High overall RMCI values are typically caused by high values regarding connectors and edges, while the value for the activities is never larger than 0.66, indicating a certain correspondence.

4 Related Work

Many contributions use (hierarchical) trace clustering for Process Discovery, intending to achieve a higher fitness and an increased precision. For example, De Medeiros et al. iteratively divide an execution log into clusters, until it is possible to discover an appropriate process model for each cluster [5]. Instead of mining a cluster hierarchy, the underlying logs are iteratively divided to mine a set of process models obtaining high precision values. Song et al. [12] motivate their approach in a similar fashion, but take additional process perspectives such as organization and data into account. They both build on previous work by Greco et al., which constructs a process model hierarchy, where higher placed models are generalizations of lower ones [13]. This differs from our hierarchy, where lower placed models cover a smaller scope than higher ones. Bose et al. [14] also construct a model hierarchy, composing a single process model into several subprocesses, as opposed to more smaller, more specific process models. The contribution by Buijs et al. [15] proposes four different approaches to mining configurable process models. Configurable models acknowledge process variation, however, they do not allow for a simple specificity evaluation in terms of the RMCI.

Ekanayake et al. [16] and García-Bañuelos et al. [17] employ a two-way approach to process discovery to control the complexity of the resulting model by defining trace clusters based on model complexity and extracting subprocesses. These works use similar techniques with an opposing objective: While they are set out to design the discovered model after a priori defined requirements, we include all possible reference models in a hierarchy and choose an appropriate design in hindsight. Recent contributions employ trace clustering for more specific applications of process discovery, such as to analyze how process models continuously evolve over a certain time period [18,19].

Despite the large number of contributions to the field of Process Discovery, the related field of Reference Model Mining is only sparsely covered. The contribution by Gottschalk et al. is explicitly set out to derive configurable reference process models from existing, well-established IT systems [20]. Existing process variants are depicted by means of a configurable reference model, i.e. models encompass several possible implementations. The contribution by Li et al. is defined on process models instead of execution traces [21]. They present approaches to reduce the distance between an existing reference models and its set of variants and to derive a new reference model from a set of process models. Both approaches are compared to traditional process mining algorithms. While both contributions provide important insights into the field of Reference Model Mining, they are not set up in a hierarchical fashion and thus do not address the dilemma of reference modeling, which is the main contribution of the article at hand.

5   Discussion and Conclusion

The contribution at hand presents a novel approach to hierarchically mining reference models from large instance data. It is set out to provide a solution to the dilemma of reference modeling, i.e. balancing a reference model's generality against its market specificity. The better a reference model fits to the requirements of an individual customer, the fewer customers are interested in applying it, hence limiting its potential sales market. This decision is formative for the design of a reference model and should thus be addressed by reference model developers and researchers alike. This contribution intends to provide a tool to facilitate business decisions. There are already many approaches employing model hierarchies to reach higher fitness and precision values and a lower complexity in Process Discovery. Our list of similar approaches is by no means complete. However, the focus of our contribution is not Process Discovery. The RMCI metric could be applied to any process hierarchy to determine an appropriate model scope.

The reference model hierarchy gives a formal structuring of the intended reference model market. This structure can be used to analyze the market itself, aligning the reference models and their scopes with the respective customer requirements and thus their intended target markets. For example, it might be a powerful tool for the realization of a platform strategy. Whereas a generic platform can be applied in a larger market, it is gradually further substantiated to fulfill the specifications of the different sub-markets. With the help of a reference model hierarchy, a reference model can be derived for both the platform and its sub-domains, addressing the varying requirements and purposes of each.

Configurable reference models are another approach to including more variability in a reference model. A configurable reference model can be seen as an integrated model, as it provides users with every possible process variant. While such models enlarge the scope and thus the market of a potential reference model, they cannot be seen as a solution to the dilemma of reference modeling. Taking the example of our evaluation in Sect. 3, it would be virtually impossible to include the details of 11297 reference models in one configurable model.

Our approach poses three requirements to the input data in order to produce a meaningful reference model hierarchy. In order to preserve the reference model character, we expect the log to contain representative—hence common—process behavior. Activities in the log can only be identified, if they are uniquely labeled and have the same level of granularity. While these requirements are quite strict, we do not assume them to be given, but suggest techniques for preprocessing the data accordingly. To our knowledge, all contributions mentioned in Sect. 4 rely on the same assumptions. Loosening them remains an open research problem in Reference Model Mining and is typically not addressed in Process Discovery.

As we have mentioned in Sect. 2, our approach is capable of deriving a reference model hierarchy from both type-level and instance-level data. One can argue that this slowly obliterates the once sharply drawn line between type models and instances. In fact, this strict separation might cause more problems than it solves. It might no longer be important to assign each instance to a type model, but to view the two as independent, yet related modeling layers [22]. Such a concept may introduce a new degree of flexibility and open a wide range of new opportunities, which will have to be further investigated by future work.

Our approach postulates a reuse-oriented conceptualization of reference models, i.e. we consider a reference model to be a special conceptual model that serves to be re-used in the design process of other conceptual models [3]. By measuring its usefulness in terms of the RMCI, we take a pragmatic view on reference model quality. Instead of consulting external measures, we focus on the situational application, stressing the reuse potential of a reference model. However, the percepted model quality, which we are able to measure and influence, is not necessarily related to the quality of the underlying process. Process quality is mainly determined by factors like cost, time, and outcome, which are not depicted in process models and only partially available in instance data. This aspect is not considered in an exclusively reuse-oriented understanding of a reference model, as no relevant data, such as process cost or duration is taken into account while constructing the reference models.

Acknowledgments. The research described in this paper was partly supported by a grant from the German Research Foundation (DFG), project name: Konzeptionelle, methodische und technische Grundlagen zur induktiven Erstellung von Referenzmodellen (Reference Model Mining), support code GZ LO 752/5-1.

References

1. Fettke, P., Loos, P.: Perspectives on reference modeling. In: Fettke, P., Loos, P. (eds.) Reference Modeling for Business Systems Analysis, pp. 1–20. Idea Group Publishing, London (2007)
2. Becker, J., Delfmann, P., Knackstedt, R.: Adaptive reference modeling: integrating configurative and generic adaptation techniques for information models. In: Becker, J., Delfmann, P. (eds.) Reference Modeling. Efficient Information Systems Design Through Reuse of Information Models, pp. 27–58. Physica-Verlag, Heidelberg (2007)
3. vom Brocke, J.: Design principles for reference modeling: reusing information models by means of aggregation, specialisation, instantiation, and analogy. In: Fettke, P., Loos, P. (eds.) Reference Modeling for Business Systems Analysis, pp. 47–75. Idea Group Publishing, London (2007)
4. Hevner, A., March, S., Park, J., Ram, S.: Design science in information systems research. MIS Q. **28**(1), 75–105 (2004)
5. Medeiros, A.K.A., Guzzo, A., Greco, G., Aalst, W.M.P., Weijters, A.J.M.M., Dongen, B.F., Saccà, D.: Process mining based on clustering: a quest for precision. In: Hofstede, A., Benatallah, B., Paik, H.-Y. (eds.) BPM 2007. LNCS, vol. 4928, pp. 17–29. Springer, Heidelberg (2008). doi:10.1007/978-3-540-78238-4_4
6. Antunes, G., Bakhshandelh, M., Borbinha, J., Cardoso, J., Dadashnia, S., et al.: The process matching contest 2015. In Kolb, J., Leopold, H., Mendling, J. (eds.) Proceedings of the 6th International Workshop on Enterprise Modelling and Information Systems Architectures. International Workshop on Enterprise Modelling and Information Systems Architectures (EMISA-15), September 3–4, Innsbruck, Austria, Köllen Druck+Verlag GmbH, Bonn, September 2015

7. Everitt, B.S., Landau, S., Leese, M., Stahl, D.: Hierarchical clustering. In: Cluster Analysis, pp. 71–110. Wiley, New York (2011)
8. Thaler, T., Ternis, S.F., Fettke, P., Loos, P.: A comparative analysis of process instance cluster techniques. In: Proceedings of the 12th International Conference on Wirtschaftsinformatik. Internationale Tagung Wirtschaftsinformatik (WI 2015), March 3–5, Osnabrück, Germany, Universität Osnabrück, March 2015
9. Rehse, J.R., Fettke, P., Loos, P.: An execution-semantic approach to inductive reference models development. In: 24th European Conference for Information Systems (ECIS). European Conference on Information Systems (ECIS 2016), June 12–15, Istanbul, Turkey, Association for Information Systems (AIS) (2016)
10. Kurpjuweit, S., Winter, R.: Concern-oriented business architecture engineering. In: Proceedings of the 2009 ACM Symposium on Applied Computing, pp. 265–272. ACM (2009)
11. van Dongen, B.: BPI Challenge 2015 (2015). http://dx.doi.org/10.4121/uuid: 31a308efc844-48da-948c-305d167a0ec1
12. Song, M., Günther, C.W., Aalst, W.M.P.: Trace clustering in process mining. In: Ardagna, D., Mecella, M., Yang, J. (eds.) BPM 2008. LNBIP, vol. 17, pp. 109–120. Springer, Heidelberg (2009). doi:10.1007/978-3-642-00328-8_11
13. Greco, G., Guzzo, A., Pontieri, L., Sacca, D.: Discovering expressive process models by clustering log traces. IEEE Trans. Knowl. Data Eng. **18**(8), 1010–1027 (2006)
14. Bose, R.P.J.C., Verbeek, E.H.M.W., Aalst, W.M.P.: Discovering hierarchical process models using ProM. In: Nurcan, S. (ed.) CAiSE Forum 2011. LNBIP, vol. 107, pp. 33–48. Springer, Heidelberg (2012). doi:10.1007/978-3-642-29749-6_3
15. Buijs, J.C.A.M., Dongen, B.F., Aalst, W.M.P.: Mining configurable process models from collections of event logs. In: Daniel, F., Wang, J., Weber, B. (eds.) BPM 2013. LNCS, vol. 8094, pp. 33–48. Springer, Heidelberg (2013). doi:10.1007/ 978-3-642-40176-3_5
16. Ekanayake, C.C., Dumas, M., García-Bañuelos, L., Rosa, M.: Slice, mine and dice: complexity-aware automated discovery of business process models. In: Daniel, F., Wang, J., Weber, B. (eds.) BPM 2013. LNCS, vol. 8094, pp. 49–64. Springer, Heidelberg (2013). doi:10.1007/978-3-642-40176-3_6
17. García-Bañuelos, L., Dumas, M., La Rosa, M., De Weerdt, J., Ekanayake, C.: Controlled automated discovery of collections of business process models. Inf. Syst. **46**, 85–101 (2014)
18. Luengo, D., Sepúlveda, M.: Applying clustering in process mining to find different versions of a business process that changes over time. In: Daniel, F., Barkaoui, K., Dustdar, S. (eds.) BPM 2011. LNBIP, vol. 99, pp. 153–158. Springer, Heidelberg (2012). doi:10.1007/978-3-642-28108-2_15
19. Hompes, B., Buijs, J., Van der Aalst, W., Dixit, P., Buurman, H.: Detecting change in processes using comparative trace clustering. In: Proceedings of the 5th International Symposium on Data-driven Process Discovery and Analysis (SIMPDA 2015), Vienna, Austria, December 9–11, 2015, pp. 95–108 (2015)
20. Gottschalk, F., Aalst, W.M.P., Jansen-Vullers, M.H.: Mining reference process models and their configurations. In: Meersman, R., Tari, Z., Herrero, P. (eds.) OTM 2008. LNCS, vol. 5333, pp. 263–272. Springer, Heidelberg (2008). doi:10. 1007/978-3-540-88875-8_47
21. Li, C., Reichert, M., Wombacher, A.: Mining business process variants: challenges, scenarios, algorithms. Data Knowl. Eng. **70**(5), 409–434 (2011)
22. Parsons, J., Wand, Y.: Emancipating instances from the tyranny of classes in information modeling. ACM Trans. Database Syst. **25**(2), 228–268 (2000)

A Framework for Interactive Multidimensional Process Mining

Thomas Vogelgesang[1](✉), Stefanie Rinderle-Ma[2], and H.-Jürgen Appelrath[1]

[1] Department of Computer Science, University of Oldenburg, Oldenburg, Germany
{thomas.vogelgesang,appelrath}@uni-oldenburg.de
[2] Faculty of Computer Science, University of Vienna, Vienna, Austria
stefanie.rinderle-ma@univie.ac.at

Abstract. The emerging concept of multidimensional process mining adopts the ideas of data cubes and OLAP to analyze processes from multiple views. Analysts can split the event log into a set of homogenous sublogs according to its case and event attributes. Process mining techniques are used to create an individual process model for each sublog representing variants of the process. These models can be compared to identify the differences between the variants. Due to the explorative character of the analysis, interactivity is crucial to successfully apply multidimensional process mining. However, current approaches lack interactivity, e.g., they require the analyst to re-perform the analysis steps after changing the view on the data cube. In this paper, we introduce a novel framework to improve the interactivity of multidimensional process mining. As our main contribution, we provide a generic concept for interactive process mining based on a stack of operations.

Keywords: Interactive process mining · Multidimensional process mining · Process Cubes

1 Motivation

Process mining comprises a set of techniques for the automated analysis of processes [2]. They are based on collections of events (event logs) which are recorded during the execution of the process. Each event represents the execution of an activity and refers to a case representing a particular instance of the process. Typically, the events of an event log are grouped and chronologically ordered by their case representing the trace of a process instance. Additionally, events and cases may have arbitrary attributes to describe their specific properties.

Most process mining techniques focus on process discovery which aims to generate a descriptive and representative process model from a given event log. However, there are also other kinds of process mining techniques. Process enhancement maps the additional attributes of the event log onto a given process model to add further perspectives (e.g. waiting times). Conformance checking techniques compare process models to event logs, e.g. to measure its quality.

© Springer International Publishing AG 2017
M. Dumas and M. Fantinato (Eds.): BPM 2016 Workshops, LNBIP 281, pp. 23–35, 2017.
DOI: 10.1007/978-3-319-58457-7_2

The notion of multidimensional process mining (MPM) [6,13,15] is an emerging concept to analyze variants of the same process. It adopts the idea of data cubes and Online Analytical Processing (OLAP) [7] from the data warehouse domain to the field of process mining and organizes the event data by its attributes which are considered as dimensions. While traditional process mining techniques typically only consider a single event log resulting in one overall process model, MPM extracts an individual subset of an event log (sublog) for each variant. Analyzing these sublogs with process discovery returns a set of process models which can be compared to each other in order to identify differences between the process variants. Providing OLAP operators like drill-down and slice on the event data, MPM enables the analyst to define arbitrary views on the event log and to iteratively explore the processes.

In contrast to static reporting, OLAP aims to interactively explore the data. The analysis typically consists of an individual sequence of queries where each query results from the preceding ones. Therefore, OLAP requires flexible and effective user interfaces which are easy-to-use [7]. Due to this explorative character, interactivity is also vital for MPM [13]: To get the desired insights, the analysts need to refine the OLAP query for many times (e.g., filter or aggregate events, change dimension granularity) and apply different process mining techniques like process discovery, process enhancement and conformance checking. Even though they provide interactive user interfaces, current tools for MPM lack interactivity with regard to four aspects:

1. *Interaction with process models:* Process models as the major result of process mining are typically visualized in a static way. Though they are complex objects composed of nodes and edges, they are typically just drawn like an image. Except for scrolling and zooming, there is often no way to interact with the process models. Direct interaction like clicking on a node has the potential, to improve the exploration of processes by making the analysis faster and more intuitive.
2. *Dynamic analysis workflow:* Especially in MPM, analysts have to follow a more or less restrictive workflow. For example, they first have to define an OLAP query (selecting some dimensions to drill down the event data, adding optional filters), select and configure a process discovery algorithm, enhance the resulting process models with time information, and finally apply conformance checking to measure the quality of the models. If a previous analysis step has to be changed (e.g., adding another filter to the OLAP query), all subsequent analysis steps have to be done again. Consequently, even minimal changes may require a lot of effort for the analysts.
3. *Undo/redo of analysis steps:* It is desirable to provide support to undo and redo particular analysis steps. This enables analysts to try arbitrary analysis steps without any risks. The chance to return to the previous view on the process without any additional effort supports the explorative character of MPM because it avoids to end up in an "analysis dead end".

4. *Performance:* Even though MPM is not a time-critical application, performance is crucial for interactivity. Long processing times may disrupt the workflow preventing analysts from defining views which they expect to be less promising.

In this paper, we will present a novel framework which addresses the first three aspects as discussed above regarding the interactivity of MPM. We will identify different concepts and use cases to improve the interactivity of MPM. As the main contribution of our paper, we will provide a generic concept for interactive process mining based on a stack of operations relevant for MPM. We have to point out that performance (4) is only considered with respect to the workflow of MPM whereas the performance of particular analysis steps (e.g., executing OLAP queries or process discovery) is out of the scope of this paper.

The structure of the paper is as follows. Section 2 briefly discusses related work. In Sect. 3, we first introduce the assumptions of our work, before we elaborate on the general ideas and concepts of our approach. The proof-of-concept implementation of our approach is presented in Sect. 4. Finally, we conclude our work in Sect. 5 and give an overview of future work.

2 Related Work

The process mining manifesto [4] gives an overview of the field of process mining. For a comprehensive introduction to this topic, we refer to van der Aalst [2]. Event Cubes [12] are an initial but very specific approach for MPM based on a multidimensional adoption of the Flexible Heuristics Miner algorithm [16]. In contrast, Process Cubes [3,6] and PMCube [15] are generic frameworks that are not limited to a particular mining algorithm or process model representation. Moreover, they are able to incorporate process enhancement and conformance checking. PMCube also introduces advanced concepts to MPM like process model comparison using difference models and the process model consolidation which provides the automatic pre-selection of possibly more relevant process models. Even though they enable the analysts to explore the processes from a multidimensional perspective, Process Cubes as well as PMCube are limited in their interactivity. For example, they do not provide interactive process representation. The lack of interactivity was also identified during the case study presented in [15] and the systematic literature review provided in [13].

In the general context of process mining, interactivity has been initially addressed by the Fuzzy Miner algorithm [9] which adopts the basic idea of interactive maps to create so-called process maps. This representation enables the analyst to dynamically abstract from less relevant parts of the process model by clustering activities in order to focus on the major relationships of the process. Similar interactivity is also implemented in commercial tools like Fluxicon Disco [8] which considers process enhancement, too. However, we aim for a more comprehensive and generic approach which considers aspects of MPM (e.g., OLAP, difference calculation for process models, process model consolidation) and supports arbitrary process discovery algorithms and process model notations. In [1], van der Aalst extends the idea of process maps by the metaphor of navigation systems to provide interactive recommendation and prediction, e.g. to estimate the remaining execution time of a process instance. In contrast to that, we only consider historical data as we aim to improve the interactivity of the overall MPM workflow. The application of MPM to real-time event data has not been investigated in research so far.

Interactivity is also considered in the intersection of process mining and visual analytics. As stated in the process mining manifesto [4], the combination of both fields has the potential "to extract more insights from event data". An example for such a visual process mining is the EventExplorer [5] which allows to browse an event log in order to visually explore and assess it. Visual analytics techniques for process mining algorithms realized in ProM[1] along the control flow, organizational, case, and time perspective are evaluated in [10]. However, the multidimensional perspective has not been considered yet.

RapidProM [11] allows analysts to interactively model scientific workflows for process mining by connecting operators in order to define complex analysis scenarios combining several techniques which should be repeated using different parameter settings or data sets. In contrast, we focus on supporting explorative ad-hoc analysis by interactivity instead of predefining an analysis workflow.

3   Approach

In this section, we will present the basic ideas and the concept for an interactive process mining. Section 3.1 introduces the underlying assumptions for this work. In Sect. 3.2, we discuss direct interaction with process models and present an operator framework for supporting interactivity in Sect. 3.3.

3.1   Assumptions

For this work, we assume that events are stored in a data cube which is accessible by OLAP queries that are expressed by a query language or a graphical user interface. This query is expected to return a set of sublogs in the usual event log structure (e.g., as defined in [2]). We do not assume a limitation to particular algorithms for process discovery, conformance checking, process enhancement, and consolidation. For the process models, we assume an arbitrary graph-based representation consisting of nodes and edges. However, there are different kinds and styles of nodes possible. Furthermore, we only focus on postmortem analysis. Process mining techniques combining historical data with real-time events like recommendation and prediction are not in the scope of this work.

3.2   Direct Interaction with Process Models

To make the process analysis as intuitive as possible, it is desirable to allow for a direct interaction with the process models (cf. first aspect *Interaction with*

---

[1] www.promtools.org.

process models introduced in Sect. 1). While modeling tools, for example, enable the user to directly interact with the presented model and its elements, process mining tools (especially tools for MPM) usually only provide a static representation. As the models form the central subject of the analysis, MPM and process mining in general can also benefit from direct interactions with the process models. Therefore, we aim for an approach that allows the analysts to directly interact with the models. An example for such an interaction is to click on a node to highlight all model elements that are part of a trace containing the selected node. Which action to perform for a particular interaction can be selected from a toolbox. Alternative actions may be the presentation of additional information in a dialog or a sidebar. Furthermore, it is possible to enrich the process models' nodes and edges with additional user controls like buttons or context menus. This enables the model to provide different interactions for the same object at the same time.

Nonetheless, the direct process model interaction can be complemented by an interaction with external user controls, e.g. sliders, buttons etc. to perform actions on the model. E.g., the Fuzzy Miner [9] provides sliders to adjust filter thresholds and update the presented process model. The interaction with the process model should also incorporate the selection of different perspectives to dynamically add additional information to the model. To provide a generic solution, we define variation points of the process model elements that can be used for the visualization of additional data. Examples are the edge labels, the border and background colors of nodes, or the line thickness of arcs. Which information (e.g. frequencies, waiting times) a variation point will visualize is determined by the user selecting a particular perspective of the process model.

3.3   Operator Framework

In order to address the second and the third aspect (*Dynamic analysis workflow* and *Und/redo of analysis steps*, cf. Sect. 1), our approach introduces an operator framework. It maps the interactions with the software – especially each analysis step – onto operators which are organized at different levels according to their position in the analysis workflow and their data dependencies. Therefore, the operators of a particular level only consume the results of lower levels as input while their results are only available for the levels above. Figure 1 shows the stack of operators defined by the framework and the data items forming their input and output. In the following, we will explain the operator levels in more detail.

OLAP. An OLAP operation of the framework represents a query that extracts data from the data cube and returns a set of sublogs as a result. Parameters of the operation are the granularity of the considered dimensions, filter predicates etc. Note that each of this operations comprises multiple low-level OLAP operations on the data cube like roll-up, drill-down, slice, and dice. As the extraction of event data from the cube is necessary to conduct a process analysis, this operation level is mandatory.

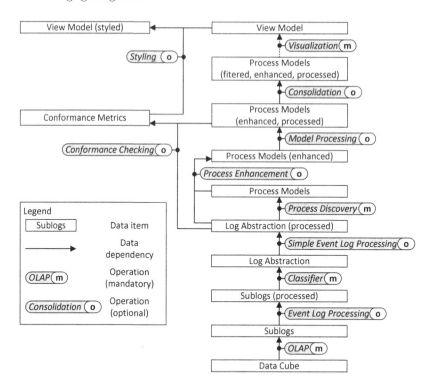

Fig. 1. Stack of operator levels with data items and their dependencies

Event Log Processing. The optional event log processing enables the analysts to manipulate the extracted sublogs, e.g., by filtering or aggregating events. It is also possible to derive new attributes from other attributes (e.g., calculating event durations from the events' start and end time-stamps). The event log processing operations are functionally overlapping with the OLAP operators. However, providing filters directly on the sublogs may significantly reduce the waiting times as loading data from the data cube is the most expensive operation in MPM in matters of performance. Therefore, applying these operations can be more efficient than changing OLAP queries.

Classifier. The classifier operations select a classifier which is a function that defines how to map the event data onto a node label necessary for process discovery. Using this classifier, these operations create a log abstraction which consists of a number of classifier-dependent log metrics (e.g., relations between events and event counts). The log abstraction also comprises a more abstract representation of the sublog (similar to simple event logs, cf. [2]) which is derived using the selected classifier. It consists of a set of (unique) traces and the frequency of their occurrence. Additionally, they also maintain references to the original cases and events to be able to use them for process enhancement. Classifier operations are mandatory, because all process discovery algorithms require a classifier to select the node labels.

Additionally, many process discovery algorithms like the Fuzzy Miner [9] or the Flexible Heuristics Miner [16] rely on these log metrics. Furthermore, other techniques like conformance checking (e.g., token replay) can benefit from the simple event log with respect to performance.

Simple Event Log Processing. Similar to the sublogs, the simple event logs contained by the log abstraction can be manipulated, e.g., traces can be filtered by the number of occurrence. As the simple event logs provide a more condensed representation of the data giving the frequency of each trace, they can be processed more efficiently than normal sublogs (which may have duplicate traces), especially if they only contain few trace variations. However, manipulations of simple event logs may effect the log metrics, so they have to be recalculated after all simple event log operations were executed.

Process Discovery. Based on the log abstractions, the process discovery operations create a process model for each cell using an arbitrary algorithm. Parameters of the operations are the selected algorithm and its individual settings (e.g., algorithm-specific thresholds). Note that changing the settings creates a new process discovery operation even if the selected algorithm remains the same. This operation level is mandatory, because the resulting process models are required for subsequent operation levels.

Process Enhancement. These operations apply arbitrary process enhancement algorithms to the previously discovered process models. As different algorithms may add different perspectives (e.g., organizational or time perspective), this operation level is chainable.

Model Processing. These optional operations manipulate process models, e.g., filtering nodes and edges or converting process models from one representation (e.g., petri nets) to another (e.g., BPMN). As multiple manipulations of the models may be desired, these operations are chainable.

Conformance Checking. Based on the log abstractions and the process models, these operations measure the quality of the discovered process models. Examples for this operation are a token replay or a trace alignment in order to measure the fitness of the process models. As the quality metric may incorporate additional perspectives as well, the enhanced and possibly manipulated process models are used.

Consolidation. The optional consolidation operations select a subset of potentially interesting process models while hiding irrelevant models in order to reduce the result complexity of MPM. An example for such an operation is the clustering consolidation (cf. [15]) which creates clusters of similar process models and selects a representative for each cluster.

Visualization. The visualization operations translate the process model into a corresponding view model to be displayed to the analysts. In contrast to the previous operations, the visualization is not automatically executed for all process models but only for the models currently presented to the user.

Styling. These optional operations manipulate the view model by changing its style (e.g., highlighting nodes by color) and linking the variation points of the visual elements to additional information (e.g., mapping the frequency of an edge onto its label). Like the visualization operations, they are only executed for the currently shown models. Furthermore, they are chainable, because multiple perspectives may be mapped to the view model.

Some levels are required to have at least one operation to be executed to be able to execute the level above. We call these levels mandatory (marked with **m** in Fig. 1). However, there are also optional levels (marked with **o**) that do not need to have an operation to be executed. For example, process discovery operations are mandatory because the resulting process model is required for the execution of other levels like process enhancement or conformance checking. In contrast, enhancing the process model is optional.

For each level, the framework manages all executed operators in a separated list to keep track of the analysis history. If the analyst performs an interaction, an operation representing this interaction is added to the list of its corresponding level. Then, the operations of this level and the levels above are consecutively executed to propagate the changes to the final analysis results. The operations of the levels below are not executed again as their results remain unchanged. This avoids unnecessary data processing and results in shorter waiting times and a faster system response. By managing the operations at different levels, it is possible to undo or redo operations separately. This enables the analyst to, for example, go back to the previous OLAP query without discarding the other performed analysis steps like process enhancement or conformance checking. The change propagation ensures that the perspective on the process remains the same, while the underlying data is updated.

The framework keeps track of all performed interactions. Besides providing an advanced undo/redo functionality, this also prevents the analysts from repeating analysis steps when they change the underlying OLAP query. The operators also form an intermediate layer decoupling user interaction and data processing which makes it easier to link direct interactions (cf. Sect. 3.2) to particular analysis step.

The operations only store the parameters required for their execution, e.g., the selected process discovery algorithm and its settings. The data items that should be processed are centrally managed by the framework and only passed for execution, so the operators are always considering the current data. For each level, the framework manages a reference to an operation marking the latest operation to execute. We call this reference the *latest active operation* of the level. There are two different kinds of operator levels:

Chainable levels. All operations up to the latest active operation of this level are consecutively executed in the respective order of their position in the operation list. All other operations of that list are considered as inactive and consequently not executed. Operation levels are only chainable if their input and the output structure are similar so the results of an operation can be passed to its successive operation as input. This can be useful, e.g., to execute multiple filter operations on the sublogs, where each operation filters by a different attribute.

Non-chainable levels. For this kind of operation levels, only the latest active operation is executed. This only applies to operator levels where the operators transform the input to a differently structured output. An example for this kind of operator levels is process discovery which takes an event log as input in order to create a process model as output.

The separation of styling and visualization operations makes it possible to differentiate between the calculation and the visualization of enhanced perspectives. This way, the analyst can show or hide specific information without any expensive recalculations. The styling operations link the operator framework to the direct interaction (cf. Sect. 3.2) by binding event handlers to particular visual elements. This way, it is possible to provide different reactions, e.g., when the user clicks on a node, depending on the selected styling operation.

As a view model should provide an integrated view onto to the process which incorporates multiple perspectives, it is necessary that the styling operations are chainable. However, it may happen that the user selects two operations that affect the same visual element, e.g., that map different values to the same label. To avoid confusion and misinterpretations by overriding the result of previous styling operations, such conflicts need to be resolved. Therefore, each operation provides a list of all properties of the visual elements it affects. Before adding a new styling operation, all existing operations are checked if their affected properties overlap with the affected properties of the new operation. Each operation that is in conflict with the new operation, will be discarded.

The described concepts contribute to the four aspects introduced in Sect. 1. The first aspect (*Interaction with process models*) is directly addressed by the concept for direct interaction introduced in Sect. 3.2. The operator framework contributes to the second aspect (*dynamic analysis workflow*). It allows the user to dynamically add and execute operations on an arbitrary operation level. Consequently, the users are less restricted during the workflow, because they do not have to follow a step-by-step configuration of the analysis. The change propagation also ensures that the users do not have to repeat previously performed analysis steps (like process enhancement) when applying changes to the underlying analysis steps (e.g., by adding filters to the OLAP query).

The operator framework also directly contributes to the third aspect (*Undo/redo of analysis steps*; cf. Sect. 1): Changing the reference to the latest active operation to the previous operation will revert it, because all operations following the reference will be ignored. However, as the undone operation is kept in the list, a redo can be easily achieved by setting the latest active operation reference to the subsequent operation. Finally, the operator framework also contributes to aspect four (*performance*). It ensures that changes to the analysis by adding, undoing or redoing an operation will only affect the subsequent operations. Operations on lower operation levels do not have to be repeated. This significantly reduces the waiting times for the user because the extraction of data from the cube is typically the most expensive task in MPM in matters of processing time.

4   Implementation

We implemented our approach in a prototype called Interactive PMCube Explorer[2] using C# and the .NET framework. It is based on the PMCube Explorer tool [14] and replaces its original user interface. The operator framework (cf. Sect. 3.3) is managed by a central operation manager component while the operators are (mainly) integrated as plug-ins. Only the visualization and styling operations are managed by a separated visualization manager component due to their asynchronous execution (triggered by events) and specific challenges like the conflict resolution.

Figure 2 shows a screenshot of the Interactive PMCube Explorer tool. As they form the main subject of the analysis, the process mining results are presented at the center of the screen using tabs. Tool boxes, options, and additional settings are flexibly arranged around them to keep the focus on the analysis results. In this example, one can see the options for defining the OLAP operation (1) and the available styling operations for the presented process model (2). The styling operations for annotating the process model with the frequencies of nodes and edges (left highlighted button) and adding direct interaction to show additional information (right highlighted button) are activated. The effect of these style operations on the visualization can be seen in the process model (3), where additional labels are attached to nodes and edges indicating their frequencies. The event handlers for direct interaction are linked to the activity nodes and to additional buttons of the edges, because clicking on thin edges might be difficult and hindering. Clicking on a node or the edge's button opens the additional information in a separated view aside of the process model (4). Above that, the

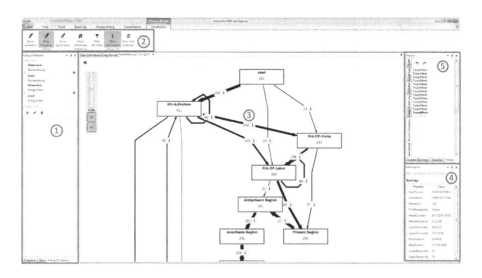

Fig. 2. Screenshot of the Interactive PMCube Explorer tool

---

[2] Screencast and tool download available at http://www.uol.de/pmcubeexplorer.

history view separately shows all performed operations for each level (5). In this example, one can see the history of process discovery operations highlighting the latest active operation while the discarded operation is toned down.

The history view allows the user to undo or redo any operation separately for each level. This enables the user, e.g., to safely try a different parameter setting of the process discovery algorithm and to return to the previous settings without repeating the configuration of other analysis steps like process enhancement. The only exception are the styling operations, which can be directly activated or deactivated without following the order that they were performed.

To show the feasibility of the approach, the prototype provides several operators for most levels. A plug-in system allows for the easy integration of additional operators. The direct interaction is demonstrated by a number of styling operations, e.g., attaching different information like frequencies and metrics to the model, color-coding of difference models, and adding event handlers to show time differences and the additional information view. Besides, it is possible to bind event handlers to visual elements in order to trigger operations on other levels. For example, one interaction filters the simple event log to all traces having a certain activity just by clicking the respective activity node in the process model.

The implementation shows that the operator framework (cf. Sect. 3.3) can be directly applied to the PMCube approach. However, the operator framework is a generic concept. Except for the assumptions from Sect. 3.1, it is not restricted to a particular approach for MPM. In principle, it can be also applied to other MPM approaches like Process Cubes [3,6] which meets our assumptions for MPM. For instance, Process Cubes uses OLAP queries to extract sublogs from the cube and is able to apply different process mining techniques like process discovery, process enhancement and conformance checking. However, some special analysis steps (e.g., consolidation of process models) are not considered in Process Cubes. However, our concepts can still be applied to it if the correspondent operation levels will be removed. On the contrary, the operator framework cannot be applied to the Event Cube because it defines MPM in a very different way and does not meet our assumptions, e.g., it does not create sublogs and defines the OLAP operations like roll-up and drill-down as manipulations of the process model.

The concept for direct interaction is generally applicable to other MPM approaches (like Process Cubes) as well as to non-multidimensional process mining. The basic idea of the operator framework can also be adopted for process mining tools in general because MPM can be considered as a generalization of traditional process mining. The differences of non-multidimensional process mining approaches are mainly related to the number of data items (typically one log and process model is processed at a time) and some operation levels have to be skipped (e.g., OLAP) as they are only meaningful in MPM context. Therefore, also the process mining community in general may benefit from this concept.

5   Conclusions and Future Work

In this paper, we presented a novel research towards interactive process mining. Its key points are the direct interaction with the process models and the operator framework which aims to avoid the unnecessary repetition of analysis steps. We implemented our approach based on the PMCube Explorer tool [14] as a proof-of-concept prototype which showed the feasibility of the approach. We plan to evaluate our approach by a user study to investigate the contribution of our approach to the interactivity of process mining.

References

1. van der Aalst, W.M.P.: Using process mining to generate accurate and interactive business process maps. In: Abramowicz, W., Flejter, D. (eds.) BIS 2009. LNBIP, vol. 37, pp. 1–14. Springer, Heidelberg (2009). doi:10.1007/978-3-642-03424-4_1
2. van der Aalst, W.M.P.: Process Mining - Discovery, Conformance and Enhancement of Business Processes. Springer, Heidelberg (2011)
3. van der Aalst, W.M.P.: Process cubes: slicing, dicing, rolling up and drilling down event data for process mining. In: Song, M., Wynn, M.T., Liu, J. (eds.) AP-BPM 2013. LNBIP, vol. 159, pp. 1–22. Springer, Cham (2013). doi:10.1007/978-3-319-02922-1_1
4. van der Aalst, W., et al.: Process mining manifesto. In: Daniel, F., Barkaoui, K., Dustdar, S. (eds.) BPM 2011. LNBIP, vol. 99, pp. 169–194. Springer, Heidelberg (2012). doi:10.1007/978-3-642-28108-2_19
5. Bodesinsky, P., Alsallakh, B., Gschwandtner, T., Miksch, S.: Exploration and assessment of event data. In: Bertini, E., Roberts, J.C. (eds.) EuroVis Workshop on Visual Analytics (EuroVA). The Eurographics Association (2015)
6. Bolt, A., van der Aalst, W.M.P.: Multidimensional process mining using process cubes. In: Gaaloul, K., Schmidt, R., Nurcan, S., Guerreiro, S., Ma, Q. (eds.) CAISE 2015. LNBIP, vol. 214, pp. 102–116. Springer, Cham (2015). doi:10.1007/978-3-319-19237-6_7
7. Golfarelli, M., Rizzi, S.: Data Warehouse Design: Modern Principles and Methodologies, 1st edn. McGraw-Hill Inc., New York (2009)
8. Günther, C.W., Rozinat, A.: Disco: discover your processes. In: Lohmann, N., Moser, S. (eds.) Proceedings of the BPM 2012 Demonstration Track. CEUR Workshop Proceedings, vol. 940, pp. 40–44. CEUR-WS.org (2012)
9. Günther, C.W., van der Aalst, W.M.P.: Fuzzy Mining – adaptive process simplification based on multi-perspective metrics. In: Alonso, G., Dadam, P., Rosemann, M. (eds.) BPM 2007. LNCS, vol. 4714, pp. 328–343. Springer, Heidelberg (2007). doi:10.1007/978-3-540-75183-0_24
10. Kriglstein, S., Pohl, M., Rinderle-Ma, S., Stallinger, M.: Visual analytics in process mining: classification of process mining techniques. In: 7th International Eurovis Workshop on Visual Analytics. Groningen (2016, accepted for publication)
11. Mans, R., van der Aalst, W.M.P., Verbeek, H.: Supporting process mining workflows with rapidprom. In: Limonad, L., Weber, B. (eds.) Proceedings of the BPM Demo Sessions 2014. CEUR Workshop Proceedings, vol. 1295, p. 56. CEUR-WS.org (2014)

12. Ribeiro, J.T.S., Weijters, A.J.M.M.: Event cube: another perspective on business processes. In: Meersman, R., et al. (eds.) OTM 2011. LNCS, vol. 7044, pp. 274–283. Springer, Heidelberg (2011). doi:10.1007/978-3-642-25109-2_18
13. Vogelgesang, T., Kaes, G., Rinderle-Ma, S., Appelrath, H.: Multidimensional process mining: questions, requirements, and limitations. In: Proceedings of the CAiSE 2016 Forum, pp. 169–176 (2016)
14. Vogelgesang, T., et al.: Multidimensional process mining with PMCube explorer. In: Daniel, F., Zugal, S. (eds.) Proceedings of the BPM Demo Session 2015. CEUR Workshop Proceedings, vol. 1418, pp. 90–94. CEUR-WS.org (2015)
15. Vogelgesang, T., Appelrath, H.-J.: PMCube: a data-warehouse-based approach for multidimensional process mining. In: Reichert, M., Reijers, H.A. (eds.) BPM 2015. LNBIP, vol. 256, pp. 167–178. Springer, Cham (2016). doi:10.1007/978-3-319-42887-1_14
16. Weijters, A.J.M.M., Ribeiro, J.T.S.: Flexible heuristics miner (FHM). Technical report, Technische Universiteit Eindhoven (2011)

Parallelization of Transition Counting for Process Mining on Multi-core CPUs and GPUs

Diogo R. Ferreira[✉] and Rui M. Santos

Instituto Superior Técnico (IST), University of Lisbon, Lisbon, Portugal
{diogo.ferreira,rui.miguel.santos}@tecnico.ulisboa.pt

Abstract. Many process mining tools and techniques produce output models based on the counting of transitions between tasks or users in an event log. Although this counting can be performed in a forward pass through the event log, when analyzing large event logs according to different perspectives it may become impractical or time-consuming to perform multiple such passes. In this work, we show how transition counting can be parallelized by taking advantage of CPU multi-threading and GPU-accelerated computing. We describe the parallelization strategies, together with a set of experiments to illustrate the performance gains that can be expected with such parallelizations.

1 Introduction

Transition counting is the basis of many process mining techniques. For example, the α-algorithm [1] uses $a >_W b$ to denote the transition between two consecutive tasks in a trace, and the HeuristicsMiner [2] uses the count of such transitions $|a >_W b|$ to derive a dependency graph from the event log. Popular process mining tools, such as Disco [3], also display the discovered model as a graph where the arcs between activities are labeled with transition counts. Transition counting is therefore an essential task in several control-flow algorithms.

Also in the organizational perspective, transition counting plays a central role in extracting sociograms based on metrics such as *handover of work* and *working together* [4,5]. The handover of work metric requires counting the transitions between successive users who participate a case. The working together metric (also known as *joint cases*) counts the number of cases in which a given pair of users have worked together (not necessarily in direct succession). This too can be regarded as a problem of transition counting, where both direct and indirect transitions between users in a case are considered.

Parallel computing [6] offers tremendous possibilities to improve the performance of process mining techniques, especially when event logs become increasingly large. The BPI Challenge 2016[1] is the first in its series (since its inception in 2011) where the event logs to be analyzed exceed 1 GB in size. But as early as

---

[1] http://www.win.tue.nl/bpi/doku.php?id=2016:challenge.

© Springer International Publishing AG 2017
M. Dumas and M. Fantinato (Eds.): BPM 2016 Workshops, LNBIP 281, pp. 36–48, 2017.
DOI: 10.1007/978-3-319-58457-7_3

2009, we were already experiencing some difficulties in processing a set of event logs that amounted to 13 GB in total size [7]. Clearly, for such large event logs, it becomes impractical to explore and analyze them by running multiple passes over the entire event log in a single-threaded fashion.

The availability of multi-core CPUs and the trend towards powerful GPUs (Graphics Processing Units) that can be used for general-purpose computing create the opportunity to leverage those technologies to accelerate the processing of large event logs. There are certainly many techniques that can potentially benefit from such parallelization, but in this work we focus on the essential task of counting transitions. Although the problem might appear to be simple, we will see that its parallelization involves some challenges and trade-offs. In particular, the parallelization on the GPU is fundamentally different from that on a multi-core CPU. We present a viable strategy to perform both.

It should be noted that we are not the first to attempt such parallelization. In the BPI Workshop 2015, there was a work on the parallelization of the α-algorithm [8]. However, such work was based on a single, high-level construct provided by MATLAB (specifically, the parallel for-loop). Here, we go much deeper into the parallelization by controlling the execution of threads at the lowest level of detail. On the CPU, we use POSIX threads [9], and on the GPU we use NVIDIA's CUDA technology and programming model [10].

Section 2 introduces a sample process and event log. Section 3 describes the algorithms that are to be parallelized. Section 4 describes the parallelization in the CPU, and Sect. 5 describes the parallelization on the GPU; both sections include experiments and results. Finally, Sect. 6 concludes the paper.

2   An Example Process

As a running example, and in order to generate variable-size event logs for testing purposes, we use the purchase process from [11]. This process is illustrated in Fig. 1. Basically, there are two main branches: if the purchase request is not approved, it is archived and the process ends; if the purchase request is approved, the product is ordered from the supplier, the warehouse receives the product, and the accounting department takes care of payment.

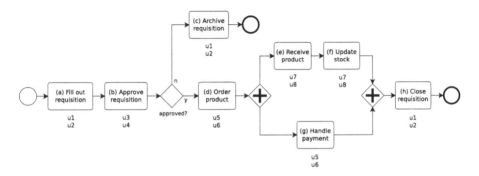

Fig. 1. A simple purchase process

In the lower branch, some activities are performed in parallel, meaning that their execution order is non-deterministic. In addition, each activity is performed by one of two users, depending on who is first available to pick the task at runtime. There are eight users involved in this process, and they are shared by multiple activities, as indicated in Fig. 1.

Table 1 shows a sample event log that has been generated from a simulation of this process. In this small example there are only three cases, but for testing purposes we will use many more (up to 10^7 cases). Our analysis will be focusing on the *case id*, *task* and *user* columns.

Table 1. Sample event log

Case id	Task	User	Timestamp
1	a	u_1	2016-04-09 17:36:47
1	b	u_3	2016-04-11 09:11:13
1	d	u_6	2016-04-12 10:00:12
1	e	u_7	2016-04-12 18:21:32
1	f	u_8	2016-04-13 13:27:41
1	g	u_6	2016-04-18 19:14:14
1	h	u_2	2016-04-19 16:48:16
2	a	u_2	2016-04-14 08:56:09
2	b	u_3	2016-04-14 09:36:02
2	d	u_5	2016-04-15 10:16:40
2	g	u_6	2016-04-19 15:39:15
2	e	u_7	2016-04-20 14:39:45
2	f	u_8	2016-04-22 09:16:16
2	h	u_1	2016-04-26 12:19:46
3	a	u_2	2016-04-25 08:39:24
3	b	u_4	2016-04-29 10:56:14
3	c	u_1	2016-04-30 15:41:22

In this example, the events have been sorted by case id and timestamp, as required by the algorithms to be described below. We assume that this sorting has already been done, or can be done as a one-time preprocessing step. When using a common log format such as MXML [12] or XES [13], such sorting step is unnecessary because the events are already grouped by case id.

3   Algorithms

In this work we consider two basic algorithms that are useful in the control-flow perspective and in the organizational perspective of process mining. The first algorithm counts direct transitions between tasks (the basis for a control-flow model), and the second algorithm counts joint cases between users (the basis for the working together metric [5]). Both are explained in more detail below.

The first algorithm can also be used to count direct transitions between users, which is the basis for the handover of work metric [4]. We will describe this variant only briefly since the algorithm is essentially the same.

3.1 The *flow* algorithm

The purpose of the flow algorithm is to count the transitions between consecutive tasks within each case. For example, if we look at the three cases in the event log of Table 1, we find that there are three occurrences of (a, b), two occurrences of (b, d), one occurrence of (b, c), two occurrences of (e, f), etc.

Let $T = \{a_1, a_2, ..., a_{|T|}\}$ be the set of distinct tasks that appear in the event log, and let (a_i, a_j) denote a transition between two tasks that appear consecutively in the same case id. A transition counting is defined as a function $f : T \times T \to \mathbb{N}_0$ which gives the number of times that each transition has been observed in the event log. The flow algorithm finds all the values for this function.

For convenience, these values can be stored in a matrix F of size $|T|^2$, which is initialized with zeros. Every possible transition in the form (a_i, a_j) has a value in this matrix, which can be found at row i and column j. Algorithm 1 goes through the event log, and every time a transition (a_i, a_j) is observed, it increments the value at position (i, j) in the matrix.

Algorithm 1. Flow

1: Let F be a matrix of size $|T|^2$
2: Initialize $F_{ij} \leftarrow 0$ for every i, j
3: **for** each case id in the event log **do**
4:     **for** each transition (a_i, a_j) in the case id **do**
5:         $F_{ij} \leftarrow F_{ij} + 1$

3.2 The *handover* algorithm

Let $U = \{u_1, u_2, ..., u_{|U|}\}$ be the set of distinct users that appear in the event log, and let (u_i, u_j) denote a transition between two users that appear consecutively in the same case id. Then substituting T by U and (a_i, a_j) by (u_i, u_j) in Algorithm 1 yields the handover of work matrix H.

3.3 The *together* algorithm

Working together is a metric to extract a social network from an event log. The goal is to find, for each pair of users, how many cases those users have worked together in. For example, in Table 1 it is possible to see that u_1 and u_2 have worked together in all three cases, u_1 and u_3 have worked together in two cases, u_1 and u_4 have worked together only once, etc. The together algorithm calculates this count for every pair of users in the event log.

As in the previous algorithms, this count can be stored in a matrix W of size $|U|^2$, which is initialized with zeros. The values in this matrix are incremented as the algorithm goes through the event log. However, these increments require a little bit more work than in the previous algorithms. Specifically, the algorithm needs to collect the set of users who participate in a case id, and then increment the edge count *for every pair of users in that set.*

Algorithm 2 shows how this can be done. Each pair of users can be denoted as (u_i, u_j). Since (u_i, u_j) and (u_j, u_i) refer to the same pair, the algorithm needs to consider only those pairs in the form (u_i, u_j) with $j > i$. As a result, W will be a triangular matrix.

Algorithm 2. Together

1: Let W be a matrix of size $|U|^2$
2: Initialize $W_{ij} \leftarrow 0$ for every i, j
3: **for** each case id in the event log **do**
4:     Let S be a set of users, initialize $S \leftarrow \emptyset$
5:     **for** each user u_i in the case id **do**
6:         $S \leftarrow S \cup \{u_i\}$
7:     **for** each user $u_i \in S$ **do**
8:         **for** each user $u_j \in S$ such that $j > i$ **do**
9:             $W_{ij} \leftarrow W_{ij} + 1$

---

4   Parallelization on the CPU

The parallelization on a multi-core CPU is based on the idea of dividing work across a number of threads. According to the description of Algorithms 1 and 2 above, the most natural division of work is by case id, where each thread receives a subset of case ids for processing.

For this processing to be as much independent as possible between threads, each thread will have a local matrix to count the transitions observed in its own subset of case ids. Then, at the end of each thread, it will be necessary to bring these local counts together into a common, global matrix, which stores the combined results of all threads, as illustrated in Fig. 2.

Since all threads will be updating the global matrix concurrently, it is necessary to employ thread synchronization to avoid race conditions. In this work, we use a mutex lock on the global matrix to ensure that it is updated by one thread at a time. This effectively reduces parallelism in that section of the code. However, if a thread has a lot of case ids to process, in principle the impact should be reduced, because the time spent on updating the global matrix becomes much shorter than the time spent on processing the case ids.

Figure 2 illustrates the parallelization of the flow algorithm in particular. The parallelization of the together algorithm follows the same approach, with the difference being that it works on users rather than tasks, and the calculation

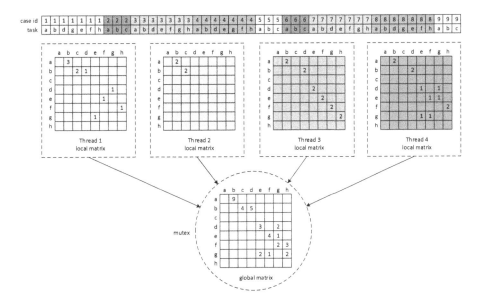

Fig. 2. Multi-threaded parallelization of the flow algorithm on the CPU

of the transition counts is slightly different (see Algorithm 2 vs. Algorithm 1). In any case, at the end of each thread it is necessary to update the global matrix with the transition counts stored in the local matrix.

4.1   Increasing the Number of Threads

For illustration purposes, Fig. 2 shows only four threads but, naturally, this can be extended to an arbitrary number of threads. Increasing the number of threads brings more parallelism, but also more concurrent updates to the global matrix. Since these updates are controlled via a mutex lock, having too many threads may lead to a longer waiting time for the mutex to be released and, eventually, to a decrease in overall performance.

To investigate how far the number of threads can be increased, we carried out an experiment on a machine with two Intel Xeon E5-2630v3 CPUs @ 2.4 GHz with 8 physical cores each, for a total of 16 physical cores. Hyper-threading was enabled [14], so the operating system (Ubuntu) saw 32 virtual cores.

Figure 3 shows the results obtained when running the three algorithms on an event log with 10^6 cases. The dashed line indicates the run-time of the single-threaded version, and the solid line the run-time of the multi-threaded version, as the number of threads is increased. All run-times were averaged over 100 runs of the same algorithm with the same number of threads.

The results in this and other similar experiments indicate that, for the flow and handover algorithms, the ideal number of threads is close to (or even a bit less than) the number of physical cores available in the machine. Any increase

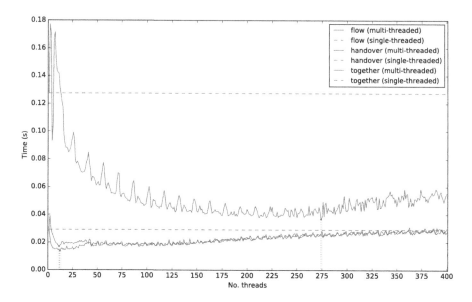

Fig. 3. Run-times of the multi-threaded versions with increasing number of threads

beyond this point results in a decrease in performance. On the other hand, for the together algorithm, there seems to be a benefit in overloading the CPU with a lot of threads. In this algorithm, there is more work to be done for each case id, so the impact of thread synchronization is lower, and the number of threads is allowed to increase up to several times the number of cores.

4.2   Increasing the Log Size

In another experiment, we increased the log size to investigate its impact on the performance of the multi-threaded versions. For this purpose, the number of cases in the event log was increased in powers of 10, from 10^1 to 10^7 cases (resulting in a log file of 1.8 GB). The number of threads was kept at 10 for the flow and handover algorithms, and at 250 for the together algorithm.

Figure 4 shows a plot of the resulting run-times. Again, the dashed line indicates the run-time of the single-threaded version, and the solid line the run-time of the multi-threaded version, as the number of cases is increased.

The single-threaded versions of the flow and handover algorithms have the same run-times (the two dashed lines are coincident), but in their multi-threaded versions the flow algorithm is slightly faster, because the flow matrix is sparser than the handover matrix.

As for the together algorithm, its run-time drops significantly in the multi-threaded version. For 10^7 cases, the performance gain was 4.0× and it was still growing as the number of cases was being increased. For comparison, the flow and handover algorithms achieved a maximum performance gain of 2.3× and 1.7×, respectively, when compared to their single-threaded versions.

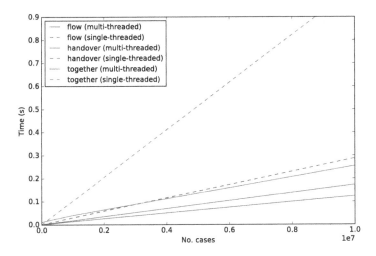

Fig. 4. Run-times of the multi-threaded versions with increasing number of cases

5   Parallelization on the GPU

The parallelization on the GPU requires a completely different approach, because it must match the hardware architecture of modern GPUs. Currently, GPUs have hundreds or even thousands of cores that operate in a thread-synchronous way, with every core executing the same instruction at the same time, but on a different piece of data. This paradigm is usually referred to in the literature as *single instruction, multiple data* (SIMD) [6]. Despite having, in general, a lower clock speed than CPUs, GPUs have so many cores that they can largely outperform the CPU in certain parallelizable tasks.

The Compute Unified Device Architecture (CUDA) [10] is a technology introduced by NVIDIA to make the parallel computing capabilities of GPUs accessible for general-purpose programming. In CUDA, each thread takes care of a single piece of data, and it finishes as soon as that work is done. The idea is to have as many short-lived threads as possible in order to distribute them across the large number of cores available in the GPU.

The work to be done by each thread is programmed into a special function called a *kernel* [15]. The kernel executes on the GPU and is replicated into as many threads as necessary in order to handle all the data that is to be processed. The data must be stored in GPU memory as input and output arrays. Typically, each thread works on a different element (or elements) of these arrays. Complex processing can be done with multiple kernels, where the output array of one kernel becomes an input array to the next, as we will see below.

5.1   Parallelization of the Flow Algorithm

The GPU parallelization of the flow algorithm is based on the idea of having each thread analyzing a single transition between two consecutive tasks. All

transitions will be collected in parallel and at once for the whole event log. In this
scenario, each thread must check if the two consecutive tasks being considered
actually belong to the same case id. If this is not the case, then the thread writes
a special value to the output array, meaning "no transition".

Figure 5 illustrates this step and also what happens afterwards. Once the
transitions have been collected, they are sorted and then counted. Both the
sorting and the counting are performed on the GPU, with the help of the Thrust
library [16]. The sort() routine of the Thrust library makes an efficient use of the
GPU to sort large arrays in GPU memory.

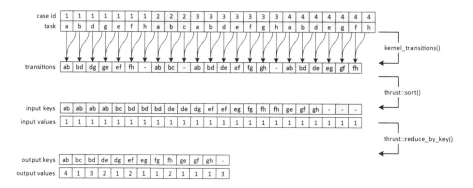

Fig. 5. Parallelization of the flow algorithm on the GPU

After sorting, the counting of transitions can be achieved through a parallel
reduction algorithm [10], namely the reduce_by_key() routine available in the
Thrust library. For this purpose, it is necessary to have an array with input keys,
and another array with input values. Basically, the reduce_by_key() routine sums
the values that correspond to the same key. Since the keys are the transitions
and the values are all 1, the sum by key yields the transition counts.

Parallelization of the handover algorithm follows the same strategy, with the
difference being that threads analyze the transitions between users rather than
tasks. The sorting and counting are performed in exactly the same way.

5.2 Parallelization of the Together Algorithm

The together algorithm is more challenging to parallelize on the GPU because
of the need to find the set of distinct users for each case id, and also the need to
consider all possible pairs of those users, in the form (u_i, u_j) with $j > i$. We do
this in two separated stages, with two different kernels.

As illustrated in Fig. 6, the first kernel marks the participants of each case
in an output array. For each case id, there is a user mask (corresponding to the
set of all users U) that is initialized with zeros. The first kernel is launched with
one thread per event, to analyze the case id and user of each event. If the thread

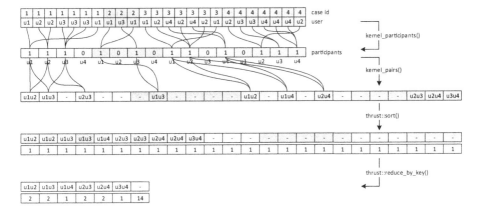

Fig. 6. Parallelization of the together algorithm on the GPU

sees a case id k and a user u_i, it writes 1 at the position $k|U|+i$ in the output array, to mark that u_i participates in case id k.

If the same user appears multiple times in a case id, there will be multiple threads writing 1 at the same position in the output array. However, this does not pose a problem since the operation is idempotent (multiple writes of 1 do not change the result). Also, in Fig. 6 we assume that there are only four users (i.e. $|U|=4$) to keep the figure under a manageable size.

The second kernel has a thread for each pair of users in the form (u_i, u_j) with $j > i$. There are $|U|(|U|-1)/2$ such pairs for each case id. If the mask is 1 for both users, then the thread writes the pair $u_i u_j$ to an output array. This pair is written to the position $k|U|(|U|-1)/2 + i|U|+j$ in the output array.

After this, the pairs are sorted and counted as before.

5.3   Results and Performance Gains

To assess the performance of these GPU parallelizations, we carried out an experiment with a GeForce GTX Titan X with 3072 cores @ 1.08 GHz. Again, we increased the number of cases in powers of 10, from 10^1 to 10^7 cases. Table 2 shows the results, and compares them to the results obtained earlier with the CPU parallelizations in the experiment of Sect. 4.2.

In Table 2, t_{CPU}, $t_{\text{CPU}*}$ and t_{GPU} denote the run-times of the single-threaded CPU version, multi-threaded CPU version, and GPU version, respectively.

From these results, it becomes apparent that the parallelizations provide a performance gain for events logs with 10^5 cases (around 6×10^5 events) or more. Also, the performance gain of the GPU version is noticeably higher than the multi-threaded CPU version, especially for the flow and handover algorithms.

These measurements were made assuming that all data are already in GPU memory, as is common in practice. If memory transfers between CPU and GPU are considered, then the performance gain of the GPU version drops down to roughly the same level as the multi-threaded version.

Table 2. Run-times and performance gains of CPU and GPU parallelizations (The data and source code used to generate these results are available at: http://web.tecnico. ulisboa.pt/diogo.ferreira/bpi2016/)

No. cases	Algorithm	t_{CPU}	t_{CPU*}	t_{CPU}/t_{CPU*}	t_{GPU}	t_{CPU}/t_{GPU}
10	flow	0.000001	0.000359	0.003×	0.000437	0.002×
10	handover	0.000001	0.000312	0.003×	0.000439	0.002×
10	together	0.000008	0.000330	0.024×	0.000443	0.018×
100	flow	0.000005	0.000280	0.018×	0.000440	0.011×
100	handover	0.000007	0.000307	0.023×	0.000442	0.016×
100	together	0.000022	0.003294	0.007×	0.000457	0.048×
1000	flow	0.000070	0.000308	0.227×	0.000357	0.196×
1000	handover	0.000058	0.000306	0.190×	0.000364	0.159×
1000	together	0.000190	0.010535	0.018×	0.000516	0.368×
10000	flow	0.000439	0.000491	0.894×	0.000671	0.654×
10000	handover	0.000431	0.000647	0.666×	0.000676	0.638×
10000	together	0.001430	0.011355	0.126×	0.000757	1.889×
100000	flow	0.003819	0.001957	1.951×	0.000942	4.054×
100000	handover	0.003827	0.003446	1.111×	0.000948	4.037×
100000	together	0.011601	0.014514	0.799×	0.002455	4.725×
1000000	flow	0.031203	0.014213	2.195×	0.004015	7.772×
1000000	handover	0.030080	0.018111	1.661×	0.004028	7.468×
1000000	together	0.102830	0.040930	2.512×	0.020586	4.995×
10000000	flow	0.287948	0.124074	2.321×	0.034346	8.384×
10000000	handover	0.287560	0.173223	1.660×	0.034632	8.303×
10000000	together	1.028615	0.254814	4.037×	0.198020	5.195×

5.4   BPI Challenge 2016 Event Logs

The largest event log in the BPI Challenge 2016 – the click-data for non-logged in customers[2] – has about 9.3×10^6 events. For testing purposes, we used the SessionID as case id, and the PAGE_NAME both as task and as user. It should be noted that there are 1381 distinct page names ($|T| = |U| = 1381$), so the transition matrix is much larger than in our previous experiments.

When running the flow algorithm, the multi-threaded CPU version provides little or no gain over the single-threaded version because the transition matrix is very large and takes a long time to be updated by each thread. However, the GPU version provides a consistent performance gain of 7.4×.

---

[2] https://data.3tu.nl/repository/uuid:9b99a146-51b5-48df-aa70-288a76c82ec4.

When running the together algorithm, the tables are turned. Here, the multi-threaded CPU version provides a performance gain of 5.3×, while the GPU version runs out of memory due to the size of the intermediate arrays.

6   Conclusion

Process mining relies on transition counting procedures which can be accelerated through parallel computing. Parallelization on the CPU provides limited gains due to the need for thread synchronization and merging at some point. Parallelization on the GPU follows a different strategy which avoids thread synchronization and provides higher performance gains, but is limited by GPU memory and memory transfers between CPU and GPU. In any case, both CPU and GPU parallelization provide a promising avenue for accelerating some of the essential tasks that are common to several process mining techniques.

References

1. van der Aalst, W.M.P., Weijters, A.J.M.M., Maruster, L.: Workflow mining: Discovering process models from event logs. IEEE Trans. Knowl. Data Eng. **16**, 1128–1142 (2004)
2. Weijters, A.J.M.M., van der Aalst, W.M.P., de Medeiros, A.K.A.: Process mining with the HeuristicsMiner algorithm. Technical Report WP 166, Eindhoven University of Technology (2006)
3. Günther, C.W., Rozinat, A.: Disco: Discover your processes. In: BPM 2012 Demonstration Track, CEUR Workshop Proceedings, Vol. 940 (2012)
4. van der Aalst, W.M.P., Song, M.: Mining social networks: Uncovering interaction patterns in business processes. In: Desel, J., Pernici, B., Weske, M. (eds.) BPM 2004. LNCS, vol. 3080, pp. 244–260. Springer, Heidelberg (2004). doi:10.1007/978-3-540-25970-1_16
5. van der Aalst, P.W.M., Reijers, A.H., Song, M.: Discovering social networks from event logs. Comput. Support. Coop. Work **14**(6), 549–593 (2005)
6. Rauber, T., Rünger, G.: Parallel Programming for Multicore and Cluster Systems. Springer, Heidelberg (2013)
7. Veiga, G.M., Ferreira, D.R.: Understanding spaghetti models with sequence clustering for ProM. In: Rinderle-Ma, S., Sadiq, S., Leymann, F. (eds.) BPM 2009. LNBIP, vol. 43, pp. 92–103. Springer, Heidelberg (2010). doi:10.1007/978-3-642-12186-9_10
8. Kundra, D., Juneja, P., Sureka, A.: Vidushi: Parallel implementation of alpha miner algorithm and performance analysis on CPU and GPU architecture. In: Reichert, M., Reijers, H.A. (eds.) BPM 2015. LNBIP, vol. 256, pp. 230–241. Springer, Cham (2016). doi:10.1007/978-3-319-42887-1_19
9. Butenhof, D.R.: Programming with POSIX Threads. Addison-Wesley, Reading (1997)
10. Nickolls, J., Buck, I., Garland, M., Skadron, K.: Scalable parallel programming with CUDA. ACM Queue **6**(2), 40–53 (2008)
11. Ferreira, D.R., Vasilyev, E.: Using logical decision trees to discover the cause of process delays from event logs. Comput. Ind. **70**, 194–207 (2015)
12. van Dongen, B.F., van Der Aalst, W.M.P.: A meta model for process mining data. In: EMOI-INTEROP 2005, CEUR Workshop Proceedings, Vol. 160 (2005)

13. Verbeek, H.M.W., Buijs, J.C.A.M., van Dongen, B.F., Aalst, W.M.P.: XES, XESame, and ProM 6. In: Soffer, P., Proper, E. (eds.) CAiSE Forum 2010. LNBIP, vol. 72, pp. 60–75. Springer, Heidelberg (2011). doi:10.1007/978-3-642-17722-4_5
14. Magro, W., Petersen, P., Shah, S.: Hyper-threading technology: Impact on compute-intensive workloads. Intel Technol. J. **6**(1), 1–9 (2002)
15. Nickolls, J., Dally, W.J.: The GPU computing era. IEEE Micro **30**(2), 56–69 (2010)
16. Bell, N., Hoberock, J.: Thrust: A productivity-oriented library for CUDA. In: GPU Computing Gems, 359–371. Jade Edition. Morgan Kaufmann (2011)

Multi-objective Trace Clustering: Finding More Balanced Solutions

Pieter De Koninck[(✉)] and Jochen De Weerdt

Faculty of Economics and Business, Research Center for Management Informatics,
KU Leuven, Leuven, Belgium
{pieter.dekoninck,jochen.deweerdt}@kuleuven.be

Abstract. In recent years, a multitude of techniques has been proposed
for the task of clustering traces. In general, these techniques either focus
on optimizing their solution based on a certain type of similarity between
the traces, such as the number of insertions and deletions needed to transform one trace into another; by mapping the traces onto a vector space
model, based on certain patterns in each trace; or on the quality of a
process model discovered from each cluster. Currently, the main technique of the latter category, *ActiTraC*, constructs its clusters based on a
single objective: fitness. However, a typical view in process discovery is
that one needs to balance fitness, generalization, precision and simplicity. Therefore, a multi-objective approach to trace clustering is deemed
more appropriate. In this paper, a thorough overview of current trace
clustering techniques and potential approaches for multi-objective trace
clustering is given. Furthermore, a multi-objective trace clustering technique is proposed. Our solution is shown to provide unique results on a
number of real-life event logs, validating its existence.

Keywords: Trace clustering · Process mining · Process model quality ·
Multi-objective learning

1 Introduction

Trace clustering is the partitioning of process instances into different groups,
called trace clusters, based on their similarity. A wide variety of trace clustering
techniques have been proposed, differentiated by their clustering methods and
biases. Table 1 contains an overview of these techniques, the data representation
used, the optimization method and the clustering bias. Two main categories of
trace clustering techniques exist: those that map traces onto a vector space model
or quantify the similarity between two traces directly, and those that take the
quality of the underlying process models into account [8,10]. The driving force
behind these proposed techniques is the observation that real-life event logs are
often quite complex and contain a large degree of variation. Since these event
logs are usually the basis for further analysis like process model discovery or
compliance checking [17], partitioning dissimilar process instances into separate
trace clusters is deemed appropriate.

© Springer International Publishing AG 2017
M. Dumas and M. Fantinato (Eds.): BPM 2016 Workshops, LNBIP 281, pp. 49–60, 2017.
DOI: 10.1007/978-3-319-58457-7_4

Table 1. Available trace clustering techniques and their characteristics

Author	Data representation	Clustering technique	Clustering bias
Greco et al. [14]	Propositional	k-means	Instance similarity: alphabet k-grams
Song et al. [16]	Propositional	Various	Instance similarity: *profiles*
Ferreira et al. [11]	Event log	First order Markov mixture model by EM	Maximum likelihood
Bose and van der Aalst [3]	Event log	Hierarchical clustering	Instance similarity: string edit distance
Bose and van der Aalst [4]	Propositional	Hierarchical clustering	Instance similarity: *conserved* patterns
Folino et al. [12]	Event log	Enhanced Markov cluster model	Maximum likelihood
De Weerdt et al. [8]	Event log	Model-driven clustering	Combined process model fitness (ICS)
Ekanayake et al. [10]	Propositional	Complexity-aware clustering	Instance similarity + repository complexity
Delias et al. [9]	Event log	Spectral	Robust instance similarity

Most of the techniques in Table 1 differentiate themselves from previous work in one or more of three different dimensions. Some techniques are unique in a data representation sense: what do they consider as input for a trace clustering? Over time, this has converged to event logs: a well-known grouping of process instances, where each instance is a sequence of activities or events. Secondly, most techniques propose a traditional or unique approach for clustering the traces: this can be based on hierarchical clustering such as in [3] or model-driven such as in [8]. The main distinction between these techniques, however, is the clustering bias, or objective they propose. An example could be the instance similarity metric based on conserved patterns from [4].

The interest in this paper lies mainly in the third differentiator, the clustering bias. Typically, clustering approaches consider a single objective. There can be situations, however, in which one is interested in combinations of these objectives, in which case one would want to deploy a multi-objective trace clustering approach.

Several different strategies are possible for dealing with these multiple objectives, both on the level of the algorithmic clustering technique as on the level of the clustering bias. Therefore, the main objective of this paper is to provide an outlook on multi-objective trace clustering. This outlook contains an elaboration of potential objectives, solution strategies and evaluations.

In light of this objective, this paper is structured as follows: in Sect. 2, an overview of potential objectives for trace clustering are detailed. In Sect. 3, we provide an new trace clustering technique, $ActiTraC_{MO}$[1], which is used as a hook for elaborating directions for future improvement and comparison to the existing trace clustering field. Finally, a small demonstration is included in Sect. 4 to highlight the distinctiveness of our technique.

2 Multi-objective Trace Clustering

2.1 Objectives for Trace Clustering

When it comes to trace clustering, several possible objectives exist. Four of them are envisioned and described here. A first objective is to optimize **process model quality**, as proposed by [8]. In that publication, trace clusters are constructed based on the Improved Continuous Semantics [15], a fitness measure for heuristic nets. This is done by mining a process model for each cluster, and evaluating the fitness of this discovered process model. Rather than purely optimizing based on fitness, process discovery techniques typically should make a trade-off between fitness, precision, generalization and simplicity [5,17]. Fitness is defined as the extent to which a process model is able to replay the behaviour in the event log that was used to discover the model. A precise model is a model that does not underfit: it should not allow too much behaviour that is unrelated to the behaviour in the event log. Generalization is related to overfitting: a process model should allow behaviour that is not present in the event log, but likely given the behaviour in the log. Finally, simplicity is conceptually related to Occam's razor: given two models that score equal on fitness, precision and generalization, the simpler model of the two is preferred.

A second objective is denoted as **similarity**, and is related to the traditional clustering objective: maximizing intra-cluster similarity and inter-cluster dissimilarity. This objective is usually conveyed using instance similarity metrics in trace clustering. Two subgroups of these metrics exist: on the one hand, a subgroup where the similarity between two instance is calculated directly, with approaches such as counting the number of insertions and deletions one would need to transform one process instance into another (Levenshtein edit distance, [3]); on the other hand, approaches where the process instances are mapped onto a vector space model based on the activities they contain, or patterns in these traces [4,14,16]. This vector space model can then be clustered in a more traditional way, using k-means or a hierarchical clustering approach, for example.

A third objective relates to the **justifiability** of a clustering solution, and is related to the expectations of a domain expert. Intuitively, the end user of the clustering may be more interested in trace clustering solutions that are justifiable given his expert knowledge. To our knowledge, there are no techniques that incorporate expert knowledge directly yet in the trace clustering domain. Potential methods for incorporating expert knowledge in an objective could be found in constraint clustering [18], if the expert has perceptions about cases that

---

[1] The approach is implemented as a ProM 6-plugin, which can be found on *http://www.processmining.be/multiobjective*.

must be grouped together or cannot be clustered together, or could be used to initialize the clustering solution based on a small set of label instances obtained from the expert (i.e. semi-supervised clustering, [2]).

A final objective one might want to optimize is the **stability** of a clustering solution. If the event log is likely to contain errors or to be incomplete, one might prefer a clustering solution that is resistant to this and produces more reliable results. In the trace clustering context, stability has been used to determine an appropriate number of clusters [6].

2.2 Combining Multiple Objectives

Apart from deciding which objectives are relevant to a certain clustering or event log, a decision needs to be made on how to combine these different objectives. Three distinct approaches are envisaged: one based on a hierarchy between the objectives, one based on a weighting of the objectives, and one based on consensus clustering.

The first possibility is based on a hierarchy between the different objectives: given that one objective is more important than others, the first objective could be strictly optimized, and the other objectives can than be used as tie-breakers. An example can be found in the discovered process model quality metrics: generally, one wants to make a trade-off between fitness, generalization, precision and simplicity. However, one is often only interested in the simplest model if the other quality measures are not affected (Occam's razor). Therefore, one could treat simplicity as having a lower hierarchical role. If no such hierarchy is present, multi-objective learning techniques usually search for several Pareto-optimal solutions, i.e. solutions for which no other solution can be found that scores better on one of the objectives without scoring worse on any of the other objectives. Typically, multiple solutions will be Pareto-efficient, therefore these algorithms usually return a set of solutions. An example of Pareto-efficient process discovery can be found in [5].

A second approach is to use a weighted objective function. The different objectives receive a weight, and this composed objective can than be used in a single-objective clustering technique. Several drawbacks exist to this approach, the main being that determining the weights can be non-trivial, especially if one prefers to work with unnormalized objectives [5]. A specific version of this type of multi-dimensional trace clustering is possible when one only uses features that can be mapped onto a vector space model, such as in [4]. By simply including more features into the vector model, one is effectively including multiple objectives with an equal weighting.

A third route could stem from co-clustering, which is an approach where cluster solutions are created for each objective separately, and these solutions are then combined to result in a final clustering. This combination is typically based on the frequency with which two traces appear in the same cluster across the different solutions: the more often they appear together, the higher the likelihood that they should be clustered together. An example of co-clustering applied in the process mining domain can be found in [1]. The main drawback of this approach is that there is no guarantee about the objective results on the combined solution: while separate clusterings might obtain very decent scores for each of the criteria, the co-clustered solution will not necessarily maintain this high quality, especially if the objectives are process model quality metrics. Another drawback is that combining cluster ensembles may not be an easy task, especially for a low number of objectives. This can be seen as follows: consider a situation where, in a two-metric solution, clustering 1 puts trace A and trace B in the same cluster, and clustering 2 puts them in a separate cluster, how will the consensus be reached? As before, a resolution could come from hierarchies or weightings.

3 A Multi-objective Trace Clustering Technique

3.1 Approach

As a first step towards acceptable multi-objective trace clustering, we propose Algorithm 1. It is built as an adaptation of ActiTraC [8], a sequential algorithm that constructs clusters based on a process quality metric, namely fitness. It is sequential since it constructs clusters one by one, by adding all traces for which a certain quality threshold is met together, before creating a new cluster. The approach proposed here, however, works parallel rather than sequential meaning that multiple clusters are being constructed at the same time, adding traces to whichever cluster would lead to the best score on a certain quality metric. Furthermore, the approach is hierarchical in the sense that a hierarchy between the cluster metrics is presumed: only if this best score is reached for multiple clusters, the next metric is inspected to test which one of these clusters should be chosen. Although the approach can be adapted to be valid for any metric, it has been conceived with two specific metrics in mind: the weighted F-score metric, which balances fitness and precision [7]; and the place/transition connection degree (P/T-CD) [8], which is a measure for the simplicity of the model. Thus, it combines two approaches from Sect. 2.1, the weighting approach and the hierarchical approach. Observe that the hierarchical approach makes sense in this case, as simplicity is often treated as an objective that is considered to be less of principal interest than the other process quality dimensions, such as fitness, precision or generalization: given two solutions with equal precision, for example, one prefers the simpler one (Occam's razor), but given two solutions with different precision, one would typically prefer the more precise model.

3.2 Algorithm

In this section, the algorithmic structure of our approach is described. The algorithm consists of three phases, each of which is discussed here.

Input. Firstly, our algorithm requires six inputs: a grouped event log, this means all process instances that are similar are grouped together in a single distinct process instance or *dpi*. Whenever the algorithm handles a trace, it is handling all traces that belong to this group of distinct process instances. Furthermore, this grouped event log is ordered, which means that the most frequent dpi's will be treated first.

Secondly, a desired number of clusters is required. If one is uncertain about an appropriate number of clusters, the approach of [6] can be used.

The next input is a list of metrics, ordered in decreasing importance. The algorithm is inspired by the approach proposed in [8], so it is conceived with process model quality metrics in mind. This means that they are evaluated based on a process model that is discovered for each cluster.

Furthermore, two thresholds are needed: a *cluster value threshold cvt*, which is the minimal quality expected from the discovered process model per cluster, it is used while assigning traces to clusters. It is called the cluster value threshold because it implies that the value of the quality metric is calculated based on all the traces that are in the cluster at that time. This is what separates it from the *trace value threshold tvt*, which is also based on this same discovered process model, but the value is calculated solely based on the traces or traces that one is testing. This difference can be important when the metric under scrutiny is a replay-related metric, such as fitness or precision metrics.

When the algorithm attempts to add traces to a cluster, it checks the value of the quality metric on the process model discovered from this cluster. If the *trace metric value tmv* or the *cluster metric value cmv* are below their respective threshold, this trace will be dubbed *unassignable*. Depending on the final input *SeperateBoolean*, these unassignable traces will either be grouped into a separate cluster (if *SeperateBoolean* is true), or added to the best possible cluster otherwise.

Initialization. The first phase of the algorithm is the initialization-phase. Since our approach is centroid-based, each cluster needs to be initialized. This is done as follows: a random trace is taken from the top 10% of the event log in terms of frequency. Then, a process model *PM* is discovered from this trace using Heuristics Miner [19]. Using this process model, the trace and cluster metric value (*tmv, cmv*) are calculated using the most important metric. Observe that at this stage, tmv and cmv are the same thing, since the cluster only contains this trace we are testing. If this trace satisfies the desired quality captured in the thresholds (*tvt and cvt*), it is added to the cluster, removed from the log and the next cluster can be initialized. If it does not satisfy the threshold, the search continues. An iteration counter is included to prevent non-terminating behaviour, which could occur if the thresholds have been set unrealistically high.

Trace assignment. In the following phase, traces are assigned to a cluster, or included in the list of unassignable traces *U*. Since the log is ordered from most frequent to least frequent, the traces will also be assigned in this order. Two variables are used to store information about the clustering process: *multiple* is a boolean that denotes whether the highest value has been reached for more than one cluster and *Check[c]* is an array of booleans which are true if cluster *c* still needs to be checked in future iterations.

The algorithm then starts to evaluate the results of the first hierarchical metric on each of the clusters. For each cluster, 4 situations are possible: (1) the results of the *trace metric value or cluster metric value* are not above the threshold, in which case this cluster is currently not assignable and calculations end here for this cluster and metric; (2) its results are above both thresholds and the *cluster metric value* is either higher than the current best or equal to the current best and the *trace metric value* is higher than the current best: in that case, the current cluster is temporarily the best one, and all the previous ones will not have to be checked in potential following rounds for this trace; (3) if the thresholds are met, but both values are exactly equal to the current best, this cluster will have to be checked in a following round, and there are now multiple clusters that are optimal, so *multiple* becomes true; (4) the thresholds are met, but the values are lower than the currently best solution. The cluster does not have to be checked again in a potential next round, so $Check[c]$ is set to false.

After a metric is checked, the trace is assigned to a cluster and removed from the log, if there were no ties found. If there is a tie, then all clusters for which $Check[c]$ is still true (i.e. the tied clusters) are evaluated again on the next metric, until a unique solution is found, or all metrics have been evaluated, in which case the trace is added to the cluster with the lowest cluster number for which $Check[c]$ is still true. If the loop ends without a cluster to assign the trace to because it did not pass the thresholds, the trace is added to the set of unassignable traces U.

Unassignable resolution. In a final phase the unassignable traces either get added to an extra, separate cluster (if *SeparateBoolean* is true), or they get assigned to the best possible existing cluster, according to a similar strategy as utilized in phase 2. There are two main differences: the thresholds are no longer checked (since these traces are the unassignable ones), and the process model used for calculating the values is not rediscovered each time a trace is added. Rather, the process models are discovered using only the 'not-unassignable' traces, to prevent bias.

3.3   Possible Enhancements

In this section, an overview is given of possible future enhancements to our algorithm. They are divided into two interrelated categories: improvements to the initialization process and window-based extensions.

Considering that our algorithm constructs clusters in parallel, an **intialization** phase is needed. Currently, this is done by randomly selecting a trace from the top 10% of the log in terms of frequency, that are confirmed to have a quality above the imposed thresholds. Intuitively, it seems sensible to start of with a solution that represents a reasonable amount of behaviour (high frequency), and a sufficiently elevated result on the main imposed quality metric. However, other initialization strategies are possible. Firstly, one could argue that rather than initializing from the most frequent traces, one might want to take seeds from the least frequent traces, since it may be likely that

Algorithm 1. Multi-objective Trace Clustering

Input: $L :=$ grouped and ordered event log, $n_b :=$ number of clusters, $M() :=$ list of metrics, $cvt :=$ cluster value threshold, $tvt :=$ trace value threshold; $SeperateBoolean :=$ true if unassignable traces should be grouped in a separate cluster;

Output: $CS :=$ A set of clusters;

Phase 1: Initialization
1: $c := 0$ % Cluster counter
2: $it := 0$ % Iterations counter
3: $CS := [\{\}, \{\}, ..., \{\}]$ % Empty cluster set, array of nb lists of traces
4: **while** $(c < n_b) \wedge (it < |L|)$ **do**
5:     $t := getRandomTrace(L)$ % Get random trace from the top 10% of log in terms of frequency
6:     $PM := HM(CS[c] \cup t)$ % Mine a process model from cluster
7:     $tmv := getMetricValue(M(0), PM, t)$ % Get result of metric on just this trace
8:     $cmv := getMetricValue(M(0), PM, CS[c] \cup t))$ % Get result of metric on entire cluster
9:     **if** $(tmv >= tvt) \wedge (cmv >= cvt)$ **then**
10:         $CS[c] := CS[c] \cup t$ % Add trace to cluster
11:         $L := L \setminus t$ % Remove trace from log
12:         $c := c + 1$ % Increment cluster counter
13:     **end if**
14:     $it := it + 1$ % Increment iteration counter to prevent non-terminating behaviour
15: **end while**
16: **if** $it \geq |L|$ **then**
17:     **return** $[\{\}, \{\}, ..., \{\}]$ % In case of failed initialization: return empty cluster solution
18: **end if**

Phase 2: Trace assignment
19: $U := \{\}$ % List of unassignable traces
20: **for** $t \in L$ **do** % Loop over the distinct traces
21:     $bestCluster := -1$ % Temporary values for assignment
22:     $multiple := false$ % Boolean that indicates multiple clusters have same scores
23:     $Check[] := [true, true, ..., true]$ % Boolean for each cluster: should we still check it?
24:
25:     **for** $m \in M$ **do** % Loop over the metrics
26:         $bestCMV := -1; bestTMV := -1;$ % Temporary values for optimization
27:         **for** $c := 0; c < nb; c := c + 1$ **do** % Inspect each possible cluster
28:             **if** $\neg Check[c]$ **then** Continue % Skip if we don't have to check it anymore
29:             **end if**
30:             $PM := HM(CS[c] \cup t)$ % Mine a process model
31:             $tmv := getMetricValue(m, PM, t)$ % Get result of metric on just this trace
32:             $cmv := getMetricValue(m, , PM, CS[c] \cup t))$ % Get result of metric on entire cluster
33:             **if** $(tmv >= tvt) \wedge (cmv >= cvt)$ **then** % Check thresholds
34:                 **if** $cmv > bestCMV \vee (cmv = bestCMV \wedge tmv > bestTMV)$ **then**
35:                     $bestCMV := cmv; bestTMV := tmv; bestCluster := c$
36:                     $Check[i] := false \ \forall i < c$ % Temporary best cluster; no need to check previous clusters in potential next round
37:                 **else if** $(cmv = bestCMV \wedge tmv = bestTMV)$ **then**
38:                     $multiple = true$ % Ties detected
39:                 **else** $Check[c] := false$ % Worse than current best; don't check again
40:                 **end if**
41:             **end if**
42:         **end for**
43:         **if** $\neg multiple \wedge bestCluster >= 0$ **then**
44:             Break
45:         **end if**
46:     **end for**
47:     **if** $bestCluster >= 0$ **then**
48:         $CS[bestCluster] := CS[bestCluster] \cup t$ % Add trace to cluster
49:         $L := L \setminus t$ % Remove trace from log
50:         Break % Exit for loop
51:     **else**
52:         $U := U \cup t$ % Add trace to unassignable
53:         $L := L \setminus t$ % Remove trace from log
54:     **end if**
55: **end for**

Phase 3: Unassignable resolution
56: **if** $SeperateBoolean$ **then**
57:     $CS[n_b + 1] := U$ % Add remaining traces to a new cluster
58: **else**
59:     Add each trace to the cluster in CS using the same procedure as in phase 2, without checking the thresholds any more. Furthermore, the trace and cluster metric values are now calculated without rediscovering a process model each time.
60: **end if**
61: **return** CS

these seeds will then be more separated than when they are taken from the selection of most frequent traces. Another alternative could come from instance similarity metrics: one could map traces onto a vector space model, e.g. using conserved patterns [4], and make sure that these seeds are sufficiently distant from each other in this space. In such a case, extra objectives (namely the features used to create the vector space) are incorporated indirectly through the seeding process, hence rendering the technique extra 'multi-objective'. Finally, one could leverage yet another category of objectives as they were described in Sect. 2.1, expert knowledge. Instead of randomly selecting the seeds, an expert could select several traces which are deemed distinctive enough to be included in separate clusters. This is expected to raise the objective of attaining a justifiable clustering solution.

A second category of enhancements is related to so-called **window**-based extensions. Similar to the approach in [8], an extra sub-phase could be added to phase 2, the trace assignment phase. The main purpose of such a window is to increase the scalability of an algorithm. This window is defined as a group of traces with two possibilities: on the one hand, it can be the top p% of traces in terms of frequency, on the other hand, it can consist of the p% traces that are closest to the current trace in terms of distance in a vector space model, like a model created with the Maximal Repeat Alphabet [4]. It could be used as follows: each time a trace is added to a cluster (line 49 in Algorithm 1) all traces in the window are checked based on trace metric value, on a process model that is discovered using a log with only the traces that are already in the cluster. If their trace metric value exceeds the trace value threshold (for the most important metric in the hierarchy), it is then added to the cluster as well. Clearly, this greedier assignment step can increase the performance of an algorithm, although a trade-off in terms of quality might be present. If the window is based on a metric not included in the main evaluation objectives, such as the MRA, then it could potentially increase the validity of the trace clustering solution as well.

Apart from implementing a window-based approach or adapting the initialization, one could also look into different process discovery techniques for the mining of a process model from each cluster. Currently, Heuristics miner is used for this, a technique that mines heuristic nets, that are converted to Petri nets if needed for the calculation of certain quality metrics. The main advantages of Heuristics miner are its scalability and its relative robustness to noise. An alternative possibility would be to use a process discovery technique that is itself multi-dimensional in nature, such as the one proposed in [5]. That technique even returns a collection of process models, so a strategy for averaging or selecting the appropriate process model from this collection for the calculation of the quality of the discovered collection of process models would then be needed as well.

4   Experimental Evaluation

The main purpose of this short evaluation is to show the general applicability of our technique, and its distinctness from existing approaches. Therefore, our approach is tested on multiple real-life datasets and compared with a variety of other trace clustering techniques.

Four real-life event logs [7] are subjected to our approach. The number of process instances, distinct process instances, number of distinct events and average number of events per process instance are listed in Table 2. The starting point is that applying process mining methods such as process discovery techniques on the entire event log leads to undesirable results [7].

With regards to trace clustering techniques, we have calculated results using 6 different methods: 3 methods based on 'process-model aware' clustering techniques: our new Act_{MO}; $ActFreq$, a sequential version of ActiTrac that does not use a MRA-based window, and $ActMRA$, [8], a version that does. Furthermore, the comparison is made with 3 'instance-level similarity'-methods (MRA [4]; GED [3]; and K-gram [16]).[2] Each of these techniques is evaluated at a number of clusters of 4 and a number of clusters of 8.

Table 2. Characteristics of the real-life event logs used for the evaluation: number of process instances (**#PI**), distinct process instances (**#DPI**), number of different events (**#EV**) and average number of events per process instance ($\frac{\#EV}{PI}$).

Log name	#PI	#DPI	#EV	$\frac{\#EV}{PI}$
MOA	2004	71	49	6.20
ICP	6407	155	18	5.99
MCRM	956	212	22	11.73
KIM	1541	251	18	5.62

For the comparison of the results, the Normalized Mutual Information (NMI, [13]) shared by two clustering solutions is used. It is a measure for the extent to which two clusterings contain the same information, conceptually defined as the extent in which two process instances are clustered together in both clusterings. This NMI is averaged across the four event logs: L is a set of event logs and P^a and P^b are two clustering solutions:

$$aNMI(P^a, P^b) = \frac{1}{|L|} \sum_{l \in L} (NMI(P^a, P^b)) \tag{1}$$

The results are included in Figs. 1 and 2. From these figures, it is clear that our technique leads to unique clustering results, given its low average similarity

---

[2] The second and third methods are implemented in the ProM-framework for process mining in the *ActiTrac*-plugin. The latter three methods can be found in the *GuideTree-Miner*-plugin.

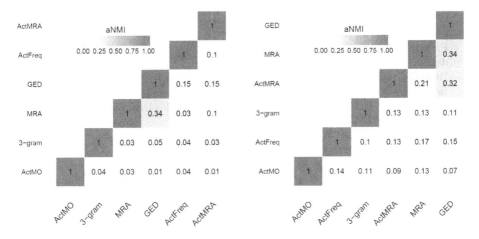

Fig. 1. *aNMI* at 4 clusters   **Fig. 2.** *aNMI* at 8 clusters

with the clustering results obtained using other techniques. Its uniqueness is higher for a cluster size of 4 than for a cluster size of 8. Observe, however, that the similarity between clustering solutions is higher in general at 8 clusters. *GED*, *MRA* and *ActMRA* appear to lead to quite similar solutions.

5   Conclusion

In this paper, a novel approach was presented for the multi-objective clustering of traces. Furthermore, an extensive overview of existing approaches, possible objectives and possible alternative optimization techniques were given. In a short evaluation, the results of our technique where shown to be distinct from those of current trace clustering techniques. Two main elaborations should be made in future work: on the one hand, a comparison of the approach presented here to approaches that include the possible enhancements described in Sect. 3.3, these were the extension with regards to distinct initialization strategies, window-based approaches and utilization of other discovery techniques. On the other hand, these approaches should be thoroughly evaluated based on their runtime, scalability, and the obtained results in terms of their objectives.

References

1. Appice, A., Malerba, D.: A co-training strategy for multiple view clustering in process mining. IEEE Trans. Serv. Comput. **PP**(99), 1 (2015)
2. Basu, S., Banerjee, A., Mooney, R.: Semi-supervised clustering by seeding. In: Proceedings of 19th International Conference on Machine Learning, ICML-2002 (2002)
3. Bose, R., van der Aalst, W.M.P.: Context aware trace clustering: towards improving process mining results. In: Sdm, pp. 401–412 (2009)

4. Bose, R.P.J.C., Aalst, W.M.P.: Trace clustering based on conserved patterns: towards achieving better process models. In: Rinderle-Ma, S., Sadiq, S., Leymann, F. (eds.) BPM 2009. LNBIP, vol. 43, pp. 170–181. Springer, Heidelberg (2010). doi:10.1007/978-3-642-12186-9_16

5. Buijs, J.C.A.M., Dongen, B.F., Aalst, W.M.P.: Discovering and navigating a collection of process models using multiple quality dimensions. In: Lohmann, N., Song, M., Wohed, P. (eds.) BPM 2013. LNBIP, vol. 171, pp. 3–14. Springer, Cham (2014). doi:10.1007/978-3-319-06257-0_1

6. De Koninck, P., De Weerdt, J.: Determining the number of trace clusters: a stability-based approach. In: ATAED 2016, vol. 1592, pp. 1–15. Ceur Workshop Proceedings (2016)

7. De Weerdt, J., De Backer, M., Vanthienen, J., Baesens, B.: A multi-dimensional quality assessment of state-of-the-art process discovery algorithms using real-life event logs. Inf. Syst. **37**(7), 654–676 (2012)

8. De Weerdt, J., Vanden Broucke, S., Vanthienen, J., Baesens, B.: Active trace clustering for improved process discovery. IEEE Trans. Knowl. Data Eng. **25**(12), 2708–2720 (2013)

9. Delias, P., Doumpos, M., Grigoroudis, E., Manolitzas, P., Matsatsinis, N.: Supporting healthcare management decisions via robust clustering of event logs. Knowl. Based Syst. **84**, 203–213 (2015)

10. Ekanayake, C.C., Dumas, M., García-Bañuelos, L., Rosa, M.: Slice, mine and dice: complexity-aware automated discovery of business process models. In: Daniel, F., Wang, J., Weber, B. (eds.) BPM 2013. LNCS, vol. 8094, pp. 49–64. Springer, Heidelberg (2013). doi:10.1007/978-3-642-40176-3_6

11. Ferreira, D., Zacarias, M., Malheiros, M., Ferreira, P.: Approaching process mining with sequence clustering: experiments and findings. In: Alonso, G., Dadam, P., Rosemann, M. (eds.) BPM 2007. LNCS, vol. 4714, pp. 360–374. Springer, Heidelberg (2007). doi:10.1007/978-3-540-75183-0_26

12. Folino, F., Greco, G., Guzzo, A., Pontieri, L.: Editorial: mining usage scenarios in business processes: outlier-aware discovery and run-time prediction. Data Knowl. Eng. **70**, 1005–1029 (2011)

13. Fred, A., Lourenço, A.: Cluster ensemble methods: from single clusterings to combined solutions. In: Okun, O., Valentini, G. (eds.) Supervised and Unsupervised Ensemble Methods and Their Applications. SCI, vol. 126, pp. 3–30. Springer, Heidelberg (2008)

14. Greco, G., Guzzo, A., Pontieri, L., Saccà, D.: Discovering expressive process models by clustering log traces. IEEE Trans. Knowl. Data Eng. **18**(8), 1010–1027 (2006)

15. Alves de Medeiros, A.K.: Genetic process mining (2006)

16. Song, M., Günther, C.W., Aalst, W.M.P.: Trace clustering in process mining. In: Ardagna, D., Mecella, M., Yang, J. (eds.) BPM 2008. LNBIP, vol. 17, pp. 109–120. Springer, Heidelberg (2009). doi:10.1007/978-3-642-00328-8_11

17. van der Aalst, W.M.P., Adriansyah, A., van Dongen, B.: Replaying history on process models for conformance checking and performance analysis. Wiley Interdiscip. Rev. Data Min. Knowl. Discov. **2**(2), 182–192 (2012)

18. Wagstaff, K., Cardie, C., Rogers, S., Schroedl, S.: Constrained k-means clustering with background knowledge. In: In ICML, pp. 577–584. Morgan Kaufmann (2001)

19. Weijters, A., van der Aalst, W.M.P., De Medeiros, A.A.: Process mining with the heuristics miner-algorithm. Technische Universiteit Eindhoven, Technical report WP 166, pp. 1–34 (2006)

Simulation of Multi-perspective Declarative Process Models

Lars Ackermann$^{(\boxtimes)}$, Stefan Schönig, and Stefan Jablonski

University of Bayreuth, Bayreuth, Germany
{lars.ackermann,stefan.schonig,stefan.jablonski}@uni-bayreuth.de

Abstract. Flexible business processes can often be represented more easily using a declarative process modeling language (DPML) rather than an imperative language. Process mining techniques can be used to automate the discovery of process models. One way to evaluate process mining techniques is to synthesize event logs from a source model via simulation techniques and to compare the discovered model with the source model. Though there are several declarative process mining techniques, there is a lack of simulation approaches. Process models also involve multiple aspects, like the flow of activities and resource assignment constraints. The simulation approach at hand automatically synthesizes event logs that conform to a given model specified in the multi-perspective, declarative language DPIL. Our technique translates DPIL constraints to a logic language called Alloy. A formula-analysis step is the actual log generation. We evaluate our technique with a concise example and describe an alternative configuration to simulate event logs based on an assumed partial execution as well as on properties that are intended to be checked. We complement the quality evaluation by a performance analysis.

Keywords: Simulation of business processes · Predictive analytics · Multi-perspective process mining

1 Introduction

Business process simulation supports those phases of the business process management lifecycle that aim at the analysis and improvement of processes [1]. New versions of processes are simulated in order to determine an optimal improvement. Logs produced by simulating processes are analyzed in order to predict effectiveness or efficiency of upcoming versions of processes. Besides analysis, another purpose of process simulation is learning about the meaning of a process. By simulating processes, modelers and users can learn to understand their behavior based on selected log contents [2]. From cognitive science we learn that studying and observing "good examples" of artifacts, here processes, develop their comprehension [3]. A third purpose of simulation is its support for testing process mining techniques [4]. Di Ciccio et al. [5] propose to use simulation in order to generate process logs that are used to test and improve process mining algorithms. It becomes obvious that simulation plays an important role in the lifecycle of business process management.

© Springer International Publishing AG 2017
M. Dumas and M. Fantinato (Eds.): BPM 2016 Workshops, LNBIP 281, pp. 61–73, 2017.
DOI: 10.1007/978-3-319-58457-7_5

Fig. 1. Concept of multi-perspective declarative process model simulation

Most of the available simulation techniques are tailored towards imperative languages such as *BPMN*, e.g. [6,7]. Over the last years, *declarative* process modeling languages (DPMLs) [8–10] and declarative process discovery techniques gained more and more attraction [4,11,12]. Imperative languages model the underlying process *explicitly* using flow-oriented representations. In contrast, declarative languages assume process executions which are restricted by constraints. Due to this semantic gap, the transformation of, especially multi-perspective, declarative process models to an appropriate imperative representation is still vague [13]. Consequently, simulation techniques for imperative models are not suitable for declarative models [5] which leads to a lack of simulation tools for the latter. The approach presented in [5] is the only representative that is able to generate traces based on rules that restrict the temporal ordering and the existence of activities. The simulator and the underlying modeling language consider only control-flow constraints but no other process perspectives including organizational and data-oriented aspects [10]. To the best of our knowledge a multi-perspective declarative process simulation technique is not available.

The approach visualised in Fig. 1 fills this research gap with a simulation technique for *multi-perspective* process models that is based on *Declarative Process Intermediate Language (DPIL)* [10]. We based our simulation technique on a transformation of DPIL rules to a logic language called *Alloy* [14]. An important advantage of using Alloy, in contrast to simulation tools designed for imperative models, is that DPIL rules and Alloy logic expressions have a direct correspondence. Alloy ships with an analyzer that is able to *exhaustively* produce unique examples and counter examples for a given Alloy model. It is possible to produce logs with desired characteristics like size, maximum trace length, trace contents or relative to a partial process execution trace.

This paper is structured as follows: Sect. 2 introduces DPIL. The discussion of the contribution in Sect. 3 is based on a brief description of Alloy (cf. Sect. 2.2). The evaluation is described in Sect. 4 and the paper is concluded in Sect. 5.

2   Background

In this section we introduce the foundations of our approach, i.e., declarative process modelling and DPIL as well as Alloy.

2.1   Multi-perspective Declarative Process Modeling with DPIL

Research has shown that DPMLs are able to cope with a high degree of flexibility [8]. The basic idea is that, without modeling anything, "everything is

Table 1. Basic set of multi-perspective macros of the DPIL language

Macro	Expanded pattern	Semantic
Sequence(a,b)	event(of b at :t) implies event(of a at <t)	Task b cannot be started before task a has been completed
Once(a)	event(of a at :p) implies not event(of a at <p)	Task a can be started once only
Consumes(c,i)	event(of c at :t) implies write(of i at <t)	Task c can not be started before a data object i has a value
Produces(p,o)	event(of p :t) implies write(of o at <t)	Task p can not be completed before a value for the data object o is present
Role(a,r)	event(of a by :id) implies relation(subject id predicate $hasRole$ object r)	Task a must be performed by a process participant in role r
Binding(a,b)	event(of b by :id) implies event(of a by id)	The tasks a and b must be processed by the same identity

allowed". To reduce this flexibility a declarative process model contains constraints which form a forbidden region for process execution paths. Independent from a specific modelling paradigm different perspectives on a process exist. The organizational perspective deals with the definition and the allocation of human and non-human resources to activities. Another perspective is the data-oriented one which deals with restrictions regarding the data flow. The *Declarative Process Intermediate Language (DPIL)* [10] is a multi-perspective declarative process modelling language, i.e., it allows for representing several business process perspectives, namely, control flow, data and especially resources in one model. Comparable languages are data-aware Declare [15] and ConDec-R [16]. In contrast to DPIL, the former only allows for formulating control-flow and data-constraints and the latter provides support only for control-flow restrictions and resource allocation contraints. The expressiveness of DPIL and its suitability for business process modelling have been evaluated [10] with respect to the well-known Workflow Patterns and in industry projects, e.g. the *Competence Center for Practical Process Management*. Although we selected DPIL as our example language, the principle is also applicable to other rule-based process modeling languages.

DPIL provides a *macro* based textual notation to define reusable rules, shown exemplarily in Table 1. We explain all macros using the example process model of Fig. 2 which shows a simple process for trip management in DPIL. The process model states, for instance, that it is mandatory for all applicants to produce the application document for a business trip before it can be approved (*produces* and *consumes*). Means of transport and accommodations can only be booked after the application has been approved (*sequence*). Every task except booking accommodations and means of transport can be performed at most once (*once*).

```
use group Administration

process BusinessTrip {
    task Apply for Trip
    task Approve application
    task Book means of transport
    task Book accommodation
    task Collect tickets
    task Archive documents
    document Application
    document TicketCollection

    ensure produces(Apply for Trip, Application)
    ensure produces(Collect tickets, TicketCollection)
    ensure consumes(Approve application, Application)
    ensure consumes(Archive documents,Application)
    ensure consumes(Archive documents,TicketCollection)

    ensure sequence(Approve application, Book means of transport)
    ensure sequence(Approve application, Book accommodation)
    ensure sequence(Book means of transport, Collect tickets)
    ensure sequence(Book accommodation, Collect tickets)
    ensure once(Apply for Trip)
    ensure once(Approve application)
    ensure once(Collect tickets)
    ensure once(Archive documents)

    ensure role(Approve Application, Administration)
    ensure binding(Book means of transport, Apply for Trip)
    ensure binding(Book accommodation, Apply for Trip)
    ensure binding(Collect tickets, Apply for Trip)
    ensure binding(Archive documents, Apply for Trip)

    milestone "Done": event(of Collect tickets) and event(of Archive documents)
}
```

Fig. 2. Process for trip management modeled with DPIL

The latter can be executed multiple times in order to allow, e.g., for flights with stopover and multiple accommodations per trip. The task *Approve application* must be performed by a resource with the role *Administration*. Additionally it is required that the same person – here the applicant – books the flight and the accommodation (*binding*). In the described setting there is no secretary which is why the applicant is also responsible for collecting the tickets and for archiving the collected documents. A process instance is finished as soon as the tickets are collected and all documents are archived (*milestone*).

2.2 Alloy in a Nutshell

Alloy is a declarative language for building models that describe structures with respect to desired restrictions. We first provide a concise and pragmatic description of Alloy's language features: A signature (*sig*) is similar to a class in object-oriented programming languages (OOPLs). It can be abstract and quantified. A *fact* is comparable to *invariants* in the *Object Constraint Language (OCL)* [17] and allows for specifying non-structural constraints. A function (*fun*) is a parameterizable snippet of re-usable code, that has a return type and performs computations based on the given parameter values. A predicate (*pred*) is comparable to a function but with the limitation that its return type is always a boolean expression. An additional major difference is that Alloy is able to *run* a predicate, which means that the analyzer tries to find models for which that predicate holds. An assertion (*assert*) can be used in combination with *check* commands to test model properties. The body of facts and assertion share the same syntax but in contrast to the former, the analyzer tries to find counter examples for a particular assertion. For further information about the general Alloy syntax we would like to refer to the dedicated literature [14].

3 Simulation of DPIL Models with Alloy

Due to Alloy's declarative nature, it can be used to represent a declarative process model. The correspondence between DPIL and Alloy as well as a mapping are described within this section, starting with a concise characterization.

3.1 Requirements and Functional Characteristics

Process simulation is used for the analysis of properties of business processes [1]. Our approach supports process analysis through event log generation. We identified the following requirements based on the introductory simulation purposes:

- *Distinctness.* Distinctness means to avoid redundant traces. This feature keeps the set of examples as small as possible. Without this feature a log can grow enormously without enhancing information content; its growth then worsen its performance and clarity.
- *Exhaustiveness.* This feature guarantees that all possible process execution paths of a defined maximum length are considered.
- *Determinism.* Determinism says that parts of the log can be replicated according to user defined settings. This is needed to specifically weight alternative execution paths.
- *Multi-perspectivity.* Processes are constituted by multiple perspectives [10]. These perspectives must be identifiable in a process log.
- *Context-awareness.* This property allows to analyze traces taking into account particular process states. Such a process state might depict a certain (partial) execution path; the log then should be analyzed whether there are processes coinciding with that execution path. For instance, if such an execution path depicts the beginning of a process trace, this analysis ascertains whether this process will eventually terminate (i.e. a process trace must be found that shows this prefix and reaches an end state).
- *Reversibility.* It can be useful to generate traces that explicitly violate process specifications (counter examples). From cognitive science we adopt that counter examples are good for gaining understanding (here: of processes) [18].

By basing the simulation on Alloy [14] the first two properties, distinctness and exhaustiveness, are guaranteed. As a consequence, determinism is incidentally achieved, too. The remaining two characteristics are explained further in Sect. 3.4.

3.2   Process Event Chain Meta-Model

Our approach currently focuses on three process perspectives which describes
(i) the temporal and existential relations between tasks (functional and behavioral perspective), *(ii)* the involvement of resources (organizational perspective), and *(iii)* data dependencies (data perspective). Due to this limited scope we are able to treat activity executions as atomic and, therefore, do not have to take into account the usual activity lifecycle. In Alloy we defined our meta model for traces in form of *process event chains (PECs)* in three modules. Two of them are shown in Listings 1.1 and 1.2. Both of them are based on another module providing only one signature, called **sig** *AssociatedElement{}*. This signature serves as an interface for extending the meta-model with additional process elements like variables or even elements of new perspectives like operations.

Listing 1.1 is the Alloy implementation of the well known organizational meta-model introduced in [19]. The first line defines the module name. Afterwards, we make the mentioned *AssociatedElement* available by opening the containing module. Line 4–8 allows for the definition of hierarchically structured relations where process resources [20] may be involved in based on a *subject-predicate-object (spo)* notation. An example would be: *John (s) hasRole (p) Admin (o)*. In our corresponding Alloy-based process model we need four additional signatures in order to represent an instance of this relation – one for *Relation* itself and one for each of the contained fields.

```
   module orgmetamodel
2  open processEventChain_commons

4  abstract sig Relation {s: one Element, p: one RelationType, o: one Element}
   abstract sig Element extends AssociatedElement {}
6  abstract sig Identity extends Element{}
   abstract sig Group{} extends Element{}
8  abstract sig RelationType{}
```

Listing 1.1. Organizational Meta-model

The structure of PECs was mainly motivated by the log structures discussed in [21] as well as related literature and is described in Listing 1.2. After defining the module name we make the two previously described modules available (line 2 and 3). The lines 5–17 describe the structural and the remaining lines describe the non-structural properties of a PEC.

PEvent is an abstract "class" for a general discrete event, including a field declaration for the unique *(disj)* position. The latter defines the position of the event in the PEC. Alternatively, a more intuitive implementation would be a *Linked List*. However, our performance tests showed that the proposed variant is much faster. The signatures in line 7 and 8 are unique (keyword *one*) and denote the beginning and the completion event of a process execution. Line 9 introduces the more interesting *TaskEvent* denoting an activity execution and comprising an integer which is the inherited position as well as associated information like the executed *Task* (cf. line 13) and the assigned organizational resource. The *Task* signature is abstract and is extended in the actual Alloy process model in order to represent concrete tasks (cf. Table 2). In order to

```
   module processEventChain_noLifecycle_multiperspect_IntBased_new
 2 open processEventChain_commons
   open orgmetamodel

 4
   // Signatures: Process Chain Element Structure
 6 abstract sig PEvent { pos: disj Int }
   one sig StartEvent extends PEvent{}{}
 8 one sig EndEvent extends PEvent{}{}
   abstract sig TaskEvent extends PEvent { assoEl: some AssociatedElement }{
10        #(Task & assoEl) = 1 }
   sig HumanTaskEvent extends TaskEvent{}{
12        #(Identity & assoEl) = 1 }
   abstract sig Task extends AssociatedElement{}
14 abstract sig DataObject {}
   abstract sig DataAccess extends AssociatedElement{ data: one DataObject }
16 abstract sig WriteAccess extends DataAccess{}

18 // Facts: Additonal non-structural constraints
   fact { ∀ intVal: Int • intVal ≥ StartEvent.pos }
20 fact { ∀ e: (PEvent - StartEvent - EndEvent) •
             e.pos < (StartEvent.pos + #TaskEvent + 1) }
22 fact { EndEvent.pos ≤ (StartEvent.pos + #TaskEvent + 1) }
   fact { ∀ assoEls: (AssociatedElement - Group) •
24          assoEls in TaskEvent.assoEl }
   fact { ∀ do: DataObject • do in DataAccess.data }
26 fact { ∀ te: TaskEvent • #(te.assoEl & Group) = 0 }

28 // Utility Functions
   fun exist(asso: AssociatedElement): set TaskEvent {
30   { te: TaskEvent • asso in te.assoEl } }
   fun inBefore(curE: TaskEvent, asso: AssociatedElement): set TaskEvent {
32   { te: TaskEvent • te.pos < curE.pos and asso in te.assoEl } }
   fun roleOf(id: Identity) : set Group{
34   { g: Group • some r: Relation • r.s=id and r.o in Group } }
   fun dAccess(d: DataObject, type: DataAccess): one DataAccess {
36   { da: DataAccess • da in type and d in da.data } }
```

Listing 1.2. Process Event Chain Meta Model

distinguish between different activity types like manual and automated tasks, the *TaskEvent* signature is abstract, too. In line 11 *HumanTaskEvent* is used to represent a manual task and it consequently extends the *TaskEvent* signature. Both signatures have an appended fact which also could be formulated using an additional *fact* statement which is only a matter of personal preferences [14]. The appended facts ensure that a *TaskEvent* encapsulates exactly one task (line 10) and one executing resource (line 12). The lines 14–16 encode the functionality to specify data objects and write accesses to these data objects. We decided to extend a more general access type (*DataAccess*) in order to allow for extending the meta-model with different access types like read accesses.

The lines 19–21 ensure that a process event chain starts with a *StartEvent* (line 19) and ends with an *EndEvent* (lines 20–21) and consequently force all *TaskEvents* to occur in between. The third fact ensures that the position increment between two consecutive tasks is 1. The remaining three facts ensure that the solver only generates process elements that are "used" in at least one event (lines 23–25) and prevents all events from containing information about organizational structures (line 26), since the organizational structures can be defined using the organizational meta-model shown in Listing 1.1.

The first two utility functions collect all *TaskEvents* that involve the overall execution of a given task (lines 29–30) or *before* (lines 31–32) a given event. The function *roleOf* calculates all roles a particular resource has. The last function identifies the concrete *DataAccess* signature for the given *DataObject* and type.

3.3 Transformation of DPIL Models to Alloy

Based on the process event chain meta-model presented above, we now discuss how to transform a DPIL model into an Alloy model that contains all restrictions for *valid* process event chains. This involves two major steps: *(i)* Creating signatures for tasks, roles and identities that fulfill these roles, data objects and access objects and *(ii)* translating the DPIL rules to Alloy facts (cf. Table 2).

Table 2. Mapping: DPIL - Alloy

DPIL	Alloy
task T	**sig** T **extends** Task{}
use group G	**sig** G **extends** Group{} **one sig** HasRole **extends** RelationType {} **abstract sig** IsG **extends** Relation {} { o =G ∧p =HasRole }
document d	**sig** d **extends** DataObject{} **sig** Write_d **extends** DataAccess{}{ data =d }
sequence(T,U)	**fact**{ ∀e: TaskEvent•U **in** e.assoEl →#inBefore[e,T]>0 }
produces(T,d)	**fact**{ ∀e: TaskEvent•T **in** e.assoEl →dAccess[d,WriteAccess] **in** e.*assoEl }
consumes(T,d)	**fact**{ ∀e: TaskEvent•T **in** e.assoEl →#inBefore[e,dAccess[d,WriteAccess]] > 0 }
once(T)	**fact**{ **lone** e: TaskEvent•T **in** e.assoEl }
role(T,r)	**fact**{ ∀e: TaskEvent•e.task=A →r **in** roleOf(e.executor) }
binding(T,U)	**fact**{ ∀e,f: TaskEvent• T **in** e.assoEl **and** U **in** f.assoEl →#((e.assoEl & Identity) & f.assoEl) =1 }
milestone event(T)	**fact** { ∀e: TaskEvent•#exist[T]=1 **and** #(T&e.assoEl)=0 →**not**(e.pos>exist[T].pos) }

Tasks are modeled by extending the existing Task signature from the meta model. In a similar way DPIL's *use group* and *documents* are mapped by extending the Group and DataObject signatures, respectively. In order to type data accesses, we additionally extend the *DataAccess* signature. Additionally a new *Relation* signature is created to be able to easily assign a role to the desired resources (*Identity*). Using this mapping it is only possible to represent flat resource-role associations. However, based on the generic organizational meta-model shown in Listing 1.1 it is possible to model hierarchical structures, too.

DPIL rules are modeled as Alloy *facts*. Alloy rules are declarative and first select atoms belonging to particular signatures the rule shall be applied to. Using the logical *implication* ($\rightarrow$) operator one can specify rule activation (left part) and validity conditions (right part). In order keep the rules concise, we make use of the functions contained in the process event chain meta-model like *inBefore*. The current simple milestone transformation considers milestones that can be reached when executing particular activities. Since *facts* are connected via conjunction we can generate one fact per activity execution that is observed by a milestone rule.

3.4 Simulation Configuration

There are two simulation parameters that are required in most cases [5]: *(i)* The *number of simulated traces (N)* and *(ii)* the *maximum trace length (L)*. Restricting the log size in terms of the number of traces is necessary to be able to provide a reproducible setting for trace generation. The number of events per trace should be restricted because of potential infinite activity loops. Furthermore, the aspect of reproducibility is also influenced by the trace length. Beside these essential simulation boundaries additional parameters may be useful, dependent on the simulation purpose. Though it is impossible to guess the particular simulation purpose (cf. Sect. 1), this section describes three different configurations: *(i)* Trace generation, *(ii)* context-aware simulation and *(iii)* property testing.

Using Alloy trace generation can be implemented by introducing an empty predicate (*sim*) and configuring a *run* command. This can be done according to the following template: **run** *sim* **for** [*L*] *TaskEvent*, [*B*] *Int*. The introduced *length* parameter *L* can be configured directly through a scope restriction for *TaskEvents*. Since we identify the position of an event in the process event chain via an index, we also have to provide the number of integer values to generate. This is done via the *bitwidth* parameter *B*. The Analyzer then generates integer values in the codomain of $\left[\frac{-2^B}{2} + 1, \frac{2^B}{2} \right]$. Hence, *B* can be calculated directly according to $B = \lceil \operatorname{ld} L \rceil$. Via collecting all unique results produced by the Alloy analyzer the desired amount of traces can be obtained.

Here, a *context-aware* simulation means that the simulation is not started at the beginning of a particular process but "somewhere between" the start and the end of the process. An example application is to check the satisfiability assuming a particular process state and to generate all traces that remain. This can be implemented by adding a *fact* for each assumed event that already happened and assigning a fixed position as well as *AssociatedElements* to an event at this position. The position can be calculated generically based on the position of the *StartEvent*. The simulation can be started using a run command, too.

A *hypothesis* is an assumption regarding structure and contents of a trace. In order to check hypotheses they have to be transformed into *predicates*. A predicate can be checked in an *assertion*. Instead of using a *run* command the *check* command has to be used but the parameters are the same. Running the analyzer results either in counter examples proving that a hypothesis is wrong or does not provide any result and, thus, corroborates a hypothesis. With this mode selected properties of the source model can be tested.

4   Implementation and Evaluation

In order to evaluate the simulation approach efficiently, we implemented a *model-to-text* transformation using *Acceleo*[1] which automatically translates DPIL models into Alloy. Acceleo is an implementation of the *MOF Model to Text Transformation Language (MOFM2T)*[2] defined by the OMG. The transformation is currently based on the macros discussed in the paper at hand. The generated Alloy file is then used in our simulator implementation[3] to generate traces of a configurable length and amount and serializes them in the *eXtensible Event Stream (XES)* [22] standard format. In order to evaluate the correctness of the generated traces regarding the source process model we make use of the same evaluation principle as in [5]. This means that we use a previously evaluated process mining technology and try to reproduce the original process model. For the paper at hand we utilized *DPILMiner* [4]. As evaluation example we used the DPIL process model shown in Fig. 2. We configured DPILMiner with the same set of rule templates like the simulation approach. After applying transitive reduction techniques on the extracted model, DPILMiner reproduced exactly the source model. Additionally, we performed property tests for all generated facts which is comparable to unit testing. These property tests have been implemented using *assert*ions and the *check* command. Another aspect of the evaluation is the performance of the proposed simulation technique. Since the simulation time increases with higher parameterizations for the number of traces (N) and their maximum lengths (L), we have performed several simulations of the continuous process model example with different configurations and results shown in Table 3.

The performance analysis shows that the computation is mainly influenced by the trace length. Furthermore, as a minor detail, we have no increase of computation time between the second and the third configuration (the time measurements in parentheses). The reason was that with a maximum trace length of 10 there are less then 100 different process event chains. For the performance analysis, we used a Dell Latitude E6430 (Core i7-3720QM with 8 × 2.6 GHz, 16 GB memory, SSD drive and Windows 8 64 Bit). The simulator is implemented in Java and we used a 64-Bit JVM with a maximum memory allocation pool of 4096M. We decided to present the performance analysis without a comparison to the technique discussed in [5] because there are large functional differences. First, the approach presented in the paper at hand considers multiple perspectives, which is not possible with the technique proposed in [5]. Secondly, our approach guaranties to simulate *all unique* traces of a defined maximum length. Additionally our simulation technique can be used in three different modes (cf. Sect. 3.4). These major functional differences result in an increase of computation time and in a significant decrease in terms of scalability. Thus, we can say that the approach presented in [5] should be used if you need event logs with longer traces that reflect the plain control flow. If the particular application involves multiple

---

[1] Download: http://www.eclipse.org/acceleo, last access: July 22, 2016.

[2] Standard: http://www.omg.org/spec/MOFM2T/1.0/, last access: July 22, 2016.

[3] Screenshot and Download (incl. example data): http://mps.kppq.de.

Table 3. Performance analysis

L	N	Time in s	L	N	Time in s
10	10	1.9	50	10	364.9
10	100	(2.4)	50	100	389.9
10	1000	(2.4)	50	1000	555.8
20	10	17.4	60	10	627.3
20	100	21.3	60	100	649.5
20	1000	52.8	60	1000	871.5
30	10	65.8	70	10	1167.1
30	100	71.8	70	100	1271.5
30	1000	122.3	70	1000	1697.0
40	10	159.0	80	10	2038.8
40	100	180.0	80	100	2194.3
40	1000	300.0	80	1000	2733.5

perspectives, and either the trace length is rather low or the computation time is not a main concern we suggest to use the presented technique.

5   Conclusion and Future Work

The paper at hand describes a process simulation technique which can be used to generate exemplary execution traces for a given process model in order to support business process management. There is only one comparable approach and this considers only plain control-flow models. Our proposed simulation approach primarily focuses on models that consider the behavioral, the organizational, *and* the data-oriented perspective. Additionally to the generation of exemplary traces, the simulation can be used in two additional modes, i.e. *(i)* context-aware simulation and *(ii)* property testing. Both modes can be used for targeted process analysis or gaining a deeper general understanding of the underlying process. A generic meta-model for process event chains and an independent logic framework called Alloy opens the opportunity for extensions. An open issue is the rather low simulation performance and scalability in the case of longer process event chains. Similar to general purpose programming languages, the same functionality can be developed more or less efficiently, dependent on the programming style. Consequently, there is a huge potential for performance optimization, e.g. the order of set joins which is a known issue in databases. Hence, we are currently planning a major evaluation study in order to get a better idea of the driving factors for scalability. Another limitation is the small set of supported rule templates (macros). In order to check Alloy's applicability we formed the set as heterogeneous as possible. Thus, extending this initial set of macros should be rather straightforward. The presented technique focuses on trace generation rather than process performance analysis. Conventional simulation tools emulate variability concerning activity durations and human decisions based on probability distributions which cannot be modeled using Alloy. Hence, we are currently developing a post-processing step which is able to compensate this limitation.

Acknowledgments. The authors would like to thank Prof. Westfechtel, Felix Schwägerl (University of Bayreuth) and Prof. Daniel Jackson (MIT) for providing tips and literature about modeling and analysis with Alloy.

References

1. van der Aalst, W.M.P.: Business process simulation revisited. Enterp. Organ. Model. Simul. **63**, 1–14 (2010)
2. Frank, U.: Multi-perspective enterprise modeling (memo) conceptual framework and modeling languages. In: HICSS, pp. 1258–1267 (2002)
3. Brown, A.L., Kane, M.J.: Preschool children can learn to transfer: learning to learn and learning from example. Cogn. Psychol. **20**, 493–523 (1988)
4. Schönig, S., Cabanillas, C., Jablonski, S., Mendling, J.: Mining the organisational perspective in agile business processes. In: BPMDS, pp. 37–52 (2015)
5. Di Ciccio, C., Bernardi, M.L., Cimitile, M., Maggi, F.M.: Generating event logs through the simulation of declare models. In: Barjis, J., Pergl, R., Babkin, E. (eds.) EOMAS 2015. LNBIP, vol. 231, pp. 20–36. Springer, Cham (2015). doi:10.1007/978-3-319-24626-0_2
6. De Medeiros, A.A., Günther, C.W.: Process mining: using cpn tools to create test logs for mining algorithms. In: Proceedings of CPN, vol. 576 (2005)
7. Burattin, A., Sperduti, A.: PLG: a framework for the generation of business process models and their execution logs. In: Muehlen, M., Su, J. (eds.) BPM 2010. LNBIP, vol. 66, pp. 214–219. Springer, Heidelberg (2011). doi:10.1007/978-3-642-20511-8_20
8. Fahland, D., Lübke, D., Mendling, J., Reijers, H., Weber, B., Weidlich, M., Zugal, S.: Declarative versus imperative process modeling languages: the issue of understandability. In: Halpin, T., Krogstie, J., Nurcan, S., Proper, E., Schmidt, R., Soffer, P., Ukor, R. (eds.) BPMDS/EMMSAD -2009. LNBIP, vol. 29, pp. 353–366. Springer, Heidelberg (2009). doi:10.1007/978-3-642-01862-6_29
9. Pichler, P., Weber, B., Zugal, S., Pinggera, J., Mendling, J., Reijers, H.A.: Imperative versus declarative process modeling languages: an empirical investigation. In: Daniel, F., Barkaoui, K., Dustdar, S. (eds.) BPM 2011. LNBIP, vol. 99, pp. 383–394. Springer, Heidelberg (2012). doi:10.1007/978-3-642-28108-2_37
10. Zeising, M., Schönig, S., Jablonski, S.: Towards a common platform for the support of routine and agile business processes. In: CollaborateCom (2014)
11. Maggi, F.M., Bose, R.P.J.C., Aalst, W.M.P.: A knowledge-based integrated approach for discovering and repairing declare maps. In: Salinesi, C., Norrie, M.C., Pastor, Ó. (eds.) CAiSE 2013. LNCS, vol. 7908, pp. 433–448. Springer, Heidelberg (2013). doi:10.1007/978-3-642-38709-8_28
12. Schönig, S., Rogge-Solti, A., Cabanillas, C., Jablonski, S., Mendling, J.: Efficient and customisable declarative process mining with SQL. In: Nurcan, S., Soffer, P., Bajec, M., Eder, J. (eds.) CAiSE 2016. LNCS, vol. 9694, pp. 290–305. Springer, Cham (2016). doi:10.1007/978-3-319-39696-5_18

13. Ackermann, L., Schönig, S., Jablonski, S.: Towards simulation- and mining-based translation of resource-aware process models. In: Proceedings of ReMa (2016)
14. Jackson, D.: Software Abstractions: Logic, Language, and Analysis. MIT Press, Cambridge (2012)
15. Montali, M., Chesani, F., Mello, P., Maggi, F.M.: Towards data-aware constraints in declare. In: Proceedings of the 28th SAC, pp. 1391–1396. ACM (2013)
16. Barba, I., Weber, B., Del Valle, C., Jiménez-Ramírez, A.: User recommendations for the optimized execution of business processes. DKE **86**, 61–84 (2013)
17. Warmer, J.B., Kleppe, A.G.: The Object Constraint Language: Precise Modeling With Uml (Addison-Wesley OTS). Addison-Wesley Professional, Boston (1998)
18. Zazkis, R., Chernoff, E.J.: What makes a counterexample exemplary? Educ. Stud. Math. **68**(3), 195–208 (2008)
19. Bussler, C.: Analysis of the organization modeling capability of workflow-management-systems. In: PRIISM 1996 Conference Proceedings, pp. 438–455 (1996)
20. Object Management Group (OMG): Business process model and notation (bpmn) version 2.0. Technical report, January 2011
21. van der Aalst, W.: Process Mining: Discovery, Conformance and Enhancement of Business Processes, vol. 2. Springer, New York (2011)
22. Verbeek, H.M.W., Buijs, J.C.A.M., Dongen, B.F., Aalst, W.M.P.: XES, XESame, and ProM 6. In: Soffer, P., Proper, E. (eds.) CAiSE Forum 2010. LNBIP, vol. 72, pp. 60–75. Springer, Heidelberg (2011). doi:10.1007/978-3-642-17722-4_5

Model Checking of Mixed-Paradigm Process Models in a Discovery Context

Finding the Fit Between Declarative and Procedural

Johannes De Smedt[1]($\boxtimes$), Claudio Di Ciccio[2], Jan Vanthienen[1],
and Jan Mendling[2]

[1] Department of Decision Sciences and Information Management,
Faculty of Economics and Business, KU Leuven, Leuven, Belgium
{johannes.desmedt,jan.vanthienen}@kuleuven.be
[2] Department of Information Systems and Operations,
Vienna University of Economics and Business, Vienna, Austria
{claudio.di.ciccio,jan.mendling}@wu.ac.at

Abstract. The act of retrieving process models from event-based data logs can offer valuable information to business owners. Many approaches have been proposed for this purpose, mining for either a procedural or declarative outcome. A blended approach that combines both process model paradigms exists and offers a great way to deal with process environments which consist of different layers of flexibility. In this paper, it will be shown how to check such models for correctness, and how this checking can contribute to retrieving the models as well. The approach is based on intersecting both parts of the model and provides an effective way to check *(i)* whether the behavior is aligned, and *(ii)* where the model can be improved according to errors that arise along the respective paradigms. To this end, we extend the functionality of Fusion Miner, a mixed-paradigm process miner, in a way to inspect which amount of flexibility is right for the event log. The procedure is demonstrated with an implemented model checker and verified on real-life event logs.

Keywords: Declarative process models · Model checking · Process mining

1 Introduction

The field of process discovery [1] has witnessed the introduction of a vast amount of approaches, both of a procedural [2–4] and a declarative [5,6] nature. Both paradigms put a different emphasis on the retrieval of control flow constructs. On the one hand, the former fits activities in paths that are extended with control flow routing objects such as (X)OR- and AND-splits and -joins. Exemplary models include Petri nets and BPMN [7]. The latter, on the other hand, typically encompasses a constraint-based approach that captures behavior in the event log by fitting sequence rules over the activities. Most notably, the Declare language

© Springer International Publishing AG 2017
M. Dumas and M. Fantinato (Eds.): BPM 2016 Workshops, LNBIP 281, pp. 74–86, 2017.
DOI: 10.1007/978-3-319-58457-7_6

[8] has often been applied in this context [5,6]. Intermediate approaches exist in the form of Hybrid Miner [9] and Fusion Miner [10,11]. The latter mines process models with intertwined state spaces, whereas the former bases upon atomic subprocesses that use either paradigm, as in [9]. This type of process mining goes beyond the traditional techniques and fits different types of behavior in the log with the appropriate paradigm, i.e., the fixed sequences are captured with procedural process discovery techniques, while behavior that is hard to fit within such fixed sequence is mined with declarative techniques. Tackling a log in such a fashion yields more comprehensive though fitting and precise models, especially in the case of environments where multiple levels of flexibility exist. The validation aspect of the retrieved models, however, has still not been investigated yet. In this paper, the checking of mixed-paradigm models is elaborated by finding the common ground of both model types in the form of a global automaton. By using such a single executable model, the behavior of the discovery produce can be checked for inconsistent behavior, which can be pinpointed along the different paradigms.

The paper is organized as follows. Section 2 provides a motivation for the usefulness of the approach. Section 3 introduces the formalisms which are further used to explain the model checking and mining techniques in Sect. 4. Section 5 evaluates the approach on a real-life data log, which is followed by the conclusion and future work in Sect. 6.

2   Motivation

Mixed-paradigm models consist of a blend of procedural and declarative process models. More precisely, this entails models which on the one hand contain fixed execution paths, while on the other hand incorporate activity-based rules. Constructing such models is not always straightforward, as one has to be able to grasp the intricacies of both parts, as well as the effect they have on one another. Especially for process discovery, a consistency problem regarding the internal behavior can occur. Because both models are mined separately, though over the same alphabet, many conflicts can occur. For instance, the procedural model might allow for an activity to be enabled, while the declarative model does not, or vice versa. In this case, the activity has to conform to the most restricting model and become disabled. However, this might cause deadlocks for the other model later on, where the activity is not enabled or did not enable another activity that is needed to reach the final state(s) of the model. Consider the model in Fig. 1. A procedural Petri net model is combined with multiple Declare constraints. In order to reach a marking containing $p5$, *precedence(g,c)* and *not no-existence(f,g)* cannot reside in the model together, as b requires f to fire first, which means that c can never be executed as g cannot fire anymore and c is in a *precedence* relation with g.

The challenge is to find whether the behavior of both models is compatible and can be used as a whole, and if not, where the discrepancies reside. This might lead to insights into how the model types interact, e.g., the procedural model

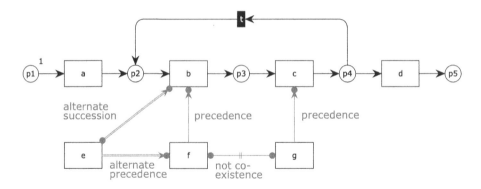

Fig. 1. An example of a mixed-paradigm model containing inconsistencies.

might be too restrictive to allow for the more flexible behavior of the declarative model. By pinpointing which constraints are not working out with the procedural model, the modeler or miner might react accordingly. This is the principle that is used by the model checking approach proposed in this paper. By incrementally matching both models, the maximal conjoined behavior is sought after, ideally yielding a full match of behavior. Furthermore, Fusion Miner is adopted to also recognize procedural models with a vast state space. This reflects the presence of either a vast model, or of a model with many routing constructs for achieving a wide array of execution paths. The latter indicates that the model is either overfitting, or tries to capture a vast deal of the flexibility still, which conflicts with the aim of the approach to capture flexibility with the declarative process model. Hence, the algorithm adapts itself automatically to revise its outcome to shift the balance of the model towards the declarative part.

3   Preliminaries

In this section, we outline the notions upon which our approach is devised. We explain the concept of a mixed-paradigm process, along with the description of the modeling languages that are utilized respectively for the declarative and the procedural part, i.e., Declare and Petri nets. Previous work of which the implementation in CPN Tools is a well-known example [12], proposes building the different execution automata and joining them on-the-fly for simulation purposes. The aim of this work, however, is not to ensure execution ex-ante, but rather to check whether a discovered mixed-paradigm process model is executable ex-post.

3.1   Mixed-Paradigm Models

Mixed-paradigm models consist of a combination of both procedural and declarative model constructs. In this paper we use Petri nets [7] and Declare [13] to represent either paradigm, because they are commonly used languages for such endeavors [14,15].

More formally, we define the activities of the model as a finite set A, for which any type of connection $F \subseteq A \times A$ can exist. There exist four types of activities:

$D \subseteq A$: the activities appearing in the declarative model,
$S \subseteq A$: the activities appearing in the procedural model,
$DD \subseteq D \setminus S$: the activities only appearing in the declarative model, and
$SS \subseteq S \setminus D$: the activities only appearing in the procedural model.

Hence, there are four corresponding connection types:

$F_D \subseteq D \times D$: the connections in the declarative model,
$F_{DS} \subseteq D \times S$: the connections from the declarative to the procedural model,
$F_{SD} \subseteq S \times D$: the connections from the procedural to the declarative model,
    and
$F_S \subseteq S \times S$: the connections in the procedural model.

The declarative model, being a constraint-based Declare model, is defined as a tuple $DM = (D, F_D)$ with F_D the set of constraints defined over D. They can be of any kind as listed in Table 1. For simplicity, we only consider unary and binary constraints, but the proposed technique would be able to incorporate any type of constraints that are expressed in the same (finite) execution semantics. All constraints can be expressed as regular expressions, which yield finite state automata (FSA) [16,17]. The automaton of a single constraint $f \in F_D$ is denoted as $\alpha(f)$. The conjunction of two or more constraints corresponds to the product $\otimes$ of the related automata. Therefore, to get the full behavior of the model, the separate automata are conjoined to build the global automaton by means of the product operation: $\Phi = \prod_{f \in F_D} \alpha(f)$.

The procedural model is defined as a Petri net, namely a tuple $PN = (P, S^{PN}, F^{PN}, L)$ with P a set of finite places, S^{PN} the set of transitions, which coincide with the aforementioned procedural activities, and $F^{PN} \subseteq (P \times S^{PN}) \cup (S^{PN} \times P)$. We consider Petri nets without weighed arcs, reset and inhibitor arcs, coloring, or stochastic extension, but with an injective labeling function $L : S^{PN} \rightarrow S \cup \tau$ with τ silent transitions used for routing purposes. For every $s \in S^{PN}$, the preset of places is defined as $\bullet s = \{p \mid (p, s) \in F^{PN}\}$ and the postset of places as $s\bullet = \{p \mid (s, p) \in F^{PN}\}$. A marking is a function $M : P \rightarrow \mathbb{N}$ which assigns a number of tokens to a place. M_0 is the initial marking of the net. A transition $s \in S^{PN}$ is said to be enabled iff $\forall p \in \bullet s, M(p) > 0$. When a transition s fires, all output places in $s\bullet$ receive an extra token, and from all input places in $\bullet s$ a token is subtracted. The new marking M' is thus such that: $\forall p \in \bullet s \quad M'(p) = M(p) - 1, \forall p \in s \bullet \quad M'(p) = M(p) + 1$, and $\forall p \in P \setminus \{s \bullet \cup \bullet s\} \quad M'(p) = M(p)$. Every marking that is generated by a sequence of firings from M is said to be reachable from M. A net is bounded when the number of reachable markings from M_0 is finite.

For a bounded Petri net, a reachability graph can be calculated [7,20]. The reachability graph of a bounded Petri net is a transition system constructed as follows. The initial marking is the initial state. Every reachable marking from M_0

Table 1. An overview of Declare constraint templates with their corresponding LTL formulas, regular expressions, and verbose descriptions.

Template	Regular Expression [18,19]	Description
Existence(A,n)	.*(A.*){n}	Activity A happens at least n times.
Absence(A,n)	[^A]*(A?[^A]*){n}	Activity A happens at most n times.
Exactly(A,n)	[^A]*(A[^A]*){n}	Activity A happens exactly n times.
Init(A)	(A.*)?	Each instance has to start with activity A.
Last(A)	.*A	Each instance has to end with activity A.
Responded existence(A,B)	[^A]*((A.*B.*)|(B.*A.*))?	If A happens at least once then B has to happen or happened before A.
Co-existence(A,B)	[^AB]*((A.*B.*)|(B.*A.*))?	If A after after A, and vice versa. happened
Response(A,B)	[^A]*(A.*B)*[^A]*	Whenever activity A happens, activity B has to happen eventually afterward.
Precedence(A,B)	[^B]*(A.*B)*[^B]*	Whenever activity B happens, activity A has to have happened before it.
Alternate response(A,B)	[^A]*(A[^A]*B[^A]*)*	After each activity A, at least one activity B is executed. A following activity A can be executed again only after the first occurrence of activity B.
Alternate precedence(A,B)	[^B]*(A[^B]*B[^B]*)*	Before each activity B, at least one activity A is executed. A following activity B can be executed again only after the first next occurrence of activity A.
Chain response(A,B)	[^A]*(AB[^A]*)*	Every time activity A happens, it must be directly followed by activity B (activity B can also follow other activities).
Chain precedence(A,B)	[^B]*(AB[^B]*)*	Every time activity B happens, it must be directly preceded by activity A (activity A can also precede other activities).
Not co-existence(A,B)	[^AB]*((A[^B]*)|(B[^A]*))?	Either activity A or B can happen, but not both.
Not succession(A,B)	[^A]*(A[^B]*)*	Activity A cannot be followed by activity B, and activity B cannot be preceded by activity A.
Not chain succession(A,B)	[^A]*(A+[^AB][^A]*)*A*	Activities A and B can never directly follow each other.
Choice(A,B)	.*[AB].*	Activity A or activity B has to happen at least once, possibly both.
Exclusive choice(A,B)	(([^B]*A[^B]*)|.*[AB].*([^A]*B[^A]*)	Activity A or activity B has to happen at least once, but not both.

is a state. Transitions between pairs of states represent the transitions that lead from a marking to another by means of a firing. A state in which no transitions are enabled anymore is called a final state. In the following, we assume that the Petri nets used to represent procedural models are bounded.

For a mixed-model, we assume to deal with closed-world models, i.e., only the activities in A can be used in the model. Nevertheless, A can be extended to also include activities that are neither constrained in the Declare model, nor included in S. The injective labeling function makes it possible to deal with duplicate activities and silent transitions, as there exists a unique element for each silent or duplicate activity in the Petri net.

4   Model Checking Approach

In the following section, we describe our approach to check the consistency of a mixed-paradigm model. Thereafter, we show that it can be used in the context of mining to reduce the required computational expensiveness.

4.1   Model Checking

The model checking approach is based on automaton multiplication, as inspired by [16]. In order to verify whether a mixed-paradigm model does not have any conflicting states, both models are brought to the same execution model, being a finite state automaton. For the Petri net, the reachability graph is calculated (Algorithm 1, line 2). In the presence of deadlocks, the reachability graph will not reach a state in which there is an end marking without remaining tokens. In this case, the conjoining would not work and the Petri net should be checked for errors on its own. Next, all the constraints in the declarative model are checked for compatibility with the reachability automaton by conjoining its executable form, also an FSA, with the procedural model (line 19–20). In case the full declarative model does not conflict, the result Φ_{MPM} will contain the full behavior of both models without the conflicting states. If not, the declarative model with the most non-conflicting constraints will be returned (line 7–11). Another possibility is to check the declarative model up front with the approach discussed in [16] and available in the MINERful framework[1], which also checks for constraint redundancy.

Iteratively conjoining constraints to get the biggest set of non-conflicting Declare constraints might be computationally inefficient in case of very big constraint sets. This can be resolved by introducing a priority scheme that checks unary and negative constraints (e.g. *exactly2* and *not succession*) last, because they often impose a high degree of interaction with other constraints [21]. Furthermore, this can also help to resolve the issue of choosing one constraint set over the other in the case of equal sizes. Note that the approach can also be tailored towards analyzing process mining results.

Algorithm 1. Model checking procedure for mixed-paradigm models.

Output: Φ_{MPM}
1: **procedure** $calculateModel(DM, PN)$
2:      $\Phi_{PN} \leftarrow calculateReachabilityGraph(PN)$
3:      $\Phi_{MPM} = \emptyset,\ b = 0,\ V \leftarrow \emptyset$
4:      **for** $f \in F_D$ **do**
5:          $\Phi_T, V \leftarrow checkConstraintForConflicts(\Phi_{PN}, F_D, f, V)$
6:          **if** $V = F_D$ **then**
7:              $\Phi_{MPM} \leftarrow \Phi_T$
8:              **break**
9:          **end if**
10:         **if** $|V| > b$ **then**
11:             $b \leftarrow |V|$
12:             $\Phi_{MPM} \leftarrow \Phi_T$
13:         **end if**
14:     **end for**
15:     **return** Φ_{MPM}
16: **end procedure**

17: **procedure** $checkConstraintForConflicts(\Phi, F_D, f, V)$
18:     $V \leftarrow f$
19:     **if** $\Phi \otimes \alpha(f) \neq \emptyset$ **then**
20:         $\Phi_{PN} \leftarrow \Phi \otimes \alpha(f)$
21:         **for** $g \in F_D \setminus V$ **do**
22:             $checkConstraintForConflicts(\Phi_{PN}, g, V)$
23:         **end for**
24:     **end if**
25:     **return** Φ_{PN}, V
26: **end procedure**

---

[1] https://github.com/cdc08x/MINERful.

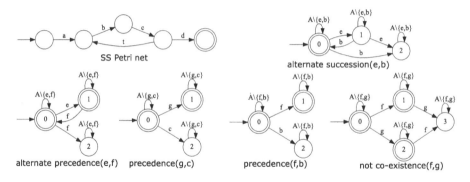

Fig. 2. The state space of the Petri net and the Declare constraints' automata from Fig. 1.

Example. Consider the model in Fig. 1 again. The reachability graph of the procedural model is relatively small and is presented in Fig. 2. The initial marking is $M_0(p_1) = 1$. The algorithm iteratively tests whether the constraints in the declarative part of the model can be intersected with this state space. Clearly, *not co-existence(f,g)* and *precedence(f,b)* cannot co-exist in this model, as c can never be executed when g is prohibited from firing through *not co-existence(f,g)*. Therefore, one of these constraints is disregarded by the model checker to provide a final execution automaton for the model. According to the priority strategy, this is the case for *not co-existence(f,g)*.

4.2 Better Mining with Model Checking Iterations

Next to checking models, the model checking algorithm can also be used to guide the discovery algorithm towards a better solution in the following way. Since calculating the reachability graph can be computationally expensive, the algorithm can be adapted as follows. Starting from the initially assigned value, the entropy level is subsequently raised to increase the size of the declarative part versus the procedural one. In the extreme case, the algorithm resorts to mining solely a declarative model, for which the model checker is guaranteed to finish. Hence, the algorithm actually checks the amount of flexibility that is in the log and adapts itself accordingly to the amount of 'declarativeness' that is needed. We call this the *self-learning* capability of the approach.

In order to achieve this, the Fusion Miner algorithm (as documented in [11]) is adapted to check whether the reachability graph of the procedural model can be calculated, containing less than a certain number of states (Algorithm 2, line 8). This number is calculated based on a threshold n, a model size multiplication coefficient, and the size of the procedural model, i.e., $n \times |S^{PN}| \times 1,000$. If the graph cannot be calculated, the entropy measure is raised by another threshold called *resilience coefficient*, r, to reiterate the process towards a model consisting of a bigger declarative part and a smaller procedural one (line 20).

Algorithm 2. FusionMINERFul algorithm.

Output: Φ_{MPM}
```
 1: procedure calculateModel(T, e, n, r)          ▷ Input: set of traces T, entropy measure e and
 2:     Φ_MPM ← ∅                                                        ▷ resilience measure r
 3:     while Φ_MPM = ∅ do
 4:         D ← getEntropicActivities(T, e)
 5:         DM ← MINERful(T, D)
 6:         PN ← HeuristicsMiner(T, A \ D)
 7:         Φ_PN ← calculateReachabilityGraph(PN, n)        ▷ Returns an empty set after
 8:         V ← ∅, b ← 0                                                  ▷ n × |S| × 1,000 states
 9:         if Φ_PN ≠ ∅ then
10:             for f ∈ F_D do
11:                 Φ_T ← checkConstraintForConflicts(Φ_PN, F_D, f, V)
12:                 if V = F_D then
13:                     Φ_MPM ← Φ_T
14:                     break
15:                 end if
16:                 if |V| > b then
17:                     b ← |V|
18:                     Φ_MPM ← Φ_T
19:                 end if
20:             end for
21:         else
22:             e ← e + r
23:         end if
24:     end while
25:     return Φ_MPM
26: end procedure
```

Example. Consider the procedural model produced by Fusion Miner depticted in Fig. 3. The procedural part of the model has to take into account many different ways of enabling d in between the other activities, introducing many silent transitions. Calculating the reachability graph will yield an enormous automaton, requiring computationally expensive conjoining operations. Because the reachability graph calculation is stopped after 32,000 ($2 \times 16 \times 1000$ when $n = 2$) states, the algorithm repeats its main procedure with an entropy level e which is increased by r. For $e = 0.4$ and $r = 0.1$, the resulting model eliminates the need for invisible transitions by removing d from the procedural workflow, as depicted in Fig. 4.

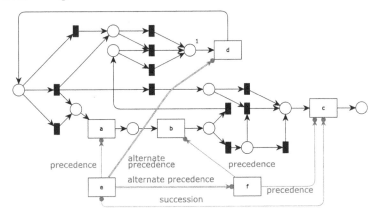

Fig. 3. Mixed-paradigm output of FusionMINERFul for an entropy level of 0.4.

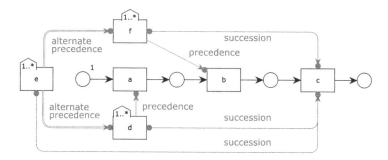

Fig. 4. Mixed-paradigm output of FusionMINERFul for the same event log after the entropy level was raised to 0.5.

5   Implementation and Evaluation

In this section, the implementation in FusionMINERFul is introduced. Next, this process mining tool is used to evaluate the approach on the 2012 BPI Challenge[2] event log.

5.1   Implementation

Mining a mixed-paradigm model with intertwined state spaces was introduced by Fusion Miner [11]. This mining algorithm uses the notion of entropy to find activity types in the log that do not fit a strict workflow well, based on the dependency information of Heuristics Miner [2]. Activities are divided into D, S, DD, and SS. F_D, F_{DS}, and F_S are mined, while F_{SD} is not considered to avoid too convoluted models that have a high risk of inconsistencies. For this work, a new version called FusionMINERFul is used. This algorithm uses MINERFul [6] to derive F_D and F_{DS} and Heuristics Miner to mine F_S. The technique also relies on the state space analysis tools which can be found in ProM[3]. The implementation is compatible with ProM and can be found at http://www.processmining.be/fusionminerful/, together with high resolution versions of the figures in this paper. The final output model is represented as a dependency graph with Declare constraints [11], which can be converted to a Petri net with Declare constraints. This serves as the basis for the model checking approach. In the output, removed constraints are colored differently, and the implementation also include the self-learning capability. Blue arcs comprise the procedural model, while black annotated arcs contain Declare constraints. Negative constraints are yellow, while constraints removed during verification are red. Declarative activities use dashed outlines, and gray and red coloring indicates *existence(A,1)* and *exactly(A,1)* respectively.

[2] DOI: 10.4121/uuid:3926db30-f712-4394-aebc-75976070e91f.
[3] http://www.promtools.org/.

5.2   Application to BPIC 2012

The approach of model checking with FusionMINERFul has been tested on the 2012 BPI Challenge event log (BPIC 2012). This log consists of three distinct subprocesses which will be treated separately, in analogy with the approach followed in [22]. The goal is to find out whether the model that is discovered is sound, and how well FusionMINERFul can determine the level of flexibility that is needed for mining an informative process model. To test the self-learning capabilities of the algorithm, the initial entropy level is always kept at 0, giving the algorithm the chance to adapt itself according to whether a procedural finite state space can be constructed. The model size multiplicator level was set to 10, and the resilience measure at 0.1.

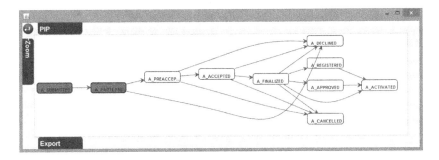

Fig. 5. Output of FusionMINERFul for the X subprocess of BPIC 2012. The set-up is: initial entropy level e of 0, resilience r of 0.1, and model size multiplicator n of 10. The entropy-level remains unchanged, indicating the log to contain procedural behavior.

The first subprocess, X, does not contain any behavior that is too unstructured to handle in a procedural model, hence the algorithm does not raise the entropy level. In this case, the full model of Heuristics Miner is outputted, as can be seen in Fig. 5. The second subprocess, subprocess Y, shows a low level of flexibility. In this case, two iterations finally churned out the process that can be seen in Fig. 6. Finally, the last subprocess, Z, reaches the maximal entropy level of 1, indicating that the process can be considered very unstructured. Only the Declare model is outputted, as can be seen in Fig. 7, which contains many negative constraints and one conflicting constraint which is removed from the model.

Overall, the approach is capable of detecting inconsistencies, although few appeared. This is in line with the intuition established in [16]. Furthermore, FusionMINERFul is also capable of finding different levels of flexibility which approaches the results from Heuristics Miner and ILP Miner. These mining techniques also need to resort to imprecise many-to-many connections to reflect the unstructuredness that is present in the event logs, especially for Y and Z.

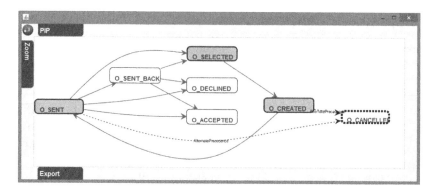

Fig. 6. Output of FusionMINERFul for the Y subprocess of the BPIC 2012. The set-up is: $e = 0$, eventually raised to 0.2, $r = 0.1$, and $n = 10$.

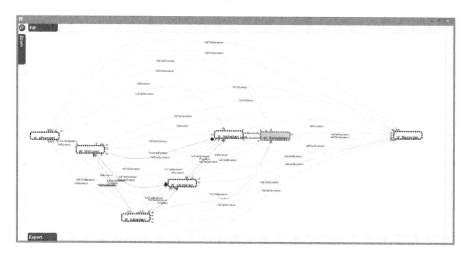

Fig. 7. Output of FusionMINERFul for the Z subprocess of BPIC 2012. The set-up is: $e = 0$, eventually raised to 1.0, $r = 0.1$, and $n = 10$.

6   Conclusion

In this paper, a model checking approach for mixed-paradigm models was proposed based on automaton multiplication. Furthermore, it was shown how this notion can be used in a mining environment to find a fit between procedural and declarative models to achieve models that accurately describe the log with the right level of granularity in terms of flexibility.

For future work, we envision to use the global automaton to apply conformance checking in order to show in which areas mixed-paradigm models can excel not only in terms of dealing with mixed-flexibility, but in terms of recall and precision. Secondly, it will be investigated how coverability graphs can be used instead of reachability graphs, in order to deal with unbounded behavior that can occur in procedural models. Finally, Heuristics Miner cannot guarantee to churn out a deadlock-free model. It is therefore in our plans to integrate the tool with other procedural process mining algorithms.

References

1. van der Aalst, W.M.: Process Mining: Discovery, Conformance and Enhancement of Business Processes. Springer, New York (2011)
2. Weijters, A., van der Aalst, W.M., De Medeiros, A.A.: Process mining with the heuristics miner-algorithm. TU Eindhoven, Technical report WP 166 (2006)
3. van der Aalst, W.M., Weijters, T., Maruster, L.: Workflow mining: discovering process models from event logs. IEEE Trans. Knowl. Data Eng. **16**(9), 1128–1142 (2004)
4. Werf, J.M.E.M., Dongen, B.F., Hurkens, C.A.J., Serebrenik, A.: Process discovery using integer linear programming. In: Hee, K.M., Valk, R. (eds.) PETRI NETS 2008. LNCS, vol. 5062, pp. 368–387. Springer, Heidelberg (2008). doi:10.1007/978-3-540-68746-7_24
5. Maggi, F.M., Mooij, A.J., van der Aalst, W.M.: User-guided discovery of declarative process models. In: CIDM, pp. 192–199. IEEE (2011)
6. Di Ciccio, C., Mecella, M.: A two-step fast algorithm for the automated discovery of declarative workflows. In: CIDM, pp. 135–142. IEEE (2013)
7. Murata, T.: Petri nets: properties, analysis and applications. Proc. IEEE **77**(4), 541–580 (1989)
8. Pesic, M., Schonenberg, H., van der Aalst, W.M.: Declare: full support for loosely-structured processes. In: EDOC, p. 287. IEEE(2007)
9. Maggi, F.M., Slaats, T., Reijers, H.A.: The automated discovery of hybrid processes. In: Sadiq, S., Soffer, P., Völzer, H. (eds.) BPM 2014. LNCS, vol. 8659, pp. 392–399. Springer, Cham (2014). doi:10.1007/978-3-319-10172-9_27
10. Smedt, J., Weerdt, J., Vanthienen, J.: Multi-paradigm process mining: retrieving better models by combining rules and sequences. In: Meersman, R., Panetto, H., Dillon, T., Missikoff, M., Liu, L., Pastor, O., Cuzzocrea, A., Sellis, T. (eds.) OTM 2014. LNCS, vol. 8841, pp. 446–453. Springer, Heidelberg (2014). doi:10.1007/978-3-662-45563-0_26
11. De Smedt, J., De Weerdt, J., Vanthienen, J.: Fusion miner: process discovery for mixed-paradigm models. Decis. Support Syst. **77**, 123–136 (2015)
12. Westergaard, M.: CPN tools 4: multi-formalism and extensibility. In: Colom, J.-M., Desel, J. (eds.) PETRI NETS 2013. LNCS, vol. 7927, pp. 400–409. Springer, Heidelberg (2013). doi:10.1007/978-3-642-38697-8_22
13. Pesic, M., Aalst, W.M.P.: A declarative approach for flexible business processes management. In: Eder, J., Dustdar, S. (eds.) BPM 2006. LNCS, vol. 4103, pp. 169–180. Springer, Heidelberg (2006). doi:10.1007/11837862_18
14. De Smedt, J., De Weerdt, J., Vanthienen, J., Poels, G.: Mixed-paradigm process modeling with intertwined state spaces. Bus. Inf. Syst. Eng. **58**, 19–29 (2016)
15. Westergaard, M., Slaats, T.: Mixing paradigms for more comprehensible models. In: Daniel, F., Wang, J., Weber, B. (eds.) BPM 2013. LNCS, vol. 8094, pp. 283–290. Springer, Heidelberg (2013). doi:10.1007/978-3-642-40176-3_24
16. Di Ciccio, C., Maggi, F.M., Montali, M., Mendling, J.: Ensuring model consistency in declarative process discovery. In: Motahari-Nezhad, H.R., Recker, J., Weidlich, M. (eds.) BPM 2015. LNCS, vol. 9253, pp. 144–159. Springer, Cham (2015). doi:10.1007/978-3-319-23063-4_9

17. Prescher, J., Di Ciccio, C., Mendling, J.: From declarative processes to imperative models. In: SIMPDA, pp. 162–173 (2014)
18. Di Ciccio, C., Mecella, M.: On the discovery of declarative control flows for artful processes. ACM Trans. Manage. Inf. Syst. **5**(4), 24:1–24:37 (2015)
19. Westergaard, M., Stahl, C., Reijers, H.A.: Unconstrainedminer: efficient discovery of generalized declarative process models. Technical report, BPMcenter (2013)
20. Desel, J., Reisig, W.: Place/transition petri nets. In: Reisig, W., Rozenberg, G. (eds.) ACPN 1996. LNCS, vol. 1491, pp. 122–173. Springer, Heidelberg (1998). doi:10.1007/3-540-65306-6_15
21. Smedt, J., Weerdt, J., Serral, E., Vanthienen, J.: Improving understandability of declarative process models by revealing hidden dependencies. In: Nurcan, S., Soffer, P., Bajec, M., Eder, J. (eds.) CAiSE 2016. LNCS, vol. 9694, pp. 83–98. Springer, Cham (2016). doi:10.1007/978-3-319-39696-5_6
22. Adriansyah, A., Buijs, J.C.A.M.: Mining process performance from event logs. In: Rosa, M., Soffer, P. (eds.) BPM 2012. LNBIP, vol. 132, pp. 217–218. Springer, Heidelberg (2013). doi:10.1007/978-3-642-36285-9_23

First Workshop on Business Process Management and Ontologies (BPMO 2016)

Introduction to the First Workshop on Business Process Management and Ontologies (BPMO 2016)

Henrik Leopold[1], Lucinéia Heloisa Thom[2],
and Pablo David Villarreal[3(✉)]

[1] Department of Computer Science, VU University Amsterdam,
Amsterdam, The Netherlands
h.leopold@vu.nl
[2] Institute of Informatics, Federal University of Rio Grande do Sul,
Porto Alegre, Brazil
lucineia@inf.ufrgs.br
[3] Center of Research and Development on Information System Engineering,
Santa Fe Faculty, National Technological University and CONICET,
Buenos Aires, Argentina
pvillarr@frsf.utn.edu.ar
http://www.henrikleopold.com/
http://www.inf.ufrgs.br/~lucineia
https://sites.google.com/site/pablodavidvillarreal/

Abstract. The BPMO 2016 workshop aims at bringing together researchers and practitioners to present, discuss, and evaluate the application of ontologies to generate new or improve existing methods, techniques, tools, and process-aware information systems that support the different phases of the business process management life-cycle. In its first edition, the workshop has received good interest from the BPM community. Several research papers of very high quality were submitted. Four of them were accepted as regular papers (with an acceptance rate of 50%). A discussion panel during the workshop with researchers and practitioners revealed several promising directions for future research integrating business process management and ontology engineering.

Keywords: Business process · Ontology · Process-aware system

1 Introduction

Business process management requires the integration and proper understanding of the organizational context, the different process representations (for analysis, modeling, and execution), the process data, and the applications/services that support the activities during the process execution. In process design, business analysts interview domain experts and translate their understanding to process models. One challenge in this context is that the vocabulary used by domain experts is often very specific and difficult to understand by process analysts. This particularly applies to specific processes from domains such as healthcare, where often profound knowledge about domain-specific terms is required to appropriately understand the overall process context. The lack of this knowledge can lead to interpretation problems, ambiguities, and misunderstandings between the process analysts and the domain experts.

Research on process design and ontologies has received increased attention in recent years. One of the reasons is that ontologies and structured vocabularies in different domains help to make data understandable by machines. Moreover, ontologies allow to add semantics to process model representations. Among others, this enables the execution of queries to make inferences about the knowledge maintained in these representations, which can then be used by methods and tools in both design time as well as execution time of business processes.

2   Goals

The goal of this workshop is to bring together researchers and practitioners to present, discuss, and evaluate the application of ontologies to generate new or improve existing methods, techniques, tools, and process-aware systems that support the different phases of the business process management life cycle. It aims to provide a high quality forum for researchers and practitioners of the communities of business process management and ontology engineering, with a focus on the application of ontologies for business process management.

The list of topics that are relevant to the workshop includes the application of ontologies to the following fields of BPM, but is not limited to:

- Foundations of business process concepts
- Analysis and design of business processes
- Business process modeling languages
- Business process reference models
- Verification and validation tools of business processes
- Process execution and monitoring
- Process mining
- Process repositories
- Process data integration and data quality
- Social BPM
- Human-centric processes and knowledge-intensive processes
- Adaptive and context-aware process execution
- Inter-organizational business process management

3   Submissions

The first edition attracted eight submissions, out of which four submissions were accepted as regular papers for presentation. Thus, the acceptance rate was 50%.

The papers presented in the workshop provided novel ideas on the application of ontologies in tools and methods for process modeling and verification. The paper "PROMPTUM Toolset: Tool Support for Integrated Ontologies and Process Models" by *Ahmet Coşkunçay, Ozge Gurbuz, Onur Demirörs* and *Erdem Eser Ekinci* presents a toolset that does not only support the joint development of business process models and domain ontologies, but also building business process models by using existing ontologies and developing ontologies by using business process model collections. The toolset enables the definition and management of labels and terms within labels of the process models and the process model elements as resources within a formal domain ontology.

The paper "Ontology-based Heuristics for Process Behavior: Formalizing False Positive Scenarios" by *Jorge Roa, Emiliano Reynares, Ma. Laura Caliusco* and *Pablo Villarreal* proposes to use ontologies-based heuristics to reason on behavioral properties of business process models. The goal is to avoid ambiguities in the definition and implementation of heuristics, which is important for improving precision of process verification. The paper presents SWRL rules along with SPARQL queries that formalize heuristics for verification of process behavior and analyzes a set of false positive scenarios for such heuristics.

The paper "Business Process Architecture Baselines from Domain Models" by *Fernanda Gonzalez-Lopez* and *Guillermo Bustos* explores the use of conceptual domain models for building a business process architecture baseline in a given business domain. In line with this idea, the paper offers guidelines for using domain models together with an entity-centric approach to derive a business process architecture.

The paper "Ontology-Based Approach for Heterogeneity Analysis of EA Models" by *João Cardoso, Marzieh Bakhshandeh, Daniel Faria, Cátia Pesquita* and *José Borbinha* proposes an ontology-based approach for heterogeneity analysis of EA business models expressed in BPMN. The proposed approach relies on encoding the models as OWL ontologies and then applying ontology matching techniques to integrate them.

4 Conclusions

The workshop consisted of four regular paper presentations and a discussion panel on the future directions of combining business process management and ontology engineering. We hope the reader will inspiring ideas in these papers on how ontologies can be applied and exploited in the area of business process management. We are looking forward to continuing discussing novel ideas on the integration of business process management and ontology engineering.

The workshop is perceived as an important start to further develop and research how ontologies can be applied to generate or improve new tools, methods, and systems for business process management.

We wish to thank all authors for their contributions and participations in the workshop, the members of the BPMO 2016 Program Committee for their support with reviewing the papers, and, finally, the workshops chairs for their support with the organization of this workshop.

5 Workshop Co-organizers

Henrik Leopold	VU University Amsterdam, The Netherlands
Lucinéia Heloisa Thom	Federal University of Rio Grande do Sul, Brazil
Pablo D. Villarreal	National Technological University & CONICET, Argentina

6 Program Committee

Agnes Koschmider	Institute AIFB, Karlsruhe Institute of Technology, Germany
Amel Bouzeghoub	Institute TELECOM SudParis, France
Andrea Delgado	Universidad de la República, Uruguay
Claudia Capelli	Federal University of Rio de Janeiro State, Brazil
Christian Meilicke	University of Mannheim, Germany
Fernanda Baiao	Federal University of Rio de Janeiro State, Brazil
Giancarlo Guizzardi	Federal University of Espírito Santo, Brazil
João Paulo Almeida	Federal University of Espírito Santo, Brazil
José Palazzo Moreira de Oliveira	Federal University of Rio Grande do Sul, Brazil
Leonardo G. Azevedo	Federal University of Rio de Janeiro State, Brazil
Mara Abel	Federal University of Rio Grande do Sul, Brazil
Manfred Reichert	University of Ulm, Germany
Marcelo Fantinato	University of São Paulo, Brazil
Marcello La Rosa	QUT Brisbane, Australia
María Laura Caliusco	National Technological University-CONICET, Argentina
Michael Fellmann	University of Rostock, Germany
Oscar Pastor	Universitat Politècnica de València, Spain
Patrick Hung	University of Ontario Institute of Technology, Canada
Samir Tata	Institute TELECOM SudParis, France
Selmin Nurcan	CRI - Centre de Recherche en Informatique, France
Walid Galoul	Institute TELECOM SudParis, France

PROMPTUM Toolset: Tool Support for Integrated Ontologies and Process Models

Ahmet Coşkunçay[1(✉)], Özge Gürbüz[1], Onur Demirörs[1], and Erdem Eser Ekinci[2]

[1] Informatics Institute, Middle East Technical University, Ankara, Turkey
{cahmet,ogurbuz,demirors}@metu.edu.tr
[2] Galaksiya Bilişim Teknolojileri, İzmir, Turkey
erdemeserekinci@galaksiya.com

Abstract. Business process models and ontologies are two essential knowledge artifacts that utilize similar information sources. In this sense, building and managing the relationships between ontologies and business process models provide benefits such as enhanced semantic quality of both artifacts and effort savings. In this study, the PROMPTUM toolset, that enables to model relations between the ontologies and the labels within the process model collections, is presented. In establishing these relations, the PROMPTUM toolset enables definition and management of labels and terms within labels of the process models and the process model elements as resources of domain ontologies. Thus, a related resource is managed as a single resource representing the same real-world object in both artifacts in both creation and maintenance. By providing the required features, the toolset supports not only building business process models and domain ontologies together but also building business process models by using existing ontologies and developing ontologies by using business process model collections.

Keywords: Business process modeling · Ontology development · Tool support

1 Introduction

Ontology development is an essential part of the knowledge management domain in terms of creating domain knowledge. Business process models, on the other hand, are important for organizations to create formal knowledge about processes [1] from knowledge management perspective. So, both activities are utilized as a part of knowledge creation.

In practice, organizations performing both process modeling and ontology building activities, allocate duplicated efforts for each development activity conducted using same or similar knowledge sources. Moreover, neither activity benefits from the knowledge created in the other, thus the resulting products are prone to be inconsistent with each other. Furthermore, the use of ontologies in process modeling would ease preventing and detecting redundancies and inconsistencies between labels of process model elements [2]. Also the process models, when used in ontology development as a source of process knowledge, would enhance the completeness of domain ontologies.

© Springer International Publishing AG 2017
M. Dumas and M. Fantinato (Eds.): BPM 2016 Workshops, LNBIP 281, pp. 93–105, 2017.
DOI: 10.1007/978-3-319-58457-7_7

Challenges of semantic process modeling, as identified in [3], include challenges that could potentially be resolved with the use of ontologies. Our study mostly focuses on the label challenges of semantic business process modeling that relate to interpretation, analysis, and improvement of the grammar and terms within the process model element labels and the process model labels. Some of the label challenges are related to the issues about identification of semantic components of labels (C1 in [3]), recognizing the meaning of terms from labels (C3), identifying homonymous or synonymous terms (C4), and assessing the similarity of labels (C6). PROMPTUM toolset described in this study presents opportunities for resolving these issues by providing means to define and manage labels and terms within labels of the process models and the process model elements as resources within a formal domain ontology. Another challenge our study relates is about "discovering a formal ontology from a collection of process models" (C24). This is a challenge PROMPTUM toolset addresses by not only enabling domain ontologies to be developed by using a set of business process models, but also by supporting business process models and domain ontologies be developed as integrated, and existing domain ontologies be used in business process modeling.

Research regarding the relations between business process modeling and ontology development has gained pace in recent years. Some [4, 5] investigate the process modeling notations based on foundational ontologies, whereas others [6, 7] focus on the importance and practical uses of process related ontologies. Mapping or transformation approaches between business process models and ontologies has also been widely studied [8–15]. However, software tools to support integrated ontology development and business process modeling is not reported in any of the surveyed academic and industrial sources. In this study, we present the PROMPTUM toolset to support integrated business process modeling and ontology building. The PROMPTUM toolset would support not only building business process models and domain ontologies together but also building business process models by using existing ontologies and vice versa. Following sections describe the PROMPTUM toolset, validation via case studies, related work and conclusions respectively.

2   PROMPTUM Toolset

Components and system interfaces of the PROMPTUM toolset are shown in Fig. 1.

Analyst is the person who models the processes via UPROM tool and/or builds the ontologies via PROMPTUM Ontology Server. Analyst also uses PROMPTUM Process Modeling Plugin via Ontology View on UPROM for associating labels in process model collections with ontology resources. The three components of the PROMPTUM toolset are described with details in the following sections.

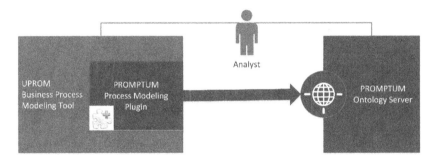

Fig. 1. Components and system interfaces of PROMPTUM toolset

2.1 UPROM Tool

UPROM tool, user interface of which is shown in Fig. 2, is the medium where business process models are developed and stored within the PROMPTUM toolset.

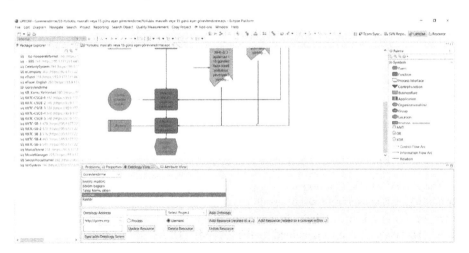

Fig. 2. User interface of UPROM tool

UPROM tool [16] is a Desktop graphical business process modeling software that provides modeling editors for several diagram types such as extended Event Driven Process Chain (eEPC), Value Chain (VC), and Function Tree (FT). UPROM possesses the common characteristics of process modeling tools such as ease of model building, formal semantics and verification of correctness, workflow patterns, resource and data perspective, and level of detail, transparency and suitability for communication [17]. UPROM provides core process modeling tool features such as a process model repository for storing and structuring the modeling projects, sub-diagram decomposition, continuous syntactic verification based on diagram meta-models, unique object assignment, and defining attributes for process models and elements. UPROM was developed based on bflow* Toolbox [18] by following Eclipse Modeling Framework (EMF) and Eclipse Graphical Modeling Framework (GMF).

2.2 PROMPTUM Ontology Server

Most ontology editors enable [19]; (1) adding/removing/modifying ontology classes, class hierarchy, object and data properties and their hierarchy, property domain and range, individuals, property individuals and literal annotations, (2) propagating a change in one part of the ontology to other parts and associated individuals, and (3) undoing a change in ontology such that previously propagated changes are also undone.

The PROMPTUM Ontology Server, which features these functionalities, is developed as a component of the PROMPTUM toolset. It serves not only as a Web-based ontology editor but also as a triple store. It is based on AngularJS application framework, uses Jetty as web server and stores ontologies in Apache Jena TDB. Web front-end of the PROMPTUM Ontology Server, shown in Fig. 3, is a simple Web-based ontology editor that enables analysts to build ontologies.

Fig. 3. Web front-end of PROMPTUM Ontology Server

The PROMPTUM Ontology Server provides web services based on RESTful APIs for adding, listing, and removing resources, data type properties and object type properties, updating resource labels, listing changes in ontologies, and getting the whole model. These services are consumed by the PROMPTUM Process Modeling Plugin.

2.3 PROMPTUM Process Modeling Plugin

The PROMPTUM Process Modeling Plugin is a plugin to the UPROM and follows the same principles in terms of the development technology used. It provides the "ontology view" shown in Fig. 4 on the UPROM's user interface and enables analysts to perform operational scenarios for integrated development of the process models and ontologies, except the regular process modeling and ontology editing operations described in Sects. 2.1 and 2.2. These operational scenarios are depicted with motivating examples from "public investment planning" and "short term assignment of academic staff members" processes and domains in following sub-sections.

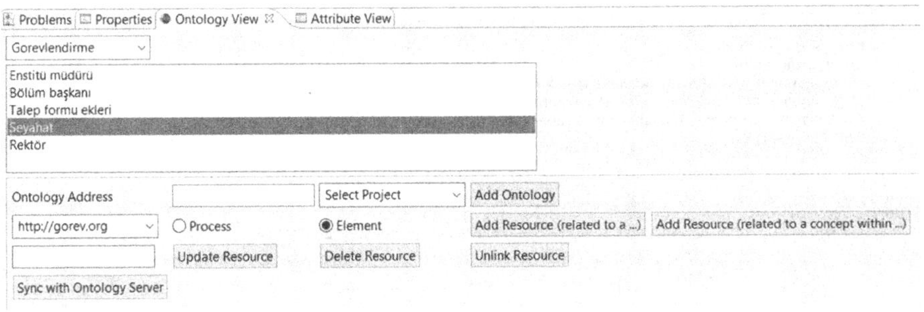

Fig. 4. Ontology view user interface in UPROM tool

Ontology selection

In PROMPTUM toolset, a label or a term in the label of a process model or a process model element within a process model collection can be added as a resource to an ontology only if a restriction is already established between the process model collection and the ontology in question. Therefore, this restriction permits that the resources can be added from only selected process model collections to the selected ontologies. A many-to-many relation between the process model collections and ontologies is possible in defining these restrictions. In other words, once the required restrictions are established, an ontology can take input from several process model collections and a process model collection can have resources defined in several ontologies.

Analysts define these restrictions using the Ontology View by entering a valid ontology address, selecting a process model collection from the "Select Project" list and clicking the "Add Ontology" button shown in Fig. 4.

Managing process model labels as ontology resources

An ontology resource and a process model label would represent the same real-world object. Figure 5 exemplifies a process model with the label "Assignment without allowance and expense and with duration of between 7–15 days" within business process models for "short term assignment of academic staff members". The object that represents this process model is also a resource that is a sub-class of the "short term assignment" class in "academic assignments" ontology.

Such ontology resources that represent the same real-world object as a process model label can be added to an ontology via Ontology View (Fig. 4). In order to add the resource, analyst clicks "Add Resource (related to a...)" button while the ontology from the ontology list, "Process" radio button, and the process model on the package explorer remain selected. Then the resource is added to the PROMPTUM Ontology Server and the label of resource is listed in the linked resources list. If the resource already exists in the ontology, which might be the case in utilizing existing ontologies, the PROMPTUM Process Modeling Plugin still keeps record of the relation between the ontology resource and the process model label.

Once a resource listed in the linked resources list, several other functions are available. Ontology resource URI and label, and the label of process model is updated by using the "Update Resource" button. The "Delete Resource" button deletes the selected

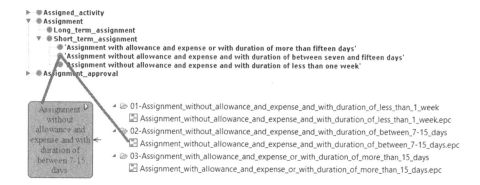

Fig. 5. Motivating example of a process model label as an ontology resource

ontology resource from the ontology and the "Unlink Resource" button removes the link between the process model and the ontology resource without deleting the resource from the ontology. And if the analyst deletes a linked resource by using the web interface of the PROMPTUM Ontology Server, upon clicking on "Sync with Ontology Server" on the Ontology View, the resource is removed from linked resources list.

Managing process model element labels as ontology resources
In Fig. 6, "Faculty member" is a role in a process model and also a class in "academic assignments" ontology. Such process model element labels can be added to an ontology via Ontology View by using the "Add Resource (related to a...)" button while the ontology from the ontology list, "Element" radio button, and the process model element are selected.

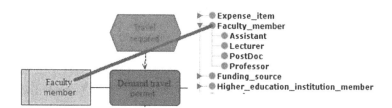

Fig. 6. Motivating example of a process element label as an ontology resource

After the resource related to a process model element label is created in ontology and listed in the linked resources list, analyst can update the process model element label by (1) clicking the "Update Resource" button and changing the label, (2) changing the process model element label in modeling area, or (3) updating the label of resource in the PROPMTUM Ontology Server and clicking the "Sync with Ontology Server" button on Ontology view. All three alternatives ensure that all instances of the process model element in the same process model collection, the linked resources list, and the resource in ontology are updated respectively. The "Sync with Ontology Server" button also ensures a deleted resource in the PROMPTUM Ontology Server is removed from the linked resources list.

"Delete resource" and "Unlink Resource" buttons perform the same functions described in previous sub-section for the resources associated with process model element labels. Moreover, if the last existing instance of a process model element is deleted from the process model collection, the associated resource is also removed from the linked resources list but it remains in the ontology.

Managing terms within process model labels as ontology resources
Not only whole labels but also the terms and phrases within the labels would represent the same real-world object that a resource in an ontology represents. Figure 7 provides a motivating example for such cases. The phrase "investment proposal" within the process model label "Gather and evaluate investment proposals" is also a resource in "public investment planning" ontology.

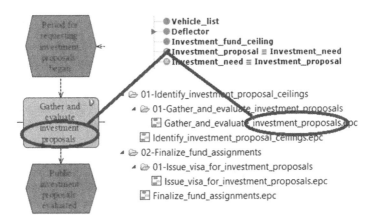

Fig. 7. Motivating example of a term within a process model label as an ontology resource

In adding a term or phrase within a process model label, analyst clicks "Add Resource (related to a concept within…)" button while the process model is selected in the UPROM package explorer, and the ontology where the resource will be added and the "Process" radio button remain selected. Then the PROMPTUM Process Modeling Plugin displays a pop-up dialog box where the analyst needs to enter the term or phrase included in process model label. The PROMPTUM Process Modeling Plugin verifies whether the input phrase is included in the process model label or not. The rest of the functionality related to managing terms within process model labels as ontology resources (i.e. updating, deleting, unlinking, and synchronizing actions) are similar to the ones described above for managing process model labels as ontology resources.

Managing terms within process model element labels as ontology resources
Similar to the terms and phrases within the process model labels, the terms and phrases within the process model element labels would also be ontology resources in the relevant domains. The motivating example depicted in Fig. 8 shows the term "sector" as a part of a process model element label (i.e. "Identify investment needs on the basis of sectors") and a resource in "public investment planning" ontology.

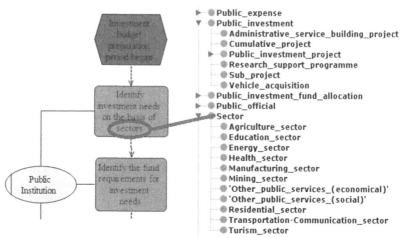

Fig. 8. Motivating example of a term within a process model element label as an ontology resource

For adding such resources to the ontology, analyst clicks "Add Resource (related to a concept within…)" button while the process model element, the ontology, and the "Process" radio button are selected and enters the valid phrase (i.e. that exists within the label of selected process model element) on the dialog box that pops up. Remaining actions related to an ontology resource that is linked to term within a process model element label are similar to those described for managing process model element labels as ontology resources.

Managing process model and process model element descriptions
Descriptions exist both for process models and process model elements in business process model collections and for resources in ontologies. For instance, the ontology resource and its related process model element, "faculty member", in Fig. 6 is described as "the professors, lecturers, assistants, and PostDocs working in higher education institutions". The PROMPTUM Process Modeling Plugin enables the descriptions of the process models and the process model elements to be added as data type properties to the related ontology resources in adding a new ontology resource via Ontology View on the UPROM tool.

As the description of a process model or process model element is revised, the value of the data type property representing the description is also changed on the PROMPTUM Ontology Server. Moreover, if the value of the data type property representing a description is revised on the PROMPTUM Ontology Server, upon clicking the "Sync with Ontology Server" button, the PROMPTUM Process Modeling Plugin updates the description of the related process model or the process model element on the UPROM tool.

3 Validation

In terms of its contribution to the state of the art, PROMPTUM toolset depends on the assumption that there are shared resources between ontologies and labels within process model collections whose relations need to be managed. In order to investigate this assumption, case studies are utilized for assessing the cohesion between process model collections and ontologies developed for the same domain. With this purpose, the PROMPTUM toolset is used retrospectively in two case studies that were performed for exploratory purposes in developing the toolset. Excerpts from these case studies are also provided as motivating examples in previous section for describing the features of PROMPTUM Process Modeling Plugin. As shown in Table 1, the cases were selected based on real-life context, resource availability, and problem complexity.

Table 1. Case selection

Case characteristics	Case 1	Case 2
Focus domain	Public investment planning	Short term assignment of academic staff members
Focus organization	Ministry of development	METU Informatics Institute
Real-life context	Good	Good
Resource availability	Fair	Good
Problem complexity	Good	Fair

First case (i.e. Case 1) includes a "public investment planning" ontology developed by using an already existing process model collection for "public investment planning" processes. Information resource availability was deficient in this case, as domain experts, who participated in modeling the processes, did not contributed to ontology building. So, the ontology was mainly built based on the documented resources.

Second case (i.e. Case 2) consists of a process model collection for "short term assignment of academic staff members" and an "academic assignments" ontology developed simultaneously. The problem space to be addressed in this case was rather unsophisticated as it covered simple work-flows, few roles, and a single organization.

Table 3 provides the number of shared resources between ontologies and labels within process model collections in each case study. For some of these resources, there are more than one instance of the same label or term within a label in process model collections, so the numbers presented do not reflect unique shared resource instances. Data about the resource descriptions are not presented in this table to avoid redundancy, as each process model and process model element label that is associated to an ontology resource has a description represented as a literal annotation in respective ontologies.

A high level of cohesion between business process models and ontologies observed from the data presented in Tables 2 and 3 highlights the importance of managing shared resources in ontologies and process model collections with tool support. For instance, 79 process model element labels and 175 terms within process model element labels are represented as ontology resources in the first case study that contains 273 process model elements and 262 ontology resources. This high cohesion might exist due to the fact that the selected domains for ontology development are highly correlated to the selected processes. Still, it demonstrates that the necessity of the PROMPTUM toolset with its aforementioned features for managing the resources shared between ontologies and process model collections is valid in many, if not all, circumstances.

Table 2. Case study summary

Metrics	Case 1	Case 2
Number of process models	9	3
Number of functions in process models	59	12
Number of process model elements	273	42
Number of ontology resources	262	88

Table 3. Relations between resources in ontology and labels in process model collections

Related resources	Case 1	Case 2
Process model labels	0	3
Terms within process model labels	5	9
Process model element labels	79	23
Terms within process model element labels	175	42

4   Related Work

Research on tools for business process management and ontologies has received increased attention in recent years. Researchers have developed approaches aiming to resolve various problems in process management and ontologies with tool support. Automated transformations from process models to ontologies is a popular research topic bringing business process modeling and ontologies together. Such transformations are described for Petri Net [10], BPMN [12, 13] and lean EPC [11, 12] languages.

Among related tool support reported in literature, SUPER [8], which is a EU funded project, aims to integrate semantic web service frameworks, an ontology infrastructure and business process management tools and techniques together. It provides tool support for annotating process models with individuals of built-in ontology classes [20].

Thomas and Fellmann [11] describe how to map process models modeled with EPC or BPMN to an ontology. They utilize a modeling tool, an ontology editor, and an ontology server alongside a script that transforms process models to ontology definition and an application that queries and validates the ontology developed within the study.

Francescomarino et al. [13] provide a BPMN based ontology, BPMNO, that defines the structural parts of process models and is populated automatically from process models via tool support. Annotations to this process ontology can be established from a domain ontology through axioms, so that process knowledge is annotated with semantics and querying is possible on the process knowledge for correct labeling, verification of semantic labeling and query answering. Later in 2011, features enabling collaborative specification of semantically annotated process models are also introduced [21].

Leopold et al. [14] presents an approach for automatic annotation of process model elements with the concepts in a taxonomy. A prototype enables to generate automatic annotations to the models.

More relevant research in terms of their approach to the problem of managing the relations between the ontology resources and labels in process model collections has suggested some matching rules (i.e. equivalence, synonymy, more general, and more specific) between concepts in process models and ontologies [22] for aligning these two artifacts for improving semantic quality of process models. In another related research [15], which follows a similar goal (i.e. reducing semantic ambiguity in process models), POBA presents a three step approach for modeling process models by using the ontology resources for labeling process model elements. However, none of these two studies report a tool support and our study differentiates itself by enabling both integrated and sequential development of ontologies and process models. Moreover, PROMPTUM toolset enables not only process model and process model element labels but also the terms within these labels be related to and managed as ontology resources.

5 Conclusion

In this study, we present the PROMPTUM toolset that is composed of a process modeling tool, an ontology editor/server, and a plugin for managing the relations between process models and ontologies. The PROMPTUM toolset provides support for integrated and/or sequential development and maintenance of business process models and ontologies. It does this by enabling the relations between ontology resources, and labels and terms within the labels in process model collections be established and managed. It is the first tool reported in literature to semantically manage the ontology resources, and the labels and terms within labels in business process models as integrated.

The PROMPTUM toolset provides features to address the label challenges in business process modeling [3] such as identifying the semantic components of labels, meaning of terms within labels, homonymous and synonymous terms, and similarity of labels. Moreover, it not only enables process models and ontologies be developed together but also process models be developed by using existing domain ontologies and ontologies be built by using the business process models. Previous studies [15, 22] have suggested that establishing relationships between ontologies and labels in process model collections even without tool support would increase the semantic quality of the process models. We anticipate that using the business process knowledge that reside in process model collections would also improve the quality of ontologies. Yet, benefits to both ontologies and process models upon managing and developing these two artifacts together by tool support requires to be discovered via further research.

Acknowledgments. The work presented was partly funded by TUBITAK (project number: 3130770).

References

1. Kalpic, B., Bernus, P.: Business process modeling through the knowledge management perspective. J. Knowl. Manage. **10**(3), 40–56 (2006)
2. van der Aalst, W.M.P.: Business process management: a comprehensive survey. ISRN Softw. Eng. **2013**, 1–37 (2013)
3. Mendling, J., Leopold, H., Pittke, F.: 25 Challenges of semantic process modeling. Int. J. Inf. Syst. Softw. Eng. Big Co. **1**(1), 78–94 (2014)
4. Santos Jr., P.S., Almeida, J., Guizzardi, G.: An ontology-based semantic foundation for ARIS EPCs. In: Proceedings of the 2010 ACM Symposium on Applied Computing, pp. 124–130 (2010)
5. Davis, I.G., Rosemann, M., Green, P.F.: Exploring proposed ontological issues of ARIS with different categories of modellers. In: Proceedings of the Australasian Conference on Information Systems (2004)
6. Höfferer, P.: Achieving business process model interoperability using metamodels and ontologies. In: Proceedings of the 15th European Conference on Information Systems (ECIS 2007), pp. 1620–1631 (2007)
7. Haller, A., Marmolowski, M., Oren, E., Gaaloul, W.: A process ontology for business intelligence. Business, p. 19, April 2008
8. Belecheanu, R., Cabral, L., Domingue, J., Gaaloul, W., Hepp, M., Filipowska, A., Kaczmarek, M., Kaczmarek, T., Nitzsche, J., Norton, B., Pedrinaci, C., Roman, D., Stollberg, M., Stein, S.: Semantics Utilized for Process Management within and between Enterprises (2007)
9. Cimpian E., Komazec S., Lintner D., Blamauer C., Evenson M.: Business Process Modeling Ontology BPMO Final Version (2008)
10. Koschmider, A., Oberweis, A.: Ontology based business process description. In: Proceedings of the CAiSE, vol. 5, pp. 321–333 (2005)
11. Thomas, O., Fellmann, M.: Semantic process modeling – design and implementation of an ontology-based representation of business processes. Bus. Inf. Syst. Eng. **1**(6), 438–451 (2009)
12. Eisenbarth T.: Semantic Process Models. University of Augsburg (2013)
13. Francescomarino, C., Ghidini, C., Rospocher, M., Serafini, L., Tonella, P.: Semantically-aided business process modeling. In: Bernstein, A., Karger, D.R., Heath, T., Feigenbaum, L., Maynard, D., Motta, E., Thirunarayan, K. (eds.) ISWC 2009. LNCS, vol. 5823, pp. 114–129. Springer, Heidelberg (2009). doi:10.1007/978-3-642-04930-9_8
14. Leopold, H., Meilicke, C., Fellmann, M., Pittke, F., Stuckenschmidt, H., Mendling, J.: Towards the automated annotation of process models. In: Zdravkovic, J., Kirikova, M., Johannesson, P. (eds.) CAiSE 2015. LNCS, vol. 9097, pp. 401–416. Springer, Cham (2015). doi:10.1007/978-3-319-19069-3_25
15. Fan, S., Hua, Z., Storey, V.C., Zhao, J.L.: A process ontology based approach to easing semantic ambiguity in business process modeling. Data Knowl. Eng. **102**, 57–77 (2016)
16. Aysolmaz, B., Demirörs, O.: Unified process modeling with UPROM tool. In: Nurcan, S., Pimenidis, E. (eds.) CAiSE Forum 2014. LNBIP, vol. 204, pp. 250–266. Springer, Cham (2015). doi:10.1007/978-3-319-19270-3_16
17. Jansen-Vullers, M., Netjes, M.: Business process simulation–a tool survey. In: Workshop and Tutorial on Practical Use of Coloured Petri Nets and the CPN (2006)
18. Laue, R., Storch, A., Höß, F.: The bflow* hive - adding functionality to eclipse-based modelling tools. In: Proceedings of the BPM Demo Session 2015 Co-located with the 13th International Conference on Business Process Management (BPM 2015), pp. 120–124 (2015)

19. Stojanovic, L., Motik, B.: Ontology evolution within ontology editors. In: EKAW02 Workshop on Evaluation of Ontology-Based Tools (EON 2002), pp. 53–62 (2002)
20. Dimitrov, M., Simov, A., Stein, S., Konstantinov, M.: A BPMO based semantic business process modelling environment. In: Proceedings of the Workshop on Semantic Business Process and Product Lifecycle Management (SBPM 2007) (2007)
21. Di Francescomarino, C., Ghidini, C., Rospocher, M., Serafini, L., Tonella, P.: A framework for the collaborative specification of semantically annotated business processes. J. Softw. Maintenance Evol. **23**(4), 261–295 (2011)
22. Cherfi, S.SS., Ayad, S., Comyn-Wattiau, I.: Aligning business process models and domain knowledge: a meta-modeling approach. In: Morzy, T., Härder, T., Wrembel, R. (eds.) Advances in Databases and Information Systems. Advances in Intelligent Systems and Computing, vol. 186, pp. 45–56. Springer, Heidelberg (2013). doi:10.1007/978-3-642-32741-4_5

Ontology-Based Heuristics for Process Behavior: Formalizing False Positive Scenarios

Jorge Roa[✉], Emiliano Reynares, María Laura Caliusco, and Pablo Villarreal

Universidad Tecnológica Nacional - Facultad Regional Santa Fe - CONICET,
Lavaise 610, S3004EWB Santa Fe, Argentina
{jroa,ereynares,mcaliusc,pvillarr}@frsf.utn.edu.ar

Abstract. Verification methods to detect errors in the behavior of process models can be formal or informal. The former are based on formal languages, whereas the latter are based on heuristics. The main advantage of informal methods with respect to the formal ones is their short run-time. However, heuristics may lead to false positives, i.e. they may detect errors in a process model even though such model is correct. In this work, we propose using ontologies to formalize heuristics that avoid false positive scenarios. With ontologies it is possible to avoid ambiguities in heuristics that may lead to inaccurate implementations and to enable their execution by ontology reasoners. To this aim, we propose a set of false positive scenarios and define SWRL rules and SPARQL queries to formalize heuristics for such scenarios by means of ontologies. In addition, we identified three requirements that should be met in order to formalize heuristics and their false positive scenarios.

Keywords: Business process model · Anti-patterns · Verification · SPARQL

1 Introduction

Business process management emerged as a means to cope with the complexity of organizations and is key to provide their services and products to customers as well as to organize their internal operational processes [4]. Since process design has direct influence on the economic efficiency of the underlying process-related project [4], it is required the verification of process behavior.

Two properties to be verified in process models are deadlocks and lack of synchronization. There are two types of verification methods to detect these properties: formal methods based on formal languages such as [1], and informal methods based on heuristics (also known as anti-patterns) such as [8,11]. Both type of methods have benefits and downsides with respect to their completeness, precision, and run-time.

A verification method has *total completeness* (or is *complete*) if it is able to verify any process model of a language and if for each process model with errors, it can detect all of such errors, i.e. there are no false negative cases. A verification

© Springer International Publishing AG 2017
M. Dumas and M. Fantinato (Eds.): BPM 2016 Workshops, LNBIP 281, pp. 106–117, 2017.
DOI: 10.1007/978-3-319-58457-7_8

method has *total precision* if each error indicated by the method is an error that exists for real in the verified process model, i.e. there are no false positive cases. *Run-time* has to do with the time it takes to return a verification result.

Both type of methods have partial completeness, since in general they are not able to verify every possible model of a language. Formal methods have total precision. However, their use is limited by their exponential run-time, since they may suffer the state space explosion problem [7]. Heuristic-based methods may have a short run-time [7,16]. However, they are promising as long as their precision and completeness can be improved. The focus of this work is on precision aspects of heuristics, whereas their completeness is left as future work.

In previous work [13], we argued that heuristics should be formalized to avoid ambiguities in their implementation and to enable the use of ontology reasoners to detect errors in process models. To this aim, we formalized heuristics with ontologies by means of SWRL rules and used an ontology reasoner to detect deadlocks and lack of synchronizations in the behavior of graph-based BPMN process models. However, an analysis of the support ontologies provide to improve precision of heuristics was out of the scope of such work. To this aim, it is necessary to consider false positive scenarios as part of the heuristics.

Therefore, the goal of this work is to make use of ontologies to formalize heuristics that avoid false positive scenarios. We explored a set of false positive scenarios for existing heuristics, and extended previous work with new SWRL rules and SPARQL queries to formalize heuristics for such scenarios by means of ontologies. This way, each heuristic considers both the combination of process elements causing the error itself, and the combination of elements that could lead to false positives cases. In addition, we identified three requirements that should be met in order to formalize heuristics and their false positive scenarios.

This work is structured as follows. Section 2 introduces the background. Section 3 describes heuristics together with false positive scenarios. Section 4 presents the formalization of heuristics with ontologies. Section 5 establishes a discussion. Finally, Sect. 6 presents conclusions and future work.

2   Background

A well-known property to verify business processes is soundness [1], which allows detecting control flow errors in process models. A sound business process is free of deadlocks and lack of synchronizations [16]. Formal approaches that verify these properties [1,7,12] mostly rely on state space exploration, which leads to a trade-off between completeness of verification methods and their run-time.

Different heuristics have been proposed to detect problems in the behavior of business processes and workflows, such as those based on anti-patterns [2,8,9,15]. However, these heuristics are not formalized, which may lead to ambiguities and wrong implementations, and hence, false positive or false negative scenarios.

In [10], authors defined heuristics by means of a set of rules with a Backus-Naur Form (BNF) grammar. However, they focus only on structured business processes, and hence, complex combinations of process model elements that could lead to false positive or negative cases are not considered. In addition, since they use a BNF grammar, rules need to be translated to a Java algorithm, and hence, it is not guaranteed that algorithms are aligned with rules.

In previous work [11], we focused on the completeness requirement of a verification method based on anti-patterns, and proposed an approach for discovery and specification of anti-patterns to detect problems in the behavior of collaborative business processes. However, heuristics were implemented by means of Java algorithms for the Eclipse platform, and no formal specification was provided.

Ontologies have traditionally been used to reason on properties related to the data perspective of business processes and to make queries to process models [3], rather than the control flow (or behavioral) perspective. In previous work [13], to verify the behavior of BPMN models, we proposed that heuristics can be formalized with a formal ontological specification of BPMN. As a result, we represented the structure of BPMN models as instantiations of an ontological specification of the ontology for BPMN presented in [14], and extended this ontology with a set of rules that specify heuristics to detect behavioral problems in process models.

Ontology is envisioned as a structure defining concepts used to represent knowledge and their relationships. An ontology model is composed of concepts, relations, a concept hierarchy or taxonomy, a function that relates concepts non-taxonomically, and a set of ontology axioms expressed in a logical language [5].

The OWL 2 Web Ontology language[1] is the latest version of an ontology implementation language with formally defined meaning. An OWL 2 ontology is a formal description of a domain of interest interpreted under a standardized semantics that allows useful inferences to be drawn. The Semantic Web Rule Language (SWRL)[2] is a proposal aimed at extending the set of OWL 2 axioms to include Horn-like rules, enabling their combination with an OWL 2 knowledge base. The rules are of the form of an implication between an antecedent and a consequent. The intended meaning can be read as: whenever the conditions specified in the body hold, then the conditions specified in the head must also hold. OWL 2 ontologies are primarily stored and exchanged as directed and labeled graphs serialized on RDF documents[3]. SPARQL[4] is a query language specifically conceived for such kind of data. It contains capabilities for querying required and optional graph patterns, supporting aggregation, filtering, subqueries, etc. The results of a SPARQL queries can be a result set or a RDF graph.

3   Heuristics for Business Process Behavior

In this section, we present three heuristics for detecting problems in process behavior. We were looking for heuristics and scenarios in which a false positive situation occurs regarding deadlocks/lack of synchronization. Heuristic 1 was

---

[1] https://www.w3.org/TR/owl2-overview/.
[2] http://www.w3.org/Submission/SWRL/.
[3] https://www.w3.org/RDF/.
[4] http://www.w3.org/TR/sparql11-query/.

defined as part of this work, whereas heuristics 2 and 3 were adapted from existing literature [8,15]. For each heuristic, we use a process model to exemplify a problem scenario with the type of problem such heuristic detects. Problem scenarios are presented together with false positive scenarios, which are process models with slight modifications to the original one that would make a heuristic fail by indicating an error in the process model when it has no error; i.e. a heuristic is satisfied by a process P and there is no error in P.

We use the concept of token to describe the behavior of process models of scenarios. A token traverses sequence flows and passes through the elements in a process model. A sequence flow is activated if it receives a token. Scenarios are represented by means of the BPMN elements start event, end event, parallel gateway, exclusive gateway, sequence flow, and activity. No loops are considered in scenarios. A detailed description of the behavioral semantics of BPMN elements and how tokens are used to describe the behavior of process models can be found in [6].

In this work, a *convergent element* can be an activity, parallel gateway, exclusive gateway, or end event that has two or more incoming sequence flows, whereas a *divergent element* can be an activity, parallel gateway, exclusive gateway, or start event that has two or more outgoing sequence flows. There is a *deadlock* in a convergent element if the element is a parallel gateway and, at a given state, at least one its incoming sequence flows cannot be activated. There is a *lack of synchronization* in a convergent element if the element is an exclusive gateway, an activity, or an end event and, at a given state, more than one of its incoming sequence flows are activated. There is a *False Positive (FP)* in a process model P if: (1) a heuristic detects a deadlock in a convergent element e of P, and for each possible execution of P every input sequence flow of e can be activated; or (2) a heuristic detects a lack of synchronization in a convergent element e, and for each possible execution of P only one of the input sequence flows of e can be activated.

3.1   Problem Scenario 1

Figure 1 shows a BPMN process model with a deadlock. In this process, from sequence flow $sf3$, it is possible to reach the exclusive gateway xs. If sequence flow $sf5$ (instead of $sf7$) receives a token, then end event $ee2$ will be activated. This means that sequence flow $sf8$ will not receive a token and gateway pj will not be activated, causing a deadlock.

Heuristic 1 (#H1). Suppose there is a process model P where its control flow diverges in parallel paths by means of a parallel gateway ps, and such paths converges into another parallel gateway pj. Suppose also that there is an exclusive gateway xs that is reachable from gateway ps, and from such exclusive gateway it is possible to reach gateway pj. If from gateway xs it is possible to reach an end event before reaching gateway pj, then the process P has a deadlock in gateway pj.

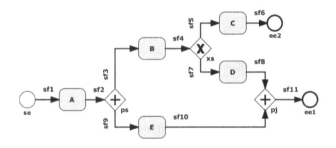

Fig. 1. Problem scenario 1

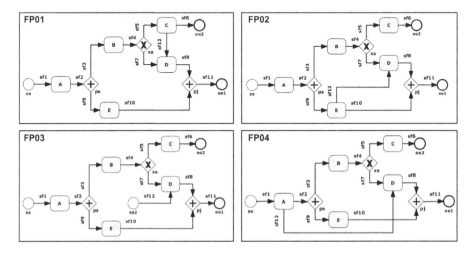

Fig. 2. False positive scenarios for #H1

False Positive Scenarios #H1. In order to determine false positive scenarios for #H1 we evaluate alternatives that can make every input sequence flow of gateway pj be activated for any possible execution. Figure 2 shows four different types of false positive scenarios where heuristic #H1 would indicate a deadlock in gateway pj (FP01, FP02, FP03, and FP04). In Sect. 4, we extend #H1 with new heuristics inferred from these scenarios as follows.

1. In FP01, if sequence flow $sf5$ is activated then activities C, D and sequence flow $sf8$ will also be activated, and hence, pj will not lead to a deadlock.
2. In FP02, activity D and $sf8$ will always be activated once activity E is executed, and hence, pj will not lead to a deadlock.
3. In FP03, activity D and $sf8$ will always be activated once the process starts, and hence, pj will not produce a deadlock.
4. In FP04, activity D and $sf8$ will always be activated once activity A is executed, and hence, pj will not lead to a deadlock.

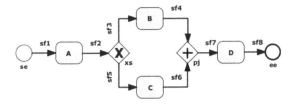

Fig. 3. Problem scenario 2

3.2 Problem Scenario 2

Figure 3 shows a BPMN process model with a deadlock. In this process, the activation of sequence flows $sf3$ and $sf5$ is mutually exclusive. Hence, if $sf3$ is activated, then there will be a deadlock since parallel gateway pj will be waiting for a token in $sf6$. On the other hand, if $sf5$ is activated, there will also be a deadlock since parallel gateway pj will be waiting for a token in $sf4$. This means there is no chance for gateway pj to be activated, causing a deadlock.

Heuristic 2 (#H2). Let P be a process model where there is a divergent exclusive gateway xs. If there is a parallel gateway pj that converges outgoing paths of xs, then the process P has a deadlock.

False Positive Scenarios for #H2. In order to determine false positive scenarios for #H2 we evaluate all alternatives that can make every input sequence flow of parallel gateway pj be activated for any possible execution. Figure 4 shows two different type of false positive scenarios where heuristic #H2 would indicate a deadlock in gateway pj (FP01 and FP02). In Sect. 4, we extend #H2 with new heuristics inferred from these scenarios as follows.

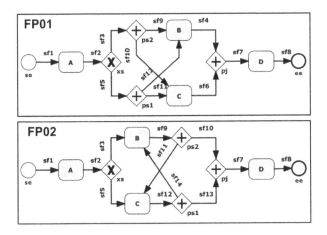

Fig. 4. False positive scenarios for #H2

1. In FP01, if either gateway $ps1$ or $ps2$ is activated, then both activities B and C, and sequence flows $sf4$ and $sf6$ will be activated, and hence, pj will not produce a deadlock.
2. In FP02, if either gateway $ps1$ or gateway $ps2$ is activated, then either activity B or C will be executed. In both cases, sequence flows $sf10$ and $sf13$ will be activated, which means that pj will not lead to a deadlock.

3.3   Problem Scenario 3

Figure 5 shows a BPMN process model with a lack of synchronization. In this process, since both activities B and C will be executed in parallel, gateway xj will be executed twice, causing a lack of synchronization in xj.

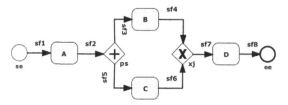

Fig. 5. Problem scenario 3

Heuristic 3 (#H3). Let P be a process model where there is a divergent parallel gateway ps. If there is an exclusive gateway xj that converges outgoing paths of ps, then the process P has a lack of synchronization.

False Positive Scenarios for #H3. In order to determine false positive scenarios for #H3 we evaluate all alternatives where for each possible executions only one of the input sequence flows of exclusive gateway xj can be activated. Figure 6 shows one false positive scenario where heuristic #H3 would indicate a lack of synchronization in gateway xj (FP01). In Sect. 4, we extend #H3 with new heuristics inferred from these scenarios as follows. In FP01, if xj is in a path from activity B, and the input sequence flows of B are part of the outgoing paths of ps, then activity B will be executed twice, and hence, B will produce a lack of synchronization instead of xj. In Sect. 4, we extend #H3 with new heuristics inferred from these scenarios.

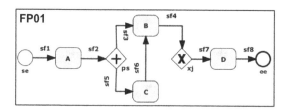

Fig. 6. False positive scenario for #H3

4   Formal Heuristics for Business Process Behavior

In Sect. 3, we showed scenarios where heuristics may fail by indicating an error in a process model that does not have such an error. This lead us to split heuristics into heuristics for problem scenarios and heuristics for false positive scenarios. The former should allow detecting errors in process models and we call it the *problem heuristics*, whereas the latter should allow detecting scenarios where heuristics may fail and we call it *false positive heuristics*. This way, to detect an error in a process model P, the problem heuristics must be satisfied by P, whereas false positive heuristics must not be satisfied by P.

In order to avoid ambiguities and improve precision of heuristics, it should be possible to formalize heuristics. We identified the following requirements that should be met by a formalization approach: (1) it should be possible to define positive facts to formalize heuristics; (2) it should be possible to define negative facts to formalize false positive cases of heuristics; and (3) it should be possible to generate the paths of process models. Requirements 1 and 2 refer to defining the problem heuristics and the false positive heuristics respectively, whereas requirement 3 is necessary to specify heuristics, since they are based on detecting the existence of paths between specific elements of a process model.

In this section, we describe how to use ontologies to formalize heuristics and try to meet the above requirements. We reused the OWL 2 ontology for BPMN proposed in [14] to formalize heuristics to detect behavioral problems in BPMN models, extending it with a set of SWRL rules and SPARQL queries. Such formalization was performed by means of Protégé - a free and open source ontology editor - and the Pellet inference engine, which provides sound-and-complete reasoning services[5]. In addition, we formalized with ontologies the concept of path, which we use in the formalization of heuristics. SWRL rules and the concept of *path* are defined in a separated ontology in the following way:

- A flow element x is connected to a flow element y if there is a sequence flow from x to y, i.e., direct connection between elements.
- A flow element x is connected to a flow element z if there is a flow element y such that x is connected to y and y is connected to z, i.e., transitive nature of the connected property.
- A *path* represents a set of sequence flows and flow nodes connecting two given flow elements.

Each problem and false positive heuristic proposed in Sect. 3 was formalized with a given SPARQL query. For a sake of space, just two SPARQL queries are depicted[6]. The below presented query is aimed to detect the *problem heuristic* #H2. The header of the query, i.e. the set of *PREFIX* sentences, states the abbreviations for the involved namespaces. The *bpmn2* prefix allows referencing

---

[5] Support, downloads and documentation about the integration of Protégé editor and Pellet inference engine can be found at http://protege.stanford.edu.

[6] Process and ontological models and SPARQL queries for all heuristics can be found at http://dx.doi.org/10.17632/xkg32p2bs6.1.

concepts defined in the BPMN2 ontology. The *hwg* prefix allows referencing the SWRL rules and the concept of path explained before. The query retrieves the names of all distinct combinations of elements $?xs$ and $?pj$ such that:

- $?xs$ is an exclusive gateway,
- $?xs$ is the source node of two different sequence flows - i.e., $?sf3$ and $?sf5$,
- $?xs$ is the source node of two different paths - i.e., $?path1$ and $?path2$,
- $?pj$ is a parallel gateway,
- $?pj$ is the target node of two different paths - i.e., $?path1$ and $?path2$, and
- each path contains one the above mentioned sequence flows.

```
PREFIX rdf: <http://www.w3.org/1999/02/22-rdf-syntax-ns#>
PREFIX owl: <http://www.w3.org/2002/07/owl#>
PREFIX xsd: <http://www.w3.org/2001/XMLSchema#>
PREFIX rdfs: <http://www.w3.org/2000/01/rdf-schema#>
PREFIX bpmn2: <http://dkm.fbk.eu/index.php/BPMN2_Ontology#>
PREFIX hwg: <http://cidisi.frsf.utn.edu.ar/emireys/ontologies/hwg#>
SELECT DISTINCT ?xsname ?pjname
   WHERE {
       ?xs a bpmn2:exclusiveGateway.
       ?xs bpmn2:name ?xsname.
       ?pj a bpmn2:parallelGateway.
       ?pj bpmn2:name ?pjname.
       ?sf3 a bpmn2:sequenceFlow.
       ?sf5 a bpmn2:sequenceFlow.
       ?path1 a hwg:Path.
       ?path2 a hwg:Path.
       ?sf3 bpmn2:has_sourceRef ?xs.
       ?sf5 bpmn2:has_sourceRef ?xs.
       ?path1 hwg:start ?xs.
       ?path1 hwg:end ?pj.
       ?path2 hwg:start ?xs.
       ?path2 hwg:end ?pj.
       ?path1 hwg:contains ?sf3.
       ?path2 hwg:contains ?sf5.
       ?path1 owl:differentFrom ?path2.
       ?sf3 owl:differentFrom ?sf5.
   }
```

The query aimed to detect the *false positive heuristic* corresponding to #H2 is depicted bellow. The query retrieves the names of all the distinct combinations of elements $?path1$ and $?path2$ such that:

- $?xs$ is an exclusive gateway,
- $?xs$ is the source node of two different sequence flows - i.e., $?sf3$ and $?sf5$,
- $?xs$ is the source node of two paths - i.e., $?path1$ and $?path2$,
- $?pj$ is a parallel gateway,
- $?pj$ is the target node of two paths - i.e., $?path1$ and $?path2$,
- each path contains one the above mentioned sequence flows, and
- both paths contains at least a common flow element

```
PREFIX rdf: <http://www.w3.org/1999/02/22-rdf-syntax-ns#>
PREFIX owl: <http://www.w3.org/2002/07/owl#>
PREFIX xsd: <http://www.w3.org/2001/XMLSchema#>
PREFIX rdfs: <http://www.w3.org/2000/01/rdf-schema#>
PREFIX bpmn2: <http://dkm.fbk.eu/index.php/BPMN2_Ontology#>
PREFIX hwg: <http://cidisi.frsf.utn.edu.ar/emireys/ontologies/hwg#>
SELECT DISTINCT ?path1 ?path2
    WHERE {
            ?xs a bpmn2:exclusiveGateway.
            ?pj a bpmn2:parallelGateway.
            ?sf3 a bpmn2:sequenceFlow.
            ?sf5 a bpmn2:sequenceFlow.
            ?path1 a hwg:Path.
            ?path2 a hwg:Path.
            ?fe a bpmn2:flowElement.
            ?sf3 bpmn2:has_sourceRef ?xs.
            ?sf5 bpmn2:has_sourceRef ?xs.
            ?path1 hwg:start ?xs.
            ?path1 hwg:end ?pj.
            ?path2 hwg:start ?xs.
            ?path2 hwg:end ?pj.
            ?path1 hwg:contains ?sf3.
            ?path2 hwg:contains ?sf5.
            ?path1 hwg:contains ?fe.
            ?path2 hwg:contains ?fe.
            ?sf3 owl:differentFrom ?sf5.
    }
```

5 Discussion

The discovery of false positive scenarios is key to improve precision of heuristics, and hence, to avoid heuristics leading to erroneous conclusions and inaccurate implementations. In this work, we found false positive scenarios by exploring alternatives to problem scenarios. We focused on a set of scenarios to know if they could be formalized with ontologies. However, there could be more false positive scenarios than the explored in this work. Discovering such scenarios is a time-consuming and complex task. An algorithm that generates process variants of a problem scenario could be used to automate such task. Each variant should be verified to determine whether it is a false positive scenario or not.

Although a false positive scenario invalidates an error indicated by a given heuristic, the scenario may have other errors that are not be detected by such heuristic. For example, in problem scenario 1, although false positive scenarios FP01, FP02, and FP03 do not have a deadlock as indicated by #H1, they have a lack of synchronization, since activity D could be executed twice. This last error should require another heuristic to detect such lack of synchronization.

In Sect. 4 we proposed that formalization of heuristics should meet three requirements. These requirements are partially met by the proposed formalization approach. Requirement 1 is met, since SPARQL queries proposed in Sect. 4 represent positive facts that formalize heuristics. Requirement 2 is not directly met, since the underlying logic formalism is not able to represent negative facts. However, to cope with this issue, we proposed heuristics should be split into a problem heuristics and false positive heuristics. This way, to detect an error in a process the former must be satisfied by the process, whereas the latter must not be satisfied. Requirement 3 is not met, since ontology reasoners cannot generate on demand the paths of a process model necessary to execute heuristics. This means that to use heuristics proposed in Sect. 4, it is first necessary to generate the paths of the process to be verified. This could be performed by making use of a graph traversal-query language such as Gremlin[7].

6   Conclusions and Future Work

In this work, we showed that the use of ontologies enables the formalization of heuristics to reason on behavioral properties of business process models, and we identified general requirements that have to be met in the formalization of heuristics. Although ontologies partially support such requirements, we showed that SWRL rules and SPARQL queries can be used to formalize different heuristics considering both problem and false positive scenarios.

The main advantage of using ontologies to formalize heuristics is to avoid ambiguities and to enable their execution by ontology reasoners. In practice, this implies that there is no need to implement heuristics as an algorithm in a given programming language. This is important, since a reasoner will not have ambiguities in its implementation, whereas an algorithm may fail due to misunderstanding of the heuristics or due to a faulty implementation. Hence, if a process model has an error, and the reasoner does not detect such error, then it will be clear that the definition of the heuristics and their formal representation must be fixed. This is specially important when formalizing false positive scenarios to improve precision of heuristics.

Future work is concerned with improving both precision and completeness of heuristics. To this aim, we will work on an approach to automate the discovery of false positive and false negative scenarios by defining variants of process models.

References

1. van der Aalst, W.M.P.: The application of petri nets to workflow management. J. Circ. Syst. Comput. **08**(01), 21–66 (1998)
2. Awad, A., Puhlmann, F.: Structural detection of deadlocks in business process models. In: Abramowicz, W., Fensel, D. (eds.) BIS 2008. LNBIP, vol. 7, pp. 239–250. Springer, Heidelberg (2008). doi:10.1007/978-3-540-79396-0_21
3. Becker, J., Pfeiffer, D., Räckers, M., Falk, T., Czerwonka, M.: Semantic business process modelling and analysis. In: Brocke, J., Rosemann, M. (eds.) Handbook on Business Process Management 1. IHIS, pp. 187–217. Springer, Heidelberg (2015). doi:10.1007/978-3-642-45100-3_9

---

[7] https://github.com/tinkerpop/gremlin/wiki.

4. Becker, J., Rosemann, M., Uthmann, C.: Guidelines of business process modeling. In: Aalst, W., Desel, J., Oberweis, A. (eds.) Business Process Management. LNCS, vol. 1806, pp. 30–49. Springer, Heidelberg (2000). doi:10.1007/3-540-45594-9_3

5. Breitman, K.K., Sampaio do Prado Leite, J.C.: Lexicon based ontology construction. In: Lucena, C., Garcia, A., Romanovsky, A., Castro, J., Alencar, P.S.C. (eds.) SELMAS 2003. LNCS, vol. 2940, pp. 19–34. Springer, Heidelberg (2004). doi:10.1007/978-3-540-24625-1_2

6. Dijkman, R.M., Dumas, M., Ouyang, C.: Semantics and analysis of business process models in bpmn. Inf. Softw. Technol. **50**(12), 1281–1294 (2008)

7. Fahland, D., Favre, C., Jobstmann, B., Koehler, J., Lohmann, N., Völzer, H., Wolf, K.: Instantaneous soundness checking of industrial business process models. In: Dayal, U., Eder, J., Koehler, J., Reijers, H.A. (eds.) BPM 2009. LNCS, vol. 5701, pp. 278–293. Springer, Heidelberg (2009). doi:10.1007/978-3-642-03848-8_19

8. Koehler, J., Vanhatalo, J.: Process anti-patterns: How to avoid the common traps of business process modeling. IBM WebSphere Dev. Tech. J. **10**(2–4) (2007)

9. Kühne, S., Kern, H., Gruhn, V., Laue, R.: Business process modeling with continuous validation. J. Softw. Maint. Evol. Res. Pract. **22**(6–7), 547–566 (2010)

10. Palma, F., Moha, N., Guéhéneuc, Y.-G.: Specification and detection of business process antipatterns. In: Benyoucef, M., Weiss, M., Mili, H. (eds.) MCETECH 2015. LNBIP, vol. 209, pp. 37–52. Springer, Cham (2015). doi:10.1007/978-3-319-17957-5_3

11. Roa, J., Chiotti, O., Villarreal, P.: Specification of behavioral anti-patterns for the verification of block-structured collaborative business processes. Inf. Softw. Technol. **75**, 148–170 (2016)

12. Roa, J., Chiotti, O., Villarreal, P.: Behavior alignment and control flow verification of process and service choreographies. JUCS **18**(17), 2383–2406 (2012)

13. Roa, J., Reynares, E., Caliusco, M.L., Villarreal, P.: Towards ontology-based anti-patterns for the verification of business process behavior. New Advances in Information Systems and Technologies. AISC, vol. 445, pp. 665–673. Springer, Cham (2016). doi:10.1007/978-3-319-31307-8_68

14. Rospocher, M., Ghidini, C., Serafini, L.: An ontology for the business process modelling notation. In: Garbacz, P., Kutz, O. (eds.) Formal Ontology in Information Systems - Proceedings of the Eighth International Conference, FOIS2014, September 22–25, 2014, vol. 267, pp. 133–146. IOS Press, Rio de Janeiro (2014)

15. Van Dongen, B., Mendling, J., van der Aalst, W.: Structural patterns for soundness of business process models. In: 10th IEEE International Enterprise Distributed Object Computing Conference, EDOC 2006, pp. 116–128, October 2006

16. Vanhatalo, J., Völzer, H., Leymann, F.: Faster and more focused control-flow analysis for business process models through SESE decomposition. In: Krämer, B.J., Lin, K.-J., Narasimhan, P. (eds.) ICSOC 2007. LNCS, vol. 4749, pp. 43–55. Springer, Heidelberg (2007). doi:10.1007/978-3-540-74974-5_4

Business Process Architecture Baselines from Domain Models

Fernanda Gonzalez-Lopez[(✉)] and Guillermo Bustos

Pontificia Universidad Catolica de Valparaiso, Valparaiso, Chile
maria.gonzalez.l@mail.pucv.cl, guillermo.bustos@pucv.cl

Abstract. Business process architectures allow to organize business processes and their relations. The entity-centric approach for business process modeling may offer new insights on this field. We present an entity-centric procedure for deriving a business process architecture baseline using the following core ideas: (i) domain model entities may be interpreted as business entities at a higher level of abstraction than usually used by entity-centric approaches, (ii) domain model relationships provide useful information for deriving a business processes architecture. We present the procedure in combination with an application example. The resulting business process architecture baseline specifies: (i) decomposition, specialization, and trigger relations, and (ii) core and support classification of business processes. These results show the potential of our approach. The main contribution of our work is providing guidelines to obtain business process architectures that may be used as reference models for companies within the same industry.

Keywords: Business process reference models · Business process architecture · Business entity life cycle · Domain model · Entity-centric

1 Introduction

Modern organizations increasingly adopt a process-oriented perspective for managing their operations. Such an approach represents the organizational business processes using business process models. These models need to be structured coherently in order to support the goals of the organization. As stated by [11], two risks arise when defining business processes in a piecemeal manner: processes might not work well together, and key processes may be left out. The concept of *business process architecture* has emerged to address these concerns by modeling the business process structure at the enterprise level. We adopt the business process architecture definition by [5]: *an organized overview of business processes that specifies their relations, which can be accompanied with guidelines that determine how these processes must be organized.* In practical terms, a business process architecture answers the following questions: (i) which business processes exist in the organization, (ii) where does one business process ends and the other begins, and (iii) how do business processes relate to each other.

© Springer International Publishing AG 2017
M. Dumas and M. Fantinato (Eds.): BPM 2016 Workshops, LNBIP 281, pp. 118–130, 2017.
DOI: 10.1007/978-3-319-58457-7_9

The works by [5,9,12,15] review available proposals for building business process architectures. Most of the proposals are framed within the traditional *activity-centric* approach to business process modeling, neglecting the newer *entity-centric* approach. An exception to this, is the *Riva* method [11], which incorporates some of the entity-centric concepts, but only in a partial way. The entity-centric approach [23] constitutes a change in perspective for business processes: while the traditional activity-centric approach focuses on *how* business processes operate, the entity-centric approach focuses on *what* is being processed, i.e. *business entities*. We do not advocate for one approach being better than the other, and in this sense we agree that they are complementary [21]. However, we believe that within the realm of the entity-centric approach lies a yet unexplored source of analysis tools for business process architecture. In particular, we believe that conceptual domain models applied in the context of the entity-centric approach, may provide a baseline for the definition of business process architectures.

The entity-centric paradigm allows to derive business process models from previously understanding the domain structure and behavior. We argue that conceptual domain models are a source of the former kind of information, since they may be interpreted as depicting business entities and their relations. Over the past decades, different organizations and industries have distilled and improved their conceptual domain models. This has led to identifying patterns that are valid for different organizations within a given business domain. For instance, [26] claims that *usually more than 50 percent of the data model consists of common constructs that are all applicable to most organizations, another 25 percent of the model is industry specific [. . .], and on average, about 25 percent of the enterprise's data model is specific to that organization.* In line with this idea, we argue that domain models may set the starting point for designing a reference business process architecture via the entity-centric paradigm, by providing information regarding the domain-specific business processes. It must be clarified that we do not propose to reuse data objects as business entities, but rather reuse domain concepts as business entities. This is a fundamental difference between our proposal and most entity-centric approaches in the literature: while typically entity-centric approaches interpret business entities as business-relevant information, we interpret them as things themselves. We understand business entities as key business-relevant dynamic concepts found in the real world (not their information registries) that are the subject matter of business process operation. When business entities are interpreted at such level of abstraction, the derived business process architecture baseline may be valid for all organizations in the business domain since it is not bounded to implementation issues. The goal of the present work is to explore the use of conceptual domain models for building a business process architecture baseline in a given business domain. We propose a five-step procedure and exemplify it via an application example showing initial insights on the potential of our approach.

The remainder of the paper is structured as follows. Section 2 reviews the entity-centric approach. Section 4 shows and discusses the proposed procedure based on the application example introduced in Sect. 3. Finally, we present the conclusions of our work.

2   Background

The *entity-centric* approach for business process modeling [23] (also *information-centric* or *artifact-centric*) is an alternative to traditional *activity-centric* approach. The central modeling construct in the entity-centric approach is the *business entity* (also *artifact* or *essential business entity*). Different proposals within entity-centric approach, interpret business entities at a different level of abstraction. Early proposals [2,16,23] define business entities at a business-relevant level. Recent proposals identify business entities with objects in the sense of object oriented paradigm [21,25], focusing on information requirements for business process enactment via software applications. In both cases, however, business entities represent information and not things themselves. This level of abstraction restricts entity-centric business process models to the particular way a given enterprise organizes its information.

Business entity identification may be achieved using two different approaches: *top-down* or *bottom-up* [22]. The top-down approach is based on subject matter experts analyzing the business system in light of a set of guiding questions that consider aspects such as what is being processed and *KPIs*. Most entity-centric literature focuses on the top-down approach, however, it is only advisable when expert knowledge is available at a reasonable cost. The less explored bottom-up approach is based on the analysis and abstraction of the data domain. The bottom-up approach is advisable when a domain model may be abstracted from a data model at a reasonable cost.

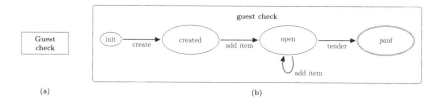

(a)   (b)

Fig. 1. Models for *guest check* business entity (a) structure, and (b) life cycle.

A (type of) business entity is represented by a structure model, e.g. entity-relationship diagram as in [2] or class diagram as in [7]. Figure 1a exemplifies a representation of a *guest check* business entity of a restaurant business system using UML class diagram notation. The generic behavior of business entity instances is represented via a *life cycle* model showing possible states, transitions, and activities during its existence in the business system, e.g. finite state machine or similar as in [16,23] or more declarative representations as the GMS in [13]. Figure 1b exemplifies a representation of the life cycle of the *guest check* business entity in which ellipses represent states, arcs represent transitions, and arc labels represent activities.

The network of interacting life cycles of the business entities in a given business system, provides a representation of the behavior of such system. Consequently, business process models may be derived by grouping sets of activities of such network, i.e. portions of the life cycles of a single or multiple business entities within a business system scope. According to [2, 23] the criteria for the activity grouping relies on a goal-based business perspective. For instance, in the restaurant example it is possible to identify a *generate sales* business process that contributes to achieving the business goal *generate revenue*. The scope of the *generate sales* business process is given by the set of activities that push the *guest check* business entity from the *created* to the *paid* state: as showed in Fig. 1b these activities include *create*, *add item*, and *tender*. The scope of the *generate sales* business process will include additional activities that may be discovered by analyzing business entities that interact with the *guest check* business entity, e.g. the *prepare items* activity for a *kitchen order* business entity.

Besides business process identification, the network of interacting life cycles of the business entities within a business system, provides information for relating business processes: the relations between the life cycles of the interacting entities may give insights for business process relations. Two kinds of life cycle interactions are identified in [23]: (i) *dependency* when a task within the life cycle of a business entity references another business entity, and (ii) a stronger *interaction* when a task within the life cycle of a business entity, needs to modify another business entity. Following the same line of work, [1] defines *correlation tasks* as activities that modify multiple business entities and can be understood as a linkage between operational goals of the business system. The work by [12] defines two types of relations between business entities: (i) *activation* when one business entity instance generates an instance of another business entity type, e.g. a *guest check* entity instance may generate a *kitchen order* entity instance, and (ii) *interaction* which is a more general relation that includes, for example, negotiation. The work by [16] defines the concept of *dominance* between business entities and proposes an algorithm that relates business entity life cycles based on such concept. Dominance refers to the relative level of participation of a business entity in an activity network that defines a process scope. We found no link between work on business process relations outside the entity-centric paradigm (e.g. [5, 20]) and life cycle relation analysis as previously described.

3   Manufacturing Industry Application Example

The business domain of the application example is the Manufacturing Industry. We consider that an organization in such domain manufactures and sells products of various types to clients that place national or international orders, which the organization delivers and invoices.

Domain model. From a structure perspective, the business domain may be represented by the domain model showed in Fig. 2a and its complementary documentation. We read the model components *Client*, *Order*, *Product type* and *Invoice*, as the business entities of the business system domain. We read associations between the components as relations between business entities, i.e. a *client Generates* multiple *orders*, an *order Includes* multiple *product types*, and an *order Invoices* an *invoice*. The *order* business entity in Fig. 2a has a hierarchy including the sub entities *national* and *international*, which represent types of *orders* to be found in the business domain. In the domain models presented throughout the work, we use an abbreviated notation for the hierarchy properties: e.g. {*d, c*} stands for {*disjoint, complete, static*} as in Fig. 2a, and {*d, c, d*} stands for {*disjoint, complete, dynamic*} as in Fig. 4. The *dynamic* property of a hierarchy allows an instance to migrate between sub classes during its existence [24]. For example, an *invoice* class may have the sub classes *sent* and *paid* arranged into a dynamic hierarchy, meaning that: (i) *status* of the invoice is the categorization criteria for the sub classes, and (ii) changes in the status of a given invoice produce its migration from one sub class to the other.

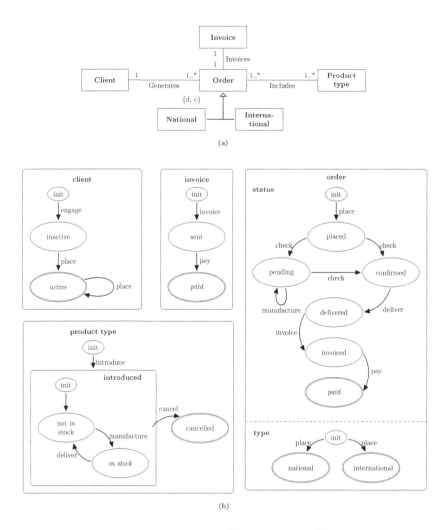

Fig. 2. Models for application example (a) domain, and (b) life cycle network.

Life cycle models. Dynamics of the business domain is represented by life cycle network showed in Fig. 2b using a notation similar to [21]. The life cycle of each business entity represents a portion of concurrent behavior within the business system. Also, the life cycle of one business entity may incorporate concurrency, as e.g. between *status* and *type* in the *order* life cycle in Fig. 2b. Within each life cycle, nodes represent *states*, arcs represent *non-instantaneous transitions*, and arc labels (e.g. *place* in Fig. 2b) represent *activities* that push business entity through their life cycles. Two kinds of special nodes may be found in a life cycle model: ellipses labeled *init* represent initial states, double-edged ellipses represent final states. We use italic font to represent desired states (e.g. *paid* in Fig. 2b). Desired states show the teleological aspect of the business system, i.e. achieving these states is part of the business system goals.

4   Proposed Procedure

The procedure we propose consists of five steps as showed in Fig. 3. It uses as input the following business system models: (i) a structure model as in Fig. 2a together with its complementary documentation, and (ii) a dynamic model as in Fig. 2b. The process analyzes and improves such models in order to produce a business process architecture baseline for the business system.

Fig. 3. Proposed procedure.

Step 1. Improve expressiveness of domain model. In [4], we discussed integration guidelines for structure and dynamic models of business entities. We argued that model expressiveness could be enhanced by incorporating dynamic hierarchy for representing possible states of the business entity in its domain model, since a business entity instance can migrate between different states along its life cycle. In line with the aforementioned, Fig. 4a shows a more expressive version of the original domain model of Fig. 2a. In the new version, each business entity has dynamic hierarchies containing its life cycle states organized in a way that is coherent to its possible state transitions. Also, the version in Fig. 4a is more expressive by making explicit the associations direction - i.e. which is the source entity and which is the target entity - using arrowheads. Such directions were derived by analyzing the semantic of each association considering the active voice verbal form for their naming. For example, the *includes* association in Fig. 4a has *order* as a source and *product type* as a target, and not the other way around.

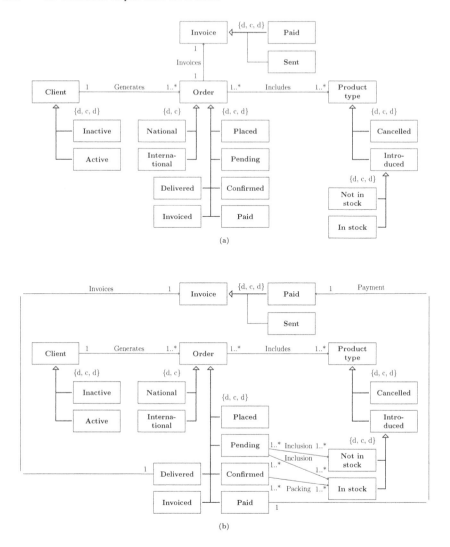

Fig. 4. Domain model for application example revisited (a) first, and (b) second iterations.

Step 2. Identify business processes. We adopt the idea of defining business processes based on a teleological view of the business system from [23] together with the idea of considering business processes as sets of valid trajectories in a state space from [3]. Consequently, it is necessary to identify which of the states of the life cycle network are desired states. A desired state is defined for each business entity within the business system. For some business entities, desired states correspond to final states as *active* for *client*, and *paid* for both *invoice* and *order*. For other business entities, desired states correspond to non-final states as *in stock* for *product type*. Business process must ensure that the business system reaches its goals, i.e. that its business entities reach their desired states. From Fig. 4b, we identify four business processes: (i) *client management* for ensuring the goal *client reaches its desired state active*, (ii) *billing* for ensuring the goal *invoice reaches its desired state paid*, (iii) *order fulfillment* for ensuring the goal *order reaches its desired state paid*, and (iv) *product management* for ensuring the goal *product type remains its desired state in stock*.

Step 3. Revisit life cycles and identify their relations. We argue that in exploring the multiplicity and semantics of associations within the domain model of the business system allows to derive life cycle interactions that will later contribute to relate business process in the business process architecture. For this purpose, it is necessary to revisit the life cycle network and the activities associated with its arcs: activity names need to consider the nature of business entities relations. In this sense, we propose the following modeling guidelines: (i) use of the same name for an activity producing a synchronized state transition in the life cycle of two or more different business entities [16], e.g. *pay* in *order* and *invoice*; and (ii) be coherent with the associations of the domain model in terms of direction (semantics) and multiplicity, e.g. since an *order* instance *includes* many *product type* instances, the life cycles of both business entities are related by the *check* activity that verifies stock availability for all product types in the given order. At this point, complementary documentation of the domain model, may allow to identify additional relations in the life cycle network. For instance, in Fig. 2b an *order* in state *pending*, may trigger the *manufacture* of one or many *product type* instances with the purpose of changing their state from *not in stock* to *in stock*. The life cycle network of Fig. 2b complies with the aforementioned modeling guidelines.

Step 4. Categorize and relate business processes. The business process categories we consider are *core* and *support*. The business process relations we consider are *decomposition*, *specialization*, and *trigger*. We present guidelines for deriving categorization and relations among the previously identified business processes, by analyzing relations between life cycles of the business entities depicted in the life cycle network, together with the relationships between business entities in the domain model.

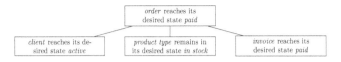

Fig. 5. Goal hierarchy for application example.

We firstly relate the business system goals in terms of a goal hierarchy as showed in Fig. 5. Goal hierarchies organize goals by grouping those that allow fulfilling higher level goals [14]. We derive this information from the domain model associations and the relations between the life cycles of the business entities. In the application example, it is possible to identify the following dependencies: (i) for an *order* instance to reach its state *placed* a *client* instance needs to

Generate it, (ii) for the *order* instance to reach its state *confirmed* it is necessary that all *product types* it *Includes* are in the desired state *in stock*, and (iii) for the *order* instance to reach its state *paid* it is necessary that the *invoice* instance it *ivoices* reaches its desired state *paid*. A goal at the top of the goal hierarchy relates to an end-to-end processes creating value for the business system customers: we categorize such processes as *core*, e.g. *order fulfillment*. On the other hand, we categorize as *support* the business processes related to goals that allow achieving a goal at the top of the goal hierarchy, e.g. *client management*, *product management*, and *billing*.

Decomposition relations may be derived between a business process and the activities within its scope in the life cycle network. Activities of the network may be interpreted as sub-processes, e.g. *deliver* in the context of *order fulfillment*.

Trigger relations may be derived by considering the decomposition relations together with the direction of the associations in the domain model. It is possible to establish a *trigger* relation between *place* in *client management* and *check* in *order fulfillment*: we say *place* triggers *check* since *client* is a source entity and *order* is a target entity within the *Generates* association in the domain model. Similarly, *check* triggers *manufacture* and *deliver* triggers *invoice*.

Specialization relations may be derived from a static hierarchy in the domain model, e.g. since *order* is linked to the *order fulfillment* business process, the *National* and *International* sub entities of *order* may be linked to *national order fulfillment* and *international order fulfillment* business process, respectively.

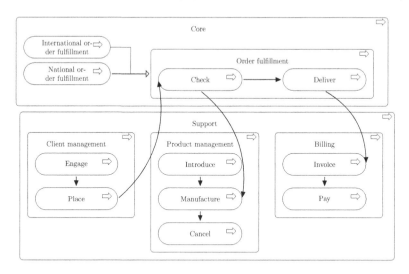

Fig. 6. Business process architecture baseline for application example.

Step 5. Create business process architecture model. The identification and categorization of business processes - into *core* and *support*-, together with the identification of business process relations - *trigger*, *specialization*, and *decomposition*-, may now be formalized using a business process architecture notation.

Figure 6 shows the resulting business process architecture for the application example. We use the Archimate [18] notation, that represents trigger relations by filled arrows, specialization relations by non-filled arrows, and decomposition by sub process nesting.

New iteration. A new iteration of the procedure will begin by improving the expressiveness of the domain model. We propose the following guideline: (re)establish associations between sub entities and other (sub) entities according to their life cycle relations. Figure 4b shows a more expressive domain model for the business system via new associations, e.g. *invoices* has been reestablished between *delivered order* and *invoice*, and *packing* has been established between *confirmed order* and *in stock product* as a specialization of *includes*.

5 Conclusions

The presented work offers guidelines for using domain models together with the entity-centric approach to derive a baseline business process architecture. The proposed procedure allowed to identify business processes, their categorization, and relations for an application example. All of which was achieved by analyzing the domain and life cycle models of the business domain in study. The analysis also served the purpose of improving the expressiveness of the domain model which accompanies and grounds the business architecture model. The topic of process relations is key to business process architecture, and we have provided guidelines to derive such relations using the entity-centric approach for business process models.

Business process relations have been recently studied by works as [5,17,20], with a focus on relation classification. The work by [16] proposes a representation of end-to-end business processes via the interconnected life cycles of dominant business entities in a given process scope, referred to as *synchronized object life cycle* by [21]. This representation acknowledges synchronization relation between business entities that simultaneously change state as a consequence of the same activity. *Artifact choreographies*, i.e. process representations that describe the life cycles of entities involved in a process and how instances of these entities interact with each other (e.g. using Petri nets as in [19] or *proclets* [29] as in [8]), provide a way of representing life cycle interactions acknowledging a variable number of business entities interacting in a given interaction. The work on business process repositories by [6] provides a formalism for representing the business process relations we explore, however this representation includes event representation, which we do not cover in this work. Our proposal lies a step before representations using artifact choreographies or event-driven business process relation formalization: we expect to develop a business oriented foundation for defining the business process architecture.

We used an application example that covers a limited set of the cases that practitioners may encounter when applying the procedure. Also, we believe that the procedure will benefit from further refinement and systematization. Another limitation of our work lies in the applicability of our proposal. Firstly, the completeness of the structure and domain models - together with other quality criteria - will affect the final outcome of the proposed procedure. In those cases where the only available structure model corresponds to a data model, it would be necessary to firstly eradicate entities that are not to be found in the real world in order to have a conceptual domain model. Secondly, we speculate that the applicability of a business process analysis based on domain models as presented in this work, may be restricted to business domains with a standardized operation and functional structure, together with a given strategy. As to the latter, for example, defining *in stock* as the desired state for *product type* is applicable to the case where the company prioritizes *service level* instead of a *just in time* policy for its inventory.

Future works are related to the application of the procedure in case studies. Currently, we are working in the analysis of the telecommunications and e-commerce business domains by comparing conceptual data models by [27] and business process architecture reference models by [10, 28]. A preliminary analysis using word coincidence search between the business entities of the former and the business process of the later provided evidence for the existence of relationships between the conceptual data models and business process architecture in both domains. We expect that applying the procedure in case studies will allow refining the proposed procedure for deriving business process architecture baselines, which might ultimately lead to its automation.

Altogether, our work contributes to the obtainment of business process architecture reference models. Despite the limitations of our work, we argue that our proposal seems promising. The entity-centric approach to business process modeling using domain models provides in fact, new insights for business process management, in particular for business process architecture baseline definition with emphasis on business processes relations.

References

1. Bhattacharya, K., Caswell, N.S., Kumaran, S., Nigam, A., Wu, F.Y.: Artifact-centered operational modeling: lessons from customer engagements. IBM Syst. J. **46**(4), 703–721 (2007)
2. Bhattacharya, K., Hull, R., Su, J.: A data-centric design methodology for business processes. In: Handbook of Research on Business Process Modeling, pp. 503–531 (2009)
3. Bider, I.: State-oriented business process modeling: principles, theory and practice. Dissertation (2002). http://www.diva-portal.org/smash/get/diva2:9168/FULLTEXT01.pdf
4. Bustos, G., Gonzalez-Lopez, F.: Integration of entity-centric models for business processes. In: II Workshop STI (2015)
5. Dijkman, R., Vanderfeesten, I., Reijers, H.A.: Business process architectures: overview, comparison and framework. Enterp. Inf. Syst. **7575**, 1–30 (2014)
6. Eid-Sabbagh, R.: Business process architectures: concepts, formalism, and analysis. Disseration (2015). https://publishup.uni-potsdam.de/opus4-ubp/frontdoor/index/index/docId/7971

7. Estanol, M., Queralt, A., Sancho, M.R., Teniente, E.: Using UML to specify artifact-centric business process models. In: BMSD 2014, pp. 84–93. SciTePress (2014)
8. Fahland, D., De Leoni, M., Van Dongen, B.F., van der Aalst, W.M.P.: Many-to-many: some observations on interactions in artifact choreographies. In: CEUR Workshop, vol. 705, pp. 9–15 (2011)
9. Fettke, P., Loos, P., Zwicker, J.: Business process reference models: survey and classification. In: Bussler, C.J., Haller, A. (eds.) BPM 2005. LNCS, vol. 3812, pp. 469–483. Springer, Heidelberg (2006). doi:10.1007/11678564_44
10. Frank, U., Lange, C.: Referenzgeschaeftsprozesse und Strategien im e-Commerce. ECOMOD (2004). https://www.wi-inf.uni-duisburg-essen.de/FGFrank/ecomod/index.php?lang=en. Accessed
11. Green, S., Ould, M.: The primacy of process architecture. In: CAiSE Workshops, vol. 2, pp. 154–159 (2004)
12. Green, S., Ould, M.: A framework for classifying and evaluating process architecture methods. Softw. Process Improv. Pract. **10**(4), 415–425 (2005)
13. Hull, R., et al.: Introducing the guard-stage-milestone approach for specifying business entity lifecycles. In: Bravetti, M., Bultan, T. (eds.) WS-FM 2010. LNCS, vol. 6551, pp. 1–24. Springer, Heidelberg (2011). doi:10.1007/978-3-642-19589-1_1
14. Kavakli, V., Loucopoulos, P.: Goal-driven business process analysis application in electricity deregulation. Inf. Syst. **24**(3), 187–207 (1999)
15. Koliadis, G., Ghose, A., Padmanabhuni, S.: Towards an enterprise business process architecture standard. In: IEEE Congress on Services, pp. 239–246 (2008)
16. Kumaran, S., Liu, R., Wu, F.Y.: On the duality of information-centric and activity-centric models of business processes. In: Bellahsène, Z., Léonard, M. (eds.) CAiSE 2008. LNCS, vol. 5074, pp. 32–47. Springer, Heidelberg (2008). doi:10.1007/978-3-540-69534-9_3
17. Kurniawan, T.A., Ghose, A.K., Lê, L.-S., Dam, H.K.: On formalizing inter-process relationships. In: Daniel, F., Barkaoui, K., Dustdar, S. (eds.) BPM 2011. LNBIP, vol. 100, pp. 75–86. Springer, Heidelberg (2012). doi:10.1007/978-3-642-28115-0_8
18. Lankhorst, M.: Enterprise architecture modelling - the issue of integration. Adv. Eng. Inform. **18**(4), 205–216 (2004)
19. Lohmann, N., Wolf, K.: Artifact-centric choreographies. In: Maglio, P.P., Weske, M., Yang, J., Fantinato, M. (eds.) ICSOC 2010. LNCS, vol. 6470, pp. 32–46. Springer, Heidelberg (2010). doi:10.1007/978-3-642-17358-5_3
20. Malinova, M., Leopold, H., Mendling, J.: A meta-model for process map design. In: CAiSE (Forum/Doctoral Consortium), pp. 25–32 (2014)
21. Meyer, A., Weske, M.: Activity-centric and artifact-centric process model roundtrip. In: Lohmann, N., Song, M., Wohed, P. (eds.) BPM 2013. LNBIP, vol. 171, pp. 167–181. Springer, Cham (2014). doi:10.1007/978-3-319-06257-0_14
22. Nandi, P., Koenig, D., Moser, S., Hull, R., Klicnik, V., Claussen, S., Kloppmann, M., Vergo, J.: Introducing Business Entities and the Business Entity Definition Language (BEDL). Data4BPM, Part 1, pp. 1–31 (2010)
23. Nigam, A., Caswell, N.S.: Business artifacts: an approach to operational specification. IBM Syst. J. **42**(3), 428–445 (2003)
24. Odell, J.J.: Dynamic and multiple classification. J. Object-Oriented Program. **4**(9), 45–48 (1992)
25. Ryndina, K., Küster, J.M., Gall, H.: Consistency of business process models and object life cycles. In: Kühne, T. (ed.) MODELS 2006. LNCS, vol. 4364, pp. 80–90. Springer, Heidelberg (2007). doi:10.1007/978-3-540-69489-2_11

26. Silverston, L.: The Data Model Resource Book. A Library of Universal Data Models for All Enterprises, vol. 1. Wiley, New York (2001)
27. Silverston, L.: The Data Model Resource Book. A Library of Universal Data Models by Industry Types, vol. 2. Wiley, New York (2001)
28. TMForum: Business Process Framework (eTOM) For The Information and Communications Services Industry, Addendum D: Process Decomposition and Descriptions (2013). http://www.tmforum.org/BusinessProcessFramework/1647/home.html. Accessed
29. van der Aalst, W.M.P., Barthelmess, P., Ellis, C.A., Wainer, J.: Workflow modeling using proclets. In: Scheuermann, P., Etzion, O. (eds.) CoopIS 2000. LNCS, vol. 1901, pp. 198–209. Springer, Heidelberg (2000). doi:10.1007/10722620_20

Ontology-Based Approach for Heterogeneity Analysis of EA Models

João Cardoso[1]([⊠]), Marzieh Bakhshandeh[1]([⊠]), Daniel Faria[2],
Cátia Pesquita[3], and José Borbinha[1]([⊠])

[1] INESC-ID, Instituto Superior Técnico, Universidade de Lisboa,
Av. Rovisco Pais 1, 1049-001 Lisboa, Portugal
{joao.m.f.cardoso,marzieh.bakhshandeh,jlb}@tecnico.ulisboa.pt
[2] Instituto Gulbenkian de Ciencia,
R. Q.ta Grande 6, 2780-156 Oeiras, Portugal
dfaria@igc.gulbenkian.pt
[3] LaSIGE, Faculdade de Ciências, Universidade de Lisboa,
Campo Grande, 1749-016 Lisboa, Portugal
cpesquita@di.fc.ul.pt

Abstract. The different needs and domains of enterprises and how they
employ EA modelling languages and tools can give rise to heterogeneity
at the syntactical, structural and semantical levels. In particular, models
dealing with the process perspective are becoming increasingly complex
and hetereogeneous. This raises difficulties in managing and reusing EA
models.

To address the heterogeneity of EA models, we propose an approach
that relies on encoding the models as OWL ontologies, then applying
ontology matching techniques to map them. By using ontology-based
techniques applied to the heterogeneity analysis of EA models.

We have applied our approach to a well known benchmarking data
set (EMISA-PMMC) encoded in BPMN. This involved the creation of
a novel ontology matching algorithm specifically designed for business
processes and their integration in a state of the art ontology matching
system, AgreementMakerLight.

Keywords: Enterprise Architecture · Heterogeneity analysis · Business
model · BPMN · Ontologies · Instance matching · Ontology alignment

1 Introduction

The main goal of enterprise architecture (EA) is to align business and business
support systems. Any architecture description should cover a wide and heterogeneous spectrum of areas, such as business processes, metrics, application
components, people, and technological infrastructure. Enterprise's needs are the
main motivator for the choice of EA modelling languages and modelling tools
that are to be adopted, as several are available. Thus, the adoption of multiple
EA modelling approaches, each based on a specific meta-model that cross-cuts

© Springer International Publishing AG 2017
M. Dumas and M. Fantinato (Eds.): BPM 2016 Workshops, LNBIP 281, pp. 131–142, 2017.
DOI: 10.1007/978-3-319-58457-7_10

distinct architectural domains, leads to heterogeneity in syntax, structure and semantics, and challenges on how to address them.

On the one hand, current model-based EA techniques have limitations in integrating multiple description languages. This occurs due to the lack of suitable extension mechanisms or because they cannot ensure consistency and traceability between the representations of the enterprise's multiple domains. On the other hand, different EA languages have similarities, so that elements of one EA language may be reused (with possible modifications) by another EA language.

Facilitating the management and reuse of EA modelling languages, and the integration of models is critical. For companies cannot afford to discard existing models and tools and start from scratch using a completely new approach [1]. This is particularly true for EA models dealing with the process perspective, which are becoming increasingly complex and heterogeneous in syntax, structure, and semantics.

Model integration aims at analysing the heterogeneity of EA models and creating links between them at both a structural and semantic level, without changing the actual meta-model and the expressiveness of the language. Model integration can be achieved in two ways [2]: direct and indirect matching. Direct matching occurs on a pair of modelling languages by having direct relations expressed between them. Thus if two models each expressed in its respective language need to be matched, the direct relations are already defined. In indirect matching, an intermediate language, such as transaction patterns from the DEMO methodology [3], is used to bridge two models. The model analysis uses information within or about the models to drive the overall analysis process [4].

Heterogeneity analysis in EA models "identifies similar entities implemented in variants or on different platforms that should be reconsidered for standardization" [5]. Ontologies and associated techniques are increasingly being recognized as valuable tools in the EA domain, as witnessed in [6,7], and can potentially be used to solve the aforementioned heterogeneity issues [8]. The use of ontologies enables interoperability between systems, humans, and systems and humans. Moreover, ontologies allow for the unique identification and disambiguation of concepts through formal semantics, which facilitates knowledge transfer, computational inference and analysis, and the detection of logical inconsistencies and deriving implicit facts.

One of the key concepts in the development of EA models is the concept of business process. Business process modelling [9] is the analytical illustration of an organization's business processes. A variety of notations have emerged in the field of business process modelling, including UML Activity Diagrams, the Business Process Modelling Notation (BPMN), Event-Driven Process Chains (EPCs), and the ArchiMate Business Layer [10]. EA models, specially the ones dealing with the process perspective, are becoming complex and heterogeneous a syntactical, structural and semantic level due to the variety of process modelling languages. These issues cause difficulties in model interoperability [11,12], which is necessary for enterprises to achieve their business goals [13]. One type of heterogeneity analysis techniques can be seen in process model matching approaches. These are concerned with supporting the creation of an alignment between process models, i.e., the identification of correspondences between their activities [14–16]. In particular, the heterogeneity analysis can be applied in two dimensions. Firstly, EA model representations made of several models but all sharing the same metamodel. Secondly, EA model representations made of several models where not all sharing the same metamodel. Table 1 shows a more detailed matrix of these two dimensions. The different dimensions are defined by two parameters metamodel and context. Context in this scenario refers to both models having the same projection on the real-world. The scope of this paper will solely focus on "AA" (one metamodel, one context) that is when, two models describe the same behavior being represented by the same modelling language, by having two architects performing the modulation, thus leading to different modelling approaches.

Table 1. Matrix dimensions

Metamodel context	A	B
A	One metamodel, One context	One metamodel, Two contexts
B	Two metamodels, One context	Two metamodels, Two contexts

In this paper we propose an ontology-based approach for heterogeneity analysis of EA business models expressed in BPMN. Said approach relies on first encoding the models as OWL ontologies, then applying ontology matching techniques to integrate them. Ontology matching is an ontology integration technique that aims to address this particular problem, through the definition of an intermediate abstraction layer that manages the heterogeneity between the ontologies to be integrated.

The remainder of the paper is organized as follows: Sect. 2 details the proposed approach; Sect. 3 describes the benchmark data set to which the proposed approach was applied; Sect. 4 details the implementation of the approach and the results it obtained; and Sect. 5 presents the main conclusions of our study.

2   Proposed Approach

In this section we'll be presenting our ontology-based approach to tackle the issue of heterogeneity analysis of EA models. In particular as will be described in Sect. 3, business models expressed in BPMN. As such, our approach follows the framework shown in Fig. 1, which includes three main processes: Ontology Transformation of EA descriptions, Analysis of each EA description and finally Consolidated Analysis. *Architectural description* can be defined as "the product used to describe the architecture of a system" [17]. In the first process the enterprise architecture description, created using specific modelling tools and conceptual languages, needs to be transformed into a logic-based ontology representation to take advantage of ontology alignment techniques. After that, the models can be used for performing analysis and can even be converted back into the original representation format if desired. Next, the Analysis of EA description and Consolidated Analysis includes analysing the similarity of EA models at the model level, with regards to process model similarity matching [18]. Process model similarity matching refers to the creation of correspondences between activities of process models.

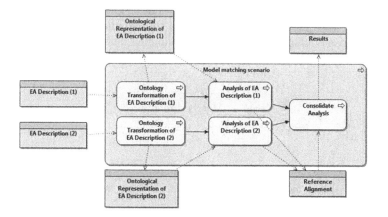

Fig. 1. Model matching framework

Performing this kind of similarity matching analysis requires:

1. Analysing each actual EA description.
2. Detecting the heterogeneities that exist, identifying the possible similarities between the two EA descriptions, and making a reference alignment for benchmarking.
3. Representing the models as ontologies.
4. Applying an appropriate ontology matching system to perform the mappings at the model level.
5. Analysing the results of the matching against the reference alignment that reflects the stakeholder's requirement.

3   Benchmark Data Set Description

A preliminary version of the proposed approach has been previously applied to a well studied benchmark. This was done through the authors participation in the Process Model Matching Contest (PMMC 2015) [19], in which data sets were evaluated through process model matching techniques. This paper builds on that effort and attempts to improve on the results obtained during the PMMC 2015, by completing the preliminary approach that was presented at the PMMC 2015 and adding a new dedicated matching algorithm (see Sect. 4.3) to increase the approach's performance.

The main goal of the PMMC 2015 was to make a comparative analysis on the results of the application of the different techniques to process model matching problems. In total, three process model matching problems were defined and published with respective data sets[1]. Three sets of process model matching problems were included in the PMCC. In this paper the focus will rest solely on the first set:

University Admission Processes (UA & UAS): This set consisted of 36 model pairs that were derived from 9 models representing the application procedure for Master students of nine German universities. The process models were available in BPMN format [20]. With respect to the reference alignment, there was a distinction between equivalence matches and subsumption matches (a general activity is matched to a more specific activity). Two versions of the gold standards were created for this data set: a strict version of the gold standard which contains only equivalence correspondences (UA), and a relaxed version which contains additionally a high number of subsumption correspondences (UAS). In the scope of this paper only the UA gold standard was considered.

The most simple representation of a business process is a directed and labelled (attributed) graph where each node represents an activity and each edge a control link between activities. In BPMN events are represented by a circle and denote something that happens, activities are represented by a rounded-corner rectangle and describe the kind of work which must be done, and finally gateways are represented by a diamond shape and determine forking and merging of paths, depending on the conditions expressed.

Table 2 summarizes the main characteristics of the selected data set. It shows the minimum, maximum, and average number of activities per model as well as the total and average number of 1:1 and 1:n correspondences. A 1:1 correspondence matches two activities A and A' such that no other correspondence in the gold standard matches A or A' to some other activity. Contrary to this, 1:n correspondences match an activity A to several other activities A1, ... , An. This can, for example, happen when an activity has to be matched to a sequence of activities. A high number of 1:n correspondences indicates that the matching task is complex and that the models describe processes on a different level of granularity.

Table 2. Characteristics of test data set.

Characteristic	UA	UAS
No. of activities (min)	12	12
No. of activities (max)	45	45
No. of activities (avg),	24.2	24.2
No. of 1:1 correspondences (total)	202	268
No. of 1:1 correspondences (avg)	5.6	7.4
No. of 1:n correspondences (total)	30	360
No. of 1:n correspondences (avg)	0.8	10

---

[1] PMMC 2015 Data Sets: https://ai.wu.ac.at/emisa2015/contest.php.

Using the PMMC 2015 data set would address the first two steps of the proposed approach, for they provided a benchmark where the similarities between models were identified and reference gold standard alignments had already been established.

4   Application of the Proposed Approach

4.1   Ontology Population with ChaosPop

In order to use an ontology matching system to match EA models, it is first necessary to represent them as OWL ontologies [21], which is the third step of the proposed approach (see Sect. 2). In the case of the PMMC 2015 data set, this implies converting the 9 models from their original ".bpmn" format into OWL.

To perform this task, we employed ChaosPop, an ontology population application under development. ChaosPop consumes raw data files and domain ontologies, and uses a mapping specification schema to correlate elements from the data files with ontological concepts from the domain ontologies. Subsequently it uses the established mappings to populate the domain ontologies. The objective of ChaosPop is to cover scenarios where the domain ontologies evolve over time, and thus allows for them to be automatically extended or instantiated. However in this particular scenario ChaosPop was used solely as a means to instantiate a BPMN 2.0 ontology[2], that could later be used with an ontology matching system.

As can be seen in Fig. 2 the process of creating a populated ontology with ChaosPop comprises four tasks:

1. Load Data Files: the data files are loaded into the ChaosPop application independently of the way the schema used for organizing the data; ChaosPop's parser interface can be extended to support previously unsupported formats.
2. Load Domain Ontology: domain ontologies are loaded into the ChaosPop application. Ontology handling within the ChaosPop application is supported by the OWL API[3] and as such ontologies can be loaded by either providing their namespace or file.
3. Create Mappings: ChaosPop has a mappings specification schema that allows for users to create mappings that correlate elements from the loaded data files with concepts from the loaded domain ontologies.
4. Populate Ontology: ChaosPop uses the created mappings to populate domain ontologies with individuals whose data comes from the data files that were loaded in the first task.

The resulting populated ontologies for all 9 models that comprised the data set can be reviewed at http://sysresearch.org/BPMO2016WorkshopPaper/ontologies/2015.07.21/. A partial view of the individual graph of the Cologne ontology can be seen in Fig. 3.

---

[2] BPMN 2.0 ontology from the DKM https://dkm.fbk.eu/bpmn-ontology.
[3] OWL API http://owlapi.sourceforge.net/.

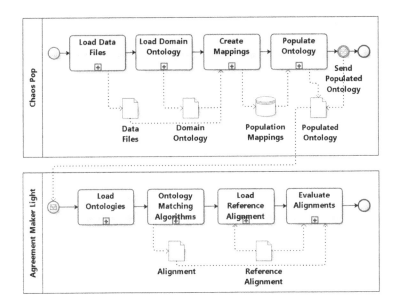

Fig. 2. BPMN model representing AML's usage applied to the selected benchmark.

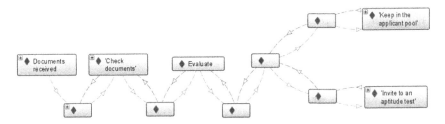

Fig. 3. Partial view of the individual graph of the cologne ontology

4.2   EA Model Matching Using an Ontology Matching System

To address the third and fourth steps of the proposed approach (see Sect. 2,
it is necessary to select an ontology matching system capable of mapping the
ontological representations of EA models, and analysing the resulting alignments
against the reference alignments provided.

AgreementMakerLight (AML) [22] is a state-of-the-art open-source ontology
matching system that has achieved top results in the last two editions of Ontology Alignment Evaluation Initiative [23,24]. It has a modular and extensible
framework, which allows for the inclusion of virtually any matching algorithm.
And while it previously focused only on mapping ontology classes and properties, it has recently been extended to map individuals as well. These features
make it an appropriate choice to handle the matching of EA models.

To match the PMMC 2015 models, a new matching algorithm was designed and implemented in AML, called Business Process Matcher, specifically designed for instance matching in the context of business processes, which is detailed in the following section.

4.3   Business Process Matcher

The Business Process Matcher algorithm comprises three steps: Lexical Similarity, Structural Similarity, and Mapping Selection.

In the Lexical Similarity step, the individuals of the two ontologies are compared with respect to their labels (which are parsed and normalized when loaded into AML to ensure uniformity). First the labels are checked for identity, then they are compared via the ISub string similarity metric [25] both before and after the removal of stop-words, and finally their word similarity is computed via a weighted Jaccard index between the words in the labels, with words weighted by length (also ignoring stop-words). Only individuals of classes "task", "intermediate catch event", and "start event" are matched. The following presents the basic algorithm for this step:

```
SET alignment to empty set
FOR each individual Is in the source ontology
    FOR each individual It in the target ontology
        IF label_Is = label_It THEN
            ADD <Is,It,1.0> to alignment
        ELSE
            COMPUTE sim1 = ISub similarity(label_Is,label_It)
            remove stop-words from label_Is, label_It
            COMPUTE sim2 = ISub similarity(label_Is,label_It)
            COMPUTE sim3 = word similarity(label_Is,label_It)
            COMPUTE final_sim = MAX(sim1,sim2*0.9,sim3)
            ADD <Is,It,final_sim> to alignment
        END IF
    END FOR
END FOR
```

The Structural Similarity step consists of finding the "best neighbour" for each mapping and averaging the similarity of the mapping with that of its best neighbour. For any given mapping, its "best neighbour" is the mapping with the highest similarity score among mappings between individuals that both precede or succeed the source and target individual in the model. In cases where the individual directly preceding or succeeding the source or target individual is not of one of the three classes above, the next individual in that direction is tested.

```
FOR each mapping <Is,It,sim> in alignment
    SET sim_struct = 0
    FOR each individual i preceding Is in the model
        FOR each individual j preceding It in the model
            IF sim(i,j) > sim_struct
                SET sim_struct = sim(i,j)
            END IF
        END FOR
    END FOR
    FOR each individual i following Is in the model
        FOR each individual j following It in the model
            IF sim(i,j) > sim_struct
                SET sim_struct = sim(i,j)
            END IF
        END FOR
    END FOR
    SET sim = (sim + sim_struct)/2
END FOR
```

The Mapping Selection step consists of selecting a subset of (high-quality) mappings from the candidate alignment, such that each individual in one of the ontologies is mapped to only one individual in the other ontology, except if there are multiple competing mappings with the same similarity score. The selection algorithm implemented by AML is a greedy heuristic. It was run with a threshold of 0.6.

```
SET alignment to empty set
SORT input_alignment by descending order of similarity
FOR each mapping <Is,It,sim> in input_alignment
    IF sim < threshold THEN
        BREAK
    ELSE IF alignment contains no mapping <Is,*,sim*> or
    <*,It,sim*> such that sim* > sim THEN
        ADD <Is,It,sim> to alignment
    END IF
END FOR
```

4.4   Results

AML's Business Process Matcher algorithm was applied to each pairwise combination of the 9 ontologies from the PMMC 2015 dataset, totaling 36 matching tasks. The alignments produced by AML were then evaluated using the gold standard alignment files[4]. The compiled results of the approach proposed in this paper can be seen in Table 3 (under "AML 2016"), together with the results of the top contestants at the PMMC 2015. A direct comparison can be made with our team's previous participation results, which are listed under "AML 2015".

---

[4] Gold standard alignment files http://sysresearch.org/BPMO2016WorkshopPaper/
goldstandard/oficial/.

In comparison with its 2015 version, AML had an increase in precision of 33.5%, an increase in recall of 10.4%, and an increase in F-measure of 29.4%. The improvement in precision was mostly due to a higher global similarity threshold (0.6 rather than 0.1) and to a stricter selection step, whereas the improvement in recall was due to refinements of the string and word-matching algorithms (namely the removal of stop-words) and the addition of the structural matcher, as described in Sect. 4.3.

In comparison with the top systems competing in the PMMC 2015, AML surpassed them with regard to F-measure (by 1.1%) while extended its lead with regard to recall (12.5% over the next best system) and improving to fourth place with regard to precision. It also had the lowest standard deviation with regard to both precision and recall among the top systems, which indicates a consistent performance.

Table 3. Results of University Admission Matching at PMMC 2015 and comparison with the results using the proposed approach

Approach	Precision			Recall			F-Measure		
	ϕ-mic	ϕ-mac	SD	ϕ-mic	ϕ-mac	SD	ϕ-mic	ϕ-mac	SD
RMM/NHCM	0,686	0,597	0,248	0,651	0,610	0,277	0,668	0,566	0,224
MSSS	**0,807**	**0,855**	0,232	0,487	0,343	0,353	0,608	0,378	0,343
OPBOT	0,598	0,636	0,335	0,603	0,623	0,312	0,601	0,603	0,300
AML (2015)	0,269	0,250	0,205	0,672	0,626	0,319	0,385	0,341	0,236
AML (2016)	0,604	0,548	0,179	**0,776**	**0,775**	0,220	**0,679**	**0,630**	0,240

5   Conclusions and Future Work

We proposed a novel approach for the heterogeneity analysis of EA models that relies on encoding them as OWL ontologies and then applying ontology matching algorithms to integrate them. This approach was evaluated on a benchmark data set, the University Admission Processes data set from the PMMC 2015, in which we had competed with a preliminary version of this approach.

The results we obtained represent a substantial improvement over the preliminary version of our approach, and more importantly, surpass the state-of-the-art in both recall and F-measure. Thus, the results attest to the soundness of the proposed approach: by encoding EA models as ontologies, we were able to harness well-established ontology matching algorithms in order to integrate them.

While the use-case on which we tested our methodology was relatively straightforward, in the sense that all EA models had the same modelling language (BPMN), we believe our approach also has merit in more complex integration scenarios, where the models are expressed in different languages. In such scenarios, encoding the models as ontologies would take more effort than it did in the present scenario, but virtually no effort would be required to then apply the ontology matching algorithms to integrate the models. By contrast, traditional approaches would have to bridge both the languages and the models themselves.

In future work, we will finish the development and release the ontology population application ChaosPop (see Sect. 4.1), so that our approach is publicly available. We will also develop ontology matching algorithms attuned to EA models, namely by exploring reasoning and pattern recognition. Finally, we will extend our approach to consider subsumption mappings in addition to equivalence mappings.

Acknowledgements. This work has been partially supported by the Fundação da Ciência e Tecnologia through funding of INESC-ID, ref.UID/CEC/50021/2013, LaSIGE Research Unit, ref.UID/CEC/00408/2013 and finally by the project SMiLaX (PTDC/EEI-ESS/4633/2014).

References

1. Lankhorst, M.M.: Enterprise architecture modelling-the issue of integration. Adv. Eng. Inf. **18**(4), 205–216 (2004)
2. Creasy, P., Ellis, G.: A conceptual graphs approach to conceptual schema integration. In: Mineau, G.W., Moulin, B., Sowa, J.F. (eds.) ICCS-ConceptStruct 1993. LNCS, vol. 699, pp. 126–141. Springer, Heidelberg (1993). doi:10.1007/3-540-56979-0_7
3. Dietz, J.: Enterprise Ontology: Theory and Methodology. Springer, Berling (2006)
4. Närman, P., Johnson, P., Nordström, L.: Enterprise architecture: a framework supporting system quality analysis. In: Enterprise Distributed Object Computing Conference, EDOC 2007, 11th IEEE International, pp. 130–130. IEEE (2007)
5. Lange, M., Mendling, J.: An experts' perspective on enterprise architecture goals, framework adoption and benefit assessment. In: Enterprise Distributed Object Computing Conference Workshops (EDOCW), 2011 15th IEEE International, pp. 304–313. IEEE (2011)
6. Azevedo, C.L., Almeida, J.P.A., van Sinderen, M., Quartel, D., Guizzardi, G.: An ontology-based semantics for the motivation extension to archimate. In: Enterprise Distributed Object Computing Conference (EDOC), 2011 15th IEEE International, pp. 25–34. IEEE (2011)
7. Antunes, G., Bakhshandeh, M., Mayer, R., Borbinha, J., Caetano, A.: Using ontologies for enterprise architecture integration and analysis. Complex Syst. Inf. Model. Q. **1**, 1–23 (2014)
8. Bürger, T., Simperl, E.: Measuring the benefits of ontologies. In: Meersman, R., Tari, Z., Herrero, P. (eds.) OTM 2008. LNCS, vol. 5333, pp. 584–594. Springer, Heidelberg (2008). doi:10.1007/978-3-540-88875-8_82
9. Dumas, M., La Rosa, M., Mendling, J., Reijers, H.A.: Fundamentals of Business Process Management. Springer, Heidelberg (2013)
10. Haren, V.: Archimate 2.0 specification (2012)
11. Sheth, A.P.: Changing focus on interoperability in information systems: from system, syntax, structure to semantics. In: Goodchild, M., Egenhofer, M., Fegeas, R., Kottman, C. (eds.) Interoperating Geographic Information Systems. The Springer International Series in Engineering and Computer Science, vol. 495, pp. 5–29. Springer, Heidelberg (1999). doi:10.1007/978-1-4615-5189-8_2

12. Haslhofer, B., Klas, W.: A survey of techniques for achieving metadata interoperability. ACM Comput. Surv. (CSUR) **42**(2), 7 (2010)
13. Lin, Y., Ding, H.: Ontology-based semantic annotation for semantic interoperability of process models. In: International Conference on Computational Intelligence for Modelling, Control and Automation, 2005 and International Conference on Intelligent Agents, Web Technologies and Internet Commerce, vol. 1, pp. 162–167. IEEE (2005)
14. Weidlich, M., Dijkman, R., Mendling, J.: The ICoP framework: identification of correspondences between process models. In: Pernici, B. (ed.) CAiSE 2010. LNCS, vol. 6051, pp. 483–498. Springer, Heidelberg (2010). doi:10.1007/978-3-642-13094-6_37
15. Klinkmüller, C., Weber, I., Mendling, J., Leopold, H., Ludwig, A.: Increasing recall of process model matching by improved activity label matching. In: Daniel, F., Wang, J., Weber, B. (eds.) BPM 2013. LNCS, vol. 8094, pp. 211–218. Springer, Heidelberg (2013). doi:10.1007/978-3-642-40176-3_17
16. Leopold, H., Niepert, M., Weidlich, M., Mendling, J., Dijkman, R., Stuckenschmidt, H.: Probabilistic optimization of semantic process model matching. In: Barros, A., Gal, A., Kindler, E. (eds.) BPM 2012. LNCS, vol. 7481, pp. 319–334. Springer, Heidelberg (2012). doi:10.1007/978-3-642-32885-5_25
17. ISO: ISO/IEC/IEEE 42010: 2011 - systems and software engineering - architecture description (2011)
18. Dijkman, R.M., et al.: A short survey on process model similarity. In: Bubenko, J., Krogstie, J., Pastor, O., Pernici, B., Rolland, C., Sølvberg, A. (eds.) Seminal Contributions to Information Systems Engineering, pp. 421–427. Springer, Heidelberg (2013)
19. Antunes, G., Bakhshandeh, M., Borbinha, J., Cardoso, J., Dadashnia, S., Francescomarino, C.D., Dragoni, M., Fettke, P., Gal, A., Ghidini, C., Hake, P., Khiat, A., Klinkmüller, C., Kuss, E., Leopold, H., Loos, P., Meilicke, C., Niesen, T., Pesquita, C., Péus, T., Schoknecht, A., Sheetrit, E., Sonntag, A., Stuckenschmidt, H., Thaler, T., Weber, I., Weidlich, M.: The process model matching contest 2015. In: EMISA 2015, pp. 127–155 (2015)
20. Allweyer, T.: BPMN 2.0: Introduction to the Standard for Business Process Modeling. BoD-Books on Demand, Norderstedt (2010)
21. Antoniou, G., Van Harmelen, F.: Web ontology language: Owl. In: Staab, S., Studer, R. (eds.) Handbook on Ontologies. International Handbooks on Information Systems, pp. 67–92. Springer, Heidelberg (2004)
22. Faria, D., Pesquita, C., Santos, E., Cruz, I.F., Couto, F.M.: Agreement maker light results for OAEI 2013. In: OM, pp. 101–108 (2013)
23. Dragisic, Z., Eckert, K., Euzenat, J., Faria, D., Ferrara, A., Granada, R., Ivanova, V., Jiménez-Ruiz, E., Kempf, A.O., Lambrix, P., et al.: Results of the ontology alignment evaluation initiative 2014. In: Proceedings of the 9th International Workshop on Ontology Matching Collocated with the 13th International Semantic Web Conference (ISWC 2014) (2014)
24. hvaiko, P., Euzenat, J., JimĂŠnez-Ruiz, E., Cheatham, M., Hassanzadeh, O.: Proceedings of the 10th International Workshop on Ontology Matching collocated with the 14th International Semantic Web Conference (ISWC 2015), vol. 1545. CEUR-WS (2015)
25. Stoilos, G., Stamou, G., Kollias, S.: A string metric for ontology alignment. In: Gil, Y., Motta, E., Benjamins, V.R., Musen, M.A. (eds.) ISWC 2005. LNCS, vol. 3729, pp. 624–637. Springer, Heidelberg (2005). doi:10.1007/11574620_45

9th International Workshop on Business Process Management and Social Software (BPMS2 2016)

Introduction to the BPMS2 Workshop 2016

Rainer Schmidt[1(✉)] and Selmin Nurcan[2,3]

[1] Faculty of Computer Science and Mathematics, Munich University
of Applied Sciences, Lothstrasse 64, 80335 Munich, Germany
Rainer.Schmidt@hm.edu
[2] Sorbonne School of Management, Paris, France
[3] CRI, University Paris 1 Panthéon-Sorbonne, Paris, France

1 Introduction

Social software [1, 2] is a new paradigm that is spreading quickly in society, organizations and economics. It enables social business that has created a multitude of success stories. More and more enterprises use social software to improve their business processes and create new business models. Social software is used both in internal and external business processes. Using social software, the communication with the customer is increasingly bi-directional. E.g. companies integrate customers into product development to capture ideas for new products and features. Social software also creates new possibilities to enhance internal business processes by improving the exchange of knowledge and information, to speed up decisions, etc. Social software is based on four principles: weak ties, social production, egalitarianism and mutual service provisioning.

Up to now, the interaction of social and human aspects with business processes has not been investigated in depth. Therefore, the objective of the workshop is to explore how social software interacts with business process management, how business process management has to change to comply with weak ties, social production, egalitarianism and mutual service, and how business processes may profit from these principles.

The workshop discussed the three topics below.

1. **Social Business Process Management (SBPM),** i.e. the use of social software to support one or multiple phases of the business process lifecycle
2. **Social Business: Social software supporting business processes**
3. **Human Aspects of Business Process Management**

Based on the successful BPMS2 series of workshops since 2008, the goal of the 9th BPMS2 workshop is to promote the integration of business process management with social software and to enlarge the community pursuing the theme.

During the workshop, four teams presented the results of their research:

In their paper "When Cognitive Biases Lead to Business Process Management Issues" Maryam Razavian, Oktay Turetken, and Irene Vanderfeesten the influence of cognitive biases on four key issues appearing during the design time of business processes.

The paper from Rüdiger Weißbach, Kathrin Kirchner, Felix Reher, Robert Heinrich titled "Challenges in Business Processes Modeling – Is Agile BPM a solution?" maps challenges in BPM projects and agile principles from software engineering, providing a foundation for selecting agile methods in a BPM project

Jan Claes, Irene Vanderfeesten, Frederik Gailly, Paul Grefen, and Geert Poels develop a smart process modeling method in their paper "Towards a structured process modeling method: Building the prescriptive modeling theory" using evidence about successful information processing techniques in cognitive psychology.

In their paper "Designing Serious Games for Citizen Engagement in Public Service Processes" Nicolas Pflanzl, Tadeu Classe, Renata Araujo, and Gottfried Vossen discusses game design as an approach to motivate, engage and change citizens' behavior to improve public services.

We wish to thank all the people who submitted papers to BPMS2 2016 for having shared their work with us, the many participants creating fruitful discussion, as well as the members of the BPMS2 2016 Program Committee, who made a remarkable effort in reviewing the submissions. We also thank the organizers of BPM 2016 for their help with the organization of the event.

Selmin Nurcan
University Paris 1 Panthéon-Sorbonne, France
Rainer Schmidt
Munich University of Applied Sciences, Germany

2 Program Committee

Renata Araujo	Department of Applied Informatics, UNIRIO, Brasilia
Jan Bosch	Chalmers University of Technology, Sweden
Marco Brambilla	Politecnico di Milano, Italy
Claudia Cappelli	UNIRIO, Brasilia
Monique Janneck	Fachhochschule Lübeck, Germany
Ralf Klamma	RWTH Aachen University, Germany
Sai Peck Lee	University of Malaya, Malaya
Michael Möhring	Munich University of Applied Services, Germany
Selmin Nurcan	Université de Paris 1 Panthéon - Sorbonne, France
Andreas Oberweis	Karlsruhe Institute of Technology (KIT), Germany
Henderik Proper	Public Research Centre Henri Tudor, Luxembourg
Hajo A. Reijers	Eindhoven University of Technology, Netherlands
Michael Rosemann	Queensland University of Technology, Australia
Gustavo Rossi	LIFIA-F. Informatica. UNLP, Italy
Flavia Santoro	NP2Tec/UNIRIO, Brasilia
Rainer Schmidt	Munich University of Applied Sciences, Germany
Miguel-Angel Sicilia	University of Alcala, Spain
Pnina Soffer	University of Haifa, Israel
Frank Termer	Bitkom e.V., Germany

References

1. Schmidt, R., Nurcan, S.: BPM and social software. In: Ardagna, D., Mecella, M., Yang, J., Aalst, W., Mylopoulos, J., Rosemann, M., Shaw, M.J., Szyperski, C. (eds.) BPM 2008 Workshops. LNBIP, vol. 17, pp. 649–658. Springer, Berlin (2009)
2. Bruno, G., Dengler, F., Jennings, B., Khalaf, R., Nurcan, S., Prilla, M., Sarini, M., Schmidt, R., Silva, R.: Key challenges for enabling agile BPM with social software. J. Softw. Maintenance Evol.: Res. Pract. 23, 297–326 (2011)

When Cognitive Biases Lead to Business Process Management Issues

Maryam Razavian, Oktay Turetken, and Irene Vanderfeesten[✉]

Department of Industrial Engineering and Innovation Sciences,
Eindhoven University of Technology, Eindhoven, The Netherlands
{M.Razavian,O.Turetken,I.T.P.Vanderfeesten}@tue.nl

Abstract. There is a broad consensus that design decision making is important for Business Process Management success. Despite many business process design approaches and practices that are available, the quality of business process analysis and design relies heavily on human factors. Some of these factors concern cognitive biases. In this paper, we explore the role of cognitive biases in four key issues regarding the design-time phases of the business process management lifecycle. We outline some research directions that may help us understand and improve the effects of cognitive biases in the design-related practices of business process management.

1 Introduction

It is well recognized in the Business Process Management (BPM) community that (re-)design decision making is an important aspect in the BPM lifecycle [1,2]. There are many approaches and best practices that aim at supporting good business process analysis and (re-)design, e.g. [3,4]. However, these approaches and best practices often lack considerations of human factors in design, such as cognitive biases, proper design reasoning and communication, and reflection; all of which can affect the design decision making [5–7]. For instance, a designer who is biased towards one solution (e.g. a reference model) can put misleading emphasis on its justification. As such, following a certain approach or best practice does not necessarily produce good business process designs and redesigns. Not only designers are prone to cognitive biases but also other BPM stakeholders such as process analists, process owners, and managers.

Human factors are a well known problem in design decision making [8]. It has been shown that people generally have cognitive biases [9]: tendencies to think in certain ways that can lead to systematic deviations from a standard of rationality or good judgment (Wikipedia). Such a pattern of deviation in judgment may be due to typical biases such as *anchoring*, *confirmation bias*, *over-confidence* and so on. These biases may lead to sub optimal results in the delineation, design, mapping, analysis and redesign of a business process. As an initial attempt to identify and overcome some of the mentioned human issues, we discuss the role of cognitive biases in design time phases of the BPM lifecycle in this paper.

M. Dumas and M. Fantinato (Eds.): BPM 2016 Workshops, LNBIP 281, pp. 147–156, 2017.
DOI: 10.1007/978-3-319-58457-7_11

In doing so, we discuss key BPM issues related to the four phases of the lifecycle: *process identification*, *process discovery*, *process analysis*, and *process redesign*. We further explore how each key issue can potentially be affected by cognitive biases. Finally, we sketch research directions we may pursue, to increase our understanding of how cognitive biases contribute to bad business process design, and ultimately improve the business processes delivered and executed.

The remainder of the paper is organized as follows: Sect. 2 presents the theoretical background in terms of cognitive biases theory and a brief description of the BPM lifecycle. Next, the research design is outlined in Sect. 3, followed by a discussion on cognitive biases in the four design time phases of the BPM lifecycle. The paper ends with a conclusion and discussion of future work.

2 Background

In this section, we first discuss about the cognitive biases as studied in the cognitive science and psychology fields. Next, we briefly present the BPM lifecycle focusing more on its design-time phases.

2.1 Cognitive Biases

A cognitive bias is the general term, introduced by Kahneman and Tversky [10], to denote human inability to reason in a rational way. They are cognition or mental behaviors that distort decision quality in a significant number of decisions for a significant number of people [9].

Although bias has been extensively studied by cognitive psychologist [10–12] and decision theory researchers [13–15], the effect of biases on business process management is relatively unknown. Anecdotal evidence of such phenomenon, however, is plentiful in the industry. As an example, some business process analysts and designers constantly underestimate business process redesign complexity and effort. These analysts and designers are biased about their ability to deliver high-quality business process designs. This is an example of a *confidence bias*. Another type of bias is *anchoring*; a first impression of a solution may become an anchor, it is difficult for some designers to adjust or change the ideas even when there is evidence to show that the initial solution is inferior.

Arnot [9], introduces a taxanomy of 37 biases, and found that decision makers suffer from these biases. Table 1 shows some of those biases, that are arranged into five categories: *Memory biases* have to do with the storage and recall of information. *Statistical biases* are concerned with the general tendency of humans to process information contrary to the normative principles of probability theory. *Confidence biases* act to increase a person's confidence in his or her prowess as a decision-maker. *Presentation biases* should not be thought of as only being concerned with the display of data. They act to bias the way information is perceived and processed, and are some of the most important biases from a decision-making perspective. *Situation biases* relate to how a person responds to the general decision situation. It is important to recognize that cognitive biases are not necessarily as discrete as described above, and that they are likely to overlap in definition and effect, and that several may apply to one situation.

Arnot also found that the use of a *debiasing* method can help decision makers to recognize their own biases and make corrections. Debiasing is a procedure for reducing or eliminating biases from the cognitive strategies of a decision-maker [9].

Table 1. Excerpt of cognitive biases proposed in [9]

Cognitive bias	Bias category	Description
Testimony	Memory biases	The inability to recall details of an event may lead to seemingly logical reconstructions that may be inaccurate
Completeness	Confidence biases	The perception of an apparently complete or logical data presentation can stop the search for omissions
Confirmation	Confidence biases	Often decision-makers seek confirmatory evidence and do not search for disconfirming information
Overconfidence	Confidence biases	The ability to solve difficult or novel problems is often overestimated
Anchoring	Adjustment biases	Adjustments from an initial position are usually insufficient
Escalation	Situation biases	Often decision-makers commit to follow or escalate a previous unsatisfactory course of action
Rule	Situation biases	The wrong decision rule may be used

2.2   BPM Lifecycle

A business process management and improvement effort in an organization goes through several stages or phases, as modeled in the BPM lifecycle by [4] (see Fig. 1).

First of all, in the *process identification* phase, an overview of the relevant business processes is created. Processes are identified, delineated, and related to each other, without giving too much detail about the contents of a process. The result of this phase is a process architecture that lists the processes and visualizes their relationships.

In the *process discovery* phase the current content and working of a process is elaborated, usually in the form of an AS-IS process model. Activities and their order are modeled, Roles and resources with their responsibilities are added. The data flow may be indicated and relevant documents and information systems may be linked to activities and resources.

Next, in the *process analysis* phase, the AS-IS process is evaluated for its current performance. Issues and bottlenecks are documented and if possible supported with quantitative performance measurements.

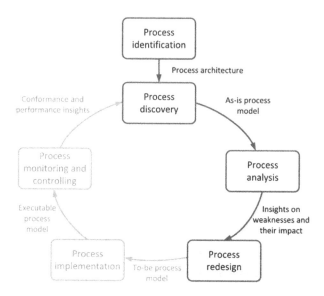

Fig. 1. The BPM lifecycle [4]. The design time phases are indicated in black, the runtime phases in grey.

In the *process redesign* phase solutions to the issues and bottlenecks, that were identified in the process analysis phase, are designed. Several solution may be considered and compared in terms of their impact and improvement potential. The redesign phase may therefore also include some process analysis. The most promising redesigns are then modeled in a TO-BE process model.

The TO-BE process model is taken as a basis in the *process implementation* phase. In this phase, the changes required to get from the AS-IS process to the TO-BE situation are prepared and implemented. This may involve organizational change as well as process automation.

When the TO-BE process is running, new information about its performance may be collected in the *process monitoring and controlling* phase. This information may again be analyzed to see if the new process is working as intended and to analyze further errors, bottlenecks, issues, etc. This may then be input to a new iteration of the BPM lifecycle, making business process management and redesign a continuous effort for improved performance.

In this paper, we focus on cognitive biases in the four design time phases of the BPM lifecycle, i.e. the process identification, process discovery, process analysis and process redesign phases. This is mainly because in all of these phases, important design decisions have to be made by human stakeholders, such as process architects, modelers and analysts. While making these decisions, these persons may suffer from particular cognitive biases that may lead to a suboptimal design of the busines process. In the next sections, we discuss the issues originating from cognitive biases in the design time phases of business process management, and we propose our view on how to overcome these.

3   Research Design

Taking the BPM academic and grey literature as our basis, we (i.e. three senior researchers working in this field) identified key issues relevant for each of the design-time phases of the BPM lifecycle (i.e. process identification, discovery, analysis and redesign). Our objective was to identify *representative* issues rather than an exhaustive list of issues relevant to all BPM lifecycle phases. Three authors identified these issues independently, after which consensus meetings were conducted to resolve disagreements, synthesize the results, and reduce the list to a single key issue per design-time phase of the lifecycle. As a result, four key issues were identified.

Next, we independently went over the list of cognitive biases as proposed by Arnott [9] and considered how each key BPM issue can potentially be affected by cognitive biases. We used Arnott's taxonomy of cognitive biases due to its extensive list compiled through a comprehensive and systematic analysis of the literature. This made it suitable to be taken as the theoretical framework underpinning the types of cognitive biases applicable to BPM lifecycle.

Similar to the previous research step, we performed a consensus meeting to discuss and integrate the results. Consequently, for each key BPM issue, a list of cognitive biases was identified. We discussed to agree on the way these biases play a role in reinforcing the underpinnings of the BPM issues under consideration.

4   Cognitive Biases in the Design-Time Phases of Business Process Management Lifecycle

In the following we introduce the key BPM issues that we identified for each of the design-time phases of the BPM lifecycle. For each BPM issue, we discuss the roles that the cognitive biases play in those issues. Cognitive biases are highlighted in brackets (e.g., [*anchoring*]). Table 2 summarizes the BPM issues and their underlying cognitive biases.

4.1   Suboptimal Process Architecture

The capability to demarcate the beginning and end point of processes is considered to be a critical success factor of BPM initiatives [1,16]. Organizations often find it difficult to clearly identify the processes and their boundaries to establish a process architecture (or process map) [4,17], leading to a suboptimal architecture. Such an architecture can be incomplete or oversimplified (lacking some essential process components), has overlapping parts among different components, or has process components with inconsistent granularity. This capability is essential particularly in the process identification phase of the BPM lifecycle. However, in practical business environments this is recognized to be a challenging undertaking. In business settings, it is sometimes difficult to delineate a coherent set of activities that satisfy a particular business goal to make up a single process or identify inputs or outputs that set the boundary between processes.

Following an existing process architecture that might have previously led to an inefficient or ineffective outcome can be a normal course of action. Identifying an interrelated set of processes and drawing a boundary between them often becomes a design choice influenced by cognitive biases. The process map designers and decision makers can perceive a process architecture to be complete and stop searching for omissions or alternative redesigns [*completeness*]. They may also underestimate the efforts in designing or redesigning a process architecture [*overconfidence*], in which case they may commit to follow an existing and possibly suboptimal process architecture that the company has previously developed [*anchoring and escalation*]. This existing (and possibly suboptimal) architecture may act as a shield that hinders analysts'/designers' capabilities in the later phases of the BPM lifecycle to detect improvement opportunities.

4.2 Incomplete Process Model

Incomplete process models miss alternative paths or business domain related exceptions, or lack sufficient level of detail in driving the subsequent efforts (for process automation or guiding process enactment). Causes of incomplete process models are diverse. Lack of sufficient capabilities of the organization in articulating end-to-end processes in a structured way is considered to be a typical cause [18]. In relation to the lack of a standard BPM methodology/approach [18], most businesses find it difficult to express and define their processes in a systematic way. This leads to the challenges in finding the right degree of balance in the level of granularity and detail in the specification, as well as in assessing the completeness of the specification. Another action that hinders completeness arises during process discovery where process information is elicited from a limited set of process participants [19]. These participants have only partial knowledge of how process runs (particularly about those parts that other process participants are performing) [20,21]. The decision to cease eliciting further process information can be influenced by cognitive biases. In particular, process modelers can stop searching for omissions in the models as they may perceive resulting models to be apparently complete and logical [*completeness*]. In cases where participants are asked to provide process information that are beyond their scope of duty or accountability, they may still offer this knowledge. They may do this despite their potential inability to recall the details and to judge the completeness of their knowledge. This has the potential to lead seemingly logical presentations that might be incomplete and inaccurate [*testimony*].

4.3 Irrelevant Bottlenecks and Weaknesses

The selection of which part(s) of a process to improve is done based on an analysis of bottlenecks and weaknesses, and their impact. There are many ways to get information on the strengths and weaknesses of the AS-IS process, e.g. by observing the process, interviewing process participants, analyzing performance indicators based on operational data [4], and by applying supporting analysis techniques to get more insights in the performance of a process, quantitative (e.g. balanced score card, queueing theory, simulation) as well as qualitative (e.g. value-added analysis, SWOT, Pareto analysis) [4]. But it is not considered an easy task to collect all the information, process it and make an evaluation and prioritization of issues. Different stakeholders may have a different view on what goes well and what goes wrong in the process. And there is a risk that important detailed data is overlooked because the collection and analysis of the data may be time-consuming. Often stakeholders (e.g. the process owner, manager or process analyst) already have an idea in mind about what goes wrong in a process and skip further systematic analysis, which may lead to an incomplete overview and insight of the AS-IS situation and to the identification of unimportant or irrelevant bottlenecks and weaknesses [1]. This may be influenced by several cognitive biases. First of all, the analyst may perceive informal information as complete and is not triggered to look for further details [completeness]. Having a preliminary idea in mind may limit the open view of the analyst and may lead to the selection of a less important weakness or bottleneck, simply because other bottlenecks where not considered nor identified [anchoring]. It may also lead to the search for evidence (figures, numbers) to support the believed bottleneck, while missing or ignoring other information that shows the contrary of indicates different problems [confirmation].

4.4   Confirmatory Redesign

Business process designers often make (re)design decisions based on their past experiences, reference solutions, or known best practices. If the current analysis of the business processes matches strongly with past situations, then the designers tend to suggest the same redesign. To justify their choice, they might seek out decision-relevant information that focuses on those aspects that confirm their redesign decision. A recent study has shown that such confirmatory thinking can happen both with professionals [22] and in academia [23].

Business process designers can make decisions based on partial information [11], but as new and relevant information becomes available, decision makers are reluctant to change their initial decisions. This is called [anchoring]. Sometimes a designer's anchor is so entrenched that he or she would not change a decision even when the decision is obviously contradictory to the new information. In a previous experiment [6], we observed that less experienced designers tend *not* to change their design solutions once they are made. In other studies [24] researchers have also observed that even with experienced designers, it is possible to anchor on the first solution that comes to mind. Past successful design decisions can lead business process designers to become increasingly confident about their judgements even when these are wrong [overconfidence]. The overconfidence bias leads BPM designers to overestimate their ability to solve novel or complex problems. Likewise, a designer might simply adopt redesign heuristics (e.g., BPR best practices [25,26]) without critically assessing whether the best practice really solves the problem [rule].

Table 2. Overview of BPM issues and cognitive biases

BPM issue	Influencing cognitive biases	Relevant BPM phases
Suboptimal process architecture	[*completeness, overconfidence, anchoring, escalation*]	Process identification
Incomplete process model	[*completeness and testimony*]	Process discovery
Irrelevant bottlenecks and weaknesses	[*completeness, anchoring, confirmation*]	Process analysis
Confirmatory redesign	[*anchoring, overconfidence, rule*]	Process redesign

5   Conclusions

Studies in cognitive psychology and design science tell us that people can make biased design decisions. Business process designers, business analysts, process owners, and many other stakeholders have to make various design decisions - particularly in the design-time phases of the BPM lifecycle. However, BPM approaches and techniques have barely addressed how human factors influence business process (re)design. One of these human factors is cognitive biases. By treating this human factor as a first class element, we explored the role that the cognitive biases can play in influencing the key issues with respect to the design-time phases of the BPM lifecycle.

Our primary objective in this research is to increase awareness and take an initial step for the future research on human factors in BPM. The initial findings and inferences from our analysis can be useful in proposing ways for 'debiasing', i.e. reducing or eliminating the effect of biases from the cognitive strategies of BPM practitioners. As a future work, we consider adopting and refining existing debiasing approaches, such as Keren's framework for debiasing medical diagnosis and prescription [27]. In brief, this requires an understanding of the environment of the bias and its triggers, studying and applying alternative means for reducing or eliminating the bias, and finally monitoring and evaluating the effectiveness of the debiasing technique. This would also involve empirical studies to give a better understanding on the degree of the influence of such human factors on the BPM issues and the underlying mechanisms in this relation. It would eventually help in developing strategies to lessen the influence of the cognitive biases.

This analysis is not based on an exhaustive list of BPM issues but only on a selected set of key issues that shows representative properties of the BPM lifecycle phases that they relate to. It should also be noted that literature identifies several issues, challenges and critical success factors of BPM that concerns organization's culture, level of process orientation and maturity, leadership, IT infrastructure, personality types, group dynamics, etc. [1,18] that may not be mapped or traced to cognitive biases. However, we believe that with this initial overview on cognitive biases in the design of business processes, we have opened a wider discussion on the influence of human factors on BPM efforts.

References

1. Parkes, A.: Critical success factors in workflow implementation. In: 6th Pacific Asia Conference on Information System (PACIS 2002), pp. 363–380 (2002)
2. Mansar, S.L., Reijers, H.A.: Best practices in business process redesign: validation of a redesign framework. Comput. Ind. **56**(5), 457–471 (2005)
3. Silver, B.: BPMN Method and Style. Cody-Cassidy Press, Aptos (2011)
4. Dumas, M., La Rosa, M., Mendling, J., Reijers, H.A.: Fundamentals of Business Process Management. Springer, Heidelberg (2013)
5. van Vliet, H., Tang, A.: Decision making in software architecture. J. Syst. Softw. **117**, 638–644 (2016)
6. Razavian, M., Tang, A., Capilla, R., Lago, P.: In two minds: how reflections influence software architecture design thinking. J. Softw. Evol. Process **6**(28), 394–426 (2016)
7. Hadar, I., Soffer, P., Kenzi, K.: The role of domain knowledge in requirements elicitation via interviews: an exploratory study. Requir. Eng. **19**(2), 143–159 (2014)
8. Kahneman, D.: Thinking, Fast and Slow. Penguin, London (2011)
9. Arnott, D.: Cognitive biases and decision support systems development: a design science approach. Inf. Syst. J. **16**(1), 55–78 (2006)
10. Tversky, A., Kahneman, D.: Judgment under uncertainty: heuristics and biases. Science **185**(4157), 1124–1131 (1974)
11. Epley, N., Gilovich, T.: The anchoring-and-adjustment heuristic. Psychol. Sci. **17**(4), 311 (2006)
12. Evans, J.: In two minds: dual-process accounts of reasoning. Trends Cogn. Sci. **7**(10), 454–459 (2003)
13. Gigerenzer, G.: Adaptive Thinking - Rationality in the Real World. Oxford University Press, Oxford (2000)
14. Klein, G.: Naturalistic decision making. Hum. Factor J. Hum. Factor Ergon. Soc. **50**(3), 456–460 (2008)
15. Parsons, J., Saunders, C.: Cognitive heuristics in software engineering: applying and extending anchoring and adjustment to artifact reuse. IEEE Trans. Softw. Eng. **30**, 873–888 (2004)
16. Murphy, F., Staples, S.: Reengineering in Australia: factors affecting success. Australas. J. Inf. Syst. 6(1) (1998)
17. Dijkman, R., Vanderfeesten, I., Reijers, H.A.: Business process architectures: overview, comparison and framework. Enterp. Inf. Syst. **10**(2), 129–158 (2016)
18. Sadiq, S., Indulska, M., Bandara, W., Chong, S.: Major issues in business process management: a vendor perspective (2007)
19. Turetken, O., Demirors, O.: Plural: a decentralized business process modeling method. Inf. Manage. **48**(6), 235–247 (2011)
20. Fleischmann, A., Schmidt, W., Stary, C., Obermeier, S., Börger, E.: Subject-Oriented Business Process Management. Springer, Heidelberg (2012)
21. Turetken, O., Demirors, O.: Business process modeling *Plural*ized. In: Fischer, H., Schneeberger, J. (eds.) S-BPM ONE 2013. CCIS, vol. 360, pp. 34–51. Springer, Heidelberg (2013). doi:10.1007/978-3-642-36754-0_3
22. Calikli, G., Bener, A., Arslan, B.: An analysis of the effects of company culture, education and experience on confirmation bias levels of software developers and testers. In: Proceedings of the 32nd ACM/IEEE International Conference on Software Engineering - Volume 2, New York, NY, USA, pp. 187–190. ACM (2010)

23. Hergovich, A., Schott, R., Burger, C.: Biased evaluation of abstracts depending on topic and conclusion: further evidence of a confirmation bias within scientific psychology. Curr. Psychol. **29**, 188–209 (2010)
24. Tang, A., Aleti, A., Burge, J., van Vliet, H.: What makes software design effective? Des. Stud. **31**(6), 614–640 (2010)
25. Reijers, H., Mansar, S.L.: Best practices in business process redesign: an overview and qualitative evaluation of successful redesign heuristics. Omega **33**(4), 283–306 (2005)
26. Vanwersch, R.J.B., Vanderfeesten, I., Rietzschel, E., Reijers, H.A.: Improving business processes: does anybody have an idea? In: Motahari-Nezhad, H.R., Recker, J., Weidlich, M. (eds.) BPM 2015. LNCS, vol. 9253, pp. 3–18. Springer, Cham (2015). doi:10.1007/978-3-319-23063-4_1
27. Keren, G.: Cognitive aids and debiasing methods: can cognitive pills cure cognitive ills? Adv. Psychol. **68**, 523–552 (1990)

Challenges in Business Processes Modeling – Is Agile BPM a Solution?

Rüdiger Weißbach[1(✉)], Kathrin Kirchner[2], Felix Reher[3], and Robert Heinrich[4]

[1] HAW Hamburg University of Applied Sciences, Hamburg, Germany
`ruediger.weissbach@haw-hamburg.de`
[2] Berlin School of Economics and Law, Berlin, Germany
`kathrin.kirchner@hwr-berlin.de`
[3] University of the West of Scotland, Paisley, UK
`felix.reher@uws.ac.uk`
[4] Karlsruhe Institute of Technology, Karlsruhe, Germany
`robert.heinrich@kit.edu`

Abstract. Agile methodologies are established in software development projects. Agility emphasizes, e.g. rapid development and facilitates communication among all stakeholders. Therefore, these principles might be useful in Business Process Modeling projects too. Till now, it is not clear how these principles could be applied in business process modeling. The contribution of this paper is a mapping of challenges in BPM projects and agile principles from software engineering that were derived from literature and own industrial projects. This comparison provides a basis of decision making which agile methods could be applied in a BPM project to face specific challenges.

Keywords: Process modeling · Agile development · Agile BPM

1 Introduction

Planning and designing under uncertainty is typical for the first steps of development of business processes as well as software systems. One way to address uncertainties during development is doing predictions for the expected behavior and quality once the process or system is realized, e.g. [1]. However, prediction is hard as often reliable data is missing. Another way is using iterative and agile development processes. Agile development principles address uncertainties and they are accepted in Software Engineering (SE) for more than a decade [2]. With a time delay, they became more important in BPM too [3], although they are not broadly established yet. We reveal challenges in the BPM process and investigate which agile principles can address which of those challenges. Thus, we aim to answer the research question: Which typical problems in BPM projects in companies can be addressed by agile principles?

The contribution of this paper is an investigation of the influences of agile development on BPM project challenges. Core principles of agility emphasize communication and collaboration as social processes [2]. This paper demonstrates how agile principles can address known challenges in business process management.

© Springer International Publishing AG 2017
M. Dumas and M. Fantinato (Eds.): BPM 2016 Workshops, LNBIP 281, pp. 157–167, 2017.
DOI: 10.1007/978-3-319-58457-7_12

The reminder of the paper structured as follows: In Sect. 2 we will describe the challenges in the elicitation and modeling process based on literature findings. Section 3 will introduce the agile methodology and derive which agile principles can provide a solution for the described challenges. Examples for own real world cases where agile principles were applied successfully to address the challenges are discussed in Sect. 4. Section 5 summarizes our findings.

2   Challenges in Business Process Modeling

Business processes are crucial for organizations [4]. Their modeling is a complex task in a social environment: Different stakeholders with different backgrounds should develop a model that will determine the way of (their) work in a formal representation. While problems of business processes itself are discussed for a long time (e.g. [5]), problems in BPM projects have been widely discussed especially within recent years. Without claiming completeness, we searched for such papers in conference proceedings and journals as media for current scientific and professional discussion. Early papers on BPM projects list practical problems based on interviews with business process analysts and/or managers. [6, 7] listed 22 pitfalls in process modeling projects related to strategy, governance, stakeholders, tools and methods, modeling practice and design as well as success and maintenance issues. [8] conducted an interview with 14 experts and point out 11 typical issues in BPM in strategic, tactical and operational levels like lack of governance likewise standards, gaps between design and execution or weak process specifications. An interview study based on a business process quality reference-model presented in [9] identified 12 potentials for process improvement in a hospital context. [10] as well as [11] discussed generic problems during modeling projects, focused on participants' competencies. With a more theoretical approach [12] reflected on the problems in empirical research and conducted several interviews using a Grounded Theory approach. Besides being able to reference some of Rosemann's pitfalls, this paper proved an awareness of challenges within companies. [13] discussed especially problems with BPMN modeling skills based on a questionnaire sent to process modelers. [14] addressed generic obstacles of BPM and propose ten principles for good BPM in addition to a scope on teaching modeling skills. A central problem in eliciting and documenting requirements is the different level of knowledge of the participants in this process. [15] analyzed the cooperation process of different stakeholders in different businesses and identified exploration and understanding of expert knowledge, building trust and mutual learning as well as the establishment of a structured and well docu-mented (software) engineering process as problems. [16] promoted the BPM model generation from natural language text as a possibility to overwhelm the knowledge acquisition bottleneck. Domain experts are normally not trained in formal description methods and write textual documentations.

Though, a holistic coverage of BPM issues, such as educational [17, 31] or cultural dimensions [18], have only emerged very recently in literature.

Interpreting the studies mentioned above we coarsely clustered the challenges to six main categories in Table 1. These identified challenges are similar to typical challenges in Software Engineering [19].

Table 1. Main challenges in business process modeling projects

CH	Challenge	References
1	Inadequate selection and usage of tools and methods	[6, 11, 12, 17]
2	Dealing with process and model complexity	[6, 7, 11, 12, 20, 21]
3	Lack of clear aims of process modeling	[6]
4	Lack of stakeholder qualification	[3, 6, 31]
5	Lack of process governance	[17, 23, 14]
6	Communication problems in the modeling process	[11, 23]

3 Agile Principles

Agile principles, already applied in software development projects, might be able to address some of the challenges in process modeling projects. These principles originally addressed dissatisfaction with classical software development approaches. The philosophy behind agile methods is stated in the agile manifesto [2]:

- Individuals and interactions over processes and tools
- Working software over comprehensive documentation
- Customer collaboration over contract negotiation
- Responding to change over following a plan
- Twelve agile principles (AP) describe the agile philosophy in more detail [24]:
- (AP1) Our highest priority is to satisfy the customer through early and continuous delivery of valuable software.
- (AP2) Welcome changing requirements, even late in development. Agile processes harness change for the customer's competitive advantage.
- (AP3) Deliver working software frequently, from a couple of weeks to a couple of months, with a reference to the shorter timescale.
- (AP4) Business people and developers must work together daily throughout the project.
- (AP5) Build projects around motivated individuals. Give them the environment and support they need, and trust them to get the job done.
- (AP6) The most efficient and effective method of conveying information to and within a development team is face-to-face conversation.
- (AP7) Working software is the primary measure of progress.
- (AP8) Agile processes promote sustainable development. The sponsors, developers, and users should be able to maintain a constant pace indefinitely.
- (AP9) Continuous attention to technical excellence and good design enhances agility.
- (AP10) Simplicity–the art of maximizing the amount of work not done–is essential.
- (AP11) The best architectures, requirements, and designs emerge from self-organizing teams.

- (AP12) At regular intervals, the team reflects on how to become more effective, then tunes and adjusts its behavior accordingly.

While the Paradigm of "Agility" influenced the development of the Software Engineering domain in the past 15 years [25], its' impact on BPM was neither such strong nor such direct. With the spread of agile approaches, especially with Scrum, the focus changed somewhat. Adapting ideas especially from Scrum, "Agility" became accepted in product development and product management as well as in project management [25]. Since a few years, "Agility" is increasingly connected with BPM too, often in the context of Social Media. But specialized papers on "Agile BPM" are rare (a search for the string "agile BPM" on the IEEExplore website lists 8 hits in May 2016).

According to [28] agility contains all the features of flexibility, plus the attributes of reactivity, proactiveness, and a positive attitude to environmental changes. [32] defines the term 'changeability' as the sum of 'flexibility' and 'responsiveness', which are (1) the "ability of a system to change status within an existing configuration of pre-established parameters" and (2) the "propensity for purposeful and timely behaviour change in the presence of modulating stimuli" [33].

Thus, communication is a pre-condition and central aspect of agility [2, 24]. Whilst the main function of communication is the exchange of information, communication also supports learning in projects. The simple discussion about processes is a kind of an informal review. Discussing the problems on BPM stated above, communication will not improve insufficient tools, however, communication can reduce the problems caused by them, at least by raising the awareness of them.

[12] illustrate that organizations used process models to train their employees and applied the agile principle of simplicity (AP10) to the modeling process. However, "incompleteness" is a characteristic of an agile business process proposal but the idea of standardization is in contradiction to the agile approach [3]. [26] claims that testing during the whole process is important for agile methods. Therefore the gap between model and process should be smaller than in traditional heavyweight approaches. Nevertheless, [23] see the focus on the customer interests as a characteristic attribute of agile processes.

From our literature findings about BPM challenges and the agile principles, we derive which challenges in BPM projects could be addressed by which agile principles. In the literature, we could not find any direct mapping of our CHs and APs, but some references gave hints what could be a good mapping.

From Table 2 it can be seen that every challenge can be addressed by at least two agile principles (marked with a cross). Looking, i.e., at challenge CH6 (communication problems), several agile principles can answer this challenge.

Thus, collaboration and communication are essential aspects of agility. Instead of treating the users as simple "objects" in the elicitation and design process, collaborative aspects are important. Therefore, we can derive that challenge CH6 (communication problems) can be addressed by AP2 (accept constant change requests), AP4 (bring all stakeholders together), AP6 (face-to-face conversation), AP8 (sustainable development) and AP12 (regular reviews).

Table 2. Agile principles (AP) and addressed challenges (CH)

AP	CH											
	1	2	3	4	5	6	7	8	9	10	11	12
1					X				X	X	X	
2			X				X		X	X		
3	X	X					X					X
4					X				X			
5					X	X		X			X	
6		X		X		X		X				X

4 Application of Agile Principles in Real World Scenarios

In the following, we report about our own experiences in four industrial projects in business process modeling, where we applied agile principles. We selected these cases because we can refer to our first-hand experience.

4.1 Farming Case (FC): Stepwise Process Refinement

Agile principles using a participatory interview technique were applied by [29] in a farming project. This project included interviews with several farming companies about how they conduct the process of cultivation planning. An interview for elicitation and modeling a process with a farmer consisted of two sessions. The first session aimed at modeling a first high level process. Thus, an early delivery of a first result was supported (AP1). In the second session, the process was refined, and open issues from the first session were clarified. This way, farmers were included in the modeling process, and early changes were allowed (AP2).

During the whole interview, paper cards were used to write down process steps and arrange and rearrange them on a table. Thus, a strong involvement of the process owner (AP4) and a better understanding of expert knowledge was achieved. Additionally, this approach helps to build trust because the farmers could see with one glance on paper cards whether the interviewer understood them or not (AP6). This case addresses our challenges (CH1) and (CH4) by using paper cards understandable for domain experts instead of formal methods and (CH6) by a strong involvement of domain experts into the modeling process.

4.2 Medical Case (MC): Involvement of Domain Experts

A similar approach that included agile principles was used for modeling clinical treatment processes [30]. At the beginning, an introduction into BPMN modeling language for medical personnel was conducted by a BPM consultant. Afterwards, the clinical treatment process was modeled using tangible BPM [22]. Thus, a structured process could be modeled using tangible BPM plastic cards by the medical personnel themselves. The model was developed in several steps: Starting with the process model on a high level of detail, process steps were refined step by step on more detailed levels. The

BPM consultant asked questions to understand the treatment process better, or corrected parts of the process based on his knowledge of BPMN modeling.

Later, the model was finally corrected by the involved BPM consultant and transferred to a BPMN model in software. Because the medical personnel was directly involved in the modeling process (AP4, AP6), they were highly engaged and could identify themselves with the modeled process. The resulting BPMN process model was understandable for all stakeholders. This case addresses our defined challenges (CH1) by using tangible BPM as a modeling language, (CH4) by providing a BPMN training and (CH6) by involving the medical personnel directly in the modeling process.

4.3   Banking Case 1 (BC1): Short Time Box

Before the agile concept was broadly disseminated, some of its principles have been successfully applied by another author in a start-up of a small private, Customer-Relationship focused bank [27]. This bank started within a span of time of 6 months from the idea to the starting of the business, including all formal and legal aspects. A real innovation in banking without any real archetype was the approach of a focal Customer Relationship Management [CRM] System for (nearly all) business processes. Typical banking transaction core processes had been operated with an externally hosted bank software system. The CRM processes had not been described concretely. A few informal user stories and some one-sentence requirements had been the only specifications. But the mission statement was clear. For supporting CRM processes the bank decided to start with the implementation of a software package that was already used in a similar bank. This "template" promised a short-term launch.

The evolution of the CRM business processes was planned with periodic iterations to be aligned to the bank's development and the learning of the employees. The implementation phase was successfully. During some months employees, managers and IT and Business Process developers collaborated intensively. The bank was able to start within a very short span of time. Some problems raised in the first time after the start of the bank: The planned regular iterations had been postponed due to market needs and so the communication flow between users, process designers and the software developers broke. During this time users lost some knowledge about the flexible handling of the CRM system. After a few months, the review workshops had been conducted and the processes had been refined.

Because the development of the business was different related to the business plan, the processes had to be adjusted. Based on the meanwhile established ongoing communication processes, these adjustments could be worked out easily. In this project the principles of Agile BPM projects as stated in [12] had been applied without a formal reflection on the method. The short time box for implementation persuaded all stakeholders to use this proceeding. This case addresses especially CH2 and CH3. Additionally, the users increased their knowledge about processes (CH4). With the implementation in a very short time box it is also a good example for the first agile principle: "to satisfy the customer through early and continuous delivery of valuable software" [2].

4.4   Banking Case 2 (BC2): Uncertain Understanding of Domain Complexity

In another - not previously described - project that was conducted by one of the authors the aspect of learning was in focus: For cost accounting there are a lot of software products in the market. But the company (a financial service company with ca. 100 employees) had no concrete specification of their requirements. To support the requirements definitionprocess the company decided to work with iterations and spreadsheet prototypes. The calculation models had been tested and expanded in several iterations. After about one year the calculation model was completed and accepted. Different to the controllers' first assumptions, the project lasted much longer than thought, because the complexity was higher. But costs for programming didn't exceed the plan that much, because most effort was on the controlling staff side, that had no concrete requirements and they learned during the project's duration about their planning lacks. Each iteration during the project made concrete values for testing the model as well as for managing the company. Additionally, the staff learned about available data and data quality. This case addresses especially CH2, CH3, and CH4.

The cases indicate that agility with its emphasis on communication between different stakeholders can help to improve the quality of BPM. However, agile principles had been used for concrete problems, not from a general management perspective. In all our case studies, it showed that the direct involvement of the domain experts (and customers) was essential for the success. Furthermore, it was useful to go in several iterations through the modeling phase to stepwise refine and review the process model.

4.5   Summary of the Empirical Results

The following table summarizes our findings from the case studies. As it can be seen from Table 3, in our case studies we experienced different challenges (column 2 in Table 3), that we derived in Table 1. These challenges could be addressed by the application of different agile principles (column 3 in Table 3) that we discussed in Sect. 3. Every challenge could be answered by the application of at least one agile principle. The results may indicate possible "good practices".

Our findings from Table 3 go in line with our derived results from literature shown in Table 2. But, from only four own case studies, we could not show all the relations that we derived from literature in Table 2.

Table 3. Applied agile principles in our industry cases and addressed/solved challenges

Casez	Addressed # of CH	# of AP	Results	Side effects
FC [29]	CH1	AP1, AP4, AP5	Usage of paper card game understandable for domain experts	CH6
FC [29]	CH4	AP5, AP6	Strong involvement of domain expert, usage of card game	
FC [29]	CH6	AP5, AP6	Stepwise development of process model, building trust by using understandable method, strong involvement of domain expert	CH1
MC [30]	CH1	AP4, AP6	Usage of tangible BPM plastic cards, increased understandability of formal models	CH6
MC [30]	CH4	AP5, AP6	Introduction for domain experts into BPMN modeling language, usage of tangible BPM	CH1
MC [30]	CH6	AP5, AP6	Including all domain experts into the modeling, face-to-face communication	
BC1 [27]	CH2	AP1, AP4 (not daily), AP6	Reduced complexity	
BC1 [27]	CH3	AP1, AP2, AP4, AP5, AP6, AP7, AP12 (planned, not done)	Elaborated project aims, better understanding of Customer Relationship processes and their possibilities	CH4
BC2 [not published]	CH2	AP3, AP4 (not really daily), AP6, AP12	Clear understanding of the domain, clear process model and algorithms	
BC2 [not published]	CH3	AP1, AP2, AP3, AP4 (not daily), AP5, AP6, AP7, AP12	Clear aims by discussion among all stakeholders	CH4, CH5
BC2 [not published]	CH4	AP2, AP3, AP5, AP6, AP12	Clear understanding of the domain, understanding of the modelling process itself	

5   Conclusion and Outlook

In this paper, we discussed and investigated the eligibility of agile principles in the process of business process modeling projects based on literature and own case studies. The main contribution of this paper is the clarification of the relationships between challenges in BPM and agile principles. In a first step, we clustered literature findings on challenges in the BPM process to six clusters (CH1 to CH6). Based on literature on BPM and agility, we developed the mapping of agile principles and addressed challenges as a foundational deduction. In addition to the foundational work we analyzed 4 industrial cases in which agile principles had been applied. These cases show that all the six challenges can be addressed by applying agile principles and confirm the theoretical deduction. Compared to the theoretical deduction the real-world situation is more intransparent. There is no direct assignment of agile principles and challenges for all project situations. Side effects are interesting, too: Communication and collaboration are essential for agile working. We observed, especially in BC2, that based on that communication and collaboration the different project participants learned from each other about the business domain, modeling process, its methods and its constraints. After this project, the business staff was able to reflect more about their ideas of requirements.

This paper does not aim at a general theory of "agile BPM". It shows the general eligibility of agile principles in the BPM process. Observed real-live projects from industry focused on practical project success, not on a rigor evaluation of the theoretical concept.

An interesting notice is that agile principles had been introduced too in the marketing of vendors. Companies like Oracle try to offer "ready-to-use" and "agile" software suites [19]. The predefinition of business process elements with its complexity reduction is praised as fast way to success in product brochures.

So this paper indicates at least three directions of further research:

1. Evaluation of the theoretical deduction in Table 2 by analyzing more cases
2. Design and evaluation of an "agile BPM" model
3. Evaluation of prototype-based products as a way of applying agile principles.

The coexistence of traditional and agile methods in industrial practice, the acceptance of agile principles in software engineering and the increasing knowledge of stakeholders concerning these principles hypothesize the relevance of this research.

References

1. Heinrich, R., Merkle, P., Henss, J., Paech, B.: Integrating business process simulation and information system simulation for performance prediction. Int. J. Softw. Syst. Model. **16**(1), 257–277 (2015). doi:10.1007/s10270-015-0457-1
2. Beck, K., et al.: Manifesto for agile software development. http://www.agilemanifesto.org/. Accessed 30 May 2016
3. Meziani, R., Magalhães, R.: Proposals for an Agile Business Process Management Methodology. In: First International Workshop on Organizational Design and Engineering, Lisbon (2009). http://archeologie-copier-coller.com/wp-content/uploads/2014/05/Meziani-Magalhaes.Lisbonne-2009.pdf. Accessed 26 April 2016
4. Dumas, M., La Rosa, M., Mendling, J., Reijers, H.A.: Fundamentals of Business Process Management. Springer, Heidelberg (2013). (pp. I-XXVII)
5. Fischermanns, G.: Praxishandbuch Prozessmanagement. Verlag Dr. Götz Schmidt (2009). ISBN: 3921313732, (in German), 28, 31, 39
6. Rosemann, M.: Potential pitfalls of process modeling: part A. Bus. Process Manag. J. **12**(2), 249–254 (2006)
7. Rosemann, M.: Potential pitfalls of process modeling: part B. Bus. Process Manag. J. **12**(3), 377–384 (2006)

8. Bandara, W., Gable, G., Rosemann, M.: Factors and measures of business process modeling: model building through a multiple case study. Eur. J. Inf. Syst. **14**(4), 347–360 (2005)
9. Heinrich, R.: Aligning Business Processes and Information Systems: New Approaches to Continuous Quality Engineering. Springer, Heidelberg (2014). ISBN 978-3-658-06517-1
10. Frederiks, P., Van der Weide, T.P.: Information modeling: the process and the required competencies of its participants. Data Knowl. Eng. **58**(1), 4–20 (2006)
11. Rittgen, P.: Collaborative modeling - a design science approach. In: 42nd Hawaii International Conference on System Sciences, HICSS 2009, Big Island, HI, pp. 1–10 (2009)
12. Malinova, M., Mendling, J.: A qualitative research perspective on BPM adoption and the pitfalls of business process modeling. In: Rosa, M., Soffer, P. (eds.) BPM 2012. LNBIP, vol. 132, pp. 77–88. Springer, Heidelberg (2013). doi:10.1007/978-3-642-36285-9_10
13. Recker, J.C.: Opportunities and constraints: the current struggle with BPMN. Bus. Process Manag. J. **16**(1), 181–201 (2010)
14. vom Brocke, J., Schmiedel, T., Recker, J., Trkman, P., Mertens, W., Viaene, S.: Ten principles of good business process management. Bus. Process Manag. J. **20**(4), 530–548 (2014)
15. Erfurth, I., Kirchner, K.: Requirements elicitation with adapted CUTA cards: first experiences with business process analysis. In: Proceedings of the 2010 15th IEEE International Conference on Engineering of Complex Computer Systems, pp. 215–223. IEEE Computer Society (2010)
16. Friedrich, F., Mendling, J., Puhlmann, F.: Process model generation from natural language text. In: Mouratidis, H., Rolland, C. (eds.) CAiSE 2011. LNCS, vol. 6741, pp. 482–496. Springer, Heidelberg (2011). doi:10.1007/978-3-642-21640-4_36
17. Bandara, W., Indulska, M., Chong, S., Sadiq, S.: Major issues in business process management: an expert perspective. In: ECIS 2007 Proceedings, Paper 89 (2007)
18. Schmiedel, T., vom Brocke, J., Recker, J.: Which cultural values matter to business process management? Results from a global Delphi study. Bus. Process Manag. J. **19**(2), 292–317 (2013)
19. Erfurth, I., Rossak, W. R.: A look at typical difficulties in practical software development from the developer perspective–a field study and a first solution proposal with UPEX. In: 14th Annual IEEE International Conference and Workshops on the Engineering of Computer-Based Systems, ECBS 2007, pp. 241–248. IEEE (2007)
20. Breuker, D., Dietrich, H.-A., Püster, J., Steinhorst, M., Becker, J., Delfmann, P.: Eine empirische studie zur strukturellen Komplexität konzeptioneller Modelle — Grundlegung eines effizienten Ansatzes zur strukturellen Modellanalyse. In: Mattfeld, D.C., Robra-Bissantz, S. (eds.), Multikonferenz Wirtschaftsinformatik 2012, pp. 1688–1701. GITO Verlag, Berlin (2012)
21. Promatis: Agile Business Process Management. Fast Implementation of Business Strategies. http://www.promatis.de/wp-content/uploads/2014/04/TS_Agile_BPM_e.pdf. Accessed 26 April 2016
22. Edelman, J., Grosskopf, A., Weske, M., Leifer, L.: Tangible business process modeling: a new approach. In: Proceedings of the 17th International Conference on Engineering Design, ICED, 9, pp. 153–168 (2009)
23. Thiemich, C., Puhlmann, F.: An agile BPM project methodology. In: Daniel, F., Wang, J., Weber, B. (eds.) BPM 2013. LNCS, vol. 8094, pp. 291–306. Springer, Heidelberg (2013). doi:10.1007/978-3-642-40176-3_25
24. Beck, K., et al.: Principles behind the Agile Manifesto (2001). http://agilemanifesto.org/principles.html. Accessed 30 May 2016
25. Dingsøyr, T., Nerur, S., Balijepally, V., Moe, N.B.: A decade of agile methodologies: towards explaining agile software development. J. Syst. Soft. **85**(6), 1213–1221 (2012)

26. Zheng, G.: Implementing a business process management system applying agile development methodology: a real-world case study. Master thesis, Erasmus Universiteit, Rotterdam (2012)
27. Weißbach, R.: Einführung von CRM-Systemen - Erfahrungen aus der Gründung einer Vertriebsbank. In: Meyer, M. (ed.) CRM-Systeme mit EAI. Konzeption, Implementierung und Evaluation, pp. 183–197. Vieweg, Wiesbaden (2002)
28. Conboy, K.: Agility from First Principles: Reconstructing the Concept of Agility in Information Systems Development. Information Systems Research, pp. 329–254. Informs, Hannover (2009)
29. Kirchner, K., Erfurth, I., Möckel, S., Gläßer, T., Schmidt, A.: A participatory approach for analyzing and modeling decision processes: a case study on cultivation planning. In: Decision Support Systems in Agriculture, Food and the Environment: Trends, Applications and Advances, pp. 138–154 (2010)
30. Kirchner, K., Malessa, C., Scheuerlein, H., Settmacher, U.: Experience from collaborative modeling of clinical pathways. In: Modellierung im Gesundheitswesen: Tagungsband des Workshops im Rahmen der Modellierung, pp. 13–24 (2014)
31. Bandara, W., Chand, D.R., Chircu, A.M., Hintringer, S., Karagiannis, D., Recker, J., van Rensburg, A., Usoff, C., Welke, R.J.: Business process management education in academia: status, challenges, and recommendations, communications of the association for information systems. Commun. Assoc. Inf. Syst. **27**(41), 743–776 (2010)
32. Milberg, J.: Produktion – Eine treibende Kraft für unsere Volkswirtschaft. In: Reinhart, G.; Milberg, J. (eds.) Mit Schwung zum Aufschwung. Münchener Kolloquium 1997, 19–39. Moderne Industrie, Landsberg/Lech (1997)
33. Bernardes, E.S., Hanna, M.D.: A theoretical review of flexibility, agility and responsiveness in the operations management literature: toward a conceptual definition of customer responsiveness. Int. J. Oper. Prod. Manage. **29**(1), 30–53 (2009)

Towards a Structured Process Modeling Method: Building the Prescriptive Modeling Theory

Regular Paper

Jan Claes[1(✉)], Irene Vanderfeesten[2], Frederik Gailly[1], Paul Grefen[2], and Geert Poels[1]

[1] Department of Business Informatics and Operations Management, Ghent University,
Tweekerkenstraat 2, 9000 Ghent, Belgium
{jan.claes,frederik.gailly,geert.poels}@ugent.be
[2] Department of Industrial Engineering and Innovation Sciences, Eindhoven University
of Technology, Postbus 513, 5600 MB Eindhoven, The Netherlands
{i.t.p.vanderfeesten,p.w.p.j.grefen}@tue.nl

Abstract. In their effort to control and manage processes, organizations often create process models. The quality of such models is not always optimal, because it is challenging for a modeler to translate her mental image of the process into a formal process description. In order to support this complex human processing task, we are developing a smart process modeling method. This paper describes how we have built the underlying prescriptive theory, which is constructed from existing evidence about successful information processing techniques in cognitive psychology.

Keywords: Business process management · Business process modeling · Human aspects of BPM · Smart BPM · Process of process modeling

1 Introduction

In an ever-increasing competitive market and in the context of globalization, mass-customization and risk management, it is currently considered important for organizations to manage and control their core processes. One of the instruments developed to support process management are process models, i.e., representations of certain aspects of the process that abstract from individual process executions [1].

Process models are typically constructed to support communication, documentation, analysis, simulation, execution, etc. [1]. Quality of process models can thus be seen as a measure of how well the model succeeds in supporting the goal: i.e. the fit-for-purpose. Hence, various process model quality variables and metrics have been studied and developed, related to different goals [2, 3].

As such, the business process management community has developed a good understanding of *what* constitutes a 'good' process model. In contrast, far less is known about *how* to build such a 'good' process model. Although in the past researchers have studied how process models are currently constructed [4, 5] and what principles to keep in mind

© Springer International Publishing AG 2017
M. Dumas and M. Fantinato (Eds.): BPM 2016 Workshops, LNBIP 281, pp. 168–179, 2017.
DOI: 10.1007/978-3-319-58457-7_13

when creating a process model [6, 7], we are not convinced that a sufficient answer to this question has been formulated.

Therefore, we started a methodological approach towards the development of a quality-oriented process modeling method. This approach consists of three phases. First, a set of more than 1.000 observations of modeling sessions was collected and analyzed [8, 9]. Second, we are currently completing the knowledge generation phase guided by the work of Gregor, et al. about theory building [10]. Third, the produced knowledge will serve as the base for developing a practical method containing concrete steps on how to successfully approach a process modeling task.

In the knowledge generation phase, we first constructed a descriptive theory that can be used to explain our observations. This is the Structured Process Modeling Theory [11], which is discussed in more detail further in this paper. Being an explanatory theory, it describes potential mechanisms about when and why people make mistakes during process modeling. In order to develop a method in the next phase of the research, this theory first needs to evolve towards a prescriptive theory. Such a theory goes beyond the explanation of the occurrence of mistakes and tries to formulate rules on how to avoid making mistakes in the first place [10]. The evolution from explanatory towards prescriptive theory is the contribution of this paper.

The development of a method that comprises the concrete implementation of these rules and that addresses the practical issues is out of the scope of this paper and has to be considered as future research. For now, the focus is on the conceptual solution, rather than on the practical method. We refer to the work of Gregor, et al. [10] for a deeper understanding about the distinction between an explanatory theory, a predictive theory and a practical method.

This paper is structured as follows. Section 2 presents related work. Section 3 provides the theoretical background. Section 5 describes the developed prescriptive theory. Section 6 concludes the paper with a discussion.

2 Related Work

Related work includes research about process model quality and about how people construct process models. State of the art of both research streams is discussed below.

2.1 Process Model Quality

Process model quality is investigated in conceptual terms by the development of *top-down quality frameworks*. They employ ontological and semiotic theories to provide a structured overview of the relations between various aspects of modeling and to identify potential quality issues for each relation. Examples are the (more general) Conceptual Modeling Quality Framework (CMQF) [12] and the (process model oriented) semiotic quality framework (SEQUAL) [13]. Both frameworks are based on the LSS framework by Lindland [14] and thus make a distinction between the correct use of symbols (syntactic quality), the correct intended meaning of the symbols (semantic quality), and the correct actual understanding of the symbols by the model readers (pragmatic quality).

Further, *bottom-up quality metrics* describe how concrete, quantifiable properties of process models are related to various quality dimensions. In the course of the years, an abundance of metrics has been defined and related to quality dimensions. Instead of discussing those metrics here, the reader is referred to the extensive literature reviews of Sánchez-González [2] and De Meyer [3].

Next, through *empirical surveys* researches have tried to gather information about the success of process modeling techniques. Recker, et al. compared modeling techniques and derived improvement opportunities about process decomposition, declarative modeling, process model lifecycle, context and modeling conventions [15]. Rosemann distracted from focus groups and semi-structured interviews a list of 22 pitfalls related to strategy, stakeholders, requirements, practicalities, future-orientation and maintenance that characterize unsuccessful process modeling [16].

Concerning *pragmatic guidelines* for process modeling, little research exists. Mendling, et al. formulated Seven Process Modeling Guidelines (7PMG), which advise to use few elements, minimize routing paths, use single start and end event, model structured, avoid OR construct, use verb-object labels, and to decompose big models [7]. Further, Becker, et al. formulated Guidelines of Modeling (GoM) about correctness, relevance, economic efficiency, clarity, comparability and systematic design of process models. The guidelines of 7PMG and GoM are criticized to lack support for the modeler on how to achieve these desired process model characteristics during modeling [17]. For such practical support, it appears that modelers have to trust on their experience and on emerged best practices [7, 18].

2.2   The Process of Process Modeling

Researchers have studied how people typically construct process models in a research stream they called "the process of process modeling" [19]. Recker, et al. identified five methods that novices apply to construct a process model: textual design, flowchart design, hybrid design, storyboard design, and canvas design [20]. Pinggera, et al. coded modeling observations in sequences of adding, deleting and laying out model elements and used clustering techniques to define three modeling styles: slower modeling/more reconciliation, slower modeling/less reconciliation, and faster modeling/less reconciliation [5]. Sedrakyan, et al. performed a process mining analysis on observational data and concluded that process modeling seems to require more effort than data modeling, and that in general it seems to be a successful tactic to start with the essence and iteratively add details to your model, which results in a quick growth of the model at the start and a slower pace of adding elements later on when thinking about the more challenging details [21]. To conclude, Claes, et al. described various patterns of process modeling, ranging from timing and spread of creation, movement and deletion of model elements to more general modeling patterns such as the process modeling styles discussed in Sect. 3.1 [9, 11].

3   Theoretical Background

The sections hereafter describe the development of the prescriptive theory, which is the main contribution of this paper. First, this section briefly presents the existing knowledge on which the developed theory is based.

3.1   Process Modeling Styles

In recent research we identified three main process modeling styles: flow-oriented, aspect-oriented and combined process modeling [11]. They are described below.

Flow-oriented process modeling is a strategy where the modeler constructs the process model following the control flow order of the process. The model elements are created, laid out and formatted in blocks from the start towards the end of the process. Every block is first completely finished, before the modeler turns to the next one. It is up to the modeler to decide what she considers a 'block'.

Aspect-oriented process modeling is a strategy where the modeler constructs the process model by focusing consecutively on different aspects of the whole model, such as content, structure, lay-out, formatting, etc. The process model is thus constructed iteratively. In each iteration another aspect of the model is considered from start to end. It is up to the modeler to decide when to address which aspects.

Combined process modeling is a strategy where the elements of the model are created in a flow-oriented way, now and then interspersed with a phase where the modeler works on a particular aspect of the partial model so far (for example interrupting the creation of elements to first layout the partial model so far).

3.2   The Structured Process Modeling Theory (SPMT)

The Structured Process Modeling Theory (SPMT) [11] is a descriptive theory that explains why mistakes are made during process modeling. Mistakes are defined as errors that are caused by cognitive failure, rather than by missing knowledge (about the process or about process modeling). Mistakes can thus be seen as cognitive imperfections during modeling that hinder a modeler to accomplish her intentions. Assuming perfect modeling intentions and perfect knowledge, mistakes would thus be the only reason a modeler doesn't create a perfect high quality model. The SPMT blames cognitive over-load of the modeler's working memory due to the complexity of modeling for the occurrence of modeling mistakes.

It states that cognitive overload can be avoided if one (i) *serializes the modeling approach* (i.e., divide the modeling task into subtasks that are handled consecutively rather than simultaneously), (ii) *in a structured way* (i.e., applying a consistent and logical approach), (iii) *that fits with the relevant properties of the modeler* (i.e., the approach matches the cognitive style and preferences of the modeler).

4 Evidence Collection

The SPMT identifies learning style, field dependency and need for structure as relevant cognitive drivers of the modeler [11]. An in-depth study of various personality factors revealed that mainly these three factors relate to the structuredness of cognitive processing during modeling. In order to transform the SPMT into a prescriptive theory, more details about the drivers - being the principal independent variables of the theory - need to be known and relative priorities in the described effects have to be determined. Therefore, we conducted a targeted literature review to collect evidence about the existent prescriptive knowledge related to these cognitive drivers, which is presented in this section. For the sake of brevity, the referenced sources are limited to a minimal amount of different papers.

4.1 Learning Style

Learning style is defined by Keefe as *"characteristic cognitive, affective, and psychological behaviors that serve as relatively stable indicators of how learners perceive, interact with, and respond to the learning environment"* [22], p. 4. A person's learning style can be classified using various dimensions such as the perception, input, organization, processing and understanding style for handling information [23]. In the context of the SPMT only the understanding dimension is considered: i.e., sequential versus global learning. The Index of Learning Styles [23] contains 11 questions that rate someone's understanding learning style. It results in an odd integer score between -11 (global learner) and +11 (sequential learner).

Table 1 presents an overview of the collected evidence related to learning style that is relevant in the context of the SPMT. From this table it can be concluded that - in contrast to sequential learners - global learners (i) work in intuitive leaps, (ii) do not use a steady pace, (iii) work on seemingly unconnected fragments, (iv) do not necessarily work in consecutive blocks, and (v) work globally.

Literature on learning style also suggests (i) that one's learning style is rather invariant, (ii) that one can apply a mismatching learning strategy, and (iii) that applying a mismatching strategy often has negative consequences. Furthermore, the literature implies that serial learners may often be field-dependent (see further).

Table 1. Overview of relevant knowledge related to learning style

Sequential learners...
"... follow *linear reasoning* processes *when solving problems*" [23], p. 679
"... learn best when material is presented in a *steady progression of complexity and difficult*" [23], p. 679
"... absorb information and acquire understanding of material in small *connected* chunks" [24], p. 289
"... progress *logically, step-by-step*" [25], p. 92
"... move to a different context *only when he or she has assimilated one portion thoroughly*" [25], p. 92
"... ask questions about much narrower relations and their hypotheses are *specific*" [26], p. 130

Global learners...
"... make *intuitive leaps* and may be unable to explain how they came up with *solutions*" [23], p. 679
"... sometimes do better by *jumping directly to more complex and difficult material*" [23], p. 679
"... take in information in *seemingly unconnected* fragments (...) in large holistic leaps" [24], p. 289
"... [are inclined to focus] upon *global*, large-predicate rules" [25], p. 93
"... are assimilating information *from many topics* in order to learn the 'aim' topic" [26], p. 130

In general
"The enigma lies in the *invariance of personal style*" [25], p. 90
"(...) yet [the students] *consistently* prefer a particular type of learning strategy (...)" [26], p. 132
"(...) competence in using a strategy *does not always* go alongside disposition to *adopt it*" [26], p. 132
"(...) some students are disposed to *act 'like holists'* (...) and others *'like serialists'* (...), with *more or less success*." [26], p. 133
"Strategic match or mismatch showed an influence upon learning - *mismatch leading to difficulty in understanding* and sometimes to complete misunderstanding of relevant topics" [25], p. 88
"The *matched consistently performed better* than the mismatched" [25], p. 88

4.2 Field Dependency

Field dependency is defined by Witkin, et al. as *"the extent to which the surrounding organized field has influenced the person's perception of an item within it"* [27], p. 6. It indicates the ease with which someone can abstract from details (i.e., the surrounding field). It is usually measured with the Hidden Figures Test [28] in which a participant has to find simple figures in a complex pattern of lines. The amount of figures that were not discovered (in the provided time) is expressed as a percentage, which quantifies someone's field dependency.

Table 2 presents an overview of the collected relevant evidence related to field dependency. From this table it can be concluded that - in contrast to field-independent people - field-dependent people (i) apply the given order, (ii) work in connected parts, and (iii) have a short attention span. It also appears that they have a high desire for structure (nfs-1) and have a high reaction to missing structure (nfs-2) (see further).

Similar to learning style literature, it is stated (i) that one's field dependency is rather stable by nature, (ii) but it can be changed, (iii) and changing it may be rather easy, (iv) that the field dependency influences the selected learning approach, (v) but the effect of the selected approach on learning success may be limited, and (vi) that these learning effects are also considered for problem solving.

Table 2. Overview of relevant knowledge related to field dependency

Field-independent learners...
"... _profit less from such a teaching approach [providing students with a plan]_" [27], p. 23
"... discern **discrete parts of the field**, distinct from the organized background" [29], p. 239
"... are also **more focused and disciplined learners**" [29], p. 239
"... are characterized by a **longer attention span** and a **greater contemplative disposition**" [29], p. 239
"... **depend more on internal than external cues**" [29], p. 239
"... are likely to **overcome the organization of the field, or to restructure it**" [27], p. 9

Field-dependent learners...
"... **profit more from 'providing students with a plan'**" [27], p. 23
"... engage a global organization of the (...) field, and **perceive parts of the field as fluent**" [29], p. 239
"... have **short attention spans**, are **easily distracted** (...)" [29], p. 239
"... **depend on the cues and structure from their environment**" [29], p. 239
"... **tend to adhere to the organization of the field as given**" [27], p. 9

In general
"(...) evidence of **self-consistency** in performance across tasks" [27], p. 6
"People are likely to be **quite stable** in their preferred mode of perceiving, even over many years" [27], p. 7
"This **does not imply that they are unchangeable**; indeed, some **may easily be altered**" [27], p. 15
"(...) it seems **easily possible** to induce FD persons to use an hypothesis-testing approach **by as simple a means as providing directions** to use such an approach" [27], p. 26
"(...) that relatively FD and FID persons tend to **favor different learning approaches**" [27], p. 27
"The approaches (...) **do not necessarily make for better achievement**" [27], p. 27
"(...) a **better learning outcome** than others seems to depend rather on the specific characteristics of the learning tasks and the particular circumstances under which learning takes place" [27], p. 27
"(...) teaching students to **use problem-solving strategies most appropriate to their styles**" [27], p. 15

4.3   Need for Structure

Need for structure is defined by Thompson, et al. as the extent to which "*an individual (...) prefers structure and clarity in most situations, with ambiguity and grey areas proving troublesome and annoying*" [30], p. 20. It falls apart in two orthogonal factors: the desire for structure (nfs-1) and reaction to missing structure (nfs-2), which are jointly measured via the 12 questions of the Personal Need for Structure Scale of [31]. For both factors an integer score between 1 (low nfs) and 6 (high nfs) is calculated as the mean of the answers to the applicable 6-point Likert-scale questions.

As can be derived from the collected relevant evidence about need for structure presented in Table 3, people with a relatively high need for structure are (i) more confident in their decisions and (ii) are biased towards information that confirms their initial thoughts.

This concludes the section that presents the theoretical background for the construction of the prescriptive theory. Based on the accumulated knowledge about process modeling styles, structured process modeling and the relevant cognitive drivers, the next section presents the developed prescriptive process modeling theory.

Table 3. Overview of relevant knowledge related to Personal Need for Structure

People with low need for structure...
"... might be <u>less confident</u> in their most accessible judgments" [30], p. 30
"... would be <u>motivated to consider alternative judgments</u>" [30], p. 30
"... their impressions are less reflective of the classic assimilation effect because they do not simply and confidently draw new information into the category that is the most readily available" [30], p. 30

People with high need for structure...
"... tend to <u>freeze on the first available explanation</u>" [30], p. 30
"... are <u>confident in their decision</u>" [30], p. 30
"... are <u>unlikely to search for further alternative judgements</u>" [30], p. 30
"... may be <u>more likely to stereotype</u> in ambiguous situations than are low-PNS individuals" [31], p. 124
"... are likely to <u>pay attention to structure-relevant and structure-consistent information</u>" [31], p. 126
"... <u>actively gather structure-consistent information</u>" [31], p. 126
"... <u>use confirmatory hypothesis-testing styles</u>" [31], p. 126
"... are particularly likely to <u>interpret ambiguous information as being structure consistent</u>" [31], p. 126
"... expend great efforts to <u>discount information perceived as being structure inconsistent</u>" [31], p. 126

5   From a Descriptive to a Prescriptive Theory

The goal of the prescriptive theory is to support the development of a quality-oriented practical process modeling method. In accordance to the Structured Process Modeling Theory, the general strategy of this method will be to divide the task of process modeling into subtasks (i.e., to serialize modeling), and to perform this approach in a consistent and logical way (i.e., in a structured way). Furthermore, the way the modeling is serialized and structured has to fit with the three identified cognitive drivers. The prescriptive theory contains the knowledge that describes how to determine which process modeling approach fits with these cognitive drivers.

The evidence presented in Sect. 4, shows that the learning style partially determines one's field dependency and in its turn the field dependency determines the need for structure. Therefore, the instructions of the prescriptive theory will consider the cognitive drivers in this order.

5.1   Learning Style

The general suggested approach is derived from the modeler's learning style. After carefully studying the identified process modeling styles (see Sect. 3.1), it was discovered that each style fits all the described needs of a certain type of learner (see Sect. 4.1).

Practically, this results in next instructions (recall that the learning style of a subject is expressed as an integer odd number between −11 and +11):

- If a modeler's learning style score is between −11 and −5 (global learner), the modeler is instructed to apply the aspect-oriented process modeling style.
- If a modeler's learning style score is between −3 and +3, the modeler is instructed to apply the combined process modeling style.
- If a modeler's learning style score is between +5 and +11 (sequential learner), the modeler is instructed to apply the flow-oriented process modeling style.

5.2 Field Dependency

The field dependency of a modeler defines additional guidelines to be implemented in combination with the general modeling style determined by the learning style.

Based on the collected evidence (see Sect. 4.2), next directions are added to the flow-oriented, aspect-oriented or combined process modeling style (recall that the field dependency is measured with a real number between 0 and 1):

- If a modeler's score is between 0.5 and 1.0 (relatively field-dependent), the modeler is instructed to take frequent short breaks and to work on smaller parts at once (because of the short attention span), to model in the provided order and to try to keep all the parts of the model connected while modeling.
- If a modeler's score is between 0.0 and 0.5 (relatively field-independent), the modeler is instructed to not be afraid to create the model in one take, to not be afraid to skip parts temporarily, to not be afraid to be working all over the model, working in big parts or working on different parts in parallel.

Note how we chose to formulate the instructions for field-independent modelers in the form of '*not be afraid to*' instead of '*try to*' because it doesn't make sense to instruct these modelers for instance to work all over the place.

5.3 Need for Structure

Whereas the evidence that was found for learning style and field dependency can be considered prescriptive because it includes references to matching and mismatching behavior (see Sects. 4.1 and 4.2), the collected evidence about need for structure seems to only describe how people behave given their personal need for structure (see Sect. 4.3). Therefore, it will be used in the prescriptive theory only to warn the modelers in order to create awareness, rather than translating it into concrete guidelines. Moreover, we feel that the evidence relates more to the desire for structure (nfs-1) variable than to the reaction to missing structure (nfs-2) variable.

Next information complements the previous instructions (recall that the desire for structure is quantified as an integer score between 1 and 6):

- If a modeler has a nfs-1 score between 1 and 3 (low desire for structure), the modeler is informed that the guidelines provided may not feel natural, but yield a high potential improvement and the modeler is requested to do an effort to apply the instructions carefully.
- If a modeler has a nfs-1 score between 4 and 6 (high need for structure), the modeler is informed that the guidelines may feel familiar, because she may already apply them and the modeler is requested to consider them as an instrument to perfect her current structured modeling style.

5.4   Summary

The prescriptive guidelines are summarized below.

6   Conclusion

The research presented in this paper focuses on human aspects of business process modeling. Real-life processes are often complex and it appears to be hard for humans to construct high quality models representing the processes. Therefore, we are developing a practical method for process modeling based on scientific theories about the human mind. The literature suggests that changing someone's modeling approach may be *"as simple (...) as providing directions to use such an approach"* [27], p. 26. One crucial step in this development is thus the formulation of the prescriptive theory that forms the basis of the method. This paper describes the transformation from the existing explanatory Structured Process Modeling Theory (SPMT) towards a prescriptive theory, which is summarized in Fig. 1.

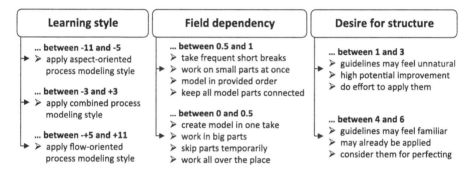

Fig. 1.   Summary of the prescriptive structured process modeling guidelines

The theory was built from the SPMT and various sources in the cognitive psychology literature. It thus combines existing knowledge in a fundamentally new way. According to Gregor, et al. such a transformation from an explanatory towards a prescriptive theory is feasible and it should be evaluated against innovativeness, utility and persuasiveness. However, in order to perform an extensive evaluation of the prescriptive theory, it first needs to be embedded in the practical method.

Therefore, a thorough validation of the theory is out of the scope of this paper. Nevertheless, we can already report on some initial results. Indeed, the method that will use the theory that is the contribution of this paper is currently under development. It consists of a digital workflow starting with the measurement of a modeler's cognitive preferences, selecting a fitting process modeling strategy based on the prescriptive theory presented in this paper, learning the strategy via a one-hour interactive digital tutorial and applying the strategy on a modeling case.

The initial results are promising and indicate (i) that it seems possible to accurately measure one's cognitive preferences in an automated digital way, (ii) that it indeed seems possible to change a modeler's modeling style with a limited intervention in the form of a digital tutorial, and (iii) that this indeed has a significant beneficial effect on modeling quality. This confirms the utility of the developed prescriptive theory.

References

1. Dumas, M., La Rosa, M., Mendling, J., Reijers, H.A.: Fundamentals of Business Process Management. Springer, Heidelberg (2013)
2. Sánchez-González, L., García, F., Ruiz, F., Piattini, M.: Toward a quality framework for business process models. Int. J. Coop. Inf. Syst. **22**(01), 1–15 (2013)
3. De Meyer, P.: Kwaliteit van procesmodellen: literatuurstudie (2015)
4. Pinggera, J., Zugal, S., Weidlich, M., Fahland, D., Weber, B., Mendling, J., Reijers, H.A.: Tracing the process of process modeling with modeling phase diagrams. In: Daniel, F., Barkaoui, K., Dustdar, S. (eds.) BPM 2011. LNBIP, vol. 99, pp. 370–382. Springer, Heidelberg (2012). doi:10.1007/978-3-642-28108-2_36
5. Pinggera, J., Soffer, P., Fahland, D., Weidlich, M., Zugal, S., Weber, B., Reijers, H.A., Mendling, J.: Styles in business process modeling: an exploration and a model. Softw. Syst. Model. **14**(3), 1055–1080 (2013)
6. Becker, J., Rosemann, M., Uthmann, C.: Guidelines of business process modeling. In: Aalst, W., Desel, J., Oberweis, A. (eds.) Business Process Management. LNCS, vol. 1806, pp. 30–49. Springer, Heidelberg (2000). doi:10.1007/3-540-45594-9_3
7. Mendling, J., Reijers, H.A., Van der Aalst, W.M.P.: Seven process modeling guidelines (7PMG). Inf. Softw. Technol. **52**(2), 127–136 (2010)
8. Claes, J., et al.: Tying process model quality to the modeling process: the impact of structuring, movement, and speed. In: Barros, A., Gal, A., Kindler, E. (eds.) BPM 2012. LNCS, vol. 7481, pp. 33–48. Springer, Heidelberg (2012). doi:10.1007/978-3-642-32885-5_3
9. Claes, J., Vanderfeesten, I., Pinggera, J., Reijers, H.A., Weber, B., Poels, G.: A visual analysis of the process of process modeling. Inf. Syst. E-bus. Manag. **13**(1), 147–190 (2015)
10. Gregor, S.: The nature of theory in information systems. MIS Q. **30**(3), 611–642 (2006)
11. Claes, J., Vanderfeesten, I., Gailly, F., Grefen, P., Poels, G.: The structured process modeling theory (SPMT) - a cognitive view on why and how modelers benefit from structuring the process of process modeling. Inf. Syst. Front. **17**(6), 1401–1425 (2015)
12. Nelson, H.J., Poels, G., Genero, M., Piattini, M.: A conceptual modeling quality framework. Softw. Qual. J. **20**(1), 201–228 (2012)
13. Krogstie, J., Sindre, G., Jørgensen, H.: Process models representing knowledge for action: a revised quality framework. Eur. J. Inf. Syst. **15**(1), 91–102 (2006)
14. Lindland, O.I., Sindre, G., Solvberg, A.: Understanding quality in conceptual modeling. IEEE Softw. **11**(2), 42–49 (1994)
15. Recker, J.C., Rosemann, M., Indulska, M., Green, P.: Business process modeling: a comparative analysis. J. Assoc. Inf. Syst. **10**(4), 333–363 (2009)
16. Rosemann, M.: Potential pitfalls of process modeling: part A. Bus. Process Manag. J. **12**(2), 249–254 (2006)
17. Moreno-Montes de Oca, I., Snoeck, M., Reijers, H.A., Rodríguez-Morffi, A.: A systematic literature review of studies on business process modeling quality. Inf. Softw. Technol. **58**, 187–205 (2014)
18. Silver, B.: BPMN: Method and Style. Cody-Cassidy Press, Altadena (2011)

19. Soffer, P., Kaner, M., Wand, Y.: Towards understanding the process of process modeling: theoretical and empirical considerations. In: Daniel, F., Barkaoui, K., Dustdar, S. (eds.) BPM 2011. LNBIP, vol. 99, pp. 357–369. Springer, Heidelberg (2012). doi: 10.1007/978-3-642-28108-2_35
20. Recker, J.C., Safrudin, N., Rosemann, M.: How novices design business processes. Inf. Syst. **37**(6), 557–573 (2012)
21. Sedrakyan, G., Snoeck, M., De Weerdt, J.: Process mining analysis of conceptual modeling behavior of novices: empirical study using JMermaid modeling and experimental logging environment. Comput. Hum. Behav. **41**, 486–503 (2014)
22. Keefe, J.W.: Learning style: an overview. In: Proceedings NASSP's Student Learning Styles: Diagnosing and Proscribing Programs, pp. 1–17 (1979)
23. Felder, R.M., Silverman, L.K.: Learning and teaching styles in engineering education. Eng. Educ. **78**, 674–681 (1988)
24. Felder, R.M.: Reaching the second tier: learning and teaching styles in college science education. J. Coll. Sci. Teach. **23**(5), 286–290 (1993)
25. Pask, G.: Learning strategies, teaching strategies, and conceptual or learning style. In: Schmeck, R.R. (ed.) Proceedings Learning Strategies and Learning Styles, pp. 83–100. Springer, New York (1988)
26. Pask, G.: Styles and strategies of learning. Br. J. Educ. Psychol. **46**(2), 128–148 (1976)
27. Witkin, H.A., Moore, C.A., Goodenough, D.R., Cox, P.W.: Field-dependent and field-independent cognitive styles and their educational implications. Rev. Educ. Res. **47**(1), 1–64 (1977)
28. Ekstrom, R.B., French, J.W., Harman, H.H.: Kit of factor-referenced cognitive tests, Princeton, New Jersey (1976)
29. Wooldridge, B., Haimes-Bartolf, M.: The field dependence/field independence learning styles: implications for adult student diversity, outcomes assessment and accountability. In: Proceedings Learning Styles and Learning: A Key to Meeting the Accountability Demands in Education, pp. 237–257 (2006)
30. Thompson, M.M., Naccarato, M.E., Parker, K.C.H., Moskowitz, G.B.: The personal need for structure and personal fear of invalidity measures: historical perspectives, current applications, and future directions. In: Proceedings Cognitive Social Psychology: The Princeton Symposium on the Legacy and Future of Social Cognition, pp. 19–39 (2001)
31. Neuberg, S.L., Newsom, J.T.: Personal need for structure: individual differences in the desire for simpler structure. J. Pers. Soc. Psychol. **65**(1), 113–131 (1993)

Designing Serious Games for Citizen Engagement in Public Service Processes

Nicolas Pflanzl[1](✉), Tadeu Classe[2], Renata Araujo[2], and Gottfried Vossen[1,3]

[1] DBIS Group, Westfälische Wilhelms-Universität Münster, Münster, Germany
nicolas.pflanzl@wi.uni-muenster.de
[2] SIGAC-CIBERDEM, Federal University of the State
or Rio de Janeiro, Rio de Janeiro, Brazil
[3] University of Waikato Management School, Hamilton, New Zealand

Abstract. One of the challenges envisioned for eGovernment is how to actively involve citizens in the improvement of public services, allowing governments to offer better services. However, citizen involvement in public service design through ICT is not an easy goal. Services have been deployed internally in public organizations, making it difficult to be leveraged by citizens, specifically those without an IT background. This research moves towards decreasing the gap between public services process opacity and complexity and citizens' lack of interest or competencies to understand them. The paper discusses game design as an approach to motivate, engage and change citizens' behavior with respect to public services improvement. The design of a sample serious game is proposed; benefits and challenges are discussed using a public service delivery scenario from Brazil.

Keywords: eGovernment · Public services process design · Serious games · Game design

1 Introduction

One of the objectives of governments lies in delivering public services to their citizens to support certain administrative tasks. To perform these services, public bodies typically execute and employ cross-organizational business processses, transactions and resources operating on Information and Communication Technology (ICT) platforms. After earlier efforts to improve access, transparency, standardization, and optimization of public services, future challenges for eGovernment initiatives have been identified as citizen involvement and participation [1–3]. In many contexts, citizen involvement in public service design through ICT is difficult to achieve, and is thus often kept to a minimum. One reason for this is that information technology (IT) is deployed internally in public organizations, and is thus not intended for interaction with citizens, as IT tools for business process management (BPM) are too difficult to be leveraged by citizens without an IT background. Citizen participation is usually addressed optimistically, with

M. Dumas and M. Fantinato (Eds.): BPM 2016 Workshops, LNBIP 281, pp. 180–191, 2017.
DOI: 10.1007/978-3-319-58457-7_14

the assumption that the availability of an interaction channel during service provision will promote participation [4]. In turn, citizens perceive public services as a boring necessity, usually want to be done with them as fast as possible, and do not develop interest in taking part in public services improvement.

Although citizens seem to have no interest in participatory involvement, preferring to assume the role of passive customers, actively involving them in the design and improvement of public services would allow governments to offer better services to its citizens. The success of such initiatives depends on the ability of the government body to motivate citizen participation. While the idea of influencing provided services should be intrinsically motivating in theory, this may not necessarily be the case in practice. Literature suggests to move towards innovative solutions for involving citizens, by creating tools that decrease the gap between public services process opacity and complexity and citizens' lack of interest and competencies to understand them [5]. We explore the use of game-based ideas and concepts in public-service BPM, based on current research that examines the potential of games to motivate, engage, and enable learning [6,7].

Games are rule-based systems that engage players in artificial conflicts with uncertain, quantifiable outcomes [8]. They take various forms, including board games, e.g., chess or go, and modern computer and video games. While games serve the primary purpose of entertainment, it has long been recognized that they also have the potential to support, for example, socialization, education, and military training [9]. It can be seen a broad adoption of digital games as objects of research whose impacts can be positive, and whose purpose may exceed that of mere enjoyment [6]. One of the research streams in this regard deals with serious games [10], i.e., actual games that are created for more serious purposes than just entertainment. This paper describes preliminary ideas on how to design and explore the use of serious games as an artifact able to motivate, engage and, to some extent, change citizens' behavior with respect to public services delivery, while enabling them to learn and understand how public administration works.

Section 2 presents the potential benefits and challenges of providing citizen participation in public service design and delivery, by illustrating a real scenario based on Brazilian public service process. Section 3 discusses the basic ideas that underline the proposal of having game-based participative public processes, and discusses related work. Section 4 proposes the concept of a serious game designed to increase citizen involvement in public services process learning and understanding. Section 5 concludes the paper and outlines future work.

2   Participative Public Service Design and Delivery

Public organizations have long focused on BPM initiatives to improve public services delivery performance and citizen satisfaction by improving the efficiency of their internal process management. Social and participative approaches to BPM – so-called Social BPM [11] – have been suggested as an organizational strategy to balance the rigidity of defined processes with the flexibility of social interaction, and as a strategy to provide innovative ways of integrating clients into process definition and execution using social software and its underlying principles. The emergence of Social BPM is envisioned as a promising approach to allow citizen participation in public services process management [5]. PFLANZL and VOSSEN performed an extensive literature review and outlined the challenges of Social BPM initiatives [12]: (i) effectively involving external stakeholders; (ii) motivating participation; (iii) training novice process modelers; (iv) providing effective tools and languages for novice users; (v) ensuring model quality; (vi) handling information overload; and (vii) coping with differences in meanings and semantics between participants. If we consider the use of Social BPM in citizen participation in public services modelling, design, delivery and evaluation, these issues can be even more challenging due to the variety of citizens' profiles, differences in cultural contexts, and the complexity of different service domains [13].

Concerning electronic participation, citizen participation improves at a progressive scale where, at each level, citizens are increasingly empowered in their possibilities for participation, discussion and decision-making in government processes and issues [14]. Through this scale, different relationships between government and citizens can be configured, in which, at the lower levels, government and citizens have very distinct responsibilities and roles, yet at higher levels roles and responsibilities are mixed and interchanged. It has been argued that in order to participate in public services process management, citizens must first understand how the service is provided, i.e., what its underlying process is [15,16]. By receiving insights into the provision of a service and its implications, citizens may find motivation to participate. Ultimately, citizens should be able to change the process, simulate distinct alternatives for its execution, or even change and adapt the process to suit their own specific needs.

As a sample scenario, we consider the public service delivery of the Brazilian Ministry of Social Security (MPS). Social security services address all citizens in Brazil and include health assurance, unemployment support, and retirement. Retirement, for example, is a right for every Brazilian citizen depending on their monetary contribution to the public social security body during their working life. Taking into account a population of est. 200 million inhabitants, a high urbanization, and shortcomings regarding services accessibility in many areas of the country, providing social security services is a costly challenge. The Brazilian government and its institutions have performed significant legal and operational efforts to improve public service delivery [17,18]. As a consequence, most public institutions nowadays provide basic information about their services online. However, the last survey performed by the Brazilian Internet Steering Committee (CGI.BR) about the use of ICT in the Brazilian Public Sector [19] showed that 66% of the local governments provide public services through websites. The type of service offered through these websites are document or forms downloads (69%); submitting forms (43%), registering for public exams or courses (38%), generating invoices (37%), checking ongoing proceedings (29%), obtaining tax payment slips (26%), obtaining official documents (22%), and scheduling appointments, assistance etc. (13%). The type of resources offered to citizens in those websites are content search engines (66%), videos (35%), requesting services (32%), real-time broadcasts (16%), and audio or web radio (11%).

Fig. 1. Services portal of the Brazilian MPS services (top) & agencies (bottom).

Information about social security services and in particular retirement claims are available on the MPS website (see Fig. 1). However, owing most likely to process complexity, information available on the website is kept to a minimum, consisting of a brief service description and facilities to schedule an appointment. Brazilian citizens have limited interface channels with the social security body and to obtain information about a service (Fig. 1). It is very common that after organizing all the documents expected to be required for a retirement claim, scheduling an appointment, and waiting to be attended, they cannot register their claim due to missing documents or lack of understanding about retirement rules. So it is reasonable to ask whether this would change if more interaction channels were provided. In addition to the service complexity and limited channels, Brazilian citizens needing access to social security are often poor, elderly, or even illiterate, and thus do not understand terms and legislation. This entails further challenges for public bodies with regard to operating effectively through social media or online channels. In practice, it is common to find private professionals specialized in helping citizens to cope with the burden of public services delivery, which may help, but at the same time facilitates corruption.

3 Game-Based, Participative Public Service Design and Delivery

One of the most prevalent areas in which games or ideas inspired by games have been applied is education and learning [6,7]. Possible approaches include the repurposed use of commercial games for education, the creation of new games specifically designed for learning, or the incorporation of game elements into non-game learning applications [7,20]. In this context, games can support knowledge acquisition, content understanding, and the development of perceptual, cognitive, and motor skills [6]. GEE argues that videogames, when designed in a way that enables critical thinking and active learning, hold the potential to introduce individuals to new domains, to let them join social groups related to the former, and to enable future problem solving and learning in these domains [21].

Based on these observations, we propose extending present research towards the conceptualization, development, and evaluation of games as a tool for supporting citizens in understanding public service processes, so as to motivate and to empower them to take part in their improvement. This is based on the assumption that playing a game with content that is built on public services allows players to actively learn about these processes, to increase their interest in understanding them, and to reflect them critically. In this way, players obtain literacy in the domain of public services, thus enabling them to correlate the contents of the game with their own, personal experiences, and to share and discuss suggestions for process improvements with related social groups. If this is done in a way as outlined in Sect. 2, this shall be understood as game-based, participative public service design and delivery. The main challenge here is how to model and design games for this purpose, considering, for instance, how to define game learning objectives, how to define critical aspects of the service to be explained, how to clarify the purpose of playing the game, how to model game dynamics, and how to introduce specific domain and cultural elements so as to allow users to quickly recognize themselves within the game.

Previous research has discussed games as a useful tool for BPM learning. RIBEIRO ET AL. described a game called ImPROVE aiming at improving BPM learning, using the real scenario of a Portuguese hospital triage system [22]. BULANDER proposed a conceptual framework to INNOV8 game (IBM's game to learn BPM), in order to measure the level of BPM learning [23]. SANTORUM proposed a method based on serious games aiming at identifying, simulating and improving organizational process, where the worker can play the game, learning with it, and suggest process improvements [24]. Using BPM to game modeling is the aim of the research presented by SOLIZ-MARTINZ ET AL., where VGPM (a modeling notation based on BPMN) was proposed for modeling game logic [25]. There is also research addressing the use of games for the involvement of citizens on civil proceedings. POPLIN evaluated the potential of serious games to support citizen participation in urban planning [26]. A game was designed through which citizens could contribute ideas for improving the urban planning of Billstedt (Hamburg, Germany). They could build their own business in a specific location and discuss issues that could arise with other citizens, helping government decisions. AHMED ET AL. discussed how serious games can support citizen participation in public services and depicted a model for designing games for that purpose, including five elements — Environment, Objectives, Goals, Rules and Players [27]. LOUNIS ET AL. evaluated the impact of two distinct game elements (incentives and community collaboration) in user participation in a gamified public service [28]. The game objective was to improve "green" behavior by motivating users to buy items with lower environmental impacts. Initial results showed that gamification improved collaboration, the users? sense of duty, and willingness to keep participating. Lastly, HORITA ET AL. described the development and deployment of a gamified social architecture to improve the management of activities and information dissemination among young volunteers during natural disasters [29]. Results indicated that participants showed more attention to the process activities and improvements in communication. Despite of the references above, research on citizen involvement in public processes through BPM using games were not found.

4   Serious Game Concept

We now describe the concept of a serious game intended to enable citizens to learn about public service processes, thus preparing them for participative service design. The concept is based on the highly successful, independent game "Papers, Please" that was developed as a commercial product, yet accomplishes a similar goal for the immigration process as outlined in Sect. 4.1. Afterwards, the design of the game to be developed itself is described in Sect. 4.2 through three lenses: mechanics, dynamics, and aesthetics (MDA) [30]. Mechanics are the components of a game that allow the player to act within the confines of its rules, dynamics represent the run-time behavior of the game, and lastly, aesthetics are the emotional responses the player exhibits while playing the game. The designer of a game can only influence its mechanics, whereas dynamics and aesthetics emerge from the former. A discussion about the expected outcomes from applying this game is provided in Sect. 4.3.

4.1   Inspiration

The concept described next is based on the highly popular and successful game "Papers, Please" released in 2013. In this game, the player is put into the role of a newly appointed immigration agent of a fictional, totalitarian country. The main task of the game lies in deciding whether to allow or deny individuals' entry depending on the legitimacy and completeness of their papers. A sample screenshot taken from the game can be seen in Fig. 2. For correct decisions, the player earns money which he can use to feed and provide for his family. Too many incorrect decisions, in turn, lead to an eventual decrease in pay.

The game is divided into multiple days (i.e., levels), and a time limit is imposed upon the player to make him decide under pressure. "Papers, Please" starts out rather simple, with the only initial rule being that citizens of the fictional country may be admitted with a valid passport. However, the complexity increases quickly, as immigration opens for citizens of other countries, new types of documents to be checked are introduced (e.g., entry permits, work passes, or diplomatic authorizations), and fingerprints need to be checked. The game becomes even more challenging when additional motivations for allowing

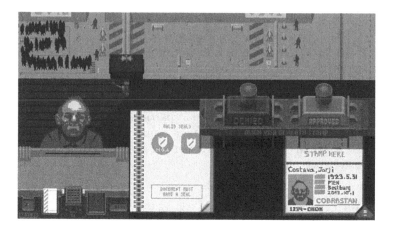

Fig. 2. Screenshot from the game "Papers, Please".

or denying a request are introduced. For instance, on one day, a woman tries to enter the country without valid documents to visit her son, asking the player to act humanely [31]. By playing "Papers, Please", players learn about the immigration process and how it allows governments to exert control over people wishing to enter a country, as well as the roles of the involved parties and the required business objects [32]. As new rules are introduced over time, they receive incremental training in executing the work processes, and an in-game rulebook serves as a reference for all decision rules that need to be applied. While "Papers, Please" was developed as a commercial product, it has also been described as a representative of a new wave of serious games that increasingly blur the lines between entertainment and education [32].

4.2 Game Design

The core idea presented in this paper is the creation of a serious game based on the gameplay of "Papers, Please" that allows citizens to learn about public service processes in an enjoyable manner while abstracting from their complexity. For that purpose, players shall be put into the role of a public official responsible for the execution of processes such as the application for a passport or the renewal of a driver?s license instead of acting as citizens.

Mechanics. The game is subdivided into workdays, each of which corresponds to a level. On each day, the player has a predetermined amount of time in order to process the requests of citizens. For each request, he must make a binary decision about whether to accept or to deny the request. Which of these two possibilities is correct depends on personal properties of the applicant, the documents that he provides, and certain rules that are currently active. Rules are prescribed in a handbook which the player may consult at any given time. For any request that the player has accepted or denied, he either earns a fixed amount of money if this was correct, or incurs a monetary penalty otherwise. The interface of the game is subdivided into two separate areas: a customer area in which a sequence of non-player citizens visualized by cartoon figures appears before the player, and a desk area, in which the player can examine documents provided by the customer. A first draft of the interface is shown in Fig. 3.

Fig. 3. Draft of the game interface with customer area and desk area.

Dynamics. Put into motion, the run-time behavior of the mechanics is expected to be as follows. At the beginning of each workday, the player may be introduced to any new rules, starting with an initial set of rules on day 1. Before the time limit of the workday is activated, the player may spend unlimited time studying old and new rules in the handbook placed on his desk. Once this preparation phase finishes, a timer presented as a wall clock which clearly denotes start and end of the workday starts running, and customers appear in front of the player in a sequence. The player must read the request of the customer and correlate it with the documents the latter presents in the desk area. At this stage, the player must form an opinion about whether to accept or deny the request. Possible reasons for denial include incomplete/faulty documents or that personal requirements not satisfied by the applicant. If the player cannot reach a decision yet, he may request additional documents, which the customer may or may not be able to provide. As the goal for the player is to maximize his earnings by processing as many customers as possible, he is forced to minimize processing times for each instance, thus acting under time pressure. The game shall start out rather simple, but becomes continuously more complex. During the first few workdays, the player will have to process rather simple requests, such as getting a pothole fixed. After occasional promotions, he will have to look at more complex service requests, e.g., applications for retirement funds. This also coincides with more complex acceptance rules that the player must adhere to. As the services become more complex, the impact that an acceptance or denial has for the customers also increases. Thus, as part of the narrative, they will increasingly try to appeal to the player to grant them their requests even if they do not fulfil all necessary requirements. Similarly to "Papers, Please", this introduces a motivation for the player to "do the right thing" even if this contradicts the rulebook and thus leads to a penalty.

Aesthetics. The emotional responses that a game invokes ultimately determine whether or not it is fun. To that extent, different games implement different aesthetic components appropriate for their design that enable fun and engaging player experiences [30]. Focusing on the outlined game design, one of the major sources of fun is reasoned to be its challenges. This means that the game presents itself as an obstacle course to the player and that the mechanics ensure that the obstacles require more and more skill to be overcome. Optimally, the difficulty of the game and the skills of the player should grow at the same scale to ensure that the game remains challenging at all times yet allows the player to overcome any challenges. This is one of the conditions for flow experience in games, which makes them enjoyable and rewarding to play [33]. Another aesthetic component of game design is discovery. Each customer can be seen as a sort of puzzle, and the player must discover the solution by finding the necessary grounds for accepting or rejecting his request.

4.3 Evaluation and Expected Outcomes

To demonstrate the effectiveness of the proposed game, an evaluation along two dimensions shall be carried out: the game itself and its game design, the effects of the game on its players. The purpose of evaluating the game itself lies in ensuring that the artifact was designed and developed in a way that actually enables it to achieve its intended outcomes. Possible choices to be considered here include user studies, survey-based measurement instruments such as the *Play Experience Scale* [34] or analysis frameworks specifically tailored to serious games such as the *Serious Game Design Assessment Framework* [35]. Focusing on the outcomes of the game, a controlled experiment shall be carried out, in which both the experimental as well as the control group will fill out the same survey, whereas the experimental group will be subjected to the proposed game first. The survey shall firstly contain a section with various statements about the realized processes whose correctness the participants must assess. As we expect the experimental group to learn about activities, decision rules, and restrictions that apply to public service processes, we assume them to be better educated about the realized processes than the control group, and thus achieve better results. As a direct result, we anticipate that members of the experimental group are able to achieve a significantly higher perceived self-efficacy, i.e., belief that they might be able to properly consume even more complex public services [36]. Thus, following [36], the survey shall contain properly constructed question items. Lastly, the depiction of public service processes from the perspective of a public servant is intended to increase the understanding of this position and thus the players? attitudes towards public service delivery in total. Following the *Theory of Reasoned Action* [37], the experimental group may thus exhibit a higher behavioral intention to use such services in the future. Accordingly, appropriate items shall be included in the survey.

In summary, we expect the players of the proposed game to be able to make correct statements about particular aspects of the implemented processes, to have a higher belief that they can consume such services, a more positive attitude towards public service delivery in general, and thus a higher intention to consume these services in the future. Results that are gained from the evaluation of the game can be used in subsequent iterations of the implementation process to improve its concept and implementation.

5   Conclusion and Outlook

This paper has presented a concept for a serious game whose purpose is to allow citizens to become literate in the domain of public service processes by enabling active learning, critical thinking, and reflection. The content of the game shall be tailored to the specific requirements of Brazil, where it will also be tested and evaluated. Ultimately, the game is intended as a first step towards game-based, participative public service design. To that extent, future developments will focus on allowing citizens to interpret the game within the context of their own experiences with public services, thus enabling them to identify and verbalize their ideas for improvement.

Presently, this research endeavor is in the concept stage. The presented approach is just one of numerous possibilities how ideas from games can enrich public services. Independently of the domain and the design approach that is used, games are becoming more and more prevalent throughout various parts of day-to-day life, thus meriting a closer examination of their potentials for public service design and delivery. We are aware that the next challenging issue in this research is the game design itself, what includes, for instance, how to clearly define the aesthetics (fun, emotional involvement) aspects of the game, and how to deal with the game complexity. However, the research is not restricted to the desing of an specific game, but on how to systematically design different games with the purpose of citizen understanding, engagement and learning. Subsequent activities in this research will be the implementation of the concept while following established game design and development processes to ensure that the result can meet its aesthetic goals. Another goal is to define a game engineering process for the context of participative public service design, taking as input business process definitions and models, thus allowing distinct public institutions to develop games for the purpose of engaging citizens with their processes.

Acknowledgments. The authors thank Prof. Geraldo Xexéo (COPPE/UFRJ) for contributions to this research and the Deutsche Forschungsgemeinschaft for financial support.

References

1. Olphert, W., Damodaran, L.: Citizen participation and engagement in the design of e-government services: the missing link in effective ICT design and delivery. J. Assoc. Inf. Syst. 8(9), Article 27 (2007)
2. Scherer, S., Wimmer, M.A.: E-participation and enterprise architecture frameworks: an analysis. Inf. Polity **17**(2), 147–161 (2012)
3. Howard, R.: Digital government will move at the speed of civic moments (2014). https://www.gartner.com/doc/2733617/digital-government-speed-civic-moments. Accessed 27 Nov 2015
4. Zheng, Y.: Explaining citizens' e-participation usage: functionality of e-participation applications. Adm. Soc. (2015)
5. Araujo, R., Taher, Y., van den Heuvel, W.J., Cappelli, C.: Evolving government-citizen ties in public service design and delivery. In: IFIP e-Government Conference, Koblenz, Germany, pp. 19–26 (2013)
6. Connolly, T.M., Boyle, E.A., MacArthur, E., Hainey, T., Boyle, J.M.: A systematic literature review of empirical evidence on computer games and serious games. Comput. Educ. **59**(2), 661–686 (2012)
7. Hamari, J., Koivisto, J., Sarsa, H.: Does gamification work? a literature review of empirical studies on gamification. In: 47th HICSS, pp. 3025–3034 (2014)
8. Salen, K., Zimmerman, E.: Rules of Play - Game Design Fundamentals, 1st edn. The MIT Press, Cambridge (2003)
9. Michael, D., Chen, S.: Serious Games - Games that Educate, Train, and Inform, 1st edn. Thomson Course Technology PTR, Boston (2005)
10. Abt, C.C.: Serious Games, 1st edn. Viking Press, New York (1970)
11. Erol, S., et al.: Combining BPM and social software: contradiction or chance? J. Softw. Maint. Evol. Res. Prac. **22**(6–7), 449–476 (2010)
12. Pflanzl, N., Vossen, G.: Human-oriented challenges of social BPM: an overview. In: 5th EMISA Workshop, St. Gallen (2013)
13. Araujo, R., Taher, Y.: Refining IT requirements for government-citizen co-participation support in public service design and delivery. In: Conference for E-Democracy and Open Government, Krems, Austria, pp. 61–72 (2014)
14. Grönlund, Å.: ICT is not participation is not democracy – eParticipation development models revisited. In: Macintosh, A., Tambouris, E. (eds.) ePart 2009. LNCS, vol. 5694, pp. 12–23. Springer, Heidelberg (2009). doi:10.1007/978-3-642-03781-8_2
15. Araujo, R., Cappelli, C., Engiel, P.: Raising citizen-government communication with business process models. In: Dolićanin, Ć., Kajan, E., Randjelović, D., Stojanović, B. (eds.) Handbook of Research on Democratic Strategies and Citizen-Centered E-Government Services, 1st edn., pp. 92–106. IGI Global, Hershey (2014)
16. Diirr, B., Araujo, R., Cappelli, C.: Encouraging society participation through conversations about public service processes. Int. J. Electron. Gov. Res. **10**(2), 22–42 (2014)
17. Brazil: Brazilian Strategy for Digital Governance in Public Administration (2016). http://www.governoeletronico.gov.br/estrategia-de-governanca-digital-egd/. Accessed 27 May 2016
18. Brazil: Open Government Partnership (2016). http://www.governoaberto.cgu.gov.br. Accessed 01 Apr 2016
19. CGI.BR: ICT Electronic Government 2013: Survey on the Use of Information and Communication Technologies in the Brazilian Public Sector (2013). http://www.cgi.br/media/docs/publicacoes/2/TIC_eGOV_2013_LIVRO_ELETRONICO.pdf. Accessed 27 Nov 2015

20. Moreno-Ger, P., Burgos, D., Martínez-Ortiz, I., Sierra, J.L., Fernández-Manjón, B.: Educational game design for online education. Comput. Hum. Behav. **24**, 2530–2540 (2008)
21. Gee, J.P.: What Video Games have to Teach us About Learning and Literacy, 1st edn. Palgrave Macmillan Ltd., New York (2007)
22. Ribeiro, C., Fernandes, J., Lourenço, A., Borbinha, J., Pereira, J.: Using serious games to teach business process modeling and simulation. In: International Conference on Modeling, Simulation and Visualization Methods, Las Vegas, USA (2012)
23. Bulander, R.: A conceptual framework of serious games for higher education: conceptual framework of the game INNOV8 to train students in business process modelling. In: International Conference on e-Business, Athens, Greece, pp. 1–6 (2010)
24. Santorum, M.: A serious game based method for business process panagement. In: International Conference on Research Challenges in IS, Gosier, France, pp. 1–12 (2011)
25. Solís-Martínez, J., Espada, J.P., García-Menéndez, N., G-Bustelo, B.C.P., Lovelle, J.M.C.: VGPM: using business process modeling for videogame modeling and code generation in multiple platforms. Comput. Stan. Interfaces **42**, 42–52 (2015)
26. Poplin, A.: Games and serious games in urban planning: study cases. In: Murgante, B., Gervasi, O., Iglesias, A., Taniar, D., Apduhan, B.O. (eds.) ICCSA 2011. LNCS, vol. 6783, pp. 1–14. Springer, Heidelberg (2011). doi:10.1007/978-3-642-21887-3_1
27. Ahmed, A.M., Mehdi, Q.H., Moreton, R.: Towards the use of serious games for effective e-government service. In: Computer Games: AI, Animation, Mobile, Multimedia, Educational and Serious Games, Louisville, KY, pp. 1–6 (2014)
28. Lounis, S., Pramatari, K., Theotokis, A.: Gamification is all about Fun: the Role of Incentive Type and Community Collaboration. In: 22nd European Conference on Information Systems, Tel Aviv, Israel (2014)
29. Horita, F.E.A., Assis, L., Castanhari, R.E.S., Isotani, S., Cruz, W.M., de Albuquerque, J.P.: A gamification-based social collaborative architecture to increase resilience against natural disasters. In: Simpósio Brasileiro de Sistemas de Informação, Londrina, Brazil (2014)
30. Hunicke, R., Leblanc, M., Zubek, R.: MDA: a formal approach to game design and game research. In: AAAI Workshop on Challenges in Game Artificial Intelligence, San Jose, CA, pp. 1–5 (2004)
31. Johnson, C.: What Games can Learn from the Engagement Layers of Papers, Please.   http://www.gamasutra.com/blogs/CoryJohnson/20141210/232045/ What_Games_Can_Learn_from_the_Engagement_Layers_of_Papers_Please.php. Accessed 13 Nov 2015
32. Fasce, F.: Beyond serious games: the next generation of cultural artifacts. In: De Gloria, A. (ed.) GALA 2014. LNCS, vol. 9221, pp. 1–4. Springer, Cham (2015). doi:10.1007/978-3-319-22960-7_1
33. Sweetser, P., Wyeth, P.: GameFlow: a model for evaluating player enjoyment in games. ACM Comput. Entertain. **3**(3), 1–24 (2005)
34. Pavlas, D., Jentsch, F., Salas, E., Fiore, S.M., Sims, V.: The play experience scale. Hum. Factors **54**(2), 214–225 (2012)
35. Mitgutsch, K., Alvarado, N.: Purposeful by design? a serious game design assessment framework. In: Foundations of Digital Games, pp. 121–128 (2012)
36. Bandura, A.: Guide for constructing self-efficacy scales. In: Self-Efficacy Beliefs of Adolescents. Information Age Publishing, pp. 307–337 (2006)
37. Ajzen, I., Fishbein, M.: Understanding Attitudes and Predicting Social Behavior. Revised edn. Prentice Hall, Englewood Cliffs (1980)

.

4th International Workshop on Decision Mining and Modeling for Business Processes (DeMiMoP'16)

Introduction to the 4th International Workshop on Decision Mining and Modeling for Business Processes (DeMiMoP'16)

Jan Vanthienen[1], Hajo A. Reijers[2], and Claudio Di Ciccio[3]

[1] Department of Decision Sciences and Information Management, KU Leuven,
Naamsestraat 69, 3000 Leuven, Belgium
jan.vanthienen@kuleuven.be
[2] Department of Computer Science, VU University Amsterdam,
De Boelelaan 1081, 1081 HV Amsterdam, The Netherlands
h.a.reijers@vu.nl
[3] Institute for Information Business, WU Vienna, Welthandelsplatz 1,
1020 Vienna, Austria
claudio.di.ciccio@wu.ac.at

Abstract. The decision process is conceptually different than the decision structure: A specific process is indeed only one possible way to model and implement a decision, including the verification of requirements, dependencies, goals, data sources, etc. There may be more possible process models and implementations for a specific decision. And the same decision could be used in multiple processes. The purpose of the workshop was thus a.o. to examine the relationship between decisions and processes, to enhance decision mining based on process data (e.g. event logs), and to find a good integration between decision modeling and process modeling. Out of the high-quality submitted manuscripts, two papers were accepted for publication, with an acceptance rate of 40%. They contributed to foster a fruitful discussion among the participants about the respective impact and the interplay of the decision perspective and the process perspective.

Keywords: Decision models · Business process models · Decision mining

1 Aims and scope

Business Process Management (BPM) and its life cycle activities – design, modeling, execution, monitoring and optimization of business processes – have become a crucial part in business management. Most processes and business process models incorporate decisions of some kind. Decisions are typically based upon a number of business (decision) rules that describe the premises and possible outcomes of a specific situation. Typical decisions are: creditworthiness of the customer in a financial process, claim acceptance in an insurance process, eligibility decisions in social security, etc. Since these decisions guide the activities and workflows of all process stakeholders (participants, owners), they should be regarded as first-class citizens in Business Process Management. The raising relevance of the role bestowed to business rules is also reflected in the recent growth of interest towards declarative process modeling, namely the rule-based behavioral representation of workflows.

Business decisions are important, but are often hidden in process flows, process activities or merely in the head of employees (tacit knowledge). As a consequence, they need to be discovered using state-of-art intelligent techniques. Decisions can be straightforward, based on a number of simple rules, or can be the result of complex analytics (decision mining). In a number of cases, a particular business process does not just contain decisions, but the entire process is about making a decision. The major purpose of a loan process or an insurance claim process, e.g., is to prepare and make a final decision. The process shows different steps, models the communication between parties, records the decision and returns the result.

It is not considered a good practice to model the detailed decision paths in the business process model. Separating rules and decisions from the process simplifies the process model (separation of concerns). Through this workshop we tried to extend the reach of the BPM audience towards the decisions and rules community and increase the integration between different modeling perspectives.

The presented papers emphasized the role of decisions in the discovery of processes, on the one hand, and on their enhancement, on the other hand. De Smedt and his colleagues discussed in their work entitled "Decision Mining in a Broader Context: An Overview of the Current Landscape and Future Directions" the interplay of the decision perspective and the behavioral perspective under the control-flow and data model dimensions during the discovery phase. They have thus categorized the existing approaches and shown that a gap exists in current literature, owing to which they delineate a research agenda. The work of Pérez-Álvarez and his colleagues entitled "Governance Knowledge Management and Decision Support using Fuzzy Governance Maps" focused on the impact that decisions, hence the consequent actions, have on the Key Process Indicators (KPIs), by proposing a formal method that can represent the relation between the KPIs and the processes.

2   Workshop Co-organizers

Jan Vanthienen   KU Leuven, Belgium
Hajo A. Reijers   VU University Amsterdam, The Netherlands
Claudio Di Ciccio   WU Vienna, Austria

3   Program Committee

Bart Baesens   KU Leuven, Belgium
Fernanda Baião   Universidade Federal do Estado do Rio de Janeiro,
                            Brazil

Josep Carmona   Universitat Politècnica de Catalunya, Spain
Guoqing Chen   Tsinghua University, Beijing, China
Johannes De Smedt   KU Leuven, Belgium
Jochen De Weerdt   KU Leuven, Belgium
Robert Golan   DBmind Technologies Inc., USA
Xunhua Guo   Tsinghua University, Beijing, China
Jae-Yoon Jung   Kyung Hee University, South Korea
Dimitris Karagiannis   University of Vienna, Austria
Ricardo M.F. Lima   Federal University of Pernambuco, Brazil
Pericles Loucopoulos   Loughborough University, UK
Jorge Munoz-Gama   Pontificia Universidad Católica de Chile, Chile
Lucinéia H. Thom   Universidade Federal do Rio Grande do Sul, Brazil
Seppe K.L.M.   KU Leuven, Belgium
   vanden Broucke
Richard Weber   Universidad de Chile, Chile
Qiang Wei   Tsinghua University, Beijing, China
Mathias Weske   Hasso Plattner Institute, University of Potsdam,
                         Germany

Decision Mining in a Broader Context: An Overview of the Current Landscape and Future Directions

Johannes De Smedt[1]([✉]), Seppe K.L.M. vanden Broucke[1], Josue Obregon[2], Aekyung Kim[2], Jae-Yoon Jung[2], and Jan Vanthienen[1]

[1] Department of Decision Sciences and Information Management, Faculty of Economics and Business, KU Leuven, Naamsestraat 69, 3000 Leuven, Belgium
{johannes.desmedt,seppe.vandenbroucke,jan.vanthienen}@kuleuven.be
[2] Department of Industrial and Management Systems Engineering, Kyung Hee University, 1 Seocheng-dong, Giheung-gu, Yongin, Gyeonggi, Republic of Korea
{jobregon,akim1007,jyjung}@khu.ac.kr

Abstract. The term *Decision Mining* has been put forward in literature to cover numerous applications in a diverse set of contexts. In the business process management community, it typically reflects the way processes and data required for decision purposes in those processes are blended into one model during discovery. However, the upcoming field of decision modeling and management requires the term to be repositioned in order to obtain a better understanding of the interplay of processes and decisions. In this paper, the different approaches that are currently available are delineated and a case is made for a new type of decision mining: one that separates the control flow and decision perspective in a less stringent form compared to existing approaches.

Keywords: Decision mining · Decision management · DMN

1 Context and Problem Statement

In recent years, decision mining has become a prevalent term in the area of business process management. It typically refers to the act of not only deriving the control flow perspective from data, the most dominating stream within process discovery [1], but also retrieving the decisions that are present in the process. Most notably, the term was introduced in the seminal work of [2], a work dedicated to deriving and describing so-called decision points in Petri nets. The approach uses a decision tree algorithm to retrieve the possible outcomes in a split node, hence explaining why a certain route in a process is taken. Since then, many variations on the decision point analysis variant of decision mining have been proposed, mostly focusing on refining the way to retrieve the decision information [3–5]. Another work paying special attention to the notion of decision mining is [6], a work dedicated to an approach that discovers the mental actions performed by decision makers, captured in a Product Data Model (PDM) [7].

© Springer International Publishing AG 2017
M. Dumas and M. Fantinato (Eds.): BPM 2016 Workshops, LNBIP 281, pp. 197–207, 2017.
DOI: 10.1007/978-3-319-58457-7_15

On the other hand, many data mining techniques exist that focus on revealing what the outcome of a decision was by making use of classification, regression, clustering, and so on [8], in order to see what data was of importance while making the decision. In very few cases, however, a dynamic approach is taken which reveals the process of making the decision and the multiple stages that might have occurred in the decision making (either in terms of parts of the input data required in the different steps or in terms of certain outputs being used as an input in a subsequent step).

There are many reasons to perform decision mining. In industry, many cases report convoluted process models that try to incorporate decisions in processes in a hard-coded and fixed form [9]. In this case, process constructs in the form of splits and joins are abused to represent typical decision artifacts such as decision tables [10]. Not only do they breach the separation of concerns between data and control flow, they also impede maintenance and reusability.

Decision mining can uncover these bad specifications, as well as resolve them by better representing the interplay between decisions and the dynamics of their inputs over control flow. Its main purpose should be aligned with the upcoming integrated approaches in business process modeling in general. A testimony of the usefulness of such an approach is the proposal of the new Decision Model and Notation (DMN) standard [11,12] as well as the clear intention to use it in the context of business process models in the form of Business Process Model and Notation (BPMN) [13]. Furthermore, other hybrid approaches surfaced, such as the Product Data Model (PDM).

The goal of this research piece is to find a framework to assess the definition of decision mining techniques by proposing a quadrant that distinguishes along the control flow and data dimensions. Furthermore, it offers a research agenda as by interpreting the framework, it can be observed that there is still a gap in the way a holistic decision model representation is combined with a holistic process model representation. It is demonstrated how to elaborate further on this quadrant by means of an example.

The paper is organized as follows. First, Sect. 2 introduces the concepts of a decision log and integrated decision-aware process model. Next, in Sect. 3 the different mining approaches are framed and defined in order to finally establish the notion of decision mining. Finally, Sect. 4 provides a research agenda for decision mining, followed by the conclusion in Sect. 5.

2   Preliminaries

In order to get a grasp of the situations in which a decision mining approach is feasible and desirable, the notion of a *decision log* is introduced, as well as the primordial way of representing a business process model conjoined with a decision model.

2.1   Decision Log

Typical inputs for data mining exist of a collection of tuples organized around one unique, identifying attribute for that tuple, as can be seen from Table 1.

Table 1. A typical data mining log

id	attr1	attr2	attr3	label
1	x1	y6	z3	l1
2	x2	y5	z2	l1
3	x1	y5	z54	l1
4	x1	y4	z7	l2
5	x2	y3	z9	l2

Depending on the problem that is tackled, some extra requirements can be posed on what data needs to be available. For supervised learning techniques, i.e. classification and regression, an extra label variable needs to be present per tuple, where the latter requires an attribute to be defined over a continuous domain. For unsupervised learning techniques such as clustering, this label is not required.

For process discovery applications, there still is the need for an instance identifier, though every instance can consist of multiple items called events or activity instances. These are marked either with a time stamp, or ordered, hence achieving a temporal aspect. An example can be found in Table 2. The instance identifier is typically referred to as the case id, with every execution of an activity being listed as an event. Besides this, often the originator or resource performing the action is included. Hence, the main differences between an event log versus a traditional, "flat" data set is the presence of grouping (where the case identifier groups tuples) and ordering (where the tuples in groups are strictly ordered). In recent years, the Extensible Event Stream (XES) standard has emerged as the main storage format for event logs; a detailed overview of the XES standard is reported in [14].

Table 2. A typical process mining log

Case id	Time	Event	Resource
1	15:20	A	r1
1	15:21	A	r1
1	16:45	B	r2
2	14:01	A	r1
2	14:58	B	r3
2	15:02	C	r2
3	9:43	A	r2
3	23:19	C	r1

A more elaborate version of an event-based log can be found in Table 3. While this type of data log is not a far stretch from a typical so-called event log, it does incorporate a non-normalized version that pertains to different aspects of the information system it originates from. Attributes can be linked not only to an event, but can also just exhibit extra information on, e.g., a particular resource. In the example, there is no difference in terms of the event that is executed, except for a sole change in *attr1* which reflects a decision regarding the resource that is used.

Table 3. A more elaborate event-based log containing extensive instance data with a particular relation between the resource and *attr1*.

Case id	Time	Event	attr
1	15:20	A	{{res=r1;attr1=x1};attr2=y6;attr3=z3}
1	15:21	A	{{res=r1;attr1=x2};attr2=y6;attr3=z3}
1	16:45	B	{{res=r2;attr1=x1};attr2=y5;attr3=z54}
2	14:01	A	{{res=r1;attr1=x2};attr2=y4;attr3=z7}
2	14:58	B	{{res=r3;attr1=x2};attr2=y3;attr3=z9}
2	15:02	C	{{res=r2;attr1=x1};attr2=y6;attr3=z54}
3	9:43	A	{{res=r2;attr1=x1};attr2=y5;attr3=z2}
3	23:19	C	{{res=r3;attr1=x2};attr2=y4;attr3=z7}

This type of data leaves the aspect of the event-first orientation and establishes the need for a so-called *decision log*.

2.2   BPMN and DMN Integration

In this paper, we choose to elaborate examples in the form of Business Process Model and Notation and Decision Model and Notation models. The former has been considered the industry standard for representing business process models. In its simplest form, it connects activities (represented as rectangles) through arcs and gateways (ranging from (X)OR/AND splits and joins) between begin and end points, typically represented as colored circles. It features numerous extensions for timing, exception handling, data, and resource/collaboration concepts. The latter is a newly proposed notation for modeling repeatable business decision processes. It represents decisions as rectangles and decision inputs, which can be reused over the entire model, as ovals. It was designed to integrate with BPMN seamlessly and can serve as its decision-oriented counterpart. For more details regarding modeling integrated models, we refer to [15].

3   The Decision Mining Quadrant

In Fig. 1, the framework with which decision mining will be defined is depicted. It covers two dimensions, the control flow and data model dimensions. First, they will be elaborated along these dimensions, next, the necessary input types and existing techniques are reviewed.

3.1   Dimensions

Decision Control Flow. Along the vertical dimension, the presence of an elaborate decision flow driving the decision making is displayed. On the one hand, data mining techniques are typically not aware of any dynamic aspects of the data, reflected in them being single-stage, or can exhibit very basic dependencies that are based on the semantics of its data structure, i.e., a composite data type

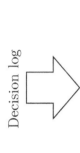

Decision log

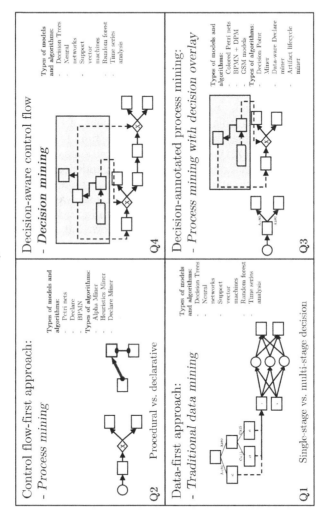

Fig. 1. The decision mining quadrant.

is used from different tables, or a previous data mining outcome is used as an input variable for a next mining exercise, making the approach multi-stage. A mix of a decision tree inserting its result into a neural network (two types of classification models) is shown in this quadrant.

Process mining and most notably the discovery pillar, on the other hand, is defined in a vast body of work on deriving a control flow of activities by fitting process models such as Petri nets [1] and BPMN [16], or declarative languages such as Declare [17,23]. On the left hand side of the quadrant, the techniques that do not include the resource, performance, or general data perspective, are out of scope and the decision is assumed to be captured within the process implicitly.

Decision Model Maturity. Along the horizontal dimension, the use of a decision model is displayed. On the left hand side, no such model is present, while on the right hand side there is. The main benefit of using such a model is the ability to structure all necessary inputs over the full life cycle of a decision. While there is little benefit for this in, e.g., a fixed decision as made by a single predictive model, it becomes crucial when many dependencies exist between decision outcomes in a dynamic artifact such as a process model.

When coupling this dimension to the first one, we see that two quadrants to the right are different in the fact that in the lower hand corner, the decision model is not connected to the whole work flow, but rather exists in parts of a control flow. As such, no holistic view of the decision is retrieved, but rather different parts of the decision model. In the right upper hand corner, the decision model is used throughout the global model and covers the full range of decisions made in the model, where some decision inputs can be reused which causes a need for keeping them consistent [15].

3.2   Relation to Data Input and Existing Techniques

The inputs for the mining approaches in Q1 and Q2 map directly onto the data logs represented in Tables 1 and 2 respectively. They either focus solely on data attributes of single instances, or on the sequential and concurrency aspect of the instances. Other approaches, i.e. [6,18,19] exist, which first mine the decisions and represent them as control flow. They rather belong to the second quadrant since they base themselves on data to find sequences to order the decision inputs. They are typically represented in either BPMN or Petri net models which are derived from PDMs. It can be argued that they fit both Q1 and Q2 at the same time, however, they nevertheless only represent the decision steps rather than the logic pertaining to making the decisions.

The approaches that can be categorized in Q3 and Q4 do actually go beyond these simpler forms of data logs and use a mix of event-based and instance-based data. The difference can be found in the way by which aspect is prioritized. On the one hand, approaches in Q3 use the control flow data to determine the overall structure of the process first, and consequently use the instance attributes to define where the data had an impact on the work flow. Indeed, this was the original setup of [2] and has later been extended in works such as [3,4,24], and [20,25].

Particularly the latter comes close to an integrated decision mining approach, but still considers fixed decision points in a log, where a fully integrated approach could consider decision (inputs) to be used throughout the model. On the other hand, approaches that favor the data perspective as the primary dimension were proposed as well and include works such as [5, 21, 22]. While the former two approaches spend a great effort in finding the most useful data variables combined with control flow variables in a log for a specific purpose, the latter constructs artifacts based on correlating the data of the events. The control flow over these artifacts is also mined and the whole is represented in a holistic model consisting of both layers. Hence, in our opinion, this is the only approach currently documented that truly belongs in Q4, as it creates an overview that both addresses the complexity of the data and the dynamic behavior of its activity generators. However, it can be argued to which extent decisions are really captured within the model, as the generation of artifacts and their life cycle rather implicitly capture the way decisions are made,

3.3   Decision Mining: The Missing Link

As illustrated before, the fourth quadrant of the definition framework for mining approaches has not been considered in numerous works. Hence, we argue that the current body of work exhibits a gap, where the current tendency is to use event-geared data as the starting point for extracting a control flow based model, which imposes the initial structure on top of which the decision perspective is placed upon, based on any additional attributes that might be available in the data set at hand. We wish to highlight herein the existence of an alternative approach, which favors the data perspective as the primary dimension, driving the discovery of a decision model, which is then enhanced with sequence based constructs mined from the data set (based on, e.g., the semantics of the data) or linked to a process model, as the fourth quadrant in our overview describes.

4   Decision Mining: A Research Agenda

The previous section has outlined a framework for classifying process, data, and decision mining approaches. Now, a research agenda is established based on the positioning of the current body of work that is available. This is done by focusing on three aspects. First of all, considerations and opportunities regarding *input data* is considered. Next, the need for more extensive support with regards to the *available approaches* is discussed. Finally, it is explained how new and better *techniques* can be constructed.

4.1   Decision Input Data: Finding the Desired Fit

Obtaining a useful discovery result from decision mining naturally depends on the input that is used for the available mining techniques. Hence, it is important to ensure a fit between the input data that is available, the results that are envisioned, as well as the technique that is used.

In order to find this fit, stakeholder interest regarding the outcome has to be clarified, as well as matched with the availability of data. E.g., process mining in a very static environment might not provide this fit, however, in case a dynamic approach is aimed for due to the business context of the problem, decision mining might provide solace as it can blend decision models with the little dynamic behavior that is present in the information system as illustrated in Fig. 2. Clearly, a lower bound is present due to the relation that exists between which approach is used and the data that is needed to perform this approach, but there also exists an upper bound in which analysis is desired. If a simple classification problem can be tackled with single case, non-temporal data, then the required quadrant one (Q1) is best, regardless of the availability of a full decision log.

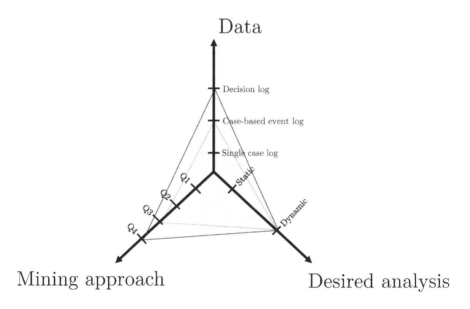

Fig. 2. Positioning data, mining approach, and desired analysis.

Future research should adopt this viewpoint and find new ways to balance these dimensions towards approaches that are clear and especially useful for the problem that is addressed. By distinguishing each new approach along these dimensions, it also becomes more straightforward to find and compare different approaches for similar use cases.

4.2 Decision Mining Support: Framework, Standards, and Adoption

We have shown that there currently exists a multitude of approaches trying to integrate control flow discovery with decision related aspects by considering the derivation of decision points as a second step following the discovery of a control flow structure. This observation, together with the upcoming popularity of DMN indicates the need for a more comprehensive framework and set of approaches to support decision mining in a more thorough manner. This aspect is heavily related to that of data requirements as indicated above, but is also directly depending on the particular use case that is to be explored. For instance, the current body of work does not consider the discovery of decision models based on decision logs only containing information on the ordering of decisions, their input attributes and output values, i.e. not containing traces or a full ordering. Instead of using control flow data to determine the overall structure of the process first, and consequently using the instance attributes to define where and how data had an impact on the work flow, one might consider the inverse approach of using decision data first to determine the overall logic of the decision, and then use sequence information to determine where and how sub-steps in the decision logic where made in terms of control flow. The fact that the latter view has been under-explored in the current body of work emphasizes the need for more encompassing approaches.

4.3 Decision Mining Techniques: A Need for Tailored Decision Mining Research

In case the ultimate goal of obtaining a full decision log is attained, there exist many opportunities for decision mining to evolve even further still. In general, it has been shown while positioning the current body of work along the dimensions of the quadrant that only a minor part of current literature actually conforms with the definition of decision mining.

Furthermore, it remains to be seen how future works will approach the theme; for now, most of the decision mining-capable approaches grew from process mining literature and verge towards incorporating data aspects into control flow-oriented approaches in order to alleviate the focus on finding process models based on routing constructs and use more appropriate ways to reflect the decision-making process that drives the control flow.

First of all, it is a challenge for researchers to further develop a decision mining approach that is driven by the construction of a decision model, rather than by the control flow containing decision points. The envelopment of a business process by its respective decision environment is a very tedious part, and can be tackled by forming, e.g., partial orders concerning decision inputs and outputs. Next, linking different techniques and decision-making types further constructs a skeleton of how the process was established. The latter is an often-overlooked or disregarded attribute of decisions and their actions, since many control flow-based discovery techniques lack an in-depth analysis of the action that is performed by the activities in general.

Secondly, many data mining techniques, although focusing on temporal aspects, often neglect the decision process as being the driver of the outcome that is analyzed. Not only does it yield more insights to know which attributes are important to a decision, it is also paramount to understand the way these attributes evolved and contributed to the general process of decision-making. In this respect, many data mining techniques can be enriched by incorporating decision models, as well as process mining techniques to not only clarify a model of attributes, but a model of processing attributes.

5 Conclusion and Future Work

This work has revised the term decision mining in a process mining context and offered a quadrant to frame approaches that exist within the domain. This allows researchers to better target the problems and opportunities that still exist in the area, and offers a research agenda for extending the sparsely-covered topic of integrated process and decision mining where both parts play an equal role and create the most holistic view out of decision logs possible.

For future work, we envision to elaborate the roadmap discussed in Sect. 4. One major topic that is currently being investigated deals with tackling decision logs both with data mining and process mining techniques to address different problems such as description, classification, and prediction at once, summarized in one decision model.

Acknowledgments. This work has been partially supported by funds from the the Flemish Fund for Science (grant FWO VS.010.14N) and from the National Research Foundation of Korea (NRF) grant (No. 2013R1A2A2A03014718).

References

1. van der Aalst, W.M.P., Weijters, T., Maruster, L.: Workflow mining: Discovering process models from event logs. IEEE Trans. Knowl. Data Eng. **16**(9), 1128–1142 (2004)
2. Rozinat, A., Aalst, W.M.P.: Decision mining in ProM. In: Dustdar, S., Fiadeiro, J.L., Sheth, A.P. (eds.) BPM 2006. LNCS, vol. 4102, pp. 420–425. Springer, Heidelberg (2006). doi:10.1007/11841760_33
3. Leoni, M., Dumas, M., García-Bañuelos, L.: Discovering branching conditions from business process execution logs. In: Cortellessa, V., Varró, D. (eds.) FASE 2013. LNCS, vol. 7793, pp. 114–129. Springer, Heidelberg (2013). doi:10.1007/978-3-642-37057-1_9
4. Mannhardt, F., Leoni, M., Reijers, H.A., Aalst, W.M.P.: Decision mining revisited - discovering overlapping rules. In: Nurcan, S., Soffer, P., Bajec, M., Eder, J. (eds.) CAiSE 2016. LNCS, vol. 9694, pp. 377–392. Springer, Cham (2016). doi:10.1007/978-3-319-39696-5_23
5. Kim, A., Obregon, J., Jung, J.-Y.: Constructing decision trees from process logs for performer recommendation. In: Lohmann, N., Song, M., Wohed, P. (eds.) BPM 2013. LNBIP, vol. 171, pp. 224–236. Springer, Cham (2014). doi:10.1007/978-3-319-06257-0_18
6. Petrusel, R., Vanderfeesten, I., Dolean, C.C., Mican, D.: Making decision process knowledge explicit using the decision data model. In: Abramowicz, W. (ed.) BIS 2011. LNBIP, vol. 87, pp. 172–184. Springer, Heidelberg (2011). doi:10.1007/978-3-642-21863-7_15
7. Vanderfeesten, I., Reijers, H.A., Aalst, W.M.P.: Product based workflow support: dynamic workflow execution. In: Bellahsène, Z., Léonard, M. (eds.) CAiSE 2008. LNCS, vol. 5074, pp. 571–574. Springer, Heidelberg (2008). doi:10.1007/978-3-540-69534-9_42
8. Fayyad, U.M., Piatetsky-Shapiro, G., Smyth, P.: From data mining to knowledge discovery in databases. AI Mag. **17**(3), 37–54 (1996)
9. Vanthienen, J., Caron, F., Smedt, J.D.: Business rules, decisions and processes: five reflections upon living apart together. In: Proceedings SIGBPS Workshop on Business Processes and Services (BPS 2013), pp. 76–81 (2013)

10. Vanthienen, J., Dries, E.: Illustration of a decision table tool for specifying and implementing knowledge based systems. Int. J. Artif. Intell. Tools **3**(2), 267–288 (1994)
11. OMG: Decision Model and Notation (2015)
12. Taylor, J., Fish, A., Vanthienen, J., Vincent, P.: Emerging standards in decision modeling (2013)
13. Chinosi, M., Trombetta, A.: BPMN: an introduction to the standard. Comput. Stan. Interfaces **34**(1), 124–134 (2012)
14. Verbeek, H.M.W., Buijs, J.C.A.M., Dongen, B.F., Aalst, W.M.P.: XES, XESame, and ProM 6. In: Soffer, P., Proper, E. (eds.) CAiSE Forum 2010. LNBIP, vol. 72, pp. 60–75. Springer, Heidelberg (2011). doi:10.1007/978-3-642-17722-4_5
15. Janssens, L., Bazhenova, E., Smedt, J.D., Vanthienen, J., Denecker, M.: Consistent integration of decision (DMN) and process (BPMN) models. In: CAiSE Forum, vol. 1612 of CEUR Workshop Proceedings, CEUR-WS.org, pp. 121–128 (2016)
16. Conforti, R., Dumas, M., García-Bañuelos, L., Rosa, M.: Beyond tasks and gateways: discovering BPMN models with subprocesses, boundary events and activity markers. In: Sadiq, S., Soffer, P., Völzer, H. (eds.) BPM 2014. LNCS, vol. 8659, pp. 101–117. Springer, Cham (2014). doi:10.1007/978-3-319-10172-9_7
17. Maggi, F.M., Mooij, A.J., van der Aalst, W.M.P.: User-guided discovery of declarative process models. In: CIDM, pp. 192–199. IEEE (2011)
18. Petrusel, R., Mican, D.: Mining decision activity logs. In: Abramowicz, W., Tolksdorf, R., Węcel, K. (eds.) BIS 2010. LNBIP, vol. 57, pp. 67–79. Springer, Heidelberg (2010). doi:10.1007/978-3-642-15402-7_12
19. Aa, H., Leopold, H., Batoulis, K., Weske, M., Reijers, H.A.: Integrated process and decision modeling for data-driven processes. In: Reichert, M., Reijers, H.A. (eds.) BPM 2015. LNBIP, vol. 256, pp. 405–417. Springer, Cham (2016). doi:10.1007/978-3-319-42887-1_33
20. Batoulis, K., Meyer, A., Bazhenova, E., Decker, G., Weske, M.: Extracting decision logic from process models. In: Zdravkovic, J., Kirikova, M., Johannesson, P. (eds.) CAiSE 2015. LNCS, vol. 9097, pp. 349–366. Springer, Cham (2015). doi:10.1007/978-3-319-19069-3_22
21. Leoni, M., Aalst, W.M.P., Dees, M.: A general framework for correlating business process characteristics. In: Sadiq, S., Soffer, P., Völzer, H. (eds.) BPM 2014. LNCS, vol. 8659, pp. 250–266. Springer, Cham (2014). doi:10.1007/978-3-319-10172-9_16
22. Popova, V., Fahland, D., Dumas, M.: Artifact lifecycle discovery. Int. J. Coop. Inf. Syst. **24**(1), 44 (2015)
23. Maggi, F.M., Dumas, M., García-Bañuelos, L., Montali, M.: Discovering data-aware declarative process models from event logs. In: Daniel, F., Wang, J., Weber, B. (eds.) BPM 2013. LNCS, vol. 8094, pp. 81–96. Springer, Heidelberg (2013). doi:10.1007/978-3-642-40176-3_8
24. de Leoni, M., van der Aalst, W.M.P.: Data-aware process mining: discovering decisions in processes using alignments. In: SAC, pp. 1454–1461. ACM (2013)
25. Bazhenova, E., Weske, M.: Deriving decision models from process models by enhanced decision mining. In: Reichert, M., Reijers, H.A. (eds.) BPM 2015. LNBIP, vol. 256, pp. 444–457. Springer, Cham (2016). doi:10.1007/978-3-319-42887-1_36

Governance Knowledge Management and Decision Support Using Fuzzy Governance Maps

José Miguel Pérez-Álvarez$^{(\boxtimes)}$, María Teresa Gómez-López,
Angel Jesus Varela-Vaca, Fco. Fernando de la Rosa Troyano,
and Rafael M. Gasca

Departmento de Lenguajes Y Sistemas Informáticos,
Universidad de Sevilla, Seville, Spain
{josemi,maytegomez,ajvarela,ffrosat,gasca}@us.es
http://www.idea.us.es

Abstract. Business process management systems incorporate the possibility of monitoring the behaviour of a company, by observing their business process indicators. Depending on the process executed, and the order of their performances, certain KPIs can be modified to render the company more competitive. This paper proposes the creation of a model-based fuzzy logic that can represent the relation between KPIs and the business processes of the companies. The use of this graph enables business experts to simulate the evolution of the business according to the decisions taken in the governance process, thereby helping in governance activities.

Keywords: Governance · Business process · Decisions · Fuzzy logic · Modelling knowledge

1 Introduction

The IT Governance Institute (2001) defines enterprise governance as the "set of responsibilities and practices exercised by the Board and Executive management Team (BET) with the goal of providing strategic direction, ensuring that objectives are achieved, ascertaining that risks are managed appropriately and verifying that the enterprise's resources are used responsibly" [1].

At management level, the BET must make decisions in order to maintain and follow the agreed business strategy, thus the right direction. The BET members that make the decisions at a specific moment must take into account the available knowledge concerning the current business processes. Performing the decision-making process implies analysing a great quantity of knowledge represented across a wide set of variables. The use of these variables in the decision-making process will improve the competitiveness of the company, which is determined by the correctness of the decided actions. The attainment of the proposed

© Springer International Publishing AG 2017
M. Dumas and M. Fantinato (Eds.): BPM 2016 Workshops, LNBIP 281, pp. 208–219, 2017.
DOI: 10.1007/978-3-319-58457-7_16

objectives in an organisation therefore implies three main activities: observing the current and heterogeneous information used and produced as process indicators in the processes of the companies; ascertaining the activities or processes that can be performed to improve the observed indicators; and making the best decision according to both aspects. Framed within this scenario, we propose a methodology to model business BET knowledge, by using fuzzy logic. Our proposal facilitates a mechanism to achieve the predicted business evolution after the execution of a set of actions. The obtained predictions help the BET make better reasoned decisions, since the team are aware of how these decisions will affect the business.

In process orientation, business processes are the main instrument for the organisation of the operations of an enterprise [2]. This implies that the overall organisation can be seen as a set of business processes, working together to achieve the objectives of the company. At organisation level, from the point of view of process orientation, lets the characterization of the operation of an enterprise using business processes [3]. Organizations can incorporate various types of business processes, and they are influenced by the business strategy that defines the objectives and goals of the organisation, but they are also influenced by the stakeholders and the information systems that support them.

Each business process can contribute towards achieving one or more business goals. In order to gain information about the business process efficiency according to the desired business goals, activities represented in controlling mechanism are performed, and KPIs of business processes are determined [3].

Certain variables can be part of the decision-making process, but others are affected by external actions in an indirect way. For example, a company can change the price of a product (variable directly determined in a decision-making process) but cannot determine the number of products sold (variable affected by the execution of other actions). However, a company normally has a set of mechanisms that can help: for example, when the price is decreased or an advertising campaign is deployed, more products will probably be sold. Some actions cannot modify these variables directly, but they can stimulate the KPIs in the desired direction.

Decision-making processes for directly determined variables have been studied previously [4]. However, to the best of our knowledge, the problem has not been extended to include variables affected by the execution of other actions.

The degree to which a business process directly or inversely affects a set of KPIs forms part of the expert knowledge of the business. Since the decision about which process should be executed is a human and manual task, the BET of the enterprise uses these indicators, typically shown on dashboards, to decide which actions to take to improve the KPIs in the future. The relation between the actions and how they can affect the variables is not always clear, since it depends on the background of the particular decision-maker whether to perform a determined action. Taking into account every item of information can be a complex task, for example if not every item of department information is included, then some profess-KPI relations can be lost. Errors can therefore be produced or decisions can fail to follow the strategy defined in the organisation.

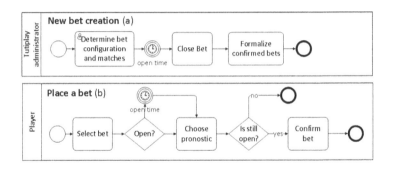

Fig. 1. Tutiplay business process for users

In order to model the indicator and action relations, in this paper we propose the creation of a model-based fuzzy logic graph that can represent the relation between KPIs and the professes of the companies, called Fuzzy Governance Maps (FGM). These FGMs help the BET in the decision-making processes, completed with a framework to simulate the various scenarios in a what-if analysis [5] supported by a tool.

The paper is organized as follows: Sect. 2 introduces an illustrative example in a real scenario; Sect. 3 shows the relation between business strategy and business processes, and how the governance helps the BET to maintain the correct direction; Sect. 4 explains the elements and structure of an FGM; Sect. 5 introduces how to use FGMs to evaluate a what-if analysis; Sect. 6 analyses an overview of related work found in the literature; and finally, conclusions are drawn and future work is proposed.

2   A Real-World Example

A real-world example is used to illustrate our proposal. It is a collaborative platform to play a football pool called Tutiplay [6]. This is a platform oriented towards allowing a set of people (usually friends) to place a betting ticket together. In each bet, each person fills in an independent row and permits the Tutiplay platform to collect every row together in one betting ticket, and formalizes the bet using the corresponding administration. In the case of economic reward, the platform also collects the winnings and divides the quantity between the participants. More than one bet can be opened for placing at the same time.

Figure 1 shows two business process models implemented to support the platform. The first model "New bet creation (a)" shows how a bet is managed by the person who administers the platform, from the creation to the close and final formalization of the bets. The second model "Place a bet (b)" shows the steps that a player must follow to place a specific bet.

The business objective of the platform is to formalize as many bets as possible in order to maximize profits, but also maximize the number of active users. To optimize these variables, Fig. 2 shows processes that allow the BET to perform certain strategies in order to ensure the proposed goal.

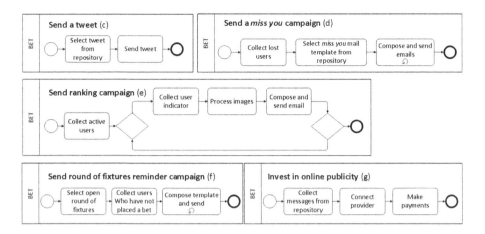

Fig. 2. Tutiplay business process for BET members

Table 1. BET strategy process

Business process	Consists of	Strategy aim
Send tweet (c)	Select a generic tweet from a repository and send it to make some noise and to connect players and followers	Increase the forecasts by making noise to followers
Execute a "miss you" campaign (d)	Send a emails to every lost user, inviting them to use the platform again	Decrease the number of lost users through the number of active users by increasing user reactivation
Execute a ranking campaign (e)	Send an individual email to every user, including indicators of the evolution of the player	Gain more bets and decrease the number of lost users
Execute a reminder campaign (f)	For a determinate round of fixtures, send an email to players that have not yet placed a bet, when the deadline is near.	Increase the number of bets for a determinate round of fixtures
Invest in online publicity (g)	Spend money on social networks, to enroll new users to the platform	Increase the number of users

Table 1 shows a small explanation regarding the business processes available for the BET and shown in Fig. 2, with the business aim that each one follows.

The correct direction of the company is based on the business strategy defined: The BET observes the evolution of the system using a dashboard, and when necessary or desired, they can decide to perform any action, that implies executing some process.

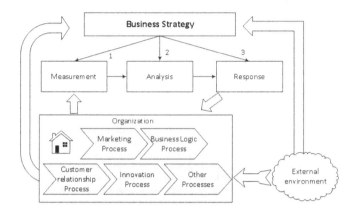

Fig. 3. Performance management process

The problem involves ascertaining which process or processes can improve the competitiveness of the company, and how they can affect the other KPIs. Our proposal includes obtaining this information by simulating the different options that the decision can produce.

3   Business Process and Business Strategy

The existing relation between certain types of processes and the capacity to modify the goals of an organisation was detected by Smith et al. [7], and depicted in Fig. 3. The alignment between the processes of an organisation, and the goals to be achieved implies performing three steps. The first (1) consists of taking measurements, which are taken from the KPIs observed from the processes defined as relevant for the organisation. (2) It is then necessary to make an analysis of these measurements in order to (3) perform possible actions that will affect the goals of the organisation. As mentioned earlier, measurement, analysis and response actions are oriented towards improving the business strategy defined, which is affected by the evolution and the status of the organisation itself, and by the external environment.

The principal aspects started above can be implemented and automated in an easy way by using BPMSs. BPMSs represent a software that supports the implementation, coordination, and monitoring of the business process execution.

The main aspect supported by BPMSs involves the handling of the business processes of the organisation. This aspect is represented in Fig. 3 in the box labelled as *Organization*. Furthermore, the *External Environment*, including the relationship with *Stakeholders*, manages input knowledge obtained from external information systems and other important sources.

In order to obtain measurements from the status of the business (edge 1 of Fig. 3), the Business Activity Monitoring (BAM) or Process Performance Measurement (PPM) tools are employed. These tools allow the expert to evaluate the defined KPIs that permit the status of the business to be ascertained at each moment. These tools require intervention from IT personnel in order to be automated. The visualization and monitoring of the status of the business by means of observations of the KPIs can easily be created through using the dashboards of these tools.

However, the following aspects cannot be automated: the first aspect is to define the business strategy or the specific KPIs to measure, since this is the responsibility of the BET and it depends on the strategies that the organisation wants to follow. This step is guided by the target markets that the company wants to cover, and the product and services offered.

Once the BET obtains the status of the business by evaluating the KPIs that can be observed on a dashboard (edge 1 in Fig. 3), the team must decide whether the status of the business is correct based on the business strategy defined (edge 2 in Fig. 3), and they must also decide whether to act (edge 3 in Fig. 3). A response can involve doing nothing, or performing a set of actions in order to archive the objectives defined as strategy. Here is where the contributions of this paper take place, by helping to model action-reaction knowledge in the process governance, and by contributing a method for the computation of this knowledge in order to make better reasoned decisions that steer the computing in the right direction to achieve its business goals.

4   Fuzzy Governance Maps (FGM)

In order to model the expert knowledge represented in Table 1, which is needed to help in the governance decision points, we propose the use of Fuzzy Governance Maps (FGMs). The use of FGMs contributes towards the effort for more intelligent governance control methods and for the development of systems that help in the governance decision process. FGM representation is a formal method that allows the BET to describe the expected behaviour of the organisation itself, and how the environment will evolve by means of the simultaneous use of stimulations of business processes and KPIs. This method is an extension of Fuzzy Cognitive Maps (FCMs) by Kosko [8], with a set of new elements for the expression of the complete semantics. As FCMs, the success of the construction of FGMs is strongly dependent on the degree of expertise held by those involved in the FCM construction [9].

A FGM is composed of $\langle IN, BPN, CE, SE \rangle$, Indicator Nodes (IN), Business Process Nodes (BPN), Causality Edges (CE), and Stimulation Edges (SE).

Indicator Nodes (IN) model the set of indicators that represents the status of the organisation: this is the set of KPIs typically included on a dashboard for the visualization of the status of the business, such as *"number of users"* or *"profits"*. On the other hand, *BPN* models the business processes or actions that a BET has available for execution, such as *"send a tweet"*, or *"invest in online publicity"* processes. Every *IN* is defined by using the name of the indicator.

The *INs* relate by means of the *Causality edges(CE)*. These edges can have a direct $(+)$ relationship in the case when they increase/decrease in the same

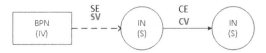

Fig. 4. Abstract FGM

direction; these edges have a indirect $(-)$ relationship when the second indicator increases/decreases in the opposite direction to the first. Another characteristic of *CE* is the velocity of causality, which represents the speed with which the second indicator is affected once the first one has changed.

It is frequently too complex to define the velocity of action-reaction between two values with accuracy, for this reason we propose the use of fuzzy logic, and we have defined five fuzzy sets, denoted as "very slow", "slow", "normal", "quick", and "very quick". Considering *IN* and *CE*, an *FGM* can be seen as a FCM considering time relationships [10].

On the other hand, *BPN* represents the set of actions in terms of business processes that can be executed by the BET.

As mentioned in Sect. 1, the business processes do not always have the capacity of modifying the indicator nodes directly, since the behaviour of these indicators depends on external factors, however, these processes can stimulate certain variables. In order to model this concept, FGMs are composed of *Stimulation Edges (SE)*. An *SE* relates a *BPN* with an *IN*, and represents that the *IN* is stimulated by the execution of a *BPN* process. In order to facilitate the modelling, four fuzzy sets have been defined, whose ranges are:

- *Greatly increase*: The execution of the business processes greatly increases the associated indicator
- *Increase*: The execution increases the associated indicator
- *Decrease*: The execution of the business process decreases the indicator
- *Greatly decrease*: The execution of the business process greatly decreases the indicator

On the other hand, the velocity at which the indicator is influenced can be modelled by using the same five fuzzy sets defined above.

Figure 4 shows how these elements are represented graphically. Formally, FGMs consist of a finite non-empty set of n *BPN* and m *IN* nodes, a finite non-empty set of o *SE* edges, and a finite set of p *CE* edges.

$BPN = \langle name, IV \rangle$ represents a business process node, and is composed of a name to represent the business process, and a set of q input variables (IV) that need to be instantiated.

$IN = \langle I, S \rangle$ represents an indicator node, and is composed of the name of the indicator (I), and a Scope (S), which represents where the indicator can be applied. The scope models the actuation ambit of *IN*, and therefore the indicator refers to a determinate ambit.

SE represents a stimulation edge between a business process node (BPN) and an indicator (I), $SE = \{se_{ij}\}$ where $i \in BPN, j \in IN$.

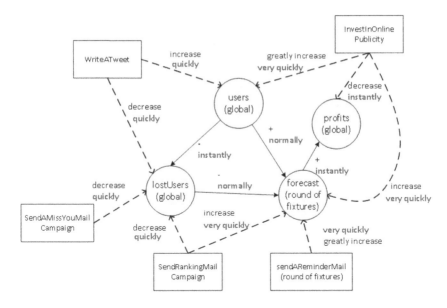

Fig. 5. FGM sample

$SE = \langle SF, SV \rangle$ is composed of a stimulation function (SF) and a stimulation velocity (SV), defined using discrete values $SV \in$ [VS, S, N, Q, VQ], which represent "very slowly", "slowly", "normally", "quickly" and "very quickly" stimulation velocities, respectively.

CE represents the set of causality relations between indicators. $CE = \{ce_{ij}\}$, $i, j \in IN$, is composed of a type of causality relation (C) and a causality velocity (CV): $CE = \langle C, CV \rangle$.

Figure 5 shows an example of an FGM for the sample shown in Sect. 2. This FGM has been designed by the BET of Tutiplay^{TM} and collects their beliefs about the operation of the enterprise.

The set of *INs* that defines the status of the enterprise for the example are:

- *users*: Number of users registered on the platform, with at least one bet placed within the last month
- *lostUsers*: Number of users registered on the platform, that have not placed a bet in the last month
- *forecast*: Number of bets per round of fixtures
- *profits*: Profits obtained by the platform

The BET considers that *forecast* and *profits* has a direct casual relation, since if the number of *forecast* is increased, the *profits* are instantly increased. They also consider the another situation, in the case where *forecast* is decreased, *profits* are instantly decreased. There is a direct casual relation between *users* and *forecast*, and an inverse casual relation between *lostUsers* and *forecast*, since in the case where *lostUsers* is increased, *forecast* is decreased (and vice versa). Finally, the

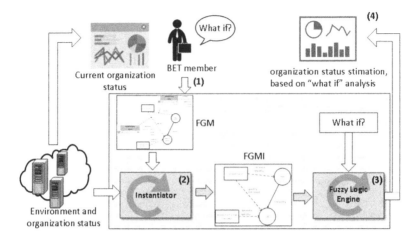

Fig. 6. Evaluation process

BET defines is an inverse casual relation between *users* and *lostUsers*, since in the case where *users* is increased, then *lostUsers* can decrease (and vice versa).

On the other hand, the processes that the BET can use to exert influence over the business, are described in Fig. 2. In this sample, the BET has modelled that the execution of the process "WriteATweet" can quickly stimulate the *users* and *lostUsers* indicator. The execution of the process "InvestInOnlinePublicity" stimulates the *users*, *profits* and *forecast* indicators. The process "SendAMissYouMailCampaign" stimulates the *lostUsers* indicator, "SendRankingMailCampaign" stimulates *lostUsers* and *forecast* indicators, and "sendAReminderMail" stimulates the indicator *forecast* for a specific round of fixtures. The degree to which these indicators are stimulated can be seen in Fig. 5.

5   Framework and Evaluation

This section describes the evaluation process of the FGM. The process is graphically shown in Fig. 6. In the case BET detects some unusual behaviour by observing the dashboard, they can decide either act in an effort to fix the problem, or they could simply remain informed as to the evolution of the organisation.

The evaluation process starts when a BET member needs to make a what-if analysis (1 of Fig. 6). The module "Instantiator" (2 of Fig. 6) explores the environment and organisational status by collecting indicators either defined by using the Process Instance Query Language (PIQL) [11], or from Business Activity Monitoring (BAM) [12] or Process Performance Measurement (PPM) [13] tools (external sources), and instances the FGM by calculating the final value for indicators and stimulation edges in order to create a Fuzzy Governance Map Instance (FGMI). An FGMI is an FGM, where the values of the *IN* are known.

Once the FGMI is obtained, it is used as the input of the Fuzzy Logic Engine module (3 of Fig. 6). This module takes the FGMI and activates the

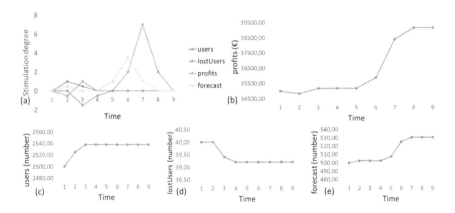

Fig. 7. Sample of estimated dashboard

actions according to the "what if" question. An FGMI is a Fuzzy cognitive map that must be instantiated in order to obtain a Fuzzy Governance Map Instance (FGMI). Once this is obtained, it can be computed by using fuzzy logic [14].

The proposal has been evaluated from an FGMI and by using simulation tools for the "Fuzzy logic engine" module. This FGMI has been mapped in the simulation tool and time relationships have been considered by introducing intermediate nodes [10]. The simulation tool has the capacity to modify the value of the nodes and propagate the results in order to obtain the degree of stimulation of each indicator. The specific evolution of each indicator can then be obtained by using these stimulation degrees and an application function.

Figure 7 shows an example of the estimated output of the dashboard obtained by using the FGM presented as a sample (Fig. 5) to simulate the evolution of the organisation from the status (*users=2500, lostUsers=40, profits=35000, forecast=[f:500]*) and the question *"what if we invest money in online publicity?"*.

Plot (a) of Fig. 7 shows the stimulation degree. The remaining plots describe the evolution of each indicator. FGM predicts that organisation will start with a loss of profits, but the action will greatly increase profits in the long-term.

6   Related Work

For many years, organisations have invested large quantities of time and money to ensure business process compliance (BPC) with policies, regulations, and legislation. A systematic selection and characterization of the literature that focuses on BPC was published in [15]. Other studies have expounded on how it is possible to utilize tactical information, knowledge and experience concerning business activities for the BPC. In previous work [16], the organisation goals are modelled using User Requirements Notation (URN) [17].

In order to react to changes in the BPC, the various techniques could be categorized in the bibliography:

1. Rule-based techniques [18] or Business Rules Management Systems (BRMS) to automate the modelling, deploying, and execution of the business rules.
2. Business Process Intelligence techniques (BPI) integrate BPMS and Business Intelligence systems [19]. Shollo [20] proposes applying "hard facts" provided by BI in the IT governance context, as a foundation for rendering arguments more convincing during decision-making discussions.
3. Goal-Oriented techniques, which extend URN to include the validation business processes by considering performance issues as compliance [21].

Many previous studies use hybrid techniques to improve the results, but from our point of view, techniques listed above provide insufficient support when dealing with the business goal models and business process execution in an integrated and efficient manner. For this reason, our research addresses these issues.

7   Conclusions and Future Work

This paper proposes a formal method to model expert knowledge through the use of Fuzzy Governance Maps (FGM). A FGM allows the Board and Executive Team (BET) to understand how the business works, and how actions can directly or indirectly affect the KPIs that define the status of the business. By computing the FGM defined by the BET, the evolution of the business according to the decisions taken in the governance process can be attained, tat is, we can ascertain what will probably happens on performing certain actions. If the evolution of the business is known according to which actions are performed, then decision-making regarding these actions becomes easier, and this help towards achieving the company's objectives.

As future work, we propose that information about regarding past instances be incorporated. By using this information, we will be able to validate the FGM designed by the BET and also to propose new stimulation relations. Furthermore, we consider the possibility of introducing dynamic stimulation edges, since the degree of stimulation sometimes depends on external factors.

Acknowledgement. This work has been partially funded by the Ministry of Science and Technology of Spain (TIN2015-63502-C3-2-R) and the European Regional Development Fund (ERDF/FEDER). Thanks to Lesley Burridge for the revision of the English version of the manuscript.

References

1. De Haes, S., Van Grembergen, W.: IT governance and its mechanisms. Inf. Syst. Control J. **1**, 27–33 (2004)
2. Hammer, M., Champy, J.: Reengineering the Corporation: A Manifesto for Business Revolution. Harper Business, New York (1993)
3. Weske, M.: Business Process Management: Concepts, Languages, Architectures. Springer, Secaucus (2007)

4. Gómez-López, M.T., Gasca, R.M., Pérez-Álvarez, J.M.: Decision-making support for the correctness of input data at runtime in business processes. Int. J. Coop. Info. Syst. **23**(2), 1–29 (2014)
5. Maio, C.D., Botti, A., Fenza, G., Loia, V., Tommasetti, A., Troisi, O., Vesci, M.: What-if analysis combining fuzzy cognitive map and structural equation modeling. In: Conference on Technologies and Applications of Artificial Intelligence, TAAI 2015, Tainan, Taiwan, 20–22 November 2015, pp. 89–96 (2015)
6. Tutiplay: Tutiplay (2015). http://www.tutiplay.com
7. Smith, P.C., Goddard, M.: Performance management and operational research: a marriage made in heaven? J. Oper. Res. Soc. **53**(3), 247–255 (2002)
8. Kosko, B.: Neural networks and fuzzy systems: a dynamical systems approach to machine intelligence/book and disk, vol. 1. Prentice Hall, Upper Saddle River (1992)
9. Tsadiras, A., Bassiliades, N.: Ruleml representation and simulation of fuzzy cognitive maps. Expert Syst. Appl. **40**(5), 1413–1426 (2013)
10. Park, K.S., Kim, S.H.: Fuzzy cognitive maps considering time relationships. Int. J. Hum. Comput. Stud. **42**(2), 157–168 (1995)
11. Pérez-Álvarez, J.M., Gómez-López, M.T., Parody, L., Gasca, R.M.: Process instance query language to include process performance indicators in DMN. In: 2016 IEEE 20th International Enterprise Distributed Object Computing Workshop (EDOCW), pp. 1–8, September 2016. doi:10.1109/EDOCW.2016.7584381
12. Bam, W.S.: Skelta business activity monitoring (BAM) and key performance indicators (kpis) for process monitoring and optimization (2015). http://www.skelta.com/products/BPM/features/BAM-KPI.aspx
13. González, O., Casallas, R., Deridder, D.: Monitoring and analysis concerns in workflow applications: from conceptual specifications to concrete implementations. Int. J. Coop. Inf. Syst. **20**(4), 371–404 (2011)
14. Margaritis, M., Stylios, C., Groumpos, P.: Fuzzy cognitive map software. In: 10th International Conference on Software, Telecommunications and Computer Networks SoftCom 2002, pp. 8–11 (2002)
15. Shamsaei, A., Amyot, D., Pourshahid, A.: A systematic review of compliance measurement based on goals and indicators. In: Salinesi, C., Pastor, O. (eds.) CAiSE 2011. LNBIP, vol. 83, pp. 228–237. Springer, Heidelberg (2011). doi:10.1007/978-3-642-22056-2_25
16. Ghanavati, S., Amyot, D., Peyton, L.: Towards a framework for tracking legal compliance in healthcare. In: Krogstie, J., Opdahl, A., Sindre, G. (eds.) CAiSE 2007. LNCS, vol. 4495, pp. 218–232. Springer, Heidelberg (2007). doi:10.1007/978-3-540-72988-4_16
17. ITU-T, I., Recommendation, Z.: 151 (11/08), user requirements notation (urn)-language definition. Geneva, Switzerland, approved November (2008)
18. Codd, E.F.: Derivability, redundancy and consistency of relations stored in large data banks. SIGMOD Rec. **38**(1), 17–36 (2009)
19. Grigori, D., Casati, F., Castellanos, M., Dayal, U., Sayal, M., Shan, M.C.: Business process intelligence. Comput. Ind. **53**(3), 321–343 (2004)
20. Shollo, A.: Using business intelligence in IT governance decision making. In: Nüttgens, M., Gadatsch, A., Kautz, K., Schirmer, I., Blinn, N. (eds.) TDIT 2011. IAICT, vol. 366, pp. 3–15. Springer, Heidelberg (2011). doi:10.1007/978-3-642-24148-2_1
21. Pourshahid, A., Amyot, D., Peyton, L., Ghanavati, S., Chen, P., Weiss, M., Forster, A.J.: Business process management with the user requirements notation. Electron. Commer. Res. **9**(4), 269–316 (2009)

2nd International Workshop on Process Engineering (IWPE 2016)

Introduction to the Second International Workshop on Process Engineering (IWPE 2016)

Mathias Weske[1(✉)] and Manfred Reichert[2]

[1] Hasso Plattner Institute, University of Potsdam, Prof.-Dr.-Helmertstr. 2-3, 14482 Potsdam, Germany
mathias.weske@hpi.de
[2] Ulm University, James-Franck-Ring, 89081 Ulm, Germany
manfred.reichert@uni-ulm.de

Motivation

After the success of the first issue of the International Workshop on Process Engineering, which was held in Innsbruck last year, this year's issue discusses novel engineering aspects related to the design and development of process-oriented information systems. While the specific topics discussed take novel development trends into account, such as microservices, the overall aim and motivation of the workshop series persists.

The workshop's motivation is based on the observation that business process management has been very successful in developing concepts, languages, algorithms, and techniques in different aspects of the domain. However, we have to admit that many of those concepts and techniques, have not yet found their way to the operational business of companies. We have identified the lack of research in engineering aspects of process-oriented information systems (process systems for short) being one of the main factors for this situation.

The goal of the workshop series is putting the engineering of process systems in the center of attention. Only if we focus on the entire value chain of system development, from the elicitation of business requirements to the engineering of suitable architectures, components and user interfaces of process systems, as well as their testing, deployment, and maintenance, we can strengthen industrial uptake.

Overview

The workshop is opened by a keynote "Case Management in the Age of Analytics and Data-driven Insights" by Boualem Benatallah from the University of New South Wales, Australia. Boualem connects two timely topics: Case management and data analytics. Doing so raises exciting opportunities, but also novel research challenges. In his abstract, Boualem characterizes his keynote as follows: "As economies undergo significant structural change, digital strategies and innovation must provide industries across the spectrum with tools to create a competitive edge and build more value into their services. With the advent of widely available data capture and management technologies, coupled with intensifying global competition, fluid business and social requirements, organizations are rapidly shifting to "data-fication" of their processes. Accordingly, they are embracing the radical changes necessary for increased productivity, added value and insights. However, while advances in big data analytics enable tremendous automation and scalability opportunities, new productivity and usability challenges have also emerged. A commonly overlooked limitation of current systems is that they do not provide effective integration of analytics and end user workspace environments (e.g., investigators, analysts user productivity tools). We discuss critical challenges in the effective integration of data-driven insights and end user-oriented case management technologies. We discuss synergies between micro-services, analytics, case management as step forward to empower end users to effectively use big data technologies, while share and collaborate on the fly, in order to generate and evolve insights."

The keynote presentation is followed by paper sessions. Heiko Beck, Marcin Hewelt, and Luise Pufahl from HPI at the University of Potsdam introduce in their paper "Extending Fragment-based Case Management with State Variables" a novel approach to case management, which is based on a dynamic combination of process fragments using state variables. This approach fits nicely to the keynote presentation, as it provides a sound framework for case management combining the traditional process modeling view with a novel data object view.

María Teresa Gómez-López, José Miguel Pérez-Álvarez, Angel Jesus Varela Vaca, and Rafael M. Gasca from the University of Seville look at process choreographies. Their paper "Guiding the creation of Choreographed Processes with Multiple Instances based on Data Models" investigates issues related to the multiple instances problem. While multiple instances of process activities are understood in the context of process orchestrations, this problem has not yet been addressed for process choreographies, i.e., interacting business processes. The authors show how choreographies can be derived from process orchestrations, data objects and conceptual data models.

In their paper "Redefining a Process Engine as a Microservice Platform", Antonio Manuel Gutierrez, Manuel Resinas, and Antonio Ruiz-Cortés, also from the University of Seville, investigate engineering challenges related to novel microservice architectures in process engine design. After introducing this novel engineering paradigm that involves technological, but also organizational aspects, this approach is followed by many organizations. The opportunities as well as the challenges related to microservices in the context of process engine design are highlighted.

The final presentation introduces the paper "Providing semantics to implement aspects in BPM" by Hércules José, Gonçalves Filipe, Claudia Cappelli, and Flavia Santoro from the Federal University of Rio de Janeiro. The authors introduce an approach that is based on crosscutting concerns to business process management. While aspects have been already discussed in the design of process languages, the engineering focus of the paper highlights novel implementation concepts and could lead to more precise and better maintainable process implementations based on web services.

Acknowledgments. The organizers express their gratitude to the members of the program committee for their detailed and timely reviews that were instrumental in compiling the program. Special thanks to Boualem Benatallah for his IWPE 2016 keynote presentation!

Potsdam and Ulm, September 2016

Mathias Weske
Manfred Reichert

Case Management in the Age of Analytics and Data-Driven Insights (Invited Talk)

Boualem Benatallah

Department of Computer Science and Engineering,
The University of New South Wales, Sydney, Australia
boualem@cse.unsw.edu.au
https://www.engineering.unsw.edu.au/computer-science-engineering/

Abstract. As economies undergo significant structural change, digital strategies and innovation must provide industries across the spectrum with tools to create a competitive edge and build more value into their services. With the advent of widely available data capture and management technologies, coupled with intensifying global competition, fluid business and social requirements, organizations are rapidly shifting to data-fication of their processes. Accordingly, they are embracing the radical changes necessary for increased productivity, added value and insights. However, while advances in big data analytics enable tremendous automation and scalability opportunities, new productivity and usability challenges have also emerged. A commonly overlooked limitation of current systems is that they do not provide effective integration of analytics and end user workspace environments (e.g., investigators, analysts user productivity tools). We discuss critical challenges in the effective integration of data-driven insights and end user- oriented case management technologies. We discuss synergies between micro-services, analytics, case management as step forward to empower end users to effectively use big data technologies, while share and collaborate on the fly, in order to generate and evolve insights.

Extending Fragment-Based Case Management with State Variables

Heiko Beck, Marcin Hewelt[(⊠)], and Luise Pufahl

Hasso Plattner Institute, University of Potsdam, Potsdam, Germany
{heiko.beck,marcin.hewelt,luise.pufahl}@hpi.de

Abstract. Modeling business processes has become standard among companies to efficiently organize their business operations. Case Management approaches have been proposed to overcome the limited flexibility of traditional business process languages as BPMN when it comes to supporting knowledge-intensive processes. One of such approaches is Chimera, in which business scenarios are captured by a set of process fragments, a domain model, and object lifecycles.

When modeling the real-world ITIL incident handling process, we observed that although Chimera is in general well-suited to capture this process, it misses the functionality to restore states of data objects. Such functionality is useful to undo errors of the case manager during process execution or to perform planned rollbacks. Therefore, we extend Chimera with state variables that memorize previous data states, making it possible to restore those states in the further course of the case. Our extension is validated using the real-world incident handling process.

1 Introduction

Business process modeling and execution has become a matter of course for many companies to reach their business goals in an efficient manner. Process models are required to achieve a common understanding of a process for all stakeholders and can be used as basis for the process implementation in an information system [12]. A business process is traditionally modeled statically in one process diagram as for example in BPMN [9]. Beginning with at least one starting point, all possible execution paths for the process are modeled until the process reaches a final state. This works well for processes with just a small number of execution paths, but gets cumbersome as the number of paths grows, e.g. for highly flexible processes.

For processes with a lot of variants but also a well known set of activities, the Production Case Management (PCM) paradigm was defined to model all execution paths in a clear way [11]. The fragment-based Case Management Chimera (for short Chimera) is a generic implementation framework for PCM [4,7]. This approach uses process fragments to model partially ordered activities in the fashion of BPMN. The complete process is then composed of all fragments, together with a domain model, object lifecycles, and a termination condition. Thus, Chimera focuses not only on activity modeling but also on data

© Springer International Publishing AG 2017
M. Dumas and M. Fantinato (Eds.): BPM 2016 Workshops, LNBIP 281, pp. 227–238, 2017.
DOI: 10.1007/978-3-319-58457-7_17

and state modeling. The state modeling concept extends the execution semantic of processes as an activity cannot only be triggered by the sequence flow but also by the states of data objects. This enables the reasonable combination of variants modeled in different process fragments.

The Chimera concept is advantageous for designing the knowledge-intensive ITIL incident handling process extended by a variant to handle mass disruptions, as the additional variant leads to many new execution paths. The ITIL process describes how a reported incident is logged and subsequently handled till the incident can be resolved and closed. In the case of a mass disruption, not all incidents must be handled separately. Instead, a parent-child relation can be used so that just the parent-incident is handled in detail and the result is propagated to its children. In interviews with our industry partner we recognized that call center employees can assign incidents to a mass disruption at different points in the process. This can be expressed as additional variant in a new process fragment which is enabled by different states of the incident. Further, call center employees have to be also able to roll back the assignment of an incident to a mass disruption, e.g. in the case of a mistakenly assignment. Chimera is designated to be used to model knowledge-intensive processes with many variants. Normally, the individual fragments can be executed at different points in the process. The execution of a wrong path and the need for a roll back is characteristic for such processes. Currently, these roll-backs cannot be handled by Chimera.

In this paper, we will close this gap by introducing a concept to handle state roll-backs with state variables. Therefore, the ITIL process will be analyzed and requirements on Chimera will be formulated. We try to fulfill the requirements with the current version of Chimera to underline the limitations of the approach before we introduce and evaluate the extension.

The remainder of this paper is structured as follows. We start with a real world example in Sect. 2. We introduce the ITIL incident process and formulate new requirements which we want to capture with case handling. In Sect. 3, the Chimera approach is introduced and in Sect. 4 we try to meet all requirements with Chimera before we introduce the new concept of state variables. Then, the extended incident process is modeled with Chimera and related work is discussed in Sects. 6 and 7 concludes the paper.

2 Running Example

The IT Infrastructure Library (ITIL) V3 [2] defines, among others, a best practice process to handle incidents in IT services. A simplified version of this process is modeled in Fig. 1. An incident is normally reported by an user in an email or a phone call. All reported incidents are logged and categorized. If the incident is a service request, then the incident is closed and a service request is created instead. Otherwise, the incident is prioritized and an initial diagnosis is conducted. Either the first level support finds a diagnosis and classifies the incident as solvable or no diagnosis is found and the incident must be escalated to the second or third level service support. In both cases, finally a solvable state is reached and the incident can be investigated, resolved and closed.

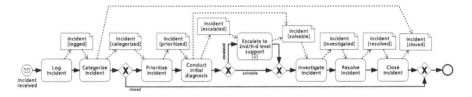

Fig. 1. Simplified version of the incident process. The incident can be closed by the "categorize incident" activity when the incident is actually a service request.

In reality it happens that several incidents with the same issue are reported by different users, for example if they have trouble with their mail programs. Such massive disruptions are not captured by the ITIL process as it is, instead each incident of a massive disruption needs to be handled separately.

Based on an interview with a german IT outsourcing company we will model an extended version of the incident process. This company logs each incident separately, however, if several incidents with the same reason are logged, one of them is marked as master incident to which the other incidents are assigned as children. Then, the master incident is handled like a normal incident and once it is solved, the solution is forwarded automatically to all children. The master incident is identified manually, as well as the children to be assigned. This is a knowledge-intensive task, as all incidents are logged by humans in natural language. Therefore, the assignment is often erroneous and rollbacks are required.

Manual assignment leads to different challenges in modeling the process. It is required that an incident is assignable to a master incident if the incident is in the state *categorized, prioritized, solvable* or *escalated*. The new state is then called *assigned to master*. Once an incident is assigned to a master, it must be either possible to wait until the master incident is resolved to apply the master result to the incident or to decouple the incident from the master and handle it separately again. If an incident is decoupled, the previous state of the incident must be restored.

To model this extended incident process with BPMN, it is necessary to insert into the process from Fig. 1 many XOR gateways to represent the different alternatives. After the activities *categorize, prioritize, conduct initial diagnosis* and *escalate* an alternative sequence flow to a separate *assign to master* activity is necessary, followed by the choice to apply the master result to the incident or to decouple the incident from the master. Separate *assign to master* activities would be necessary to ensure that in case of a decoupling the right state for the incident is restored. This approach would lead to a very complex and spaghetti-like process model with a lot of redundant activities. For this type of processes the fragment-based case management approach Chimera is more suitable, as it allows to handle incident assignment and decoupling in separate process fragments. In this paper we employ Chimera to model the extended incident handling process and analyze which capabilities of Chimera can be used to model rollbacks. In the next section, we explain the Chimera framework in more detail.

3   Fragment-Based Case Management with Chimera

This section introduces the concepts of Chimera that constitute the basis for this paper. For a thourough investigation of fragment-based case management and its formalization please refer to [4,7].

Chimera fragment-based case management is an approach to model flexible, variant-rich business processes as so-called *case models*. In contrast to traditional process modeling approaches like BPMN or Event-driven Process Chains (EPC) that capture a business process in one process model, the Chimera approach uses several process fragments to describe the process. Each fragment is modeled using standard BPMN, containing start and end events, gateways, and data nodes, and describes only a part of the activities necessary to handle the case. During case execution the fragments are dynamically combined based on data dependencies and user decisions. Hence, the execution paths of a business process are not visible as end-to-end paths in the model.

Data nodes in the process fragments refer to data classes and their states to express pre- and post-conditions of activities. For a pre-condition to be satisfied, a data object (the run-time representation of a data node) of the specified data class needs to exist in the specified state, which is denoted in square brackets below the data class name. A data object can either be *read* by an activity in a specific state (incoming data flow) or can be *written* by an activity in a specific state (outgoing data flow).

Let us assume the process model of Fig. 1 is a process fragment and *Incident* is a data class of the case model. The *categorize incident* activity reads the incident in the state *logged* and writes it either in the state *categorized* or *closed*. If an activity reads or writes more than one data object of the same class, this is logically combined with an *or*, while data objects from different classes are combined by an *and*. We call all read data objects the input set of an activity and the written data objects the output set.

The execution semantics of a fragment is defined by the control flow and the data flow. An activity gets enabled if it is control-flow enabled, i.e., the preceding activity was executed or the preceding event happened, as well as data-flow enabled, i.e., all data objects in the input set are available in the required state. The first activity of each fragment gets automatically control-flow enabled, once the case starts. As the *log incident* activity has no further preconditions, it is directly enabled. If this activity is terminated, the incident is written in the state *logged* and the *categorize incident* activity gets enabled.

A Chimera Case Management case model has two different termination concepts, the fragment termination and the process termination. If a fragment gets terminated, it is re-initialized meaning that the first activity gets again control-flow enabled. The termination of the complete business process is defined by a termination condition which references a set of data objects in a certain state. For the incident process the termination condition is *Incident[closed]*, i.e. the incident is in state *closed*.

Chimera supports link activities to represent the same activity in different fragments. Each link activity gets enabled by data and control flow like a normal activity. If several link activities are enabled which share the same output set, this activities are executed in parallel. A link activity will be used in Sect. 5.

4 Concept

The use case in Sect. 2 already introduced the requirements which we want to model with Chimera. These requirements are now formulated explicitly to find a generic solution that applies not only to our use case. Following properties must be fulfilled by the process model:

R.1 An activity a_1 gets enabled if a specific data object d is in exactly one of several states. If then a_1 is executed, it reads the data object d in the actual state.

R.2 The data object d is written by an activity a_2 in exactly the same state from which it was read in R.1 by a_1 (possibility for the roll-back).

R.3 The requirement R.2 should be feasible even if the read activity a_1 from R.1 takes part in a different fragment than the write activity a_2 from R.2.

The requirements R.1 and R.2 were directly derived from the use case. The assignment to a master incident should be possible from different data states and it must be possible to decouple an incident and to restore the previous state. R.3 is required because the methodology for Chimera suggests in [7] to start modeling with the happy path of a scenario and refine the scenario by adding new fragments. Therefore, it is not unlikely that a new fragment is added in which a former state is restored.

In Subsect. 4.1 we try to meet the requirements with Chimera. We show that the requirements cannot be completely fulfilled, with the current concepts. Thus, in Subsect. 4.2 an extension to Chimera is suggested to meet all requirements.

4.1 Range of Functions of Chimera

We start modeling a new fragment which fulfills the requirements R.1 and R.2. As the enablement of an activity in Chimera is triggered by the control flow and the states of data objects, it is simple to fulfill R.1. The activity which shall be enabled by one of several data states is put as first activity into a new fragment. All states which can trigger the activity are used as input for this activity. This is visualized in Fig. 2 with the *modify data object activity*. As described in Sect. 3, the first activity of each fragment is control flow enabled with the instantiation of a new case. Consequently, the enablement of the *modify data object activity* is controlled by the data object $\mathcal{DO}$. If this object is in state s_1 or s_2, the activity is enabled as demanded by R.1. The number of possible states for the data object $\mathcal{DO}$ could be chosen freely.

To meet R.2, a second activity is added into the same fragment. This activity can write the previously read data object in all states which were specified as input for the read activity. For the example that means, that the data object $\mathcal{DO}$ is either written in state s_1 or s_2. However, this solution is not satisfying as

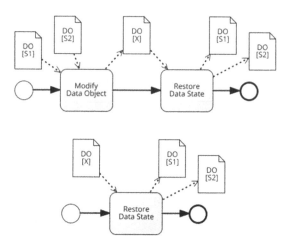

Fig. 2. The top fragment tries to fulfill the requirements R.1 and R.2 while the bottom fragment tries to fulfill R.3. However, this implementation enables invalid executions as for example reading $\mathcal{DO}$ in s_1 and writing $\mathcal{DO}$ in s_2.

it is possible to read the data object in state s_1 and then to write it in s_2. We are looking for a solution where such paths are forbidden and where it is only possible to write $\mathcal{DO}$ in s_2, if s_2 was used as input for the *modify data object activity* before.

We tried to meet R.3 in the bottom fragment of Fig. 2 by implementing a second *restore data state activity* which writes $\mathcal{DO}$ in s_1 or s_2. This produces the same problem as before. There are invalid paths possible in the process.

A way to enable just valid paths would be to split up the first fragment from Fig. 2 into several fragments. Instead of reading and writing all states in one fragment, each fragment would have just one input for the *modify data object activity* and the same state as output for the *restore data object activity*. This would lead to two fragments for our example, one for state s_1 and one for state s_2. Unfortunately, the number of states can be chosen freely and therefore a high number of almost redundant fragments would be necessary. That leads to unnecessary maintenance effort. Furthermore, R.3 cannot be fulfilled with this approach, because it is necessary that the state is restored in the same fragment it was read before to forbid invalid paths.

We will now introduce the concept of state variables and show that this small extension meets all requirements in a clear and easy to handle manner.

4.2 Extension to Chimera

The Chimera approach is quite new and still under development. The application to a real world use case showed, that there exists a gap in functionality which we will now close by the concept of state variables. This new concept can be used to enable read access to a data object that is in one of several states and memorize the state that was actually read. The memorized state can be used in the further course of the process to restore this specific state.

So far, a data object was specified in a process model in exactly one state by the notation *DataObject[state]* or in an arbitrary state by *DataObject[*]*. We extend this notation with a new set notation as follows:

Definition 1 (Set notation). *Let $\mathcal{DO}$ be a data object and s_1 to s_n possible states of $\mathcal{DO}$. Further, $\mathcal{DO}[s_1]$ specifies the data object in state s_1. We now define:*
$$\mathcal{DO}[\{s_1, s_2, ..., s_n\}] = \mathcal{DO}[s_1] \vee \mathcal{DO}[s_2] \vee ... \vee \mathcal{DO}[s_n]$$

The above definition introduces just a new notation but no new concept. However, this notation is required for the following concept of a state variable:

Definition 2 (Assign values to state variables). *Let α be a state variable to which the actual state s_i of possible states $\{s_1, s_2, ..., s_n\}$ read by an activity can be assigned. In the model, we define this by $\mathcal{DO}[\{s_1, s_2, ..., s_n\} \mapsto \alpha]$.*

The state variable enables to store a read state of a set of possible states $\{s_1, s_2, ..., s_n\}$ for a data object $\mathcal{DO}$. As a data object is always in exactly one state, the assignment of a state to a variable is unique. If an activity is executed multiple times, then the last read state is stored in the state variable. The previous assigned state of the variable is overwritten. Defined state variables are visible in the complete scenario, and hence can be accessed in all fragments.

Further, we define the use of a state variable to write a data object in a previously saved state.

Definition 3 (Use value of state variable). *Let $\mathcal{A}$ be an activity and $\mathcal{O}(\mathcal{A})$ be the output set of $\mathcal{A}$ with $\mathcal{DO}[\alpha] \in \mathcal{O}(\mathcal{A})$, then the data object $\mathcal{DO}$ is written in the state which was previously stored in α.*

The concept of state variables is now used to fulfill the previously defined requirements R.1 to R.3 with the two fragments from Fig. 3. The upper fragment fulfills R.1 and R.2 while the below fragment was created to meet R.3. The *modify data object* activity is the first activity in the upper fragment and therefore automatically control flow enabled with the instantiation of the scenario. The enablement of the activity is determined by the state of $\mathcal{DO}$. As we use the set notation defined above, the activity gets enabled if $\mathcal{DO}$ is in one of the specified states. This fulfills R.1. Moreover, the actual read state is saved in the state variable α for which we assume it is globally defined in the scenario. To meet R.2, the *restore data object* activity was implemented in the same fragment. This activity writes $\mathcal{DO}$ in the state of the referencing state variable. Thereby, it accesses the actual value of α and writes $\mathcal{DO}$ in the same state in which it was read before. The *restore data object* activity in the bottom fragment uses the same state variable α in the output set. As state variables are globally unique in a scenario, exactly the same state is referenced which was read in the top fragment by the *modify data object* activity. This fulfills R.3.

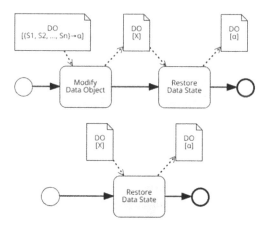

Fig. 3. The top fragment uses the state variable α to memorize the actual read state of the data object $\mathcal{DO}$ by the *modify data object activity*. The state variable is then used in the *restore data object activity* in the top fragment as well as in the *restore data state activity* in the bottom fragment to restore the previous read state. This fulfills R.1 to R.3.

To ensure a correct execution of a scenario it is necessary that each state variable, which is used in an output set was initialized before. Otherwise, if an uninitialized state variable is used the data object state to be written is undefined. Let us assume, the *restore data state* activity of the bottom fragment is executed before the *modify data object* activity, the variable α is not initialized and the new state for $\mathcal{DO}$ is unclear. To solve this problem, two alternatives are suggested: (1) Either an activity can just be enabled if all variables are initialized or (2) the user gets the responsibility to determine the correct output state. The restriction of the enablement would mean that the *restore data object* activity of the bottom fragment can just be enabled, when the *modify data object* activity was executed before. This could be achieved by integrating the initialization of state variables into the enablement semantics of activities. The second alternative would allow an execution of the *restore data state* activity in the bottom fragment, even if the *modify data object* activity was not executed before. In this case, the user will choose among all possible states $\{s_1, s_2, ..., s_n\}$ for $\mathcal{DO}$ and select the one to be written.

As case management scenarios are normally executed by experienced case managers, we recommend to use the second alternative and to give them the authorization to determine the new state of the data object. This has the advantage that the new concept can be implemented directly. In contrast, the first variant would require an additional effort to extend the enablement semantics.

We will now apply the newly introduced concept to the ITIL process from Sect. 2 to model the process with a variant to handle mass disruptions.

5   Application to the Use Case

In this section, we use the extended Chimera concept with state variables to model the incident use case with a variant to handle mass disruptions. The process consists of three fragments which are depicted in Fig. 4. Fragment (a) describes how a normal incident is handled as described in the ITIL standard. Beginning with a logging activity, the incident is analyzed step by step until it can be resolved. The fragments (b) and (c) illustrate the handling of mass disruptions. In fragment b), an incident in different states can be assigned to a master incident and subsequently the incident can be decoupled in case of a erroneous assignment. Fragment (c) describes how the result of a master incident can be applied to an assigned incident.

Fragment (b) fulfills the requirements that an incident can be assigned at different points in the process and that a roll-back after an assignment is possible. The *assign to master* activity gets enabled, if the incident is in state *categorized, prioritized, solvable* or *escalated*. The state variable α is used to memorize the read state of the incident. The incident is written by the *assign to master* activity in state *assigned to master*. As long as the incident is in this state, the *remove from master* activity is enabled. If the user executes the *remove from master* activity, the incident is written into the state which was previously memorized in the state variable α. Thus, incorrect state transitions are prevented.

Fragment (c) describes the application of a master result to a single incident. The *apply master result* activity gets enabled if the incident is in state *assigned to master* and the master incident is in state *resolved*. If the master result is applicable, then the incident is written in state resolved. Otherwise, the incident stays in state *assigned to master* and must be removed in the subsequent activity.

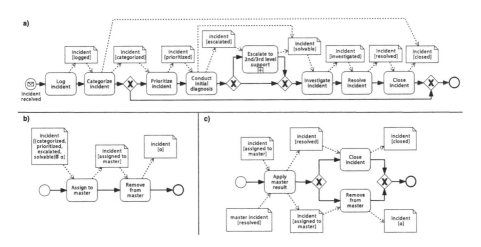

Fig. 4. The extended incident process consists of three fragments: (a) handles a normal incident, (b) enables the assignment and decoupling of an incident to a master incident. In (c) the result of a master incident can be used to resolve an assigned incident.

The *remove from master* activities from Fig. 4 (b) and (c) are linked activities. They are both executed simultaneously, if they are both enabled. As a result, fragment (b) gets always reinitialized if an incident is decoupled from a master incident. It is not possible that the *remove from master* activity from (c) is enabled and the same activity from fragment (b) is not enabled.

With the concept of state variables it is possible to model the incident process correctly and in a clear way. To this end, three fragments were required and just one activity occurs twice (*remove from master*). It was not necessary to edit the happy path of the process from Fig. 4 (a). The state variable α guarantees, that no invalid paths can occur in the process.

A further use case where the concept of state variables can be used is the temporary upgrade of a customer in the context of a marketing campaign. If several customers of different membership categories as for example, *student*, *basic* or *premium* get all a two week trial of a gold membership, it is important that they are assigned back to the right membership category after the two weeks. It is forbidden that a *student* member is categorized as *premium* member after the end of the trial. This can be captured by the state variable.

6   Related Work

The case handling paradigm was defined by van der Aalst et al. in 2005. Since then the research area of case handling is quite active. In contrast to traditional business process modeling languages, the case handling paradigm [1] is not only process-driven but also data-driven to support the modeling of flexible processes. A further well-known concept for highly flexible processes are PHILharmonic Flows [3,6]. The concept of PHILharmonic Flows builds on the case handling paradigm and is described by the authors not only as data-driven, but as being object-aware. Data objects are central in PHILharmonic Flows and all modifications of data objects during the process execution are described in a fine-grained way in micro processes. The importance of case handling was also recognized by the Object Management Group (OMG) and the first version of the Case Management Model and Notation (CMMN) [10] standard based on Guard-Stage-Milestones (GSM) [5,8] was published in 2014. This three mentioned case handling approaches are now compared to Chimera with respect to the presented extension and the capability to restore states of data objects.

The case handling paradigm defines a process mainly by activities and data objects. Whether an activity can be enabled is decided based on the values for the data objects and the termination of other activities. The permission to execute an activity is administrated by roles. The *execute* role allows an user to perform an activity. A special feature in the case handling approach is the *redo* role, which allows an user to undo an activity. This enables the reset of a data object into a specific state by undoing all modifications of the data object. However, the undo functionality is semantically different from the concept of state variables. State variables can also be used for a planned recreation of a state as in the example for the temporary upgrade of an user in Subsect. 4.2. This should not be modeled by an undo.

In PHILharmonic flows processes are defined in so-called micro and macro processes, where the former describes modifications of single attributes of data objects and the latter describes object interactions. Part of the concept is a mapping between attribute values and data states, each state is explicitly defined by a set of attribute-value relations. In a micro process a fine-grained view on this mapping is provided, as not only state transitions are modeled but even each value assignment is represented. If a state is defined by an attribute having a non-null value, a previous state could be restored by setting that attribute to null again. For example, the state *assigned to master* could be defined by the attribute *masterId* being set to an arbitrary value. The decoupling could be modeled in a micro process which sets this value to null again. This should restore the previous state. However, if two states are distinguished by the value of an attribute, rather then by the attribute being null or non-null, resetting state is not possible. This is due to the fact that the previous attribute value is not memorized.

CMMN employs the concept of a case file and case file items that can represent arbitrary data elements relevant for the case, from XML documents to folder structures. The standard states that "there is no need to know the internals of the those content objects" [10, Sect. 5.2.3], which is strange, given that sentries, controlling entry into and exit from stages, can contain conditions that evaluate over the case file. In CMMN it is possible to express that the *assign to master* activity enters a stage that represents the assignment, in which activities *remove from master* and *apply master result* can be executed. However, due to the informal data modeling concept of CMMN it is not possible to specify that exiting this stage through the *remove from master* activity resets the assigned issue to its previous state. Using the more formal GSM approach in [5] this might be possible.

7   Conclusion

With the ITIL incident handling process extended by a variant to handle mass disruptions we found a real-world process which is difficult to model with traditional process modeling languages. We explained why case management is more suitable to model the process and we used Chimera as case management notation to model the ITIL process. Like related case management approaches Chimera is not capable to model rollbacks of data objects. However, as Chimera is designated to model knowledge-intensive processes with a lot of variants it is important to support roll-backs, as the execution is very error-prone. With the concept of state variables we formally introduced a memory for data states which can be used to restore data states even after the execution of several tasks. The paper evaluates the concept by applying it to the incident process and showing that it is possible to model the handling of mass disruptions in two small fragments. A state variable was used to restore a previous data state.

This new concept is just a small extension to the actual implementation of Chimera and it should be easy to integrate into the ongoing development.

However, there is still an open question how to ensure in the fragment-based approach that roll-backs can be only executed in the case that the state variable of a data object was initialized before. The paper gives first recommendations to handle this challenge, but this should be further investigated in future work.

References

1. van der Aalst, W.M., Weske, M., Grünbauer, D.: Case handling: A new paradigm for business process support. Data Knowl. Eng. **53**(2), 129–162 (2005)
2. Cannon, D., Wheeldon, D., Taylor, S.: Office of Government Commerce UK: ITIL v3 - Service Operation. The Stationery Office (2007)
3. Chiao, C.M., Künzle, V., Reichert, M.: Enhancing the case handling paradigm to support object-aware processes. In: 3rd International Symposium on Data-Driven Process Discovery and Analysis (SIMPDA 2013), pp. 89–103, No. 1027 in CEUR Workshop Proceedings, CEUR-WS.org (2013)
4. Hewelt, M., Weske, M.: A hybrid approach for flexible case modeling and execution. In: La Rosa, M., Loos, P., Pastor, O. (eds.) BPM 2016. LNBIP, vol. 260, pp. 38–54. Springer, Cham (2016). doi:10.1007/978-3-319-45468-9_3
5. Hull, R., Damaggio, E., De Masellis, R., Fournier, F., Gupta, M., Heath III., F.T., Hobson, S., Linehan, M., Maradugu, S., Nigam, A., Sukaviriya, P.N., Vaculin, R.: Business artifacts with guard-stage-milestone lifecycles. In: DEBS, pp. 51–62. ACM (2011)
6. Künzle, V., Reichert, M.: PHILharmonicFlows: Towards a framework for object-aware process management. J. Softw. Maint. Evol. Res. Pract. **23**(4), 205–244 (2011)
7. Meyer, A., Herzberg, N., Puhlmann, F., Weske, M.: Implementation framework for production case management: Modeling and execution. In: Enterprise Distributed Object Computing Conference, pp. 190–199. IEEE (2014)
8. Nigam, A., Caswell, N.S.: Business artifacts: An approach to operational specification. IBM Syst. J. **42**(3), 428–445 (2003)
9. Object Management Group: Business Process Model and Notation (BPMN), version 2.0 (2011)
10. Object Management Group: Case Management Model and Notation (CMMN), version 1.0 (2014)
11. Swenson, K.D.: State of the Art in Case Management. White Paper Fujitsu (2013)
12. Weske, M.: Business Process Management. Springer, Heidelberg (2012)

Guiding the Creation of Choreographed Processes with Multiple Instances Based on Data Models

María Teresa Gómez-López, José Miguel Pérez-Álvarez,
Angel Jesús Varela-Vaca, and Rafael M. Gasca

Departamento de Lenguajes y Sistemas Informáticos,
Universidad de Sevilla, Seville, Spain
{maytegomez,josemi,ajvarela,gasca}@us.es
http://www.idea.us.es/

Abstract. Choreography in business processes is used as a mechanism to communicate various organizations, by providing a method to isolate the behaviour of each part and keeping the privacy of their data. Nevertheless, choreography diagrams can also be necessary inside an organization when a single instance of a process needs to interact and be synchronized with multiple instances of another process simultaneously. The description, by business experts, and the implementation, by developers, of these choreographed models are highly complex, especially when the activities involved in the processes exchange various data objects and with different cardinalities. We propose the automatic detection of the synchronization points, when a choreographed process model is needed. The choreography will be derived from the analysis of the process model, data objects consumed and generated through the process, and the data conceptual model that relates the data objects. A graphical tool has been developed to support where the synchronization points must be included, helping to decide about the patterns that describe how a single model can be transformed into a choreographed model.

Keywords: Business process choreography · Multiple instances · Conceptual model · Data model relation

1   Introduction

Process choreography is frequently related to a business contract between two or more organizations, used as a mechanism to communicate various entities, by providing a method to isolate the behaviour of each part and keeping the privacy of their data. Nevertheless, choreography diagrams can also be necessary within an organization when the relation of single and multiple instances need to be synchronized simultaneously. Unfortunately, one aspect that remains a challenge is the description and development of these choreographed business processes. The challenges that choreography establishes, such as correlation and

© Springer International Publishing AG 2017
M. Dumas and M. Fantinato (Eds.): BPM 2016 Workshops, LNBIP 281, pp. 239–251, 2017.
DOI: 10.1007/978-3-319-58457-7_18

the inclusion of heterogeneous data, once the process is choreographed, have been studied in [1,2]. However, the creation of the choreographed model remains highly complex, specially when the data objects involved have an N:M cardinality relation. Also, the choreography diagrams are typically too low-level, and the terms of the interactions between the two parties are not always clear.

Regarding our proposal, in an activity-centric paradigm, such as BPMN [3], where the process is described by an explicit order between the activities, the choreography between processes becomes difficult to model and implement in a Business Process Management System (BPMS). One example, where difficulty in choreographing a model can be observed is given by the process that manages the *Calls for Residence* to help University students that want to apply for an assignment of a room in a residence. Figure 1 depicts a reduced example that is enlarged in the next sections: (1) the call for residences is created; (2) students apply by filling out a form; (3) administrative staff of the University evaluate the proposals; (4) the administrative staff notify each student; and, finally, (5) the accepted students can formalize documents and make payments. Although the process is easy to model and understand, it is totally incorrect and incomplete, since the activities cannot be executed in the same and single instance. For example, the activity *Fill out the form* can be executed several times for each *Publish Residence Call*, and not only once per call as the model describes. Something similar occurs between *Evaluate the Proposals* and *Publish a list with the resolution*, since they are not executed the same number of times. The problem of the different number of executions of the activities cannot be solved with a loop activity, since there exists a correlation between each execution of *Fill out the form* and *Formalize documents and make payments* depending on the student. This problem can be solved by using event handlers, however how to create the pools, the activities for the synchronization and the exchanged data is a hard problem that we want to reduce in this paper.

Sometimes inconsistencies in business processes are derived from the activities that consume and produce different data objects. For the example, *Fill out the form* generates a data object *Residence Call*, but *Fill our the form* generates only one *Student Form* per instance, and since it is a competitive process *Evaluate the proposal* uses a set of *Student Forms*. Therefore, in order to discover the inconsistencies related to the data object evolutions and to help in the creation of a correct choreography, it is necessary to include the data objects in the process model. In order to ascertain the points where choreography is required, both an analysis of the expected relation between the objects described in the Data Model, and the data evolution during the process model are needed. The importance of the data perspective and its verification in business process models has been studied in previous work [4]; the challenge now is to ascertain how it can affect the choreography.

In this paper, we propose the combination of various modelling paradigms: business process models described by using BPMN, the data object evolution also included in the BPMN model, and the Data Model that supports the managed data. With these three models combined, it is possible to analyse the process

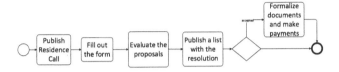

Fig. 1. Example of business process which needs to be choreographed.

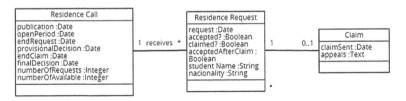

Fig. 2. A piece of the conceptual model.

correctness according to the exchanged data. If any inconsistency is found, the necessary modifications are proposed. The main advantages of our proposal are: (1) we define a model that combines various and different paradigms that are easy to describe independently, but difficult to choreograph; (2) the relations between the objects to generate a correct choreographed model are analysed providing the mechanisms to create a correct model; and, (3) the warnings and the transformation patterns have been implemented on a graphical tool as an extension of Activiti^{TM}.

The remainder of this paper is organized as follows. Section 2 presents the formalization of the concepts and the defined model, including a motivating example. Section 3 presents the suggested approach for the automatic detection of synchronization points and how they can be implemented. Section 4 discusses related work. Finally, Sect. 5 concludes and presents lines for further research.

2   Formalization of Concepts

The modelling paradigms that are combined and included in the formalization are:

– **Conceptual Model for Data Modelling.** The data management used in a process is crucial in achieving the established objectives. Various models can be used to describe the data relation, but we propose the use of a Conceptual Model (cf. Fig. 2) for a higher level of abstraction managed by using Object-Relational Mapping [5].

– **BPMN for Business Process Modelling.** The main goal of BPMN [3] is to provide a standard notation which is readily understandable by all business users, from the business analysts that create the initial drafts of the processes, to the technical developers responsible for implementing the technology, and finally, to the business people who will manage and monitor those processes.

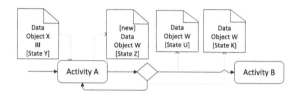

Fig. 3. Example of data object relations.

Thus, BPMN creates a standardized bridge for the gap between the business process design and process implementation.

– **Data Object Life-cycle Modelling.** Although BPMN is not primarily designed for data modelling, there exists a set of notations that enables the designer to model the data involved in a business process. The primary construct of BPMN for modelling data within the process flow is the *Data Object* element. The *Data Objects* can include the State of Data Objects at various points in a process [3]. An activity in a process can *Consume* a Data Object, which implies that the Data Object is in a particular state. When the activity is executed, the object may transit to a new state (an object in a new state is *Produced*). Figure 3 represents an example of accessing the data objects related to *Activity A*. A data flow edge from a data object to an activity describes a read access to an instance of the data object (cf. Data object X), which has to be in this state to execute the activity. Likewise, a data flow edge from an activity to a data object describes a write access (cf. Data object W), which either creates a data object instance if it does not already exist (labelled with [new] as proposed in [6]), or updates the instance if it already exists. A data flow edge connecting a data object with a sequence flow indicates the data object is flowing through that connection, and the state of the flowing data object (cf. in Fig. 3, Data Object W in State U). Data objects can be modelled as a single object or as a collection of objects (marked by three parallel bars, |||). Only the execution of an activity can imply the creation of a new object, although it is possible to represent different states of an object depending on the executed branch, for example, after the execution of activity A, an XOR-split is executed, where the data object W in state U flows through the upper branch, or in state K for the lower branch.

2.1 Model Formalization

The model applied in the formalization and automatic choreography of our problem is composed of two sub-models: (1) one model to describe the Conceptual Model (CM_{Graph}); and (2) one model to describe the components related to the business process model (BP_{Graph}).

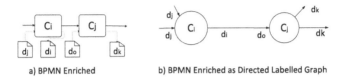

a) BPMN Enriched   b) BPMN Enriched as Directed Labelled Graph

Fig. 4. Labelled directed graph to formalize the BP_{Graph}.

Formalization of the Conceptual Model. A conceptual model is described as a connected labelled graph (CM_{Graph}) composed of nodes (EN) and edges (AS), $CM_{Graph} = \langle EN, AS \rangle$, where:

- EN is the disjoint union of the entities of the conceptual model $(e_1 \ldots e_n)$,
- AS is the set of edges that represent the association between two entities (e_i, e_j). Each edge can be labelled with two values that represent the cardinality $(card_i, card_j)$, and each one can represent simple (1 or 0..1) or multiple (0..N or 1..N) cardinality.

Formalization of the Enriched Process Model. The enriched business process is modelled as a directed labelled connected graph (BP_{Graph}), based on the annotated graph presented in [7,8], composed of nodes (C) and edges (F), $BP_{Graph} = \langle C, F \rangle$ whose equivalences are shown in Fig. 4, where:

- C is the disjoint sets of components formed of activities A, events E, and gateways G:
  - A is the set of activities.
  - E is the set of events, that can be partitioned into disjoint sets of one start event E^s, intermediate events E^i and end events E^e. Just one start event is allowed in the model, and therefore there is one and only one node whose input degree is equal to 0.
  - G can be partitioned into disjoint sets of parallel-fork gateways G^F, parallel-join gateways G^J, data-based XOR-decision gateways G^D and XOR-merge gateways G^M.
- F is the set of edges that represent the association between the components, such as activities. Each edge is labelled with two values (origin and destination of the arc) $\langle d_i, d_o \rangle$. From the point of view of the nodes (C components), $d_{Consumed}$ and $d_{Produced}$ are the incoming and outgoing data objects respectively. Each Data Object (DO), associated to the origin (d_i) or to the end (d_o) of each directed edge, is described by the tuple $\langle Entity, Cardinality, new \rangle$, where:
  - $Entity$ is one of the entities EN described in the CM_{Graph} $(\in e_1 \ldots e_n)$,
  - $Cardinality$: single or multiples (collection represented by $||||$) objects,
  - new: a Boolean that represents whether the data object is *[new]*.

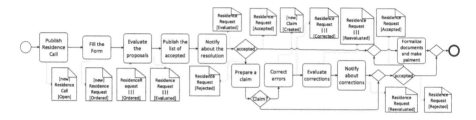

Fig. 5. No choreographed process for the management of University Residences.

2.2   Application Case Study

The business process shown in Fig. 5 extends the previous example for the assignment of rooms in the University Halls of Residences of Seville. Firstly, the call for positions is published, and students can apply by filling out a request form. Once the requests have been sent to the administration staff, a competitive analysis is performed. According to the resolution, the students are notified. Each student analyses whether the assigned residence is suitable and decides whether to claim. If a student decides to claim, then the documentation is revised again. The final decision is sent to each candidate, and the final list is published.

3   Guiding the Transformation to a Consistent Choreographed Model

As explained in the introduction, a correct business process workflow can be incorrect for the data object consumed and produced in the model. The necessity or not to transform a BPMN model into another choreographed model can be derived from the necessity to synchronize processes with multiple and different number of instances.

3.1   Choreography Derived from Data Object Relation

As explained previously, incorrectness in business processes can be detected through the data used in each activity, and therefore the choreography becomes necessary. This analysis is based on allocating the activities involved with objects with the same cardinality in the same pool, where a pool is the representation of a participant in a collaboration [3]. The cardinality and relation of the objects are described in the conceptual model. For the fragment of the Conceptual Model shown in Fig. 2, two activities that consume *Residence Call* and *Residence Request* respectively, cannot be in the same pool, since both data objects have a 1:N relation in the conceptual model. This incompatibility arises due to the fact that both activities cannot be executed the same number of times in the same instance, and for each evaluation of a *Residence Call*, several managements of *Residence Request* can be performed. Otherwise, two activities that consume *Residence Call* and a set of *Residence Requests*, respectively, can be in the same pool, since both data objects have a 1:N relation in the conceptual model. In an equivalent way, the activities that manage a *Residence Request* and a *Claim*, can be in the same pool since there exists a 1:1 relation between them in the conceptual model. The relations in the conceptual model are determined by a direct association between two entities, such as between *Residence Call* and *Residence Request*, or transitively, such as between *Residence Call* and *Claim*. Derived from this idea, we include the following definitions:

Definition 1. Data Context. A Data Object D is **in the same context** as a Data Object D' iff:

- D and D' are single objects and they have a 1:1 relation in the conceptual model,
- D is a single object and D' is a collection of objects, and they have a 1:N relation in the conceptual model,
- D is a collection of objects and D' is a single object, and they have an N:1 relation in the conceptual model,
- D and D' are collections of objects, and they have a 1:1 or N:N relation in the conceptual model.

Definition 2. Points of Choreography. Being f_i an edge that connects two C components (c_i, c_j) with the data object d_i and d_j, if d_i and d_j do not belong to the same Data Context, then f_i is a point of choreography. There are two types of Points of Choreography:

- **Many-to-one point:** In BP_{Graph}, it implies that $card_i > card_j$ in CM_{Graph}, and the cardinality of d_i is $\leq$ to the cardinality of d_j. This is, for example, the relation between *Residence Call* and *Residence Request*.
- **One-to-many point:** It implies that $card_i < card_j$ in CM_{Graph}, and $d_i \geq d_j$ in BP_{Graph}. This is, for example, the relation between *Residence Request* and *Collection of Residence Request*.

In order to model the Point of Choreography following BPMN 2.0 [3], it is necessary to take into account that each interaction has an initiator or sender (the party sending the message), and a recipient or receiver (the party receiving the message, who may reply with a return message). In order to choreograph the instances of two pools, multi-instance activities ($\equiv$) to send and receive messages are used. They are divided into:

- **Loop Activity** ($\circlearrowleft$) is a type of activity that acts as a wrapper for an inner activity that can be executed multiple times in sequence. The Loop Activity executes the inner Activity as long as the *loop Condition* evaluates to true.
- **Parallel Activity** ($|||$) is a type of activity that acts as a wrapper for an activity that has multiple instances spawned in parallel, and can be used to receive the messages of another process.

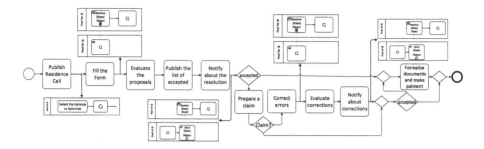

Fig. 6. Example of annotated process about University Residences.

3.2 Patterns for Synchronization Point Identification

Assuming that BP_{Graph} must be correct according to the workflow [9–11], our analysis about the data aspect correctness is centred in the existence of various pools to support the components that manage different Data Contexts than pools. In order to ascertain that a business process model formed by one or more than one pool is correct related to exchanged data, two properties must be satisfied:

1. Every $d_{Consumed}$ and $d_{Produced}$ by the components c_i of a pool P, must belong to the same *Data Context* of P (Definition 1).
2. For every point of synchronization (Definition 2) where (c_i, c_j) are involved, c_i and c_j must belong to different pools.

If it is found an incorrect synchronization point that does not satisfy the first property, or a $d_{Produced}$ that does not satisfy the second property, the following warnings (in Fig. 6) will be annotated in the original model (Fig. 5):

1. **Data Objects Consumed and Produced by an Activity** (Fig. 7.a): If $d_{Consumed}$ (d_i) and $d_{Produced}$ (d_j) by a component C_i do not belong to the same Data Context, two situations and labelled patterns can be found:
   – If d_i is a *new* data object, such as the activity *Fill out the form*, then synchronization is necessary. To synchronize the instance of d_i and d_j, a new activity to correlate the instances of different tasks must be located immediately before the component C_i (Fig. 7.a.I), such as the activity *Select the Residence Call* for the example of Fig. 6. This is necessary because for each d_i, several d_j can be created, and therefore when a d_j is created, it is necessary to determine the corresponding d_i.
   – Else, two types of synchronization can be carried out:
     • **Many-to-One Relation** (Fig. 7.a.II)) advises the creation of synchronization by means of a *Loop Activity* that forms a wrapper around for the inner activity C_i, which can be executed multiple times to send messages, and a single activity that receives the messages. In Fig. 6, an example is the activity *Notify the resolution*.

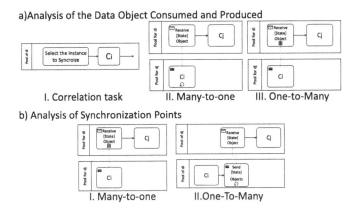

Fig. 7. Patterns of synchronization

- **One-to-Many Relation** (Fig. 7.a.III)) advises the creation of synchronization by means of the single activity C_i that also sends messages to synchronize with a *Parallel Activity* that receives these messages.

2. **Points of Synchronization between two Activities** (Fig. 7.b): The synchronization points involve components that must be in different pools. It implies that a correct choreographed process must have, at least, one pool for each Data Context. For the example, at least, two pools must be necessary to cover the activities that manage the data objects {*Residence Call, Collection of Residence Request*} and {*Residence Request, Claim*}, respectively. In order to highlight the components that can belong to the same pool, we have coloured some activities of Fig. 6 to distinguish the tasks of both pools. There exist two types of synchronization to satisfy the properties expressed before when d_i and d_o are not in the same Data Context:

   - **Many-to-One Relation** (Fig. 7.b.I) which synchronizes by means of two extra activities: a *Loop Activity*, to send messages; and an activity to receive the messages in the another pool, such as the activity *Notify the resolution*.
   - **One-to-Many Relation** (Fig. 7.b.II), which synchronizes by means of two extra activities: a *Parallel Activity*, to receive messages; and a single activity to send the messages, such as *Send Residence Request*.

3.3   Architecture of the Implementation

In order to evaluate our proposal, we have implemented a graphical tool that enables the inclusion of a connection to a database to create the Conceptual Model automatically using ORM (Hibernate in our case). It has been developed as an extension of Activiti^{TM} [12] with additional components (cf. Fig. 8), since this is an open-source distribution that offers the business expert a friendly

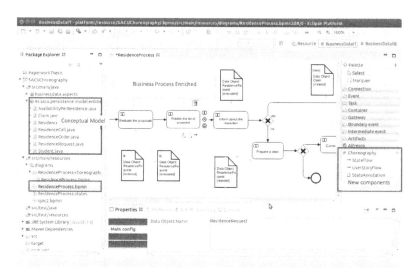

Fig. 8. Screenshot of the Activiti^{TM} process modeller plugin.

interface to model the process using BPMN elements and easy to extend with new components and functionality.

Activiti^{TM} is a light-weight workflow and Business Process Management Platform targeted at business people, developers and system administration. Since neither the data state description nor warning patterns are included in Activiti^{TM}, an extension of Eclipse-based Activiti^{TM} Designer has been developed to support the new graphical elements, and included in the graphical tool (with a video available in: http://www.idea.us.es/condchoreography/).

4   Related Work

The complexity of managing the choreography of business processes is well-known. Certain studies have been published describing the implementation of business process choreographies [13,14]. However, these only take into account the control flow, although the data flow across messages is also an important aspect. In [15], the authors introduce a first proposal about data-aware collaboration.

According to the data inclusion the process models, it is crucial to analyse the evolution of the objects [16], and to determine whether there are data dependencies and which these dependencies are [17] in order to define whose model can be aligned in BPMN model [6,18].

The automating data exchange in a process choreography has been solved in [1], where the choreography model is known. Nevertheless, we consider the difficulty encountered when faced with how to create a correct choreography model specially derived from the limitations of BPMN studied in [19], or how to perform the choreography using technologies such as BPEL [20].

[21,22] propose extensions to choreography modelling by means of interaction patterns between the involved participants. But these works do not study the choreography problem derived from the synchronization of multiple instances necessary for the data dependencies. They only face the choreography problem when the activities must be executed by different organizations, but with a one-to-one relation between the instances executed in the different pools that tend to be produced in conversation environment. Khalaf et al. have introduced an approach for the partitioning of single process models into multiple participants [2,23] including the data aspect. However this partitioning cannot detect the necessity to synchronize N instances of a process with another derived from the data that manage.

5   Conclusions and Future Work

In this paper, we propose an automatic detection of the necessary and type of synchronization points in a process models derived from the exchanged objects. The detection of the synchronization points is based on the comparison between the type and cardinality of the data objects consumed and produced by the components of a business process, and on the relation in the conceptual model. We have developed a graphical tool as an extension of Activiti, where the business process, conceptual model and data object description can be included in the same model, and the warnings related to synchronization correctness are located in the process automatically. The result has been applied to a real example of the University of Seville.

As future work, we propose an extension of the proposal to create a new synchronized model automatically. Also we consider relevant the application of user stories combined with business processes to facilitate the recovery of requirements, and the automatic creation of lanes and pools during the process of choreography.

Acknowledgement. This work has been partially funded by the Ministry of Science and Technology of Spain (TIN2015-63502-C3-2-R) and the European Regional Development Fund (ERDF/FEDER). We would like to thank SACU of the University of Seville for the valuable information that has contributed towards the development of the ideas in this paper.

References

1. Meyer, A., Pufahl, L., Batoulis, K., Fahland, D., Weske, M.: Automating data exchange in process choreographies. Inf. Syst. **53**, 296–329 (2015)
2. Khalaf, R., Kopp, O., Leymann, F.: Maintaining data dependencies across BPEL process fragments. Int. J. Coop. Inf. Syst. **17**(3), 259–282 (2008)
3. OMG: Object Management Group, Business Process Model and Notation (BPMN) Version 2.0. OMG Standard (2011)
4. Borrego, D., Gasca, R.M., Gómez-López, M.: Automating correctness verification of artifact-centric business process models. Inf. Softw. Technol. **62**, 187–197 (2015)

5. Vennam, S., Dezhgosha, K.: Application development with object relational mapping framework - hibernate. In: Proceedings of the 2009 International Conference on Internet Computing, ICOMP 2009, 13–16 July 2009, Las Vegas, Nevada, USA, pp. 166–169 (2009)

6. Meyer, A., Pufahl, L., Fahland, D., Weske, M.: Modeling and enacting complex data dependencies in business processes. In: Daniel, F., Wang, J., Weber, B. (eds.) BPM 2013. LNCS, vol. 8094, pp. 171–186. Springer, Heidelberg (2013). doi:10.1007/978-3-642-40176-3_14

7. Weber, I., Hoffmann, J., Mendling, J.: Semantic business process validation. In: 3rd International Workshop on Semantic Business Process Management at ESWC 2008, SBPM 2008, June 2008

8. Gómez-López, M.T., Gasca, R.M., Pérez-Álvarez, J.M.: Decision-making support for the correctness of input data at runtime in business processes. Int. J. Coop. Inf. Syst. 23(4), 1–29 (2014)

9. Dijkman, R.M., Dumas, M., Ouyang, C.: Semantics and analysis of business process models in BPMN. Inf. Softw. Technol. 50(12), 1281–1294 (2008)

10. Borrego, D., Eshuis, R., López, M.T.G., Gasca, R.M.: Diagnosing correctness of semantic workflow models. Data Knowl. Eng. 87, 167–184 (2013)

11. Kheldoun, A., Barkaoui, K., Ioualalen, M.: Specification and verification of complex business processes - A high-level petri net-based approach. In: Motahari-Nezhad, H.R., Recker, J., Weidlich, M. (eds.) BPM 2015. LNCS, vol. 9253, pp. 55–71. Springer, Cham (2015). doi:10.1007/978-3-319-23063-4_4

12. Rademakers, T.: Activiti Documentation (2015)

13. Decker, G., Weske, M.: Interaction-centric modeling of process choreographies. Inf. Syst. 36(2), 292–312 (2011)

14. van der Aalst, W.M.P., Lohmann, N., Massuthe, P., Stahl, C., Wolf, K.: Multiparty contracts: Agreeing and implementing interorganizational processes. Comput. J. 53(1), 90–106 (2010)

15. Knuplesch, D., Pryss, R., Reichert, M.: Data-aware interaction in distributed and collaborative workflows: Modeling, semantics, correctness. In: 2012 8th International Conference on Collaborative Computing: Networking, Applications and Worksharing, CollaborateCom 2012, Pittsburgh, PA, USA, October 14–17, pp. 223–232 (2012)

16. Herzberg, N., Meyer, A., Weske, M.: Improving business process intelligence by observing object state transitions. Data Knowl. Eng. 98, 144–164 (2015)

17. Senderovich, A., Rogge-Solti, A., Gal, A., Mendling, J., Mandelbaum, A., Kadish, S., Bunnell, C.A.: Data-driven performance analysis of scheduled processes. In: Motahari-Nezhad, H.R., Recker, J., Weidlich, M. (eds.) BPM 2015. LNCS, vol. 9253, pp. 35–52. Springer, Cham (2015). doi:10.1007/978-3-319-23063-4_3

18. Gómez-López, M.T., Borrego, D., Gasca, R.M.: Data state description for the migration to activity-centric business process model maintaining legacy databases. In: Abramowicz, W., Kokkinaki, A. (eds.) BIS 2014. LNBIP, vol. 176, pp. 86–97. Springer, Cham (2014). doi:10.1007/978-3-319-06695-0_8

19. Cornax, M.C., Dupuy-Chessa, S., Rieu, D., Mandran, N.: Evaluating the appropriateness of the BPMN 2.0 standard for modeling service choreographies: Using an extended quality framework. Softw. Syst. Model. 15(1), 219–255 (2016)

20. Decker, G., Kopp, O., Leymann, F., Pfitzner, K., Weske, M.: Modeling service choreographies using BPMN and BPEL4Chor. In: Bellahsène, Z., Léonard, M. (eds.) CAiSE 2008. LNCS, vol. 5074, pp. 79–93. Springer, Heidelberg (2008). doi:10.1007/978-3-540-69534-9_6

21. Barros, A., Dumas, M., Hofstede, A.H.M.: Service interaction patterns. In: Aalst, W.M.P., Benatallah, B., Casati, F., Curbera, F. (eds.) BPM 2005. LNCS, vol. 3649, pp. 302–318. Springer, Heidelberg (2005). doi:10.1007/11538394_20
22. Barros, A., Hettel, T., Flender, C.: Process choreography modeling. In: vom Brocke, J., Rosemann, M. (eds.) Handbook on Business Process Management. International Handbooks on Information Systems, pp. 257–277. Springer, Heidelberg (2010)
23. Khalaf, R., Leymann, F.: E role-based decomposition of business processes using BPEL. In: 2006 IEEE International Conference on Web Services (ICWS 2006), 18–22 September 2006, Chicago, Illinois, USA, pp. 770–780 (2006)

Redefining a Process Engine as a Microservice Platform

Antonio Manuel Gutiérrez–Fernández[(✉)], Manuel Resinas,
and Antonio Ruiz–Cortés

School of Computer Engineering, University of Seville, Seville, Germany
{amgutierrez,resinas,aruiz}@us.es

Abstract. In recent years, microservice architectures have emerged as
an agile approach for scalable web applications on cloud environments.
As each microservice is developed and deployed independently, they can
be developed in the platform and programming language that best suite
their purposes, using a simple communication protocol, as REST APIs
or asynchronous event-based collaborations, to compose them. In this
paper, we argue that process engines provide an excellent platform to
develop microservices whose business logic involves complex work flows
or processes so that a Business Process language can be used as high-
level language to develop these services and a process engine to execute it. We identify the requirements for integrating a process engine
in a microservice architecture and we propose how the communication
and deployment in a microservice architecture can be handled by the
process engine.

Keywords: Process engine · Event-based asynchronous communication · Microservice architecture

1 Introduction

The popularity of microservices architecture is increasing in software development. Opposite to monolithic designs or classic 3-layer web application,
microservice architectures propose decomposing application into small components around business concepts. Although distributed systems are not a nov-
elty, current technology stack with web environments, cloud platforms or persistence servers hinders monolithic application development and deployment.
In short, the microservice architectural style is an approach to develop a single application as a suit of small services around business concepts (opposite
to functional responsibilities in n-layer architectures), each running in its own
platform and communicating with lightweight mechanisms [3]. The rationale
behind the microservices architecture is that decomposing complex applications

This work has been partially supported by the European Commission (FEDER), the
Spanish and the Andalusian R&D&I programmes under grants TIN2015-70560-R,
P12-TIC-1867, and P10-TIC-5906.).

M. Dumas and M. Fantinato (Eds.): BPM 2016 Workshops, LNBIP 281, pp. 252–263, 2017.
DOI: 10.1007/978-3-319-58457-7_19

into microservices allows evolving them independently which leads to more agile developments and technological independence between them [4]. However, the decomposition into microservices increases complexity of integration so the use of event-based asynchronous communication is encouraged to also decouple integration interfaces between services. A number of organizations such as Netflix or Twitter have moved to microservice architectures as their services grew. As a matter of fact, one of the advantages of microservice approach is it allows to choose the best suitable language to develop it. The decision making regarding the platform lies on the microservice domain, developers expertise or technical requirements.

In the last two decades, business process applications have been developed using process engines. These systems currently support the development of web front-ends, REST interaction and light deployments, such as a Java library (e.g. Activiti[1] or Camunda[2]). Traditionally, process engines have performed the orchestrator role in Service Oriented Architectures. However, in a microservice based application, this role is not strictly required as services are self choreographed [10].

In this paper, we propose redefining the role of a process engine as microservice platform instead of an orchestrator of services, for the development of services whose business logic is a workflow, such as purchase orders in ERP. To this end, we analyse the main required properties of a microservice and how a process engine can be adapted to provide them as a microservice.

In the next section, we introduce an example application and introduce the main features of a microservice architecture. In Sect. 3, we propose a methodology to design the microservice and its interfaces with a process engine and, in Sect. 4, we review other works related to services and process engines. Finally, conclusions and possible work extensions are described in Sect. 5.

2 Motivation and Background

2.1 Purchase Order Management

We introduce, as an example, the development of an application that manages Purchase Orders including their shipment logistics and bank payments. The application starts when an employee fulfills a new purchase order, waits for budget responsible validation, then deals with provider to request shipment, track and receive it, and, lastly, handles the related account payments. This application manages Purchase Order lifecycle and its related shipment and payment and includes a number of decision points close to business domain. This lifecycle can be quite straight-forward modelled as a Business Process, BP depicted in the Fig. 1, and deployed it in a process engine. Therefore, a process engine can fit as development platform to implement Purchase Order lifecycle.

---

[1] http://activiti.org/.
[2] https://camunda.com/.

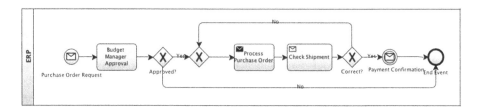

Fig. 1. Purchase order process

In current technology context, microservices architectures have provided an agile development pattern for scalable systems in cloud environments. These architectures are characterized by the following features.

Organized Around Business Capabilities. Monolithic systems become more difficult to maintain as they grow, so, following the single responsibility principle, microservice architecture approach proposes decomposing functionality into specific domain contexts to facilitate the development and testing stages and manage scalability. These separated domain contexts are named bounded contexts [2]. The decision about the granularity degree relies on the specific business domain concepts and development management (team size, expertise,...) so there are no fixed rules about decomposition scale but it should be enough small so a single team could be the only responsible for the development of a given microservice (and, e.g., in a development iteration, such as two weeks).

Decentralized Data Management. The microservice is responsible for managing the complete lifecycle of the business objects that fall in its bounded context (business logic, persistence,...) so that their details are encapsulated in the microservice and the interactions between objects in different microservices has to be performed with a public exposed interface.

Deployment in Isolation. Besides the encapsulation of the management of objects, each microservice has to be deployed in isolation, which decouples the development and execution platform between different microservices. Therefore, each microservice can be developed with the most suitable language for its purposes (considering technical or functional requirements). And, through this isolation, a microservice can be developed and updated avoiding that the related microservices (those that requires or provides operations from or to this microservice) require changes.

Interaction with Other Microservices. As we introduced, the dependencies between microservices are exposed with an application programming interface (API). Following the boundary context, this interface should be defined in terms of business concepts. In traditional service oriented architectures (SOAs), the interactions are commonly orchestrated through a mediator. However, in microservices architectures is preferred avoiding this orchestrated role so the services are self choreographed with a lightweight protocol, such as REST APIs. This choreography requires each microservice knows exact details of consumed operations (not only domain details but also endpoints or communication protocol). Furthermore, if they communicate with a synchronous pattern, microservices are dependent of consumed microservices (errors or too slow operations block execution). To avoid this coupling, event-driven asynchronous communication are preferred in microservices architectures [6]. The asynchronous approach requires a message broker which event consumers subscribe to and which handles emitted events to the related subscribers. With this approach, the blocking errors in consumed operations and the direct dependency between providers and requesters (event emitter and subscriber) are avoided.

Interaction with User Interface. Lastly, microservices architectures has to provide a proper handling of user interface. There are several approaches to develop the whole application frontend based on different microservices. On one side, we can develop a complete independent UI which uses the provided APIs by microservices to handle the business data. In this approach, UI component is highly coupled with the microservices and device-aware rendering is difficult as the provided APIs are business guided. Other approach is each microservice provides its own piece of User Interface and they are assembled to provide a full front-end. This approach requires a mechanism with templates systems or style sheets to provide a seamless appearance and it also is dependent on user client technological context where compositions could be harder to deal with (native applications or thick clients). And lastly, another approach is composing UI pieces from the different microservices in the backend, so a central server layer provides different user interfaces, for different user roles, devices, etc. This approach relies certain control in the UI management so it could handle logic it shouldn't.

2.2 Purchase Order Application as Microservice Architecture

Going back to our example, our application includes three business concepts - Purchase Orders, Shipments and Payments. Regarding the boundary contexts for a microservice architecture, we consider a decomposition into three microservices around these business concepts. The developers of these services should consider functional requirements, team expertise or technological aspects to implement them. As we introduced, business process modelling notation and process engine are, respectively, a suitable language and platform to develop Purchase Order logic. However, there are a number of challenges that have to be addressed in order for the process engine to perform as a microservice platform. First, the process engine has to support the management of the full lifecycle of Purchase Orders, including business logic and persistence (Point 1 in Fig. 2). Second, as the development and deployment is independent between different microservices, one

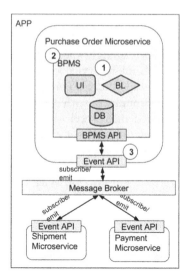

Fig. 2. Microservice architecture example

process engine has to be independently deployed and executed for each microservice where they perform as platform (opposite to traditional single deployment for a process-based application), as depicted in Point 2 in Fig. 2. Furthermore, process engines usually provide management and interaction interfaces but they have to be adapted or wrapped to communicate in asynchronous event-based ground (Point 3 in Fig. 2) and focused on business events (not to process events). And, last, the integration of user interfaces for the different microservices have to be addressed.

3   Building Microservices with a Process Engine

In this section, we argue how to address the introduced microservice features with a process engine as development platform for Purchase Order microservice.

3.1   Organized Around Business Capabilities

The example application introduced in previous section involves three business concepts: Purchase Order, Shipment and Payment. We depicted the business process for Purchase Orders which handles the full lifecycle of them. Shipment and Payment have their own processes, related to different business areas as logistics and accounting. Therefore, a decomposition into three microservices around each business concept fits with the leading single responsibility principle in the microservices architectures.

3.2   Decentralized Data Management

We have proposed developing Purchase Order microservice with a process engine as a platform. As we introduced, this implies the process engine has to manage the full lifecycle of Purchase Order objects. Process engines commonly self manage objects during business process instances and delete them after they finish (sometimes they store them just for keeping history traceability), encapsulating their persistence system (relational database, runtime memory,...). Therefore, the domain object managed by microservice has to be completely handled by one business process instance, as the Purchase Order in the Business Process in Fig. 1 (in other case, the system should be extended to manage the objects out of the process instance runtime or other development platform would be encouraged).

The properties for Purchase Order Data Object are depicted in Table 1.

Table 1. Purchase order data object

Reference	An unique string to identify purchase order (e.g.: 'A#432-2015')
Description	A string to humanly describe the content of the purchase order (e.g.: 'Monitor 22")
Date	A date value related to Order date (e.g.: '15/12/2014')
Status	An enumerated from {ordered, verified, requested, delivered, finished} to indicate the purchase order status (e.g.: 'ordered')
Payment	Payment information (bank account, price, date,...). This resource is managed by another microservice
Shipment	Shipment information (provider, receive date,...). This resource is managed by another microservice

In process engines, data are not a first-class citizen and they are simple managed as process variables without relationship between them. Therefore relating object requires manual handled references or extending the object model.

3.3   Deploying Process Engine Microservice

A microservice has to be self managed, i.e., each one is deployed and run independently. Therefore, each microservice has to be deployed with their own platform. Process engines have traditionally provided a platform to develop single full process oriented applications. However, in the last years, lightweight open source process engines, such as Activiti or Camunda, have appeared where their process engine can be deployed as a Java library in a external application.

In the context of cloud applications, a number of solutions propose deploying independent platforms through Virtual Machines or other approaches as Docker containers (chapter Deployment in [6]). Therefore, the deployment of an individual process engine to perform a single microservice (through virtualization or container), enables the required self-management. Figure 3 depicts an example stack of technologies to provide a process-engine-based microservice with a Java process engine (such as Camunda). The business process model and their related code to automate tasks (e.g. Java code in case of Camunda) is installed on a Camunda instance. This instance runs over a Java Virtual Machine and its single deployed on a Virtual Machine. Camunda stores execution data on a database which commonly would be deployed in a separate Virtual Machine. And the two Virtual Machines together form the microservice.

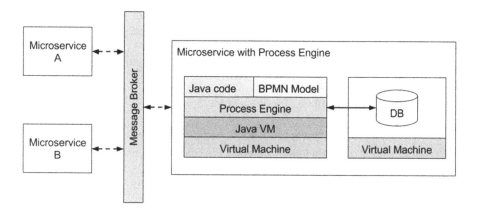

Fig. 3. Process engine microservice technology stack

As a result of this stack, there could be several process engines running simultaneously (one per each microservice) in the same application but the introduced overhead is not significant and it allows that different microservices evolve and scale independently.

We have also to point out that, in this work, we propose getting advantage of the process engine capabilities but Business Process Management Systems also commonly provide other functionality, such as a Process Modeller, that could be used in the processes definition but this is not deployed or included in the microservice.

3.4   Managing Microservice Asynchronous Communication

In order to develop an asynchronous event-based mechanism with a message broker, such as RabbitMQ -which fully support enterprise integration patterns- is proposed, although a similar simpler approach could be based on Atom standard and HTTP.

We study how to communicate with the process engine through a message broker to provide an asynchronous event-driven communication protocol, as depicted in Fig. 4.

In the figure, related to our example Purchase Order, an event "New Purchase Order (PO)" is emitted by an external service. Then it's propagated to the Purchase Order microservice (Step 1 in Fig. 4), which is subscribed to this event. This microservice, developed with a process engine, requires handling the event message as an operation provided by the process engine (Step 2 in Fig. 4). This operation is mapped to a process interaction, such as the creation of a "New Process Instance" (Step 3 in Fig. 4). The handling of "New Purchase Order" event message as a "New Process Instance" operation in the Process Engine is provided by an Event Mapper component. This component is part of the process engine microservice and would responsible for: (i) Subscribing to the microservice related events (e.g. "New Purchase Order" or "Payment Confirmation"), (ii) Map the incoming event message to the own operation provided by the process engine (e.g. "New Purchase Order" event message into "New Process Instance" operation and the outcoming requests from the process engines into event messages (e.g. "Process Purchase Order" request into "Process Purchase Order" event message). This map includes the routing and adapting of messages content.

Some process definition languages, such as BPMN, provide an event-based explicit mechanism to communicate process from different parties. There are a number of existing solutions to develop this Event Mapper component, such as Spring Integration or Mule, which provide mappers of different message protocols, so we can use them to different process engines (which are compatible with the supported message protocols). Modern process engines provide a synchronous management interface, commonly as a REST API, to create or interact with process instances, so it could also be used for several purposes (as support to user interface), but in this work, we focus on providing an asynchronous communication.

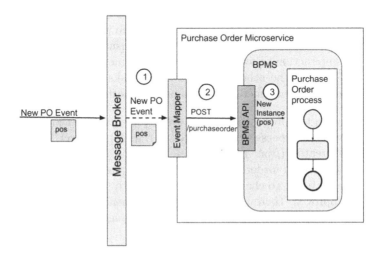

Fig. 4. Purchase order microservice with event-driven communication

Regardless the Event Mapper we use, to define a generic Event API and required mappings in the Event Mapper, we identify the required interactions with process engine that should be provided as events. These events relate with objects, regarding the possible states in their lifecycle: (i) Propagation of Object information, (ii) Creation of objects, (iii) Object updates. Therefore, the Event API has to meet the following requirements:

R1: Object information. Relevant changes in objects managed by the microservice process/es has to be notified through an event message.
R2: Creation of Objects. As without process instances there are no Objects, the API has to enable process instance creation.
R3: State-Machine awareness. The object state machine is provided by business process execution so the Event API has to enable interacting with process instances in those points that are waiting for external data, so the process instance can advance (and business object changes its state).

According to R1, all relevant changes in objects managed by business process have to be notified. This can be explicitly included in the process definition with the process language mechanisms or develop a generic mechanism of notifying any change on Data Objects. To support R2 and R3, we analyse tasks and data flow in business process for Purchase Orders. Regarding to R2, when a process instance starts there is no existing Data Objects so using events different to "New Purchase Order" should have no effect. If the process instance starting is related to the creation of a Data Object, then it is invoked by a New Data Object Event (i.e.: New Purchase Order Event). Regarding to R3, possible interactions with Object workflow are:

– An explicit business process event. If the business process explicitly waits for an external event, the object flow activates at the reception of a message (e.g., as intermediate event or as part of a reception task in BPMN).
– Implicit waiting task. On the one side, some tasks in process require from user interaction (User tasks). Modern process engines usually provide a web interface for this interaction so we can directly communicate with these interfaces through HTTP. On the other side, automatic tasks can also include active waiting for events (performed by programmatic mechanisms).

These stages in the proposed example are depicted with BPMN in Fig. 5. In the example, when a process is instantiated (1), a Data Object for a Purchase Order (PO) is created (from required Reference and Description values) and its Status is initiated to 'ordered'. After budget manager checks PO, the Data Object is updated in (2) to change the Status value to 'approved' or 'cancelled'. If the PO is approved, there is a request for shipment and when shipment is received in (3), the PO Object is updated again to change the Status to 'received' and to relate it to shipment Data Object.

Considering the previous discussion about Message, these three events can be related to external Messages. Purchase Order creation in (1) relates to message "New Purchase Order". Purchase Order update in (2) and (3) relates to

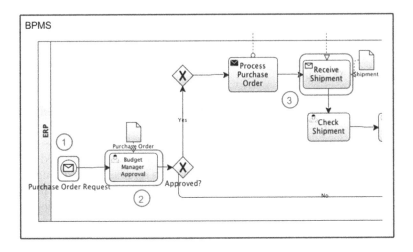

Fig. 5. Process flow interactions

Table 2. Mapping from events to process instance stage

Business object event	Process stage
New purchase order event	Start Instance
Budget manager approval	User Task
Shipment received	Receive Event
Purchase order update	Any stage

"Budget Manager Approval" and "Shipment Received" messages. After any creation or update of Purchase Order objects, corresponding event is notified with a "Purchase Order Update" message. A summary map for these relationships is depicted in Table 2. As deleting a Data Object is not consistent with a process workflow, we do not consider this event in our proposal.

3.5   Developing User Interface

In an event asynchronous communication pattern, the frontend can be developed using this event interface. Therefore, the frontend has to be aware of Purchase Order lifecycle to provide a consistent interface (e.g.: Interface to create a new Purchase Order has to provide all the required data fields for a new Purchase Order). To support any screen related to purchase order management, the user interface should subscribe to all "Purchase Order Update" events and emit an event after each "Purchase Order" updating ("Shipment Received" or "Budget Manager Approval" events) or creation ("New Purchase Order" event) through the user interface.

Process engines already feature user interface mechanisms to manage process instances, interact with tasks, authorization, etc. However, these facilities are thought to be used in a single process engine application and not integrated in a composed application. Therefore, the frontend has to be fully developed.

4   Related Work

As far as we know, there is no approach related to use a process engine as a platform to implement microservices. A number of research works have explored managing business processes as services. In [9], a REST API is proposed to consume a business process as a REST Resource. This idea is extended in [5] to provide a scalable architecture for business process services, considering runtime aspects. While both of them provide a REST API to interact with business processes, their goal is managing them as resources while our goal is to use business process model as the 'programming language' to manage -other- resources.

Other papers propose to use business processes as services orchestrators[3]. In this respect, Pautasso et al. [8] propose using BPEL to compose REST services and Bellido et al. [1] define control-flow patterns for REST services managed by business process. The base of these proposals is to use processes for their capabilities to provide a specific function, orchestration, in a service architecture while our proposal focuses on providing a generic purpose platform to develop microservices.

And, at last, Overdick [7], proposes extending BPEL to manage the state machine of resources in REST way. They get advantage of BPEL as event handler to control HTTP operations on the resources. So, while our approach is to fully develop suitable microservices with processes, this work only provides an extension to BPEL in order to capture and manage REST requests.

5   Conclusions and Future Work

In this paper, we have shown how to encapsulate business processes and process engines as programming language and platform to develop a microservice. Specifically, we first analyse the characteristics of microservice architectures to define the requirements for a process engine to implement a microservice. With these requirements, our proposal proposes different approaches to handle these requirements.

This work is a first approach to support a novel platform for the industry microservice architectures but there are several lines of future work. First, this approach has to be extended to more complex aspects such as interactions between processes in the same microservice or responsibility delegation events. Second, we focus on the interface to handle business objects, but common microservices also usually provide a management interface to monitor execution or to control configuration properties. And, last, this approach has to be applied to real scenarios to validate its feasibility.

---

[3] http://www.bpm-guide.de/2015/04/09/orchestration-using-bpmn-and-microservices/.

References

1. Bellido, J., Alarcón, R., Pautasso, C.: Control-Flow patterns for decentralized RESTful service composition. ACM Trans. Web (TWEB) **8**(1), 5 (2013)
2. Evans, E.J.: Domain-Driven Design: Tacking Complexity in the Heart of Software. Addison-Wesley Longman Publishing Co., Inc., Boston (2003)
3. Fowler, M.: Microservices, March 2014. http://martinfowler.com/articles/microservices.html
4. Fowler, M.: Microservices trade-offs, January 2015. http://martinfowler.com/articles/microservice-trade-offs.html
5. Gambi, A., Pautasso, C.: RESTful business process management in the cloud. In: 2013 ICSE Workshop on Principles of Engineering Service-Oriented Systems (PESOS), pp. 1–10, May 2013
6. Newman, S.: Building Microservices. O'Reilly Media, Incorporated, Sebastopol (2015). https://books.google.es/books?id=1uUDoQEACAAJ
7. Overdick, H.: Towards resource-oriented BPEL. In: Gschwind, T., Pautasso, C. (eds.) Emerging Web Services Technology, Volume II. Whitestein Series in Software Agent Technologies and Autonomic Computing, pp. 129–140. Birkhäuser, Basel (2008)
8. Pautasso, C.: RESTful web service composition with BPEL for REST. Data Knowl. Eng. **68**(9), 851–866 (2009)
9. Pautasso, C., Wilde, E.: Push-enabling RESTful business processes. In: Kappel, G., Maamar, Z., Motahari-Nezhad, H.R. (eds.) ICSOC 2011. LNCS, vol. 7084, pp. 32–46. Springer, Heidelberg (2011). doi:10.1007/978-3-642-25535-9_3
10. Richards, M.: Microservices vs Service-Oriented Architecture. O'Reilly Media, Incorporated, Sebastopol (2015)

Providing Semantics to Implement Aspects in BPM

Hércules S.S. José[(✉)], Filipe Esteves Gonçalves, Claudia Cappelli,
and Flávia Maria Santoro

Programa de Pós-Graduação em Informática, Universidade Federal do Estado do Rio de Janeiro,
Av. Pasteur, 458 – Urca, Rio de Janeiro, RJ, Brazil
{hercules.jose,filipe.goncalves,claudia.cappelli,
flavia.santoro}@uniriotec.br

Abstract. Crosscutting concerns in business processes have been addressed, among other forms, under the aspects orientation paradigm. The goal is reducing visualization complexity, allowing reuse and improving maintainability. Literature presents techniques that address aspects in BPM lifecycle stages of modeling and implementation. However, those techniques adopt different semantical representations, making the integration between those stages very difficult. This paper proposes a service identification method to select an implementation for aspects in order to meet goals set in the modeling stage. We describe an artifact produced with this purpose within an application scenario where Web Services are discovered and selected during a process execution. We conclude that aspects' behavior can be flexible and adaptable at runtime .

Keywords: Aspect orientation · Business process · Semantics

1 Introduction

Separation of crosscutting concerns is a technique currently used in systems to address complex phenomena by breaking them down into smaller, meaningful and manageable parts. This reduces the complexity and improves the reusability of the resulting subsystems [1]. Such concerns, so-called crosscutting concerns, become scattered and interwoven in different parts of one or more systems. The separation of crosscutting concerns has been addressed using the aspect orientation paradigm [2].

Aspect-Oriented Business Process Management (AO-BPM) has been investigated in both the modeling phase [3] and in the implementation phase [4] of the BPM cycle [15]. Bastos et al. [5] identified the existence of different representation semantics for aspects and its related elements in business processes. This difference creates a gap between the layers (conceptual model and executable model), making the integration between them difficult in the business process lifecycle.

Moreover, it is important that the process meets previously set goals in order to allow assessment of the operational quality and to facilitate the understanding of organizational changes. However, business experts do not always know during the modeling phase how a particular aspect should be implemented; in turn, developers are frequently unaware of the objectives defined by the experts during the modeling phase. The issue

© Springer International Publishing AG 2017
M. Dumas and M. Fantinato (Eds.): BPM 2016 Workshops, LNBIP 281, pp. 264–276, 2017.
DOI: 10.1007/978-3-319-58457-7_20

is more evident in a scenario where each service may represent a crosscutting concern and its implementation is not linked to the particularities of the business domain.

This work extends the research of Bastos et al. [5] creating a method to identify services able to select an implementation for an aspect that meets established objectives. The method uses Web Service Modeling Ontology (WSMO) [6] to incorporate the necessary semantics for invoking Web services that meet the goals set during the modeling phase. In this paper, we describe a software artifact produced to support that method enabling a Workflow Management System (WfMS) or a Business Process Management System (BPMS) to interact with an execution environment for WSMO. An initial evaluation was applied in an application scenario where it was observed the discovery and selection of web services that dynamically met the goals set for the aspect. The use of the method and artifact produced shows that the behavior of the aspect becomes more flexible and adaptable at runtime to allow a service to be replaced by another that meets the same goals.

The paper is organized as follows: Section 2 contextualizes the problem and research question; Sect. 3 presents related works; Sect. 4 describes the method and the artifact; Sect. 5 illustrates the use of the method and the artifact in an application scenario; Sect. 6 provides conclusions and future work.

2 Problem Contextualization

In software development, crosscutting concerns such as security, auditing, authentication, logging, persistence, usability and transparency can be encapsulated in modules using the aspect-oriented paradigm. Kiczales et al. [2] state that the incorporation of an aspect to the main flow must be through pointcuts and advices. Pointcuts are the elements of an aspect that describe at what points (join points) within the software that aspect should act. Advices define the actions to be taken when a join point is reached. Advices may be configured to act before (before advice), during (around advice) and after (after advice) the join point.

The same principle has been applied in business processes [8]. For example, activities related to the concept of logging can be modularized in one aspect. The logging aspect can provide advices that represent a set of activities that perform information logging of the process. Figure 1 shows the information logging during the review process of academic articles. Existing approaches to represent aspects in the stages of process modeling and implementation differ in their semantics, hindering an interaction between these phases. The problem is evident when comparing the representation of aspects adopted in Fig. 1, with the representation adopted in Fig. 2. In Fig. 1, the aspect is represented as a single activity and arrows indicate its join points. In the Fig. 2, the aspect and its elements are represented as XML tags.

Another point that further highlights the issue of semantic problems between modeling and implementation of aspects is when business experts do not know the details about the implementation for that aspect. Taking the logging example, experts know that the data exchanged during the process execution will be stored and recorded in a database, but they do not know how this operation will take place, or if it will be in accordance with the goals set in the process. Once separated from the process, the aspect becomes an independent module, and may contain its own implementation encapsulated in a Web Service, for example. The developer who implements this service is unaware of the business experts' intentions, who designed the service to fulfill a certain goal, which might not meet the goals set for the process.

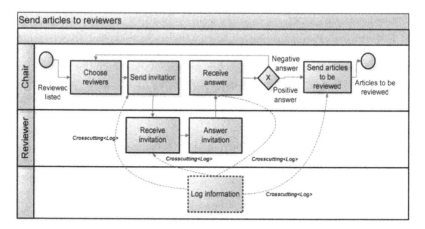

Fig. 1. "Choose reviewers" process model with log information aspect [9].

```
1    <aspect name="PerformanceMonitor">
2     <partnerLinks>
3      <partnerLink name="AuditingWS" partnerLinkType="AuditingPLT"
4              myRole="caller" partnerRole="measurer"/>
5     </partnerLinks>
6     <variables>
7      <variable name="startTimerRequest" messageType="startTimerInput"/>
8      <variable name="startTimerResponse" messageType="startTimeOutput"/>
9      <variable name="stopTimerRequest" messageType="stopTimerInput"/>
10     <variable name="stopTimerResponse" messageType="stopTimeOutput"/>
11    </variables>
12    <pointcutandadvice>
13     <pointcut name="monitored activities" contextCollection="true">
14       //invoke[@operation="findAFlight"] | //invoke[@operation="findARoom"]
15     </pointcut>
16     <advice type="around">
17      <sequence>
18       <assign>
19        <copy>
20            <from variable="ThisJPActivity" part="name"/>
21            <to variable="startTimerRequest" part="activityName"/>
22        </copy>
23        </assign>
24      <invoke partnerLink="AuditingWS" portType="AuditingPT"
25        operation="startTimer" inputVariable="startTimerRequest"
26        outputVariable="startTimerResponse"/>
27      <proceed/>
28      <assign>...</assign>
29      <invoke partnerLink="AuditingWS" portType="AuditingPT"
30        operation="stopTimer" inputVariable="stopTimerRequest"
31        outputVariable="stopTimerResponse"/>
32      </sequence>
33     </advice>
34    </pointcutandadvice>
35   </aspect>
```

Fig. 2. Aspect to monitor the execution time [3].

Bastos et al. [5] evidenced the existing problem in representing aspects of business processes, and they extended an Aspect Ontology [10] to address aspects in the business processes domain. According to the extended ontology, three items establish the behavior of aspects: process, base code and goal. The authors emphasize the Class Goal since the aspect must have an operational goal to be accomplished.

The concept of goal was the same used by Santos et al. [11]. The use of operational goals as part of the description of the aspect during the process modeling allows design and code base to be linked during the process execution. Moreover, when defining a goal to be achieved, we have the possibility to choose the best implementation for the aspect, without having to know in advance its implementation.

3   Related Work

Cappelli et al. [4] proposed an approach to separate crosscutting concerns in business process models in order to solve the scattering and interweaving problem. The approach was implemented as an extension of the Oryx[1], called CrossOryx. The approach is limited to the modeling phase. Comparing with the approach proposed in this paper, we provide the link of the model with its implementation, defined by the goals to be achieved.

Charfi et al. [3] applied the aspect-orientation in the composition of services to increase the reusability, reduce maintenance costs, and facilitate the handling of complexity within a system. Such approach was implemented as an extension to the Business Process Execution Language (BPEL) [13], called Aspect-Oriented BPEL (AOBPEL). BPEL does not provide a graphical representation, thus the AOBPEL is restricted to the implementation/execution phase of the BPM lifecycle. Our approach also enables the modeling of aspects.

Jalali et al. [14] proposed a generic solution to support execution of aspect oriented business process models in runtime on a Workflow Management System (WfMS). It is based on the principle behind dynamic weaving of aspects, and it was formally specified using Coloured Petri Nets. From this solution, two artifacts were built: first, the Aspect Service enables a WfMS to support dynamic weaving; second, the Pointcut Editor, enables the definition of aspects, their advices that will be weaved, and pointcuts and their conditions. The implementation of this solution described in [7] enables the YAWL system[2] to support aspect-oriented business processes in practice. Like in Cappelli et al. [4], the solution has kept its focus on the modeling phase. Despite the aspect has been enabled to run on a WfMS, no mention was made about a possible link between the model and its implementation. Still, the approach was promising and thus it supported our proposal to enable YAWL to make dynamic weaving using AspectService.

---

[1] Additional information about Oryx available in http://bpt.hpi.uni-potsdam.de/Oryx.

[2] YAWL stands for Yet Another Workflow Language (http://www.yawlfoundation.org/).

4   A Proposal for Semantically Integration in AO-BPM

According to the ontology proposed by Bastos et al. [5] the behavior of an aspect could be represented in three ways: process, base code or goal. We argue that for a greater alignment between the phases of modeling and implementation, all the three representations must compose the definition of the aspect behavior: Process - using a graphical workflow notation such as Petri- Nets or BPMN; Codebase - functionality to be performed at runtime, written using a programming language, such as Java; and Goal - abstract objective definition expressing the desired behavior of the aspect within the business process. The Goal could be described using the concepts of Web Service Modeling Ontology (WSMO) [6].

The WSMO describes resources and data exchanged during the call to a service. Ontologies provide formal definition that are processable, and through them other components and applications take into account the real meaning of the data. Therefore, the concepts inherent to the data exchanged during the execution of a process are formally defined from ontologies described using the Web Service Modeling Language (WSML).

Figure 3 illustrates the behavior of the aspect composed of the process model, goal and base code. An ontology describing the data domain ensures that its meaning remains the same during the execution of the aspect. In the modeling phase, the aspect is modeled in a workflow tool. The data domain ontology and operational goal related to the aspect are described in WSML. In the implementation phase, possible source codes for the aspect are available through Web Services and hosted by a service provider. Available Web Services receive their respective semantically description using WSML. The semantic description of ontologies, goals, and Web Services are managed by a Semantic Execution Environment (SEE).

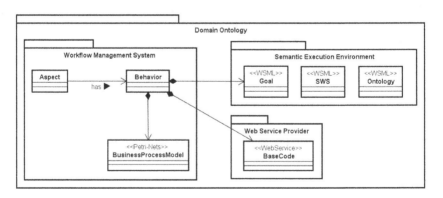

Fig. 3. Process behavior representation.

Bastos et al. [5] chose the Web Service Modeling eXecution (WSMX) [12] as the SEE to perform the services described on the basis of operational goals. WSMX allows the discovery, selection, mediation, invocation and interaction of Semantic Web Services (SWS). This environment implements a subset of WSMO and contains repositories to maintain the service entities, ontologies, goals, mediators and data. Combining the description of Goals and Web Services through criteria matching, the environment can discover and invoke the service that best addresses the desired operational goal. Based on Fig. 3, goals, ontologies and Web Services written in WSML are stored in WSMX.

The GoalService artifact was developed (Fig. 4) to allow a WfMS to interact with WSMX. It is responsible for enabling a WfMS to interact with WSMX in order to invoke the WebService according to the operational goals set out during the execution of the aspect. The GoalService was implemented as a custom component of the WfMS where the communication with the workflow engine is made by a service interface. The communication of GoalService with WSMX is made through two service interfaces: (i) to discover candidate services that meet the goals defined in aspect; and (ii) to invoke the service chosen among the candidate services. The GoalService makes use of the WfMS rule repository to identify the goals assigned to the aspect. The use of rule repository is useful, because it allows that a change can be addressed at runtime, without causing any interruption of the process, and neither changes in modeling. Algorithm 1 illustrates the interaction of GoalService with WSMX to discover and invoke the WebService based on goals stored on rule repository.

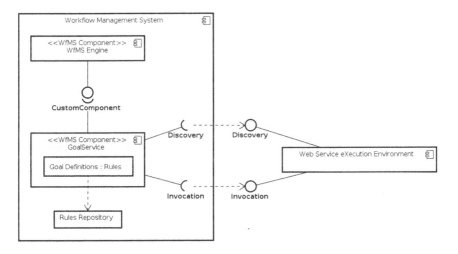

Fig. 4. GoalService artifact scheme.

Algorithm 1 – Discovery and invocation service based on operational goal.

```
1. connect_to(WfMS_Engine);
2. actual_Aspect = WfMS_Engine.get_Actual_Actitivy();
3. WSMX_Connection = WSMX_Server.get_Active_Connection();
4. if (WSMX_Connection not exists) then
5.   WSMX_Server.start_New_Connection();
6.   WSMX_Server.load_Ontologies();
7.   WSMX_Server.load_Semantic_Web_Services();
8.   WSMX_Server.load_Goals();
9.   WSMX_Connection = WSMX_Server.get_Active_Connection();
10. end if
11. aspect_Goal =
find_operational_goal_from_rule_repository(actual_Aspect);
12. if (aspect_Goal not exists) then
13.   return_Control_to_Engine(WfMS_Engine, actual_Aspect);
14.   end_of_method;
15. end if
16. candidate_Web_Service =
WSMX_Connection.run_Discovery_Service(aspect_Goal);
17. if (candidate_Web_Service exists) then
18.   WSMX_Connection.invoke_Web_Service(candidate_Web_Service);
19. else
20.   return_Control_to_Engine(WfMS_Engine, actual_Aspect);
21.   end_of_method;
22. end if
23. return_Control_to_Engine(WfMS_Engine, actual_Aspect);
24. end_of_method;
```

5   Evaluation of the Proposal

The method presented in the previous section was applied in an application scenario to evaluate the artifact produced. The scenario used was adapted from the process "Send Articles to Review" described by Cappelli et al. [9] (depicted in Fig. 1), comprising the Aspect Log Information. The process chosen deals with the chair of a conference selecting and inviting a reviewer to review articles. The objective of this evaluation was to investigate if the approach proposed is able to select an adequate service to implement the aspect according to the goal defined in the modeling phase. Besides, by using the setting already provided and discussed by [9], we could compare both approaches and highlight the differences and contributions.

The main aspect and the process were modeled in YAWL (Fig. 5a). To allow the aspect Log Information to be incorporated in the main flow, we used the concepts explained in [7, 14], where a special activity called "PROCEED" shall be inserted in the process, indicating the aspect join point. According to [7, 14], Fig. 5b shows a join point of the type after. Moreover, in the activity "Log Information", the custom service

GoalService is set, indicating to YAWL that the execution of the activity will be done by a custom service.

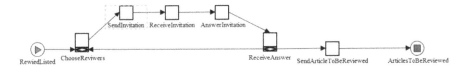

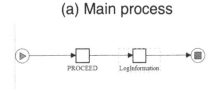

(a) Main process

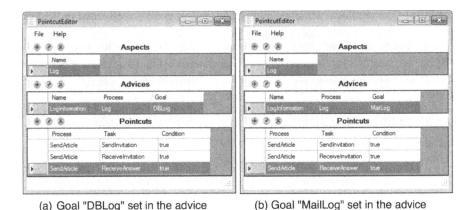

(b) Aspect "Log Information"

Fig. 5. Process "Send Articles to Review" and the aspect "Log Information", adapted from [9].

The definition of the aspect, its advice and pointcuts for AspectService was made using the extended version of Pointcut Editor, where the "Goal" column was added on advice section (Fig. 6). Also, in the definition of the advice, we have set the goal that the aspect must achieve. The name of the goal is the same that is available in WSMX during process execution.

(a) Goal "DBLog" set in the advice   (b) Goal "MailLog" set in the advice

Fig. 6. Pointcut editor with the change of goals during the aspect execution

Three Web Services were created using Eclipse IDE for Java EE[3] configured with Apache Axis2[4], each one with a different strategy log record: the first is a log file; the

---

[3] Eclipse IDE available in https://eclipse.org/downloads/.
[4] Apache Axis2 available in http://axis.apache.org/axis2/java/core/.

```
wsmlVariant _"http://www.wsmo.org/wsml/wsml-syntax/wsml-flight"
namespace { _"http://www.uniriotec.br/aspect#",
    onto _"http://www.uniriotec.br/aspect#",
    wsml _"http://www.wsmo.org/wsml/wsml-syntax#",
    dc _"http://purl.org/dc/elements/1.1#",
    discovery _"http://wiki.wsmx.org/index.php?title=DiscoveryOntology#"
  }

webService MailLog
    importsOntology {
        onto#LogOntology
    }
    capability MailLogCapability
        nfp
            discovery#discoveryStrategy hasValue discovery#LightweightDiscovery
            discovery#discoveryStrategy hasValue discovery#NoPreFilter
        endnfp

        postcondition
            definedBy
                ?logEffect memberOf onto#LogEffect
                    and ?logEffect[messageLogged hasValue _boolean("true")]
                    and ?logEffect[mailSent hasValue _boolean("true")].
```

Fig. 7. Mail log Web Service described in WSML.

second sends the log by e-mail; and the third registers the log in a database. Each Web Service is semantically described using the concepts of WSMO (Fig. 7) using Web Service Modeling Toolkit (WSMT)[5]. The WSMT is justified because it provides syntax checking of the WSML files. The three goals to be achieved by each of the services were defined using WSML (Fig. 8). The log ontology used in Web Services and Goals is the same in [5]. The ontology, semantic Web Services and goals were loaded in WSMX, and they become available during the process execution.

At the beginning of the execution of process "Send Articles to Review", the goal "DBLog" has been set for the activity "Log Information" (Fig. 6a). After the execution of activities "Send Invitation" and "Receive Invitation", the aspect was activated using the implementation of Jalali et al. [7], and activated the "Log Information" activity, as a result GoalService invoked WSMX discovery method, and received as the outcome the Web Service that made the log to be recorded in the database (Fig. 9), so as to invoke it to represent the aspect implementation for this instance.

---

[5] Web Service Modeling Toolkit available in https://sourceforge.net/projects/wsmt/.

```
wsmlVariant _"http://www.wsmo.org/wsml/wsml-syntax/wsml-flight"
namespace { _"http://www.uniriotec.br/aspect#",
    onto _"http://www.uniriotec.br/aspect#",
    wsml _"http://www.wsmo.org/wsml/wsml-syntax#",
    dc _"http://purl.org/dc/elements/1.1#",
    discovery _"http://wiki.wsmx.org/index.php?title=DiscoveryOntology#"
}
goal MailLogGoal
    importsOntology {
        onto#LogOntology
    }
    capability MailLogGoalCapability
        nfp
            discovery#discoveryStrategy hasValue discovery#LightweightDiscovery
            discovery#discoveryStrategy hasValue discovery#NoPreFilter
        endnfp
        postcondition
            definedBy
                ?logEffect memberOf onto#LogEffect
                    and ?logEffect[messageLogged hasValue _boolean("true")]
                    and ?logEffect[mailSent hasValue _boolean("true")].

ontology MailLogGoalOntology
    importsOntology onto#LogOntology
    instance logMail memberOf onto#LogRequest
        severity hasValue onto#WARN
        message hasValue "Test mail log using SWS."
```

Fig. 8. Mail log goal described in WSML.

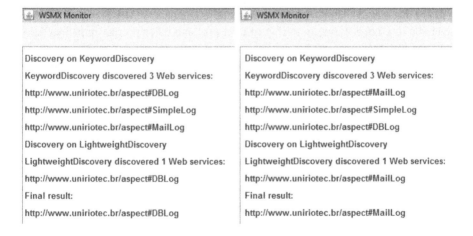

Fig. 9. WSMX monitoring (services discover and selection).

To evaluate the flexibility and adaptability in selecting an implementation at run time, the goal was then changed to "MailLog" (Fig. 6b). When activity "Receive Answer" was performed, GoalService sent the new goal for WSMX, and obtained as the outcome the Web Service that met this new goal without interrupting the execution of the aspect and the main process flow.

Figure 9 shows the screen with the monitoring of WSMX and the three calls in order to discover the Web Services according to the goals set.

5.1 Discussion

The results of this preliminary evaluation showed that it is possible to successfully carry out a dynamic weaving process, since the aspect "Log Information" was incorporated into the main process at runtime. Moreover, it was possible to verify the change in the behavior of the aspect when another Web Service has been selected to meet the new goal defined.

The results can be considered sound. During process execution, the aspect was inserted into the main process, with no changes in the main process model. From this evaluation, we could see in practice that Jalali et al. [7] research supported YAWL to perform the weaving process, and to check the evidence presented by Bastos et al. [5]. Moreover, our proposal allows an expert to focus on the construction of the main process without worrying about how crosscutting concerns will be incorporated.

This study was limited to only one aspect as an example in order to simplify this preliminary evaluation. The scenario was based on Bastos et al. [5] where the ontology of logging, the semantic Web Services and goals are described using WSML The development of Web Services were made by the authors of this work, so they can be not so precise, since the ontology and semantic description served as the basis for its implementation. Details in the description of Web Services could do WSMX return any or more Web Services during the discovery process. However, this work evidenced that the use of this method, supported by an artifact running on a WfMS, potentially creates a semantic link between the modeling of an aspect and its respective implementation from a defined goal.

The research of Jalali et al. [7] sought to enable a WfMS to incorporate aspects to the main process in practice. Our method has made progress on this issue while bringing the concept of operational goal for the aspect, so different implementations can be used to perform the activities of each advice. Unlike the findings presented in [7] relating to limitation of the use of the "Proceed" activity, the assignment of the GoalService can be done in more than one activity using the same advice, and may be included in loops.

In addition, other issues were raised in [14]. One of them is that sometimes different modeling options are possible, for example, an activity called "Archive" in an advice can be considered as various aspects such as logging. Our study moves forward with this issue by allowing different behaviors for the same advice simplifying the definition of an operational goal to be achieved, achieving flexibility and adaptability to the aspect.

6   Conclusions and Future Work

The aspect orientation in business processes aims to reduce complexity and enable reusability. The literature has demonstrated the advantages of this approach in different stages of the BPM life cycle. However, in order to adopt a unique approach to both modeling and implementation stages of BPM, it is necessary that the representations of aspects share a common vocabulary.

This work has extended the research of Bastos et al. [5] to set the behavioral characteristics of the aspects in order to establish a link between modeling and implementation. This link intends to decrease the gap between the phases. The artifact proposed in this paper allow that a WfMS environment selects the best implementation for an aspect, so that, during process execution, an implementation for the aspect can be discovered and invoked according to the operational goal set. Thus, we obtain flexibility in achieving the goals, and make aspects more adaptable during the implementation phase.

As future work, we aim to further investigate discovery and selection of candidate services returned by WSMX by GoalService, and with the support of simulations tools determine the degree of accuracy in the selection of services, in case the WSMX return two or more Web Services that it complies with the same goal. Furthermore, we will investigate a particular issue on how the method could address conflict between different aspects (e.g., between personal information privacy, and information gathering strategies, or between personal information privacy and logging policies), and it is also planned to extend the solution developed to other WfMS.

References

1. Tarr, P., Ossher, H., Harrison, W., Sutton Jr., S.M.: N Degrees of separation: multidimensional separation of concerns. In: Proceedings of the 21st International Conference on Software Engineering, pp. 107–119. ACM, New York (1999)
2. Kiczales, G., Lamping, J., Mendhekar, A., Maeda, C., Lopes, C.V., Loingtier, J.-M., Irwin, J.: Aspect-oriented programming. Comput. Sci. **1241**(1997), 220–242 (1997)
3. Charfi, A., Mezini, M.: AO4BPEL: an aspect-oriented extension to BPEL. World Wide Web **10**, 309–344 (2007)
4. Cappelli, C., Santoro, F.M., Do Prado Leite, J.C.S., Batista, T., Medeiros, A.L., Romeiro, C.S.C.: Reflections on the modularity of business process models: The case for introducing the aspect-oriented paradigm. Bus. Process Manag. J. **16**, 662–687 (2010)
5. Bastos, A., Santoro, F.M., Siqueira, S.W.M.: Bringing semantics to aspect-oriented business process management. In: Lohmann, N., Song, M., Wohed, P. (eds.) BPM 2013. LNBIP, vol. 171, pp. 291–302. Springer, Cham (2014). doi:10.1007/978-3-319-06257-0_23
6. Fensel, D., Facca, F.M., Simperl, E., Toma, I.: Web service modeling ontology. In: Fensel, D., Facca, F.M., Simperl, E., Toma, I. (eds.) Semantic Web Services, pp. 107–129. Springer, Heidelberg (2005). doi:10.1007/978-3-642-19193-0_7
7. Jalali, A., Wohed, P., Ouyang, C., Johannesson, P.: Dynamic weaving in aspect oriented business process management. In: Meersman, R., et al. (eds.) OTM 2013. LNCS, vol. 8185, pp. 2–20. Springer, Heidelberg (2013). doi:10.1007/978-3-642-41030-7_2
8. Jalali, A.: Foundation of Aspect Oriented Business Process Management (2011). http://aobpm.blogs.dsv.su.se/files/2011/10/Thesis.pdf
9. Cappelli, C., Leite, J., Batista, T., Silva, L.: An aspect-oriented approach to business process modeling. In: Proceedings of the 15th Workshop on Early Aspects - EA 2009, Charlottesville, VA, USA, p. 7 (2009)
10. van den Berg, K., Conejero, J.M., Chitchyan, R.: AOSD Ontology 1.0 - Public Ontology of Aspect-Orientation (2005). http://www.aosd-europe.net/deliverables/d9.pdf

11. Santos, F.J.N., Cappelli, C., Santoro, F.M., do Prado Leite, J.C.S., Batista, T.V.: Using goals to identify aspects in business process models. In: Proceedings of the 2011 International Workshop on Early Aspects - EA 2011, p. 19. ACM Press, New York (2011)
12. Haller, A., Cimpian, E., Mocan, A., Oren, E., Bussler, C.: WSMX - a semantic service-oriented architecture. In: IEEE International Conference on Web Services (ICWS 2005), vol. 1, pp. 321–328. IEEE (2005)
13. Andrews, T., Curbera, F., Dholakia, H., Goland, Y., Klein, J., Leymann, F., Liu, K., Roller, D., Smith, D., Thatte, S.: Business Process Execution Language for Web Services. Version 1.1 (2003)
14. Jalali, A., Ouyang, C., Wohed, P., Johannesson, P.: Supporting aspect orientation in business process management. Softw. Syst. Model., 1–23 (2015). http://xml.coverpages.org/BPELv11-20030505-20030331-Diffs.pdf
15. Dumas, M., La Rosa, M., Mendling, J., Reijers, H.A.: Fundamentals of Business Process Management. Springer, Heidelberg (2013)

First International Workshop on Process Querying (PQ 2016)

Introduction to the First International Workshop on Process Querying (PQ 2016)

Artem Polyvyanyy[✉] and Arthur H.M. ter Hofstede

Queensland University of Technology, Brisbane, Australia
{artem.polyvyanyy,a.terhofstede}@qut.edu.au
http://www.processquerying.com

Abstract. *Process querying* studies methods for automated management of real-world and envisioned processes, process models, process repositories, and process knowledge within modern organizations. The main objective of the first edition of the workshop was to provide a forum for researchers and practitioners to exchange findings and ideas on process querying research and practices. The program of the workshop included an invited talk and three accepted research papers (50% acceptance rate). The invited talk entitled "Process Querying with Uncertainty" was given by Associate Professor Avigdor Gal of the Faculty of Industrial Engineering & Management at the Technion, Israel. The three presentations of the accepted research papers on the workshop day addressed the topics of querying data in event logs and measuring similarity between business process models. The workshop took place on September 19, 2016, in Rio de Janeiro, Brazil.

Keywords: Process querying · Process · Process knowledge · Process management · Process model · Collection of process models · Process model repository

1 Aims and Background

Process-related information grows exponentially in organizations via workflows, guided procedures, business transactions, Internet applications, real-time device interactions, and other coordinative applications underpinning commercial operations. Event logs, application databases, process models, and business process repositories capture a wide range of process data, e.g., activity sequences, document exchanges, interactions with customers, resource collaborations, and records on product routing and service delivery. *Process querying* studies methods for automated management of information that concerns models of real-world or envisioned processes as well as executions of processes with the goal of converting this information into decision making capabilities.

The International Workshop on Process Querying 2016 (PQ 2016) is the first workshop for researchers and practitioners in the area of process querying. The workshop aims to provide a high quality forum for exchanging research findings and ideas on process querying technologies and practices. Process querying research spans a range of topics from theoretical studies of algorithms and the limits of computability of process querying techniques to practical issues of implementing process querying technologies in software. Process querying is an interdisciplinary area which makes use of existing knowledge (e.g., methods, techniques) but also contributes novel findings to research areas such as business process management, analysis and design of information systems, model checking, formal methods, process mining, information retrieval, big data, formal languages, programming languages, etc.

2 Main Topics

The main topics of the Process Querying 2016 workshop include:

– Behavioral and structural methods for process querying
– Imperative and declarative process querying
– Exact and approximate process querying
– Expressiveness of process querying
– Decidability and complexity of process querying
– Process query languages and notations
– Indexing for fast process querying
– Empirical evaluation and validation of process querying
– Label management in process querying
– Information retrieval methods in process querying
– Process querying of big (process) data
– Automatic management of process models, e.g., process model repair
– Automatic management of process model collections
– Event log querying
– Process performance querying
– Multi-perspective process querying, e.g., resource-, data-perspective, etc.
– Process querying exploiting rich ontology annotations
– Applications of process querying: process compliance, standardization, reuse, etc.
– Experience reports from implementations of process querying tools
– Case studies in process querying

3 Invited Talk

Note that information that follows in this section was provided by Associate Professor Avigdor Gal prior to the workshop day.

Title: Process Querying with Uncertainty.

Abstract: This talk will focus on aspects of uncertainty when querying process repositories to offer, for instance, re-use driven modelling support, harmonisation of model variants, and model-based system validation. Many of these techniques share reliance on the identification of correspondences between the entities of different models, also termed process matching. Taking up recent advances in matching process models in particular, we shall provide an overview of concepts and matching techniques in this domain. The talk will include a brief review of similarity measures for the textual, structural, and behavioural dimensions of process models, which form the basis of matching techniques. The talk will then present a vision of using uncertainty management tools to support effective querying of process repositories.

Speaker: Associate Professor Avigdor Gal of the Faculty of Industrial Engineering & Management at the Technion is a Technion graduate and an expert on information systems. His research focuses on effective methods of integrating data from multiple and diverse sources, which affect the way businesses and consumers seek information over the Internet. His current work zeroes in on schema matching – the task of providing communication between databases, and connecting such communication to real-world concepts. Another line of research involves the identification of complex events such as flu epidemics, biological attacks, and breaches in computer security, and its application to disaster and crisis management. He has applied his research to European and American projects in government, eHealth, and the integration of business documents. Prof. Gal has published more than 100 papers in leading professional journals, conferences, and books. He authored the book Uncertain schema Matching in 2011.

4   Workshop Organizers

Artem Polyvyanyy	Queensland University of Technology, Australia
Arthur H.M. ter Hofstede	Queensland University of Technology, Australia

5   Program Committee

Ahmed Awad	Cairo University, Egypt
Hyerim Bae	Pusan National University, South Korea
Joos Buijs	Eindhoven University of Technology, The Netherlands
Massimiliano de Leoni	Eindhoven University of Technology, The Netherlands
Jochen De Weerdt	Katholieke Universiteit Leuven, Belgium
Gero Decker	Signavio, Germany
Claudio Di Ciccio	Vienna University of Economics and Business, Austria
Remco Dijkman	Eindhoven University of Technology, The Netherlands
Dirk Fahland	Eindhoven University of Technology, The Netherlands
Luciano García-Bañuelos	University of Tartu, Estonia
David Knuplesch	Ulm University, Germany
Matthias Kunze	Zalando, Germany
Marcello La Rosa	Queensland University of Technology, Australia
Henrik Leopold	VU University Amsterdam, The Netherlands
Jorge Munoz-Gama	Pontificia Universidad Católica de Chile, Chile
Chun Ouyang	Queensland University of Technology, Australia

Artem Polyvyanyy	Queensland University of Technology, Australia
Stefanie Rinderle-Ma	University of Vienna, Austria
Andreas Rogge-Solti	Vienna University of Economics and Business, Austria
Minseok Song	Pohang University of Science and Technology, South Korea
Arthur ter Hofstede	Queensland University of Technology, Australia
Seppe vanden Broucke	Katholieke Universiteit Leuven, Belgium
Boudewijn van Dongen	Eindhoven University of Technology, The Netherlands
Matthias Weidlich	Humboldt University of Berlin, Germany

6 Additional Reviewers

Raffaele Conforti	Queensland University of Technology, Australia
Johannes Pflug	University of Vienna, Austria

Process Model Search
Using Latent Semantic Analysis

Andreas Schoknecht[✉], Nicolai Fischer, and Andreas Oberweis

Karlsruhe Institute of Technology, Institute AIFB,
Building 05.20, 76128 Karlsruhe, Germany
{andreas.schoknecht,andreas.oberweis}@kit.edu,
nicolai.fischer@student.kit.edu

Abstract. Process model similarity measures can be utilized for searching process model collections, which is also called similarity-based search. While there are quite a lot of approaches, most of them base on an underlying alignment between the activities of the compared process models. Yet, according to the results of the process model matching contests conducted in recent years, such an alignment seems to be quite difficult to achieve. The Latent Semantic Analysis-based Similarity Search approach described in this paper circumvents the matching challenge by not requiring such a matching. Instead, it uses a Latent Semantic Analysis-based Similarity Measure to query model collections and retrieve similar models. An evaluation with a collection of 80 models resulted in very good results in terms of Precision, Recall, and F-Measure. The best F-Measure value obtained during the experiments was 0.92.

Keywords: Process model querying · Process model search · Process model similarity · Latent Semantic Analysis

1 Introduction

Calculating the similarity between business process models has been the focus of many research publications in the past years due to its importance regarding the management of large collections of process models. These collections can contain hundreds or even thousands of models nowadays (see, e.g., the collections mentioned in [1,2]), which makes sophisticated operations like conformance checking, duplicate detection, or the reuse of (parts of) models hard to conduct without automated support. Besides these application areas, process model similarity measures are also utilized for searching process model collections, which is also called similarity-based search [3]. In this context a process model is used as query model, the input, with the aim to find similar models as output. These output models are then usually ranked by a decreasing similarity value to the input model. Hence, such approaches differ from another popular stream of research which uses query languages specifically designed for querying process model collections [4].

Regarding the automatic similarity measurement between process models used in similarity-based search techniques, many approaches have been published

© Springer International Publishing AG 2017
M. Dumas and M. Fantinato (Eds.): BPM 2016 Workshops, LNBIP 281, pp. 283–295, 2017.
DOI: 10.1007/978-3-319-58457-7_21

during the last years (see, e.g., the survey in [5] for an overview). But while there are quite a lot of approaches, most of them base on an underlying alignment between the activities of the compared process models, i.e., before calculating a final similarity value these approaches conduct some kind of process model matching. Yet, such a matching – and even more so a high quality matching – seems to be quite difficult to achieve according to the results of the process model matching contests conducted in recent years [6]. With respect to the results of the Process Model Matching Contest 2015 [6] the participating matching techniques performed quite poorly. Even the best Recall values were below 0.7, which means that each matching technique did not detect at least 30% of the correct matches. For one dataset the authors of the matching contest publication even stated that 36% of the correct matches were not detected at all.

The *Latent Semantic Analysis-based Similarity Search* (LS3) approach described in the following circumvents the matching challenge by not requiring such a matching. Instead, it uses a *Latent Semantic Analysis-based Similarity Measure* (LSSM) which treats a whole process model as a text corpus and calculates a similarity value based on the contained words. It leaves aside the structure and the behavior of process models as these aspects are regarded less relevant with respect to the application in a search function. As an example, the order of activities might arguably be more important in the context of conformance checking, in which legal regulations might prohibit certain execution sequences than in a search function, which focuses on the content of a process model. First evaluation results of the *LS3* are very promising as they yield an F-Measure of 0.92 in the best case.

The rest of the paper is organized as follows: In Sect. 2 basic terminology and definitions are introduced while in Sect. 3 related work with regard to similarity-based search of process model collections focusing on the semantic similarity of labels is described. Afterwards, the newly proposed *Latent Semantic Analysis-based Similarity Search* is presented in Sect. 4 followed by an evaluation and discussion of the evaluation results in Sect. 5. Finally, Sect. 6 summarizes the results of the paper and provides an outlook on future research.

2   Basic Terminology and Definitions

2.1   Business Process

The term business process refers to sequences of manual or (partly) automated activities executed in a company or organization according to specific rules with a certain aim. Such processes are typically represented as process models with the help of business process modeling languages like EPC, BPMN, or Petri Nets for various reasons like documentation or process analysis. While the *LS3* approach could generally be applied for all the mentioned languages, the process models in the paper at hand are represented as Petri Nets as the models used in the evaluation were available in this language. The following definition provides a formal description (for the definition of Petri Nets see, e.g., [7]).

Definition 1 (Labelled Petri Net). *A labelled Petri Net is a 4-Tuple* $PN = (P, T, F, \ell)$ *with*

- $P = \{p_1, p_2, \cdots, p_m\}$ *being a finite set of places,*
- $T = \{t_1, t_2, \cdots, t_n\}$ *being a finite set of transitions, i.e., the activities,*
- $F \subseteq (S \times T) \cup (T \times S)$ *being a set of arcs representing the control flow and*
- $\ell : P \cup T \to L$ *being a labeling function which assigns a label* $l \in L$ *to each place* $p \in P$ *and to each transition* $t \in T$ *with* L *being a set of labels.*
- *Additionally, it holds that* $P \cap T = \emptyset$ *and* $P \cup T \neq \emptyset$.

Note that we specifically included a labeling function for the transitions and places as the Latent Semantic Analysis-based Similarity Measure used in the *LS3* search approach utilizes these labels to determine a similarity value between process models.

2.2   Latent Semantic Analysis

Latent Semantic Analysis (LSA) is both a theory and a mathematical/statistical method for capturing the meaning of words and documents. It was developed to improve document retrieval based on user queries [8]. LSA goes one step further than ordinary indexing techniques – which are based on a syntactical comparison of terms of the query and the documents (so called word matching approach) – by adding a semantic aspect to these comparisons. As a consequence meanings of terms and documents can be used to improve retrieval results [9].

LSA as a theory is based on the assumption that the semantics of terms is determined by the meanings of documents in which they appear and vice versa that the semantics of a document is determined by the meanings of the terms contained in it. Therefore, LSA aims at analyzing the choice of words, which reflects a hidden respectively latent semantic structure of a document [10].

LSA as a mathematical/statistical method is based on the vector space model, which can be represented by a term-document matrix. In this context a document is a text passage with a predefined length and a term is a word or a meaningful unit (e.g. statue of liberty). An entry in the term-document matrix represents the weight of a specific term in a specific document such as the term frequency. LSA generates knowledge by analyzing large collections of text [11]. By applying singular value decomposition, an $m \times n$ term-document matrix A with rank r is decomposed into three matrices T, Σ and D^T. The rows of the orthogonal matrix T describe the term vectors whereas the columns of the orthogonal matrix D^T describe the document vectors. Matrix Σ contains the corresponding singular values sorted in decreasing order. For analyzing semantics, the LSA uses a so called semantic space. It is created by keeping only the k largest singular values and the corresponding values of the term and document vectors thereby effectively executing a dimension reduction. This step is supposed to reduce noise within the document collection, i.e., handling problems related to synonymy and polysemy of words. The product A_k of the generated matrices T_k, Σ_k and D_k^T is the best least squares approximation of the original term-document matrix A. The explained singular value decomposition and

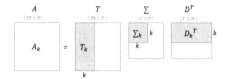

Fig. 1. Singular value decomposition and dimension reduction

the reduction on k dimensions is shown in Fig. 1. The four main steps of the LSA-based retrieval procedure according to [11] are then as follows:

1. **Extraction of Terms for the Term-Document Matrix:** In this process step terms are extracted from each document of the examined collection. It can be useful to apply natural language processing concepts before constructing the term-document matrix (e.g. removal of stop words). The result of this step is a term-document matrix containing term frequencies.
2. **Transformation of the Term-Document Matrix:** In this step the entries of the term-document matrix (term frequencies) are transformed by applying a weighting scheme, which enables better differentiation of documents.
3. **Singular Value Decomposition and Dimension Reduction:** This step decomposes the term-document matrix and truncates the resulting matrices as shown in Fig. 1. The result of this step is the so called semantic space.
4. **Retrieval in Semantic Space:** Term-Term, Document-Document and Term-Document comparisons are computed in the semantic space by applying an appropriate similarity measure for vectors, such as the cosine similarity.

3   Related Work

Two major streams of research can be identified which are related to the *LS3* approach. The first one focuses on querying languages for process model collections, while the second one uses existing process models for a similarity-based search. Works falling into the first category include, e.g., the specification of query languages for process models in BPMN [12] or BPEL [13]. Going one step further, APQL has been published recently, which can be used independently of a particular process modeling language [4]. Yet, common to all these approaches is the need to formulate a query with the respective query language, whereas existing process models cannot be used as a query. In this respect the *LS3* approach differs from the aforementioned works as it uses existing models as queries. Hence, it is closer to the second research stream.

An early work on similarity-based search is described in [14]. The presented algorithms focus on the graph structure of process models to determine a similarity value, while requiring similarity values of labels, i.e., an alignment of process model elements. The *LS3* approach, however, explicitly focuses on the textual labels and does not require such an alignment. The approach described in [15] does also require an alignment while focusing on the behavior of process models. And it does not calculate a similarity value, but requires models to include all of the behavior of a query model. The approach by Gater et al. [16] uses an index based on behavioral characteristics of a model to speed up the query answering. Speeding up the querying process is also the aim of [17] by using simple but representative abstractions of process models, so called features. [18] focuses on the structural aspects and clustering of workflows for similarity-based search, which the *LS3* approach does not use. The approach described by Awad et al. [19], however, is closer to our approach. They use an Enhanced Topic-based Vector Space Model, which is based on an ontology, for the comparison of a BPMN-Q query against process models. The *LS3* approach in contrast does not rely on an ontology but on LSA to calculate a similarity value, hence there is no effort required for constructing such an ontology. Another approach by Qiao et al. [20] uses topic language modeling, specifically the Latent Dirichlet Allocation (LDA), combined with structural similarity calculation and clustering to retrieve similar models. The LDA aspect is closely related to our Latent Semantic Analysis approach. But while the LSA compares vectors in a k dimensional space to determine a similarity value, the LDA-based approach requires the estimation of the probability that a process model can generate a certain query. Another recent approach described in [21] performs a similarity search based on the structure of process models and by calculating an alignment of model elements using WordNet [22] to calculate the semantic relatedness of labels. In contrast, the *LS3* approach does not require such an alignment nor WordNet.

We acknowledge that there exist further publications describing a similarity measure for process models which could be used in a search approach. But due to space restrictions, we limit this section to the aforementioned works. To our knowledge, these relate most closely to our approach of similarity-based search. For an overview of further related publications see, e.g., the survey in [5].

4   LS3: Latent Semantic Analysis-Based Similarity Search

In this section we present the *Latent Semantic Analysis-based Similarity Search* (LS3). The *LS3* retrieves results for a query model by calculating similarity values between process models of a process model collection using a *Latent Semantic Analysis-based Similarity Measure* (LSSM) and by returning all models above a certain threshold. The LSA utilized in the *LSSM* originally aims at analyzing the word choice of a given text passage with a predefined length, a so called document. Therefore, it is necessary to delimit word sets in a meaningful and understandable manner to define such a document. For our purpose – measuring the similarity between process models – whole process models serve as a basis to define a document. Thus, a document vector of a process model is defined as follows (adapted from [23]):

Definition 2 (Document Vector of a Process Model). *Let M be a set of process models, W_{all} be a set of terms containing all distinct terms of*

M, and w(m) be a function, which returns the set of terms (bag-of-words)
W_m of a process model $m \in M$, whereby $W_m \subseteq W_{all}$. The vector $d_m =$
$(w_{1m}, w_{2m}, ..., w_{tm})$ then represents the document vector of the process model
m, whereby each index t represents a term of the set of all terms contained in
the process model collection $W_{all} = \bigcup w(m)$ for all $m \in M$. The entries w_{tm}
reflect a weight of the term frequency, which describe how often a certain term
appears within a model.

The similarity-based search is then conducted according to the following five
steps (see Fig. 2). Thereby, the first four steps constitute the calculation of the
LSSM similarity while query results are determined in the fifth step. As an
existing process model might be used as a query q, we treat q as any other
process model m. To calculate term frequencies correctly, we include q in the set
of process models M containing n models. Therefore, the set M contains both
the query q and all process models of the collection ($n + 1$ models in total).

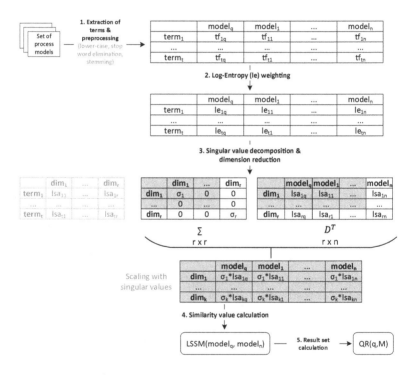

Fig. 2. Conceptual LS3 querying approach

Step 1 - Extraction of terms for the term-document matrix: For creating the
term-document matrix each process model and the query has to be represented
by its document vector (the columns of the matrices in Fig. 2), which is specified
in Definition 2. In our case of Petri Net-based process models we first extract
the distinct terms of all transition and place labels L from each process model

$m \in M$ (note that M now also contains q). Then, each (multi-)set of terms is transformed into lower-case letters, stop words are removed and the terms are stemmed with a Porter-Stemmer [24]. This preprocessing is conducted to reduce the vector space, i.e., we want to eliminate terms that do not contribute to the semantics of a process model. For example, the term respectively stop word "the" does not contribute any semantics to a model. Result of this step is a term-document matrix containing absolute term frequencies. These term frequencies tf_{ij} indicate how often a term i appears in process model j ($1 \leq i \leq t$, with t being the amount of distinct terms and $0 \leq j \leq n$).

Step 2 - Transformation of the term-document matrix: As the usage of absolute term frequencies is discouraged in studies [25], we apply the *log-entropy* weighting scheme on the values of the term-document matrix generated in step 1. The log-entropy weighting scheme [10] is defined as follows:

$$le_{ij} = log_2(tf_{ij} + 1) \cdot (1 + \sum_{j=1}^{n+1} \frac{p_{ij} \cdot log_2 p_{ij}}{log_2 j}), \quad \forall tf_{ij} > 0. \tag{1}$$

Thereby, $p_{ij} = \frac{tf_{ij}}{gf_i}$ is the quotient of the term frequency tf_{ij} and the global frequency gf_i. The global frequency indicates how often a term i appears in the whole process model collection. See also the second step of Fig. 2.

Step 3 - Singular value decomposition and dimension reduction: In this step the most important part of the LSA is executed: the singular value decomposition and dimension reduction to reduce noise and create latent, semantic concepts, thus generating the so called semantic space. As shown in Figs. 1 and 2, the (log-entropy transformed) term-document matrix is decomposed into three matrices. Only the matrices Σ and D^T are relevant for the purpose of determining the similarity between process models. The matrix D^T contains the calculated process model vectors, matrix Σ contains the corresponding singular values sorted in descending order. The amount of singular values greater than zero corresponds to the rank r of the original term-document matrix. Therefore, the upper boundary for choosing certain dimensions is set by the rank r. By choosing r dimensions the calculated similarity values are the same as if they were calculated with the original (weighted) term-document matrix.

Step 4 - Similarity value calculation: Before calculating the similarity between two column vectors, i.e., two process models with formula 2, it is necessary to scale the entries of these vectors with their corresponding singular values [8]. For the actual determination of the similarity between a query model q and another business process model m, we then use the cosine similarity [26, p. 25] of their vector representations transformed on the interval $[0, 1]$ as specified in formula 2 with 0 indicating dissimilarity and 1 indicating similarity.

$$LSSM(q, m) = \frac{cos_{sim}(q, m) + 1}{2} \tag{2}$$

Hence, the semantics of a process model (and a query) is represented by the direction of its document vector whereas the length of a vector might be misleading. To be independent of the vector length, we prefer the cosine similarity over distance measures (e.g. the euclidean distance) [27]. See also steps 3 and 4 in Fig. 2.

Step 5 - Retrieval of query results: After calculating the $LSSM$ similarity values between a query and all other models, the query results for a query model q are then determined by applying a threshold value θ. I.e., all models with an $LSSM$ similarity value equal to or higher than θ are included in the result set $QR(q, M) = \{m \mid m \in M \wedge LSSM(q, m) \geq \theta\}$ (step 5 in Fig. 2.).

5   Evaluation and Discussion

For the empirical evaluation of the $LS3$ we used the process model collection introduced by Vogelaar et al. [28]. This collection contains models of eight different business processes performed by 10 different dutch municipalities. Hence, there exist 80 process models in the collection in total. This model collection is linguistically harmonized, i.e., node labels are unambiguous and consistent. Thus, same activities are labeled in the same manner. For our purpose, we transformed these models from the original YAWL representation into Petri Nets and stored them in the XML-based format PNML, which enables an automatic processing of the models. A prototype[1] of the $LS3$ has been implemented in JAVA and the evaluation has been conducted on a laptop with the following specifications: Intel(R) i5-4300U CPU, 8 GB RAM, Windows 10 and JAVA Version 1.8.

5.1   Results

To evaluate the $LS3$ we determined Precision, Recall and F-Measure on the basis of the described process model collection. Thereby, Precision is defined as the fraction of relevant and received results (true positives TP) to all received results (B); Recall is defined as the fraction of relevant and received results to all relevant results (A); and F-Measure is defined as the harmonic mean of Precision and Recall. Formally, these values are calculated as follows:

$$P = \frac{|TP|}{|B|}, \quad R = \frac{|TP|}{|A|}, \quad F = 2 \cdot \frac{P \cdot R}{P + R}$$

These measures were calculated for each possible dimensionality k (the maximum is 69 in our case) by querying the collection with each process model as input. This means that we executed 80 queries for each dimensionality. A process model m is returned if $LSSM(q, m) \geq 0.75$, which is a manually determined threshold. Returned process models are relevant if they describe the same business process as the input model. Hence, ideally, for each query model the result

---

[1] The JAVA API can be downloaded from `butler.aifb.kit.edu/asc/LS3/ls3.html`.

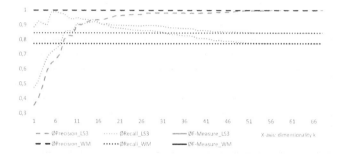

Fig. 3. Results of the empirical evaluation: Precision, Recall and F-Measure

set would only contain the nine other models which describe the same process as the query model. Figure 3 shows the average values of Precision, Recall and F-Measure dependent on the chosen dimensionality. As described in Sect. 4 the upper boundary for the choice of dimensions is set by the original term-document matrix. For the aforementioned model collection the rank of the corresponding term-document matrix is 69. Therefore, Fig. 3 presents the average Precision, Recall and F-Measure values for the whole 80 queries of all 69 possibilities. Note that we excluded dimensionality 1 as it only contains the similarity values 1 and 0 since there exists only one latent concept.

The values of the traditional word matching approach serve as a benchmark. These values are calculated by applying the (on the interval [0,1] transformed) cosine similarity measure on the weighted term-document matrix after step 2 of the presented conceptual approach. The term "word matching" refers to a vector space model whose dimensions are formed by the set of unique terms W_{all} and not by latent, semantic dimensions as generated through dimension reduction [11]. Consequently – in case of word matching – only syntactical comparisons of document vectors are made. The reference average values are: $\varnothing$ Precision$_{WM}$ = 1, $\varnothing$ Recall$_{WM}$ = 0.77 and $\varnothing$ F-Measure$_{WM}$ = 0.85.

First of all, Fig. 3 shows the typical development for Precision and Recall curves in Information Retrieval: if Precision is high, Recall is low and vice versa. The highest $\varnothing$ Recall$_{LS3}$ is achieved for dimensions 6 and 7 (0.99) whereas the highest $\varnothing$ Precision$_{LS3}$ is achieved for dimensions above 56 (1). Dependent on the number of dimensions the *LS3* Precision curve gradually approximates the Precision$_{WM}$ value from below. The higher the dimension, the smaller is the difference between $\varnothing$ Precision$_{LS3}$ and $\varnothing$ Precision$_{WM}$. As far as the *LS3* Recall curve is concerned, there is a different trend visible. The $\varnothing$ Recall$_{LS3}$ values approximate from above the corresponding Recall$_{WM}$ value.

Regarding the development of the F-Measure curve it is obvious that for dimensions lower than 9 the *LS3* performance is poorer than the performance of the traditional word-matching approach. Yet, the highest overall $\varnothing$ F-Measure value is calculated for dimension 16 (0.92). For dimensions above 16 the F-Measure$_{LS3}$ curve gradually approaches its corresponding F-Measure$_{WM}$ value. This means that for all dimensions below 9 the *LS3* F-Measure performance is at least equal to the word-matching performance.

5.2 Discussion

The previously described results are typical for LSA based applications as far as the F-Measure is considered as a measure for overall performance [11]: for small dimensions performance is poor compared to the word-matching approach, for a certain dimension interval the performance is better (for the presented process model collection interval [9, 43]), and for all other dimensions the performance difference of the two approaches is negligible small. The Recall curve in Fig. 3 is a typical property for LSA based applications since the LSA is supposed to increase Recall compared to the word matching approach. The Precision reference value ($\varnothing$ Precision$_{\mathrm{WM}} = 1$) is remarkable: two process models describing two different business processes of the collection were modeled in a way (by using "enough" syntactical different terms) that no similarity values above the defined threshold ($\theta = 0.75$) were calculated, i.e., no false positives were calculated. The described approximation of $\varnothing$ Precision$_{\mathrm{LS3}}$, $\varnothing$ Recall$_{\mathrm{LS3}}$ and $\varnothing$ F-Measure$_{\mathrm{LS3}}$ to their corresponding reference values is typical for LSA based applications as well [11]. Besides these LSA-based aspects, the results show that the *LS3* seems to be very well suited for identifying similar process models. The F-Measure values are very high for dimensions higher than 8, which means that the query results mostly contain true positives. Additionally, the average standard deviation values over all queries for all dimensions for Precision, Recall and F-Measure are quite low (0.09; 0.20; 0.19). This implies that the query results have a consistently high quality.

Strengths: The *LS3* enables the similarity calculation only by analyzing the text corpus itself and not by looking up "external" information, such as ontologies or thesauri (e.g. WordNet [22]). Hence, no effort must be spent on constructing those ontologies, which might also not be able to reliably measure the semantic similarity between words which are specific to an organization. Furthermore, the *LS3* is not dependent on an underlying alignment between the activities of the compared process models. Thus, this complex problem is completely avoided. Finally, the run time (overall 3.23 s; only calculating similarity values and data storage in csv-files without computing the SVD 1.30 s) of our prototype is promising. Therefore, the *LS3* has a high potential for an application in real world contexts.

Limitations: As the *LS3* is based on the LSA they have the same shortcomings in common, especially determining the optimal dimension or the interpretation of the created latent dimensions. While the interpretation of latent semantic dimensions might not be of importance in the context at hand, finding the optimal dimension does have a significant impact on the F-Measure values, which are comparably low for dimensions smaller 9. Further details on these shortcomings can, e.g., be found in [29]. Furthermore, for applying the *LSSM* it is crucial that the process model collection is sufficiently large and that each process model contains sufficient labels as the LSA gains its knowledge by analyzing large text corpora. If process models are textually described by only a few labels, it might be difficult to detect semantics for the *LSSM*.

6   Conclusion

The *Latent Semantic Analysis-based Similarity Search* (LS3) approach described in the paper at hand utilizes the Latent Semantic Analysis (LSA) concept from document retrieval for searching process model collections. Based on a newly developed similarity measure, the *Latent Semantic Analysis-based Similarity Measure*, the *LS3* approach returns all process models in a collection that have a similarity value above a certain threshold. In a first evaluation with 80 process models from Dutch municipalities [28] the *LS3* showed very promising results. When choosing the best possible number of dimensions the F-Measure achieves a value of 0.92.

Regarding future work, further empirical studies are needed to assess the strengths and weaknesses of the *LS3* more precisely. The inclusion of other model data sets with different characteristics would be desirable. As the models used in the evaluation have quite harmonized labels, it would be interesting to assess the *LS3* performance with more heterogeneous labels. Besides, a comparison with other similarity-based search approaches should be conducted to compare the *LS3* approach not only against a basic technique from document retrieval but also against other approaches specifically designed for process model search. Furthermore, the run time of the *LS3* might be improved by not requiring the calculation of a term-document matrix for each query. E.g., calculating the matrix once for a model collection and updating it on model insertions/deletions might be more efficient.

References

1. Lau, C.K., Fournier, A.J., Xia, Y., Recker, J., Bernhard, E.: Process model repository governance at suncorp. Technical report, Queensland University of Technology (2011)
2. Song, L., Wang, J., Wen, L., Wang, W., Tan, S., Kong, H.: Querying process models based on the temporal relations between tasks. In: Proceedings of the 15th IEEE International EDOCW Workshops, pp. 213–222 (2011)
3. Dumas, M., García-Bañuelos, L., Dijkman, R.M.: Similarity search of business process models. IEEE Data Eng. Bull. **32**(3), 23–28 (2009)
4. ter Hofstede, A.H.M., Ouyang, C., Rosa, M.L., Song, L., Wang, J., Polyvyanyy, A.: APQL: a process-model query language. In: Proceedings of the 1st AP-BPM Conference, pp. 23–38 (2013)
5. Becker, M., Laue, R.: A comparative survey of business process similarity measures. Comput. Ind. **63**(2), 148–167 (2012)
6. Antunes, G., Bakhshandeh, M., Borbinha, J., Cardoso, J., Dadashnia, S., Francescomarino, C.D., Dragoni, M., Fettke, P., Gal, A., Ghidini, C., Hake, P., Khiat, A., Klinkmüller, C., Kuss, E., Leopold, H., Loos, P., Meilicke, C., Niesen, T., Pesquita, C., Péus, T., Schoknecht, A., Sheetrit, E., Sonntag, A., Stuckenschmidt, H., Thaler, T., Weber, I., Weidlich, M.: The process model matching contest 2015. In: Proceedings of the 6th EMISA Workshop, pp. 127–155 (2015)
7. Reisig, W.: Understanding Petri Nets - Modeling Techniques, Analysis Methods, Case Studies. Springer, Heidelberg (2013)

8. Deerwester, S.C., Dumais, S.T., Landauer, T.K., Furnas, G.W., Harshman, R.A.: Indexing by latent semantic analysis. J. Am. Soc. Inf. Sci. **41**(6), 391–407 (1990)

9. Dumais, S.T., Furnas, G.W., Landauer, T.K., Deerwester, S., Harshman, R.: Using latent semantic analysis to improve access to textual information. In: SIGCHI Conference on Human Factors in Computing Systems, pp. 281–285 (1988)

10. Landauer, T.K., Foltz, P.W., Laham, D.: An introduction to latent semantic analysis. Discourse Process. **25**(2–3), 259–284 (1998)

11. Dumais, S.T.: LSA and information retrieval: getting back to basics. In: Handbook of Latent Semantic Analysis, pp. 293–321. Lawrence Erlbaum Associates (2007)

12. Awad, A.: BPMN-Q: a language to query business processes. In: Proceedings of the 2nd EMISA Workshop, pp. 115–128 (2007)

13. Beeri, C., Eyal, A., Kamenkovich, S., Milo, T.: Querying business processes with BP-QL. Inf. Syst. **33**(6), 477–507 (2008)

14. Dijkman, R., Dumas, M., García-Bañuelos, L.: Graph matching algorithms for business process model similarity search. In: Dayal, U., Eder, J., Koehler, J., Reijers, H.A. (eds.) BPM 2009. LNCS, vol. 5701, pp. 48–63. Springer, Heidelberg (2009). doi:10.1007/978-3-642-03848-8_5

15. Kunze, M., Weidlich, M., Weske, M.: Querying process models by behavior inclusion. Softw. Syst. Model. **14**(3), 1105–1125 (2015)

16. Gater, A., Grigori, D., Bouzeghoub, M.: Indexing process model flow dependencies for similarity search. In: Meersman, R., Panetto, H., Dillon, T., Rinderle-Ma, S., Dadam, P., Zhou, X., Pearson, S., Ferscha, A., Bergamaschi, S., Cruz, I.F. (eds.) OTM 2012. LNCS, vol. 7565, pp. 128–145. Springer, Heidelberg (2012). doi:10.1007/978-3-642-33606-5_9

17. Yan, Z., Dijkman, R., Grefen, P.: Fast business process similarity search. Distrib. Parallel Databases **30**(2), 105–144 (2012)

18. Kastner, M., Wagdy Saleh, M., Wagner, S., Affenzeller, M., Jacak, W.: Heuristic methods for searching and clustering hierarchical workflows. In: Moreno-Díaz, R., Pichler, F., Quesada-Arencibia, A. (eds.) EUROCAST 2009. LNCS, vol. 5717, pp. 737–744. Springer, Heidelberg (2009). doi:10.1007/978-3-642-04772-5_95

19. Awad, A., Polyvyanyy, A., Weske, M.: Semantic querying of business process models. In: Proceedings of the 12th EDOC Conference, pp. 85–94 (2008)

20. Qiao, M., Akkiraju, R., Rembert, A.J.: Towards efficient business process clustering and retrieval: combining language modeling and structure matching. In: Rinderle-Ma, S., Toumani, F., Wolf, K. (eds.) BPM 2011. LNCS, vol. 6896, pp. 199–214. Springer, Heidelberg (2011). doi:10.1007/978-3-642-23059-2_17

21. Li, S., Cao, J.: A new similarity search approach on process models. In: Cao, J., Wen, L., Liu, X. (eds.) PAS 2014. CCIS, vol. 495, pp. 11–20. Springer, Heidelberg (2015). doi:10.1007/978-3-662-46170-9_2

22. Miller, G.A.: WordNet: a lexical database for English. Commun. ACM **38**(11), 39–41 (1995)

23. Malinova, M., Dijkman, R., Mendling, J.: Automatic extraction of process categories from process model collections. In: Lohmann, N., Song, M., Wohed, P. (eds.) BPM 2013. LNBIP, vol. 171, pp. 430–441. Springer, Cham (2014). doi:10.1007/978-3-319-06257-0_34

24. Porter, M.F.: An algorithm for suffix stripping. Program **14**(3), 130–137 (1980)

25. Zaman, A.N.K., Brown, C.G.: Latent semantic indexing and large dataset: study of term-weighting schemes. In: Proceedings of the 5th ICDIM Conference, pp. 1–4 (2010)

26. van Rijsbergen, C.J.: Information Retrieval, 2nd edn. Butterworth-Heinemann, Newton (1979)

27. Turney, P.D., Pantel, P.: From frequency to meaning: vector space models of semantics. J. Artif. Intell. Res. (JAIR) **37**, 141–188 (2010)
28. Vogelaar, J.J.C.L., Verbeek, H.M.W., Luka, B., Aalst, W.M.P.: Comparing business processes to determine the feasibility of configurable models: a case study. In: Daniel, F., Barkaoui, K., Dustdar, S. (eds.) BPM 2011. LNBIP, vol. 100, pp. 50–61. Springer, Heidelberg (2012). doi:10.1007/978-3-642-28115-0_6
29. Hu, X., Cai, Z., Wiemer-Hastings, P., Graesser, A.C., McNamara, D.S.: Strengths, limitations, and extensions of LSA. In: Handbook of Latent Semantic Analysis, pp. 401–426. Lawrence Erlbaum Associates (2007)

Everything You Always Wanted to Know About Your Process, but Did Not Know How to Ask

Eduardo González López de Murillas[1(✉)], Hajo A. Reijers[1,2],
and Wil M.P. van der Aalst[1]

[1] Department of Mathematics and Computer Science,
Eindhoven University of Technology, Eindhoven, The Netherlands
{e.gonzalez,h.a.reijers,w.m.p.v.d.aalst}@tue.nl
[2] Department of Computer Science, VU University Amsterdam,
Amsterdam, The Netherlands

Abstract. The size of execution data available for process mining analysis grows several orders of magnitude every couple of years. Extracting and selecting the relevant data to be analyzed on each case represents an open challenge in the field. This paper presents a systematic literature review on different approaches to query process data and establish their provenance. In addition, a new query language is proposed, which overcomes the limitations identified during the review. The proposal is based on a combination of data and process perspectives. It provides simple constructs to intuitively formulate questions. An implementation of the language is provided, together with examples of queries to be applied on different aspects of the process analysis.

Keywords: Process mining · Databases · Event logs · Query languages

1 Introduction

One of the main goals of process mining techniques is to obtain insights into the behavior of systems, companies, business processes, or any kind of workflow under study. Obviously, it is important to perform the analysis on the right data. Being able to extract and query some specific subset of the data becomes crucial when dealing with complex and heterogeneous datasets. In addition, the use of querying tools allows finding specific cases or exceptional behavior. Whatever the goal, analysts often find themselves in the situation in which they need to develop ad-hoc software to deal with specific datasets, or use existing tools that might be difficult to use, too general, or just not suitable for process analysis.

Different approaches exist to support the querying of process data. Some of them belong to the field of Business Process Management (BPM). In this field, events are the main source of information. They represent transactions or activities that were executed at a certain moment in time in the environment under study. Querying this kind of data allows to obtain valuable information about the behavior and execution of processes. There are other approaches originating

© Springer International Publishing AG 2017
M. Dumas and M. Fantinato (Eds.): BPM 2016 Workshops, LNBIP 281, pp. 296–309, 2017.
DOI: 10.1007/978-3-319-58457-7_22

from the field of data provenance, which are mainly concerned with recording and observing the origins of data. This field is closely related to scientific workflows, in which the traceability of the origin of experimental results becomes crucial to guarantee correctness and reproducibility. In each of these fields, we find query languages and techniques that focus on the particularities of their input data. However, none of these approaches succeeds at combining process and data aspects in an integrated way. In addition, the development of a query mechanism that allows to exploit this combination, while being intuitive and easy to use, represents an additional challenge to overcome.

In order to make the querying of process event data easier and more efficient, we propose a new query language that exploits both process and data perspectives. Section 2 presents a systematic literature review, analyzing existing solutions and comparing them. Section 3 presents our approach together with examples of its use. Section 4 provides information about the implementation and, finally, Sect. 5 concludes the paper.

2 Systematic Literature Review

In order to get an overview of existing approaches, we first concluded a systematic literature review [1]. Figure 1 shows an overview of the procedure. First, a coarse set of candidate papers needs to be obtained from a scientific publications database or through a search engine (*Query*). Afterward, a *relevance screening* is performed in order to identify with papers are actually within the scope. To do so, a set of criteria are defined. Only papers that fulfill these criteria pass to the next phase. Next, a *quality screening* is conducted on the relevant papers. This is done by defining some minimum quality criteria that the papers must satisfy. Finally, with the selected papers that are relevant and have sufficient quality, a detailed review is performed.

Fig. 1. Pipeline of the systematic review process

In accordance with the procedure described in Fig. 1, as a first step, we performed a search of related papers. To do so, we chose Scopus[1], one of the largest abstract and citation database of peer-reviewed literature, including scientific journals, books, and conference proceedings. This database provides a search engine that, by means of queries, allows to specify different kinds of criteria to filter the results.

In our case, we are interested in papers that refer to *business processes* or *workflows*, that relate to *queries* and that analyze *events*, *logs*, *provenance*, *data* or *transactions*. In addition, we want to filter out any work that does not belong to the area of *Computer Science*, or that is not written in English. The exact query as executed in the search engine can be observed in Listing 1.

---

[1] http://www.scopus.com.

Listing 1. Query as executed in Scopus

```
TITLE−ABS−KEY("business process"OR "workflow") AND
TITLE−ABS−KEY("query"OR "querying") AND
TITLE−ABS−KEY("event"OR "log" OR "provenance"OR
        "data" OR "transaction") AND
( LIMIT−TO(SUBJAREA,"COMP" ) ) AND (
LIMIT−TO(LANGUAGE,"English") )
```

The query above yielded 835 results, from the years 1994 to 2016, with the distribution depicted in Fig. 2.

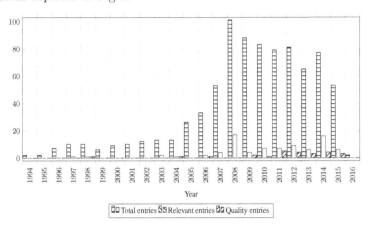

Fig. 2. Distribution per year of the related work as a result of the first query on Scopus.

However, not all of these proved relevant to our topic. When performing the *relevance screening*, we make a distinction between inclusive (I) and exclusive (E) criteria. In this way, candidates that satisfy *all* the inclusive criteria and do not satisfy any of the exclusive will be included in the next phase of the review. The ones that do not satisfy all the inclusive criteria or satisfy at least one of the exclusive ones will be discarded. The specific criteria used in this review are listed below:

1. Does the study consider process data as input? (I)
2. Does the study use only process models as input? (E)
3. Does the study propose a language to query the data? (I)

As a result of the *relevance screening* phase, the whole set of 835 entries was reduced to a set of 83 relevant works. These 83 entries are considered to be related and in the scope of process and provenance data querying. However, to guarantee a minimum level of detail, we defined the following criteria for the *quality screening* phase:

1. Does the study provide a sufficient description of the language?
2. If the language already exists, are the extensions, modifications, adaptations sufficiently explained?
3. Does the study include concrete examples of application of the language?

As a result of this phase, the set of 83 relevant works was reduced to a set of 25 papers with enough quality to be analyzed in detail. At the final stage, these papers have been analyzed to identify their most important features, and then compared to our approach. To do so, the content of each paper was reviewed, closely considering the main characteristics of the approach they describe. These characteristics refer to the kind of input data that is used by each approach (*Input data aspects*), qualities related to provenance (*Provenance aspects*), business processes (*Business process aspects*), database or artifact environments (*Database or Artifact aspects*), and the nature of the questions that can be queried with them (*Query aspects*). Table 1 presents the main characteristics of the remaining 25 references and how they can be classified when looking at the features listed below:

Input data aspects

– **Event data:** The approach allows to query and analyze event data.
– **Model-based:** The approach takes into account execution models such as workflows, BPEL, or Petri nets.
– **Storage model:** A meta-model for provenance or event data storage is proposed by the approach.
– **Complex event processing:** Different sources of events are analyzed by the approach to infer new kinds of more complex events.
– **Streams:** It is possible to query streams of events instead of complete logs.

Provenance aspects

– **Provenance-oriented:** The approach is provenance oriented or allows to record and query provenance information on the execution of workflows, scientific workflows, business processes, etc.
– **OPM-compliant:** The storage model used by the approach complies with the Open Provenance Model [2].
– **Data lineage:** The language allows to query about the life cycle of data, its origins and where it moves over time.
– **Dependency graphs/relations:** Relations between data entities are considered by the approach. For example dependency graphs, common in the provenance area, are used.

Business Process aspects

– **Business process oriented:** The approach is applied to the field of business processes management. In addition, it considers business process specific aspects while querying, e.g., using concepts such as activities, cases, resources, etc.

Database or Artifact aspects

- **Entities/artifacts:** The approach captures information about the objects, data entities or artifacts related to the event or provenance data. This information can be queried as well.
- **Database-oriented:** The approach captures database specific data such as schema, tables, column information, keys, objects, etc., in case the event information is extracted from that kind of environment.
- **State-aware:** The state of the analyzed system is considered by the approach at the time of capturing and querying data. This can be related to process state, data state, versioning, etc.

Query aspects

- **Graph-based:** Queries and results are expressed as graphs, in which edges are relations and nodes can be data objects, events, activities, etc.
- **Relevance querying:** It is possible to query about data relevance, i.e. relations between data that do not only reflect data origin.
- **Semantic querying:** The query language is based on or compatible with semantic technologies such as RDF, OWL, SPARQL, etc.
- **Regular path queries (RPQ):** The language allows to make queries that select nodes connected by a path on a graph based database.
- **Projection queries:** It is possible to query cases that fit a partial pattern using projection.
- **Temporal properties/querying:** The language or technique makes it possible to query temporal properties related to the data. It can also refer to temporal logic languages such as LTL, CTL, etc.
- **Event correlation:** The approach does not consider events in isolation, but allows to query the correlation between them, e.g. querying evens related to the same artifact.
- **Multi process:** The approach allows to query aspects related to several processes at the same time on the same dataset.
- **Multi log:** Several event logs can be queried at the same time in order to combine results for a single query.
- **Multi data schema:** Several data schemas can be considered in a single query.

Looking at Table 1, it can be seen that most of the approaches can be categorized in one of two big groups: *provenance-oriented* and *business process oriented*. The *provenance-oriented* approaches [3–12] usually support some kind of provenance model, data lineage or so. However, not all the approaches under this category support every aspect of data provenance. Only one of them [7] is database-oriented and considers states and artifacts. Most of the *business process oriented* approaches [13–26] seem to ignore data provenance aspects, and focus mainly on capturing causal relations of business activities and supporting different ways to query the data. There is an outlier [27] that focuses only on the temporal analysis of event logs using temporal logic checking. However, this solution ignores all other aspects of the data. As can be seen, none of the existing

Table 1. Comparison of features for the references at the end of the systematic review.

Ref	Title	Input data					Provenance				BP	DB				Query aspects								
		Event data	Model-based	Storage model	Complex event processing	Streams	Provenance-oriented	OPM-compliant	Data lineage	Dependency graphs/Relations	Business process oriented	Entities/Artifacts	Database-oriented	State-aware	Graph-based	Relevance querying	Semantic querying	Regular path queries (RPQ)	Projection queries	Temporal properties/querying	Event correlation	Multi process	Multi log	Multi data schema
[3]	Answering regular path queries on workflow provenance	✓					✓									✓		✓						
[4]	Capturing and querying workflow runtime provenance with PROV: A practical approach	✓	✓				✓			✓														
[5]	Modeling and querying scientific workflow provenance in the D-OPM	✓	✓					✓								✓		✓						
[6]	Towards a scalable semantic provenance management system	✓					✓	✓	✓								✓							
[7]	Towards integrating workflow and database provenance	✓	✓				✓	✓	✓			✓	✓	✓										
[8]	MTCProv: A practical provenance query framework for many-task scientific computing	✓	✓				✓	✓	✓															
[9]	OPQL: A first OPM-level query language for scientific workflow provenance	✓					✓	✓	✓						✓									
[10]	Storing, reasoning, and querying OPM-compliant scientific workflow provenance using relational databases	✓	✓				✓	✓	✓				✓											
[11]	XQuery meets Datalog: Data relevance query for workflow trustworthiness	✓					✓									✓								
[12]	A model for user-oriented data provenance in pipelined scientific workflows	✓	✓				✓	✓	✓				✓											
[13]	A knowledge driven approach towards the validation of externally acquired traceability datasets in supply chain business processes	✓					✓	✓							✓									
[14]	Process query language: A way to make workflow processes more flexible	✓	✓				✓				✓													
[15]	Workflow History Management	✓	✓				✓																	
[16]	The HIT model: Workflow-aware event stream monitoring	✓	✓		✓	✓											✓							
[17]	Semantic Enabled complex Event Language for business process monitoring	✓			✓	✓					✓					✓	✓			✓	✓			
[18]	A framework supporting the analysis of process logs stored in either relational or NoSQL DBMSS	✓	✓								✓						✓							
[19]	Business impact analysis: a framework for a comprehensive analysis and optimization of business processes	✓	✓								✓	✓	✓	✓										
[20]	Model-driven event query generation for business process monitoring	✓	✓	✓							✓													
[21]	Querying process models based on the temporal relations between tasks		✓																	✓		✓		
[22]	A query language for analyzing business processes execution	✓	✓				✓	✓	✓		✓				✓					✓		✓		
[23]	Top-k projection queries for probabilistic business processes	✓	✓	✓							✓								✓					
[24]	Integration of Event Data from Heterogeneous Systems to Support Business Process Analysis	✓	✓	✓							✓											✓		
[25]	Enabling semantic complex event processing in the domain of logistics	✓			✓						✓						✓							
[26]	Optimizing complex sequence pattern extraction using caching	✓			✓						✓													
[27]	Log-based understanding of business processes through temporal logic query checking	✓																		✓				
	Our approach	✓	✓				✓		✓	✓	✓				✓		✓			✓	✓	✓	✓	✓

approaches succeeds at combining data provenance and business processes with a good support for querying aspects.

The insight from this literature review is that in the field of process data querying, there is a need for a solution that combines business process analysis with the data perspective, which also allows to query all this information in an integrated way. Taking into account that, in most cases, the execution of business processes is supported by databases, the consideration of the data perspective becomes specially relevant.

3 Data-Aware Process Oriented Query Language

To illustrate our approach for process data querying, we propose the following running example. Consider that we want to study a process related to ticket sales and the organization of concerts. To support this, a database is used, which stores all the information relating to customers, bookings, tickets, concert halls, seats, bands, and concerts. The simplified schema of such a database is depicted in Fig. 3.

The analysis of such an environment presents many challenges. The most prominent one is the lack of explicit process execution information. The data

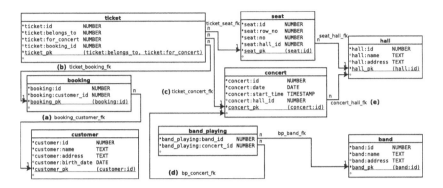

Fig. 3. Data schema of the example database

schema in Fig. 3 does not allow to record events or execution trails like cases or process instances. However, this issue has been dealt with in [28], where we explore several methods to obtain events from database settings and a meta model is proposed to store all the information about the process and data perspectives. From now on, whenever we refer to this example, we will assume the existence of a data set that complies with the meta model proposed in [28]. This meta model, that can be observed in Fig. 4, allows to combine the traditional process view (events, instances and processes), with the data perspective (data models, objects and object versions).

In order to provide a solution that enables the querying of event data considering process and data perspectives, we propose a new query language. The *Data-Aware Process Oriented Query Language* (DAPOQ-Lang), built as an interface on top of the meta model proposed in [28], allows to query not only process execution data such as events and cases, but related objects, its versions and data schemas as well, together with temporal properties of all these elements. The combination of all these aspects makes it possible to get better insights and improve the user experience. The process-oriented nature of this language improves query development time and readability, which business analysts will, hopefully, appreciate. A detailed definition of the syntax and additional documentation on the language can be found on the website of the project[2]. In the subsections below we provide some examples of use of the language related to the different characteristics.

3.1   Input Data Aspects

Event data: Our query language is based on the meta model proposed in [28]. An ER diagram of this meta model is depicted in Fig. 4. As can be observed, logs, cases, and events are part of it. Listing 2 provides an example of a query to obtain a certain subset of events.

---
[2] https://www.win.tue.nl/~egonzale/projects/dapoq-lang/.

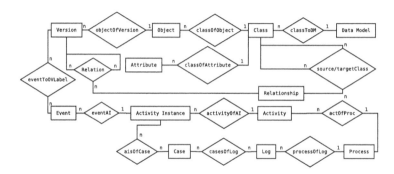

Fig. 4. ER diagram of the meta model

Listing 2. DAPOQ-Lang query to get all the events for which attribute "address" contains the substring "Madrid".

```
return allEvents where at.address CONTAINS "Madrid"
```

Model-based: Being model based, it is possible to make queries that combine event data with model information such as activities. For example, Listing 3 shows a query to obtain events that belong to a specific activity of a certain process.

Listing 3. DAPOQ-Lang query to get all the events corresponding to activities which label contains the substring "update"of the process named "concert-organizing".

```
return eventsOf(activitiesOf(allProcesses where name == "concert-organizing")
              where name CONTAINS "update")
```

Storage model: As has been mentioned before, the language builds on top of the meta model described in [28], and the ER diagram in Fig. 4. This meta model represents a valid storage model for event data combining the process and data perspectives on a single structure.

3.2   Provenance Aspects

Provenance-oriented: Data provenance aims at tracing the origins of data and enabling reproducibility of experiments. An example of tracing the origins of such data is presented in Listing 4, in which we query all the events that affected a specific object in our dataset.

Listing 4. DAPOQ-Lang query to get all the events that affected a specific object.

```
return eventsOf(allObjects where id == "43")
```

Data lineage: Some of the data lineage aspects, such as the origins of data, have been covered in the example in Listing 4. However, data lineage also deals with the lifecycle of objects. Listing 5 is an example of lifecycle and data history, querying all the existing versions that existed through time of a certain object.

Listing 5. DAPOQ-Lang query to get all the versions of objects affected by a specific case.

```
return  versionsOf(allCases  where  id  ==  "77")
```

Dependency graphs/Relations: Data relations are a fundamental aspect of our approach. Relations hold between object versions during certain periods of time. Listing 6 shows that it is possible to query all the objects that have ever been related to another specific object.

Listing 6. DAPOQ-Lang query to get booking objects related to the customer "Customer35".

```
var _bookings = objectsOf(allClasses  where  name == "booking");
var _objectsC = objectsOf(versionsRelatedTo(
                          versionsOf(allClasses  where  name == "customer")
                          where at.CUSTOMER_NAME == "Customer35"));
return _objectsC INTERSECTION _bookings
```

3.3   Business Process Aspects

Business process oriented: Our language is able to query elements specific of business processes, such as processes, activities, logs, traces and events. Listing 7 demonstrates how to query concurrent activities for a certain process.

Listing 7. DAPOQ-Lang query to get all the activities of the same process that have ever happened concurrently with any "insert" activity for a process called "concert-organizing".

```
var _actCon = activitiesOf(allProcesses  where  name == "concert-organizing");
return concurrentWith(_actCon where name CONTAINS "insert") INTERSECTION _actCon
```

3.4   Database or Artifact Aspects

Entities/Artifacts: Being based on the meta model depicted by Fig. 4, the language considers objects as first class elements. Listing 8 shows how to obtain the objects affected by events in a certain period of time.

Listing 8. DAPOQ-Lang query to get all the objects affected by events that happened between a certain period of time.

```
return  objectsOf(allEvents  where  (timestamp > "2014-01-31 13:45:23"
                             and   timestamp < "2015-01-31 13:45:23"))
```

Database-oriented: Our language is essentially database oriented, and that is evident due to the fact that it considers data models, objects and versions. Listing 9 demonstrates how to obtain classes of a data model.

Listing 9. DAPOQ-Lang query to get all the classes of a data model named "concert-portal".

```
return  classesOf(allDatamodels  where  name == "concert-portal")
```

State-aware: Given that the language is state-aware, it is possible to query the values of objects at certain moments in history (versions). Listing 10 shows how to get the state of a database at a certain moment in time.

Listing 10. DAPOQ-Lang query to get all the versions of objects of the class "customer"as they where at a certain point in time.

```
versionsOf(allClasses where name == "customer")
        where (start_timestamp =< "2014−01−31 13:45:23"
        and   end_timestamp => "2014−01−31 13:45:23")
```

3.5   Query Aspects

Relevance querying: Relevance querying refers to relations between data elements that not only reflect data origin. Our language is able to query relations of any nature. Listing 11 queries objects that were related at some point with certain object through a very specific relation.

Listing 11. DAPOQ-Lang query to get all the objects that have been related at some point to a certain ticket (35) by the relationship "ticket_concert_fk".

```
var _ticketVers = versionsOf(allClasses where name == "ticket") where at.id == "35";
return objectsOf(versionsOf(allRelationships where name == "ticket_concert_fk")
        INTERSECTION versionsRelatedTo(_ticketVers))
```

Temporal properties/querying: The meta model on which our language is based considers time attributes of different elements of the structure. This makes it easy to formulate queries with temporal properties with our language. For example, Listing 12 shows how to obtain activities executed during a certain period of time.

Listing 12. DAPOQ-Lang query to get all the activities that where executed during a period for the process "ticket-selling".

```
return activitiesOf(createPeriod("2014−01−31 13:45:23" , "2014−01−31 14:45:23"))
        INTERSECTION activitiesOf(allProcesses where name == "ticket−selling")
```

Event correlation: Events are not considered as single and isolated elements in DAPOQ-Lang. They can be correlated to cases, logs, versions, objects, etc. Listing 13 shows a query to obtain events correlated to the same object.

Listing 13. DAPOQ-Lang query to get all the events that affected any version of the object corresponding to the customer named "Customer35".

```
return eventsOf(objectsOf(versionsOf(allClasses where name == "customer")
        where at.CUSTOMER_NAME == "Customer35")))
```

Multi process/Multi log/Multi data schema: An advantage of our approach not found in other approaches is the support to query properties of several processes, logs, and data schemas at the same time. Listings 14, 15 and 16 show examples of querying all the processes of several activities, the logs of some processes and the data schemas that contain a certain relationship.

Listing 14. DAPOQ-Lang query to get the processes with an activity which name contains the substring "insert".

```
return processesOf(allActivities where name CONTAINS "insert")
```

Listing 15. DAPOQ-Lang query to get all the logs of the processes named "ticket-selling".

```
return logsOf(allProcesses where name == "ticket-selling")
```

Listing 16. DAPOQ-Lang query to get the data models with a relationship "ticket_seat_pk".

```
return datamodelsOf(allRelationships where name == "ticket_seat_pk")
```

4 Implementation

The query language (DAPOQ-Lang) proposed in the previous section has been implemented as a library[3] in Java, and integrated as part of the Process Aware Data Suite[4] (PADAS). The parser for the grammar has been created using ANTLR[5], a parser generator widely used to build languages and other tools. DAPOQ-Lang queries are executed on OpenSLEX[6] files, which store all the data according to the meta model defined in [28]. In addition to the query engine,

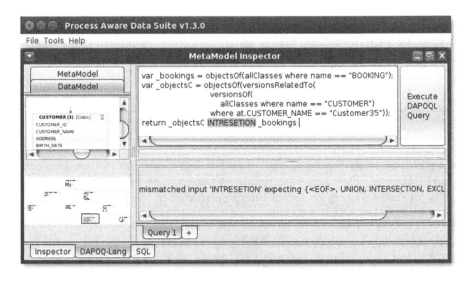

Fig. 5. Screenshot of the tool while writing a query, highlighting a misspelled word.

---

[3] https://www.win.tue.nl/~egonzale/projects/dapoq-lang/.
[4] https://www.win.tue.nl/~egonzale/projects/padas/.
[5] http://www.antlr.org/.
[6] https://www.win.tue.nl/~egonzale/projects/openslex/.

DAPOQ-Lang provides assistance on the query development by means of *predictive typing*. This functionality, implemented in the GUI of the tool, helps the user to write queries by suggesting language keywords as they are typed. Also, *syntax checking* is performed simultaneously, highlighting any problematic or incorrect parts of the query as shown in Fig. 5.

5 Conclusion

This work analyzes the existing approaches for querying execution trails such as process events, transactions and provenance data. The functionalities of these approaches have been identified and classified accordingly. As a result of this study, we have identified the need for a query mechanism that combines both process and data perspectives and helps with the task of obtaining insights about the process or workflow under analysis.

To fulfill this need, we proposed a new query language that, combining process and data perspectives, allows the analyst to ask meaningful questions in a simple way. Examples of the use of the language have been presented, covering different aspects of the data, in order to guarantee its usefulness and simplicity. In addition, we provide an implementation that not only enables the execution, but also assists on the writing and development of the queries by means of *predictive typing* and *syntax checking*.

The paper shows that it is feasible to develop a query language that satisfies the needs of process analysts, while keeping simplicity and ease of use. As future work, a full specification of the language will be provided. Also, efforts will be made to improve the performance when dealing with big datasets and to keep the language evolving, adding new functionalities and constructs.

References

1. Vanwersch, R., Shahzad, K., Vanhaecht, K., Grefen, P., Pintelon, L., Mendling, J., Van Merode, G., Reijers, H.A.: Methodological support for business process redesign in health care: a literature review protocol. Int. J. Care Pathways **15**(4), 119–126 (2011)
2. Moreau, L., Clifford, B., Freire, J., Futrelle, J., Gil, Y., Groth, P., Kwasnikowska, N., Miles, S., Missier, P., Myers, J., et al.: The open provenance model core specification (v1. 1). Future Gener. Comput. Syst. **27**(6), 743–756 (2011)
3. Huang, X., Bao, Z., Davidson, S.B., Milo, T., Yuan, X.: Answering regular path queries on workflow provenance. In: 2015 IEEE Proceedings of the 31st International Conference on Data Engineering (ICDE), pp. 375–386. IEEE (2015)
4. Costa, F., Silva, V., De Oliveira, D., Ocaña, K., Ogasawara, E., Dias, J., Mattoso, M.: Capturing and querying workflow runtime provenance with PROV: a practical approach. In: Proceedings of the Joint EDBT/ICDT 2013 Workshops, pp. 282–289. ACM (2013)
5. Cuevas-Vicenttin, V., Dey, S., Wang, M.L.Y., Song, T., Ludascher, B.: Modeling and querying scientific workflow provenance in the D-OPM. In: High Performance Computing, Networking, Storage and Analysis (SCC), pp. 119–128. IEEE (2012)

6. Sakka, M.A., Defude, B.: Towards a scalable semantic provenance management system. In: Hameurlain, A., Küng, J., Wagner, R. (eds.) Transactions on Large-Scale Data- and Knowledge-Centered Systems VII. LNCS, vol. 7720, pp. 96–127. Springer, Heidelberg (2012). doi:10.1007/978-3-642-35332-1_4
7. Chirigati, F., Freire, J.: Towards integrating workflow and database provenance. In: Groth, P., Frew, J. (eds.) IPAW 2012. LNCS, vol. 7525, pp. 11–23. Springer, Heidelberg (2012). doi:10.1007/978-3-642-34222-6_2
8. Gadelha, L.M., Wilde, M., Mattoso, M., Foster, I.: MTCProv: a practical provenance query framework for many-task scientific computing. Distrib. Parallel Databases 30(5–6), 351–370 (2012)
9. Lim, C., Lu, S., Chebotko, A., Fotouhi, F.: OPQL: A first OPM-level query language for scientific workflow provenance. In: 2011 IEEE International Conference on Services Computing (SCC), pp. 136–143. IEEE (2011)
10. Lim, C., Lu, S., Chebotko, A., Fotouhi, F.: Storing, reasoning, and querying OPM-compliant scientific workflow provenance using relational databases. Future Gener. Comput. Syst. 27(6), 781–789 (2011)
11. Liu, D.: XQuery meets Datalog: data relevance query for workflow trustworthiness. In: Research Challenges in Information Science (RCIS 2010), pp. 169–174. IEEE (2010)
12. Bowers, S., McPhillips, T., Ludäscher, B., Cohen, S., Davidson, S.B.: A model for user-oriented data provenance in pipelined scientific workflows. In: Moreau, L., Foster, I. (eds.) IPAW 2006. LNCS, vol. 4145, pp. 133–147. Springer, Heidelberg (2006). doi:10.1007/11890850_15
13. Solanki, M., Brewster, C.: A knowledge driven approach towards the validation of externally acquired traceability datasets in supply chain business processes. In: Janowicz, K., Schlobach, S., Lambrix, P., Hyvönen, E. (eds.) EKAW 2014. LNCS (LNAI), vol. 8876, pp. 503–518. Springer, Cham (2014). doi:10.1007/978-3-319-13704-9_38
14. Momotko, M., Subieta, K.: Process query language: a way to make workflow processes more flexible. In: Benczúr, A., Demetrovics, J., Gottlob, G. (eds.) ADBIS 2004. LNCS, vol. 3255, pp. 306–321. Springer, Heidelberg (2004). doi:10.1007/978-3-540-30204-9_21
15. Koksal, P., Arpinar, S.N., Dogac, A.: Workflow history management. ACM Sigmod Rec. 27(1), 67–75 (1998)
16. Poppe, O., Giessl, S., Rundensteiner, E.A., Bry, F.: The HIT model: workflow-aware event stream monitoring. In: Hameurlain, A., Küng, J., Wagner, R., Amann, B., Lamarre, P. (eds.) Transactions on Large-Scale Data- and Knowledge-Centered Systems XI. LNCS, vol. 8290, pp. 26–50. Springer, Heidelberg (2013). doi:10.1007/978-3-642-45269-7_2
17. Liu, D., Pedrinaci, C., Domingue, J.: Semantic enabled complex event language for business process monitoring. In: Proceedings of the 4th International Workshop on Semantic Business Process Management, pp. 31–34. ACM (2009)
18. Fazzinga, B., Flesca, S., Furfaro, F., Masciari, E., Pontieri, L., Pulice, C.: A framework supporting the analysis of process logs stored in either relational or NoSQL DBMSs. In: Esposito, F., Pivert, O., Hacid, M.-S., Raś, Z.W., Ferilli, S. (eds.) ISMIS 2015. LNCS (LNAI), vol. 9384, pp. 52–58. Springer, Cham (2015). doi:10.1007/978-3-319-25252-0_6
19. Radeschütz, S., Schwarz, H., Niedermann, F.: Business impact analysis: a framework for a comprehensive analysis and optimization of business processes. Comput. Sci. Res. Dev. 30(1), 69–86 (2015)

20. Backmann, M., Baumgrass, A., Herzberg, N., Meyer, A., Weske, M.: Model-driven event query generation for business process monitoring. In: Lomuscio, A.R., Nepal, S., Patrizi, F., Benatallah, B., Brandić, I. (eds.) ICSOC 2013. LNCS, vol. 8377, pp. 406–418. Springer, Cham (2014). doi:10.1007/978-3-319-06859-6_36

21. Song, L., Wang, J., Wen, L., Wang, W., Tan, S., Kong, H.: Querying process models based on the temporal relations between tasks. In: 2011 15th IEEE International Enterprise Distributed Object Computing Conference Workshops (EDOCW), pp. 213–222. IEEE (2011)

22. Beheshti, S.-M.-R., Benatallah, B., Motahari-Nezhad, H.R., Sakr, S.: A query language for analyzing business processes execution. In: Rinderle-Ma, S., Toumani, F., Wolf, K. (eds.) BPM 2011. LNCS, vol. 6896, pp. 281–297. Springer, Heidelberg (2011). doi:10.1007/978-3-642-23059-2_22

23. Deutch, D., Milo, T.: Top-K projection queries for probabilistic business processes. In: Proceedings of the 12th International Conference on Database Theory. ACM (2009)

24. Baquero, A.V., Molloy, O.: Integration of event data from heterogeneous systems to support business process analysis. In: Fred, A., Dietz, J.L.G., Liu, K., Filipe, J. (eds.) IC3K 2012. CCIS, vol. 415, pp. 440–454. Springer, Heidelberg (2013). doi:10.1007/978-3-642-54105-6_29

25. Metzke, T., Rogge-Solti, A., Baumgrass, A., Mendling, J., Weske, M.: Enabling semantic complex event processing in the domain of logistics. In: Lomuscio, A.R., Nepal, S., Patrizi, F., Benatallah, B., Brandić, I. (eds.) ICSOC 2013. LNCS, vol. 8377, pp. 419–431. Springer, Cham (2014). doi:10.1007/978-3-319-06859-6_37

26. Ray, M., Liu, M., Rundensteiner, E., Dougherty, D.J., Gupta, C., Wang, S., Mehta, A., Ari, I.: Optimizing complex sequence pattern extraction using caching. In: 2011 IEEE Proceedings of the 27th International Conference on Data Engineering Workshops (ICDEW), pp. 243–248. IEEE (2011)

27. Räim, M., Ciccio, C., Maggi, F.M., Mecella, M., Mendling, J.: Log-based understanding of business processes through temporal logic query checking. In: Meersman, R., et al. (eds.) OTM 2014. LNCS, vol. 8841, pp. 75–92. Springer, Heidelberg (2014). doi:10.1007/978-3-662-45563-0_5

28. González López de Murillas, E., Reijers, H.A., van der Aalst, W.M.P.: Connecting databases with process mining: a meta model and toolset. In: Proceedings of the 17th International Conference on Enterprise, Business-Process and Information Systems Modeling, BPMDS (2016)

A Comparative Analysis of Business Process Model Similarity Measures

Tom Thaler[1]([⊠]), Andreas Schoknecht[2], Peter Fettke[1], Andreas Oberweis[2], and Ralf Laue[3]

[1] German Research Center for Artificial Intelligence (DFKI),
Saarland University, Saarbrücken, Germany
{tom.thaler,peter.fettke}@dfki.de
[2] Karlsruhe Institute of Technology, Karlsruhe, Germany
{andreas.schoknecht,andreas.oberweis}@kit.edu
[3] University of Applied Sciences Zwickau, Zwickau, Germany
ralf.laue@fh-zwickau.de

Abstract. To work efficiently with and unlock the potentials of business process models, measuring their similarity is a basic requirement. Thus, many automatic similarity measurement approaches have been developed during the last years, which utilize very different aspects of a model. At the same time, it is unclear which measures can be meaningfully applied in which context and how they behave in general. Hence, this paper analyzes how the values of existing similarity measures correlate and how corresponding implementations perform with respect to their resource consumption. The results of our analysis show that the similarity values of most measures highly correlate while their performance prohibits the usage of more than 50% of the measures in practice.

Keywords: Process model similarity · Process model matching

1 Introduction

Regarding a practical usage of business process models, there are many scenarios, wherefore measuring the similarity between particular models is even a basic requirement. This includes checking the conformance of process models to legal regulations or to reference models, enabling the reusability of process fragments or merging process models. At the same time, a manual measurement of similarity values and differences between hundreds or even thousands of models would take an enormous effort leading to high costs [3].

Thus, many business process model similarity measures have been developed during the last years. However, the interpretation of similarity is quite different, since several dimensions of similarity can be considered. While these are not dimensions in a strong mathematical sense, they focus on different aspects of process models (cf. Fig. 1). Since the activites in *seller p. 1* and *seller p. 2* are identically labelled, the similarity value would be 1 when only considering the natural language dimension. Yet, their graph structure and behavior are slightly

© Springer International Publishing AG 2017
M. Dumas and M. Fantinato (Eds.): BPM 2016 Workshops, LNBIP 281, pp. 310–322, 2017.
DOI: 10.1007/978-3-319-58457-7_23

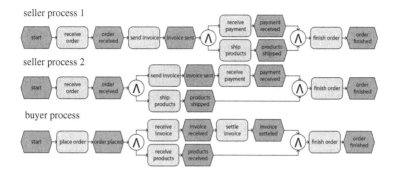

Fig. 1. Illustration of different dimensions used for similarity measurement (cf. [11]).

different. E.g., in *seller p. 1* the activity "send invoice" is always executed before "ship products" while this is not necessarily true for *seller p. 2*. Thus, the final similarity value might be less than 1. When looking at the process models *seller p. 2* and *buyer p.*, they are identical with respect to the graph structure, which would lead to a similarity value of 1 when only considering this dimension.

Considering this variety of dimensions, which are possibly addressed by existing similarity measures, it is unclear which measures can be meaningfully applied in which usage scenario and how they behave in general. Therefore, we make a first step in this direction by addressing the following questions: (1) How do the values of existing similarity measures correlate? (2) How do existing implementations perform and what does that imply for their practical usage?

The rest of the paper is organized as follows: In Sect. 2, the relevant core terms are introduced and explained in detail. Section 3 outlines related work, followed by the applied research approach in Sect. 4. Afterwards, Sect. 5 describes the results of the comparative analysis, which covers a correlation analysis of similarity values as well as a run time analysis of eight publicly available measures. The results and limitations of the paper are discussed in Sect. 6. Finally, a conclusion is given in Sect. 7.

2   Fundamentals of Business Process Model Similarity Measurement

Business process model similarity measures try to quantify the similarity between business process models in general. A similarity value is mostly expressed either on an interval or on a ratio scale, which provide the frame for the typical operationalization of business process model similarity in a metric space. A metric generally fulfills the properties non-negativity ($\forall x, y \in \mathcal{D}, d(x, y) \geq 0$), symmetry ($\forall x, y \in \mathcal{D}, d(x, y) = d(y, x)$), identity ($\forall x, y \in \mathcal{D}, x = y \Leftrightarrow d(x, y) = 0$) and triangle inequality ($\forall x, y, z \in \mathcal{D}, d(x, z) \leq d(x, y) + d(y, z)$) [24], with $\mathcal{D}$ being a domain of objects and $d : \mathcal{D} \times \mathcal{D} \to \mathbb{R}$ being a function measuring the distance between two objects. However, as shown in [14], many existing process model similarity measures do not fulfill the above mentioned properties. Only three of eleven analyzed measures met all metric properties.

The quantification of process model similarity is based on process model matches in most cases [4]. Such matches explicate correspondences between single nodes or sets of nodes of two models based on criteria such as similarity, equality or analogy [20] and are expressed using different cardinalities (1:1, 1:N, M:N). As shown in [4], at that time, most similarity measures used matching approaches creating solely 1:1 matches, while recent works tend to M:N matches [2,6]. As mentioned above, there are several dimensions of similarity in business process models, which are now briefly introduced.

Natural language: Natural language is, e.g., used for labeling the elements of a model, for naming a model or for tagging it. Such labels serve as one of the most important knowledge bases for process model similarity measurement and are analyzed with regard to syntactical and semantic aspects. While the syntactical analysis focuses on the characters of labels, the semantic analysis aims at understanding their meaning based on the used words and the grammar.

Graph structure: The relevant aspects of this dimension arise from graph theory and can be divided into general graph structure-based and business process aware control flow similarity measures. The graph structure-based similarity between models can be quantified by, e.g., quantitative metrics related to common subgraphs. Since general graph-based algorithms do not consider the meaning of any control flow connectors, they are either ignored or the existing measures are extended in order to handle them.

Behavior: This dimension focuses on the execution of business processes. Corresponding execution traces can be generated through simulation runs or during the actual execution of a process. An example for such a similarity measurement is to count the number of identical execution sequences in a trace log. Thereby, also characteristics of possible execution sequences are considered.

Human estimation: Another dimension is the human judgment on how similar process models are. One can differentiate three types of human estimation based on the knowledge of the involved people: Process experts have a grounded knowledge on the process landscape of a company, while process participants are specialists for particular processes or parts thereof. Thus, it can be assumed that process experts quantify the similarity on a general and more high level point of view, while process participants take up a detailed perspective. In contrast to that, the crowd gains its knowledge solely based on the process description (the models). Hence, the crowd quantifies the similarity with its own interpretation at the back of their mind.

Other aspects: Other aspects that are described in the literature are collected in this group, as they are infrequently used or are specific to a certain similarity measurement approach. Examples are the usage of ontology alignment techniques [5] or resources as, e.g., input and output objects [13].

3   Related Work

Currently, two structured literature surveys on process model similarity have been published [4,18]. Besides these, a few other articles give an overview on published similarity measures as well as matching techniques. One highlights open questions regarding similarity measurement [8] and another three compare different approaches in evaluation settings [2,6,9]. Becker and Laue [4] provide a detailed overview on the exact calculations used by process model similarity measures without considering their behavior in realistic scenarios. The survey conducted by Niesen and Houy [18] focuses on Natural Language Processing (NLP) techniques used in process model similarity measures. Again, the survey of Dijkman et al. [8] describes categories of problems related to similarity measurement as well as future research directions. While these surveys provide a theoretical overview on process model similarity, our article compares different similarity measures regarding their run time and similarity values. Hence, we focus on an empirical assessment point of view.

Furthermore, the evaluations described in [2,6] can be regarded as related work as process model matching techniques are summarized and compared to each other. Thus, these papers also constitute practical, empirical works. They present results of two process model matching contests, wherein several matching approaches were compared regarding their performance of finding similar activities in several process model data sets. The closest work to ours is the comparison of three process model similarity measures for process model retrieval, which are evaluated with one data set described in [9]. In contrast to this work, we additionally analyze the correlation between the similarity values of eight publicly available measures on three different data sets.

4   Analysis Objectives and Methodology

4.1   Selection of Model Data

Although a representative data set is not achievable from a statistical point of view (the overall population of process models is unknown), an experimental analysis of similarity measures is necessary to characterize their behavior in concrete application scenarios. For that purpose, one can distinguish laboratory and field investigations. In laboratory investigations the process models are (possibly synthetically) generated in a controlled environment, while in field investigations, they are generated by modelers in the real world. However, results from a laboratory investigation cannot always easily be transferred to the field. Against that background, the field should be considered as well. Hence, we use three different groups of samples with different characteristics, which are taken from a large process model corpus [19]. The data sets are described below, several corresponding model metrics are presented in [19].

1. **Field models:** No restrictions regarding the labeling of model elements are given to the modeler(s). Thus, equal or similar aspects might be modeled

in a different manner and expressed with different words. An adequate data set, containing models from the domains university admission (9 models) and birth registration (9 models), is provided in [6].

2. **Models from controlled modeling environments:** Models are created in a controlled environment, wherein different modelers independently model the same process based on a natural language text description. Since, in this way, a terminology is given by the textual description, it is assumed that it is used by the modelers. Student exercises[1] (8 models) serve as an adequate data set. An analysis based on this data set covers a laboratory investigation.

3. **Mined models:** The process models are derived using process mining techniques. Thus, the node labels are linguistically harmonized and therefore (1) unambiguous and (2) consistent over the whole collection (matching problem is faded out). The models from Dutch governance presented in [21] fulfill this requirement (80 models). However, one can discuss whether they are synthetically created in a laboratory sense or, since the processes are executed in the real world, whether they are derived from field.

4.2   Selection and Setup of Similarity Measures

In order to identify relevant similarity measurement approaches we conducted a structured literature search [22], which led to 120 relevant papers.[2] Based on this selection, all papers with an existing or known implementation, were selected. If available, the tools of the remaining 16 candidates were checked, whether either Petri Net or EPC is supported, since the models used in our study are available in these notations. Finally, the selected similarity measures address the dimensions mentioned in Sect. 2 as described in Table 1 and were set up as follows:

Similarity Score based on Common Activity Names [1]:[3] The similarity is calculated based on the number of identically labeled activities.

Graph Edit Distance Similarity [7](**see Footnote 3**): The concept of edit distance is applied to both, node labels (string edit distance) and the graph structure (graph edit distance). The Greedy algorithm is used for the (approximate) optimization of the similarity matrix. The three mentioned quotients are equally weighted.

Causal Footprints [10](**see Footnote 3**): The similarity measure was implemented in the research prototype RefMod-Miner. Although there is a proposal of a semantic node similarity measure, in the context at hand, two nodes are considered as equal (matched), iff they have the same label.

Percentage of Common Nodes and Edges [17](**see Footnote 3**): This similarity measure enhances [1] by regarding all nodes and edges of process models

---

[1] Model set "Exams" is available in the model repository at http://rmm.dfki.de.

[2] A complete list of investigated publications can be retrieved from http://rmm.dfki.de/docs/BPM2016_ProcessModelSimAnalysis_LiteratureList.pdf.

[3] http://rmm.dfki.de.

Table 1. Dimensions addressed by selected measures.

Dimension/Ref	[1]	[7]	[10]	[17]	[23]	[15]	[16]	[12]
Natural language	syn	syn	syn	syn	syn	syn	syn+sem	syn
Graph structure		x	x	x	x	x	x	
Behavior								x

(instead of activities only). Thus, the structure of a process model is considered as well. The control flow connectors are ignored such that their preceding and succeeding nodes are interpreted as directly connected through edges.

Feature-based Similarity Estimation [23](**see Footnote 3**)**:** The approach consists of syntactical and graph-structural components. Firstly, for each node pair, the Levenshtein similarity for their labels is calculated (syntactical component). As a graph-structural component, the authors define five roles characterizing a node. The thresholds are set as proposed in the original paper. The resulting similarity matrix is optimized using the Greedy algorithm.

Activity Matching and Graph Edit Distance [15](**see Footnote 3**)**:** This measure extends [7] by also considering control flow connectors. The Greedy algorithm is again used for optimizing the similarity matrix. All other settings are equal to the above mentioned graph edit distance similarity.

La Rosa Similarity [16]:[4] This similarity measure extends [15] by calculating a node matching not only based on the Levenshtein similarity but also using a linguistic similarity measure based on a lexical database. The original implementation is used with the standard parameters.

Longest Common Sets of Traces [12](**see Footnote 3**)**:** The authors propose two components expressing in how far the traces of one model are reflected by the traces of the other model. To make the similarity values comparable, the average of both components is calculated and interpreted as the similarity value. The node mapping (which is not explicated in the original paper) is calculated using the Levenshtein similarity with a minimum threshold of 0.9.

5   Analysis Results

Overall, we analyzed 3,371 model pairs with 8 similarity measures and thereby proceeded 26,968 similarity calculations plus the corresponding node matchings. Since the practical applicability is one important aspect, the analysis was executed on a standard PC with 4 cores (3 GHz each) and 4 GB main memory.

Correlation between Similarity Values: Since all analyzed techniques use node matchings as a calculation basis, they highly depend on the corresponding

---

[4] http://www.processconfiguration.com/download.html.

node matching approach. Against this background, it is assumed that the correlation between the measures' values differ based on the data sets. While mined models might reliably be matched by all approaches, quite diverging matchings between non-linguistically harmonized models are expected. Thus, the similarity values are evaluated separately on the above mentioned model categories (1) field models, (2) models of controlled modeling environments and (3) mined models. Since the underlying node matching is of major importance, they were additionally analyzed with regard to known reference matchings considering the activities only. Table 2 includes only six mapping approaches instead of eight, since [1,10,17] use the same approach. Moreover, it looks like [7,15,16] produce exactly the same matchings in all data sets. This is in fact not the case, since only the activities are considered in the reference matchings, while the named approaches also match events or connectors, which are not covered by the analysis. Furthermore, we chose the Pearson correlation coefficient since we assume a normal distribution of the similarity values. In fact, it is not possible to determine the distribution in a methodical correct way, since, as mentioned above, it is not possible to randomly select process models from the overall population of process models. Nevertheless, the Pearson correlation coefficient is suitable for interpreting the correlation behavior for the selected data sets.

Table 2. Analysis of underlying matchings with regard to reference activity matchings.

	Field						Controlled						Mined					
Approach	TP	FP	FN	P	R	F	TP	FP	FN	P	R	F	TP	FP	FN	P	R	F
[1,10,17]	152	17	962	0.9	0.14	0.24	6	0	284	1.00	0.02	0.04	9,767	0	0	1.00	1.00	1.00
[7]	289	205	825	0.59	0.26	0.36	96	33	194	0.74	0.33	0.46	9,767	9,187	0	0.52	1.00	0.68
[23]	315	906	799	0.27	0.28	0.27	125	228	165	0.35	0.43	0.39	9,554	50,354	213	0.16	0.98	0.27
[15]	289	205	825	0.59	0.26	0.36	96	33	194	0.74	0.33	0.46	9,767	9,187	0	0.52	1.00	0.68
[16]	289	205	825	0.59	0.26	0.36	96	33	194	0.74	0.33	0.46	9,767	9,187	0	0.52	1.00	0.68
[12]	175	20	939	0.90	0.16	.027	19	1	271	0.95	0.07	0.12	9,767	1,257	0	0.89	1.00	0.94

sim = similarity measure, TP = true positives, FP = false positives, FN = false negatives, P = μ-average of precision, R = μ-average of recall, F = μ-average of f-measure.

As one can see in Table 3 and contrary to the expectations, a higher correlation of the similarity values for mined models in comparison to the other two data sets cannot be identified. As one can easily see in Table 2, the mapping quality in case of the mined models is, as expected, much higher than for the other two cases. At the same time, a higher mapping quality of the controlled models to the field models cannot generally be attested. However, that does not influence the correlation between the similarity values, since the quality of the different approaches within the different scenarios are comparable in most cases. Especially the analysis of [1,10,17] in case of the mined models is very meaningful, since they have a perfect mapping (a linguistic harmonization corresponds to a label-identical mapping). Although the first approach solely considers the equally labeled activities, while the others also consider the control flow, there is a high correlation in the resulting similarity values.

Furthermore the presented heat map shows a very high correlation between the similarity values of all measures except of [23]. As one can see in Table 2, the matching approach produces at least four times more false positive matches than all other approaches, which is also the reason for the missing significance of the correspondent correlation values. That underlines the thesis of the high dependency of a similarity measure to the underlying node mapping. Particularly, a cluster consisting of [1,7,15–17] can be identified. Therein, all similarity measure pairs correlate to more than 0.95 on average. This is surprising since [1] solely considers node labels for the similarity quantification, while the other four approaches also take the structure into account.

Considering the measures [10,23], it is conspicuous that both show a comparable low correlation to most other similarity measures, but also between each other. As that is founded in the different matching approach of [10,23] seems to measure a specific aspect of similarity. In fact, [10] bases on the causal footprints, and thus considers the causal dependencies between nodes, while [23] considers the correspondences between nodes similar to [1].

However, looking at the very low mapping quality of [1,10,12,17] in the controlled scenario, a general effect on their correlation cannot be identified. As a further intermediate result, except of [10,23], there is a very high correlation between all analyzed measures. The expected lower correlation between measures focusing on the behavior of possible process executions in comparison to those focusing solely on the labels could not be experimentally verified. Hence, except of the [10,23], the analyzed measures are scarcely distinguishable based on their values and are therefore exchangeable in the demonstrated cases. Hence, except of [23], the similarity measures seem to be exchangeable as their values correlate to a very high degree. Other aspects such as the run time of similarity calculation could therefore be more important when choosing a measure for a specific application.

Computing Performance: The second analysis aspect of the comparative analysis is the computing performance in terms of run time. The importance of this aspect is founded in the practical applicability, for which a calculation time of several minutes or hours would generally not be desirable. As expected, based on the number of models within the data sets *University admission*, *Birth registration*, and *Student exercises*, they are suitable for fast calculations. Nearly all measures returned the similarity values in less than one minute, 50% of the calculations under five seconds.

Considering real model databases, perhaps containing thousands of individual models, it seems that more than 50% of the analyzed similarity measure implementations [10,12,15,16,23] are not suitable for an application in real contexts. On the other hand, there are three approaches [1,7,17] which generally provide results in short time. Especially [1,17] have high potential for an application in real contexts, since next to the short calculation time, they also constitute the highest correlation to all other measures (Table 4).

Table 3. Pearson correlation coefficients for the analyzed data sets.

		[1]	[7]	[10]	[17]	[23]	[15]	[16]	[12]
[1]	F	1.00	0.93	0.77	0.97	0.70	0.96	0.94	0.94
	C	1.00	0.94	0.80	0.98	0.43*	0.97	0.96	0.92
	M	1.00	0.96	0.85	0.98	0.60	0.95	0.93	#
[7]	F	0.93	1.00	0.91	0.98	0.62	0.97	0.98	0.98
	C	0.94	1.00	0.91	0.95	0.49*	0.98	0.94	0.97
	M	0.96	1.00	0.87	0.99	0.55	0.94	0.98	#
[10]	F	0.77	0.91	1.00	0.86	0.55	0.85	0.87	0.85
	C	0.80	0.91	1.00	0.95	0.49	0.98	0.89	0.96
	M	0.85	0.87	1.00	0.85	0.49	0.82	0.84	#
[17]	F	0.97	0.98	0.86	1.00	0.65	0.98	0.98	0.97
	C	0.98	0.95	0.95	1.00	0.56*	0.99	0.97	0.91
	M	0.98	0.99	0.85	1.00	0.57	0.94	0.97	#
[23]	F	0.70	0.62	0.55	0.65	1.00	0.75	0.63	0.63
	C	0.43*	0.49*	0.49	0.56*	1.00	0.55	0.65	0.55*
	M	0.60	0.55	0.49	0.57	1.00	0.76	0.57	#
[15]	F	0.96	0.97	0.85	0.98	0.75	1.00	0.97	0.96
	C	0.97	0.98	0.98	0.99	0.55	1.00	0.97	0.95
	M	0.95	0.94	0.82	0.94	0.76	1.00	0.93	#
[16]	F	0.94	0.98	0.87	0.98	0.63	0.97	1.00	0.98
	C	0.96	0.94	0.89	0.97	0.65	0.97	1.00	0.96
	M	0.93	0.98	0.84	0.97	0.57	0.93	1.00	#
[12]	F	0.94	0.98	0.85	0.97	0.63	0.96	0.98	1.00
	C	0.92	0.97	0.96	0.91	0.55*	0.95	0.96	1.00
	M	#	#	#	#	#	#	#	#

p-value ≤1%, F = field models, C = controlled models, M = mined models, # = calculation aborted because of memory overflow, * = p-value >1%.

In contrast to that, it was not possible to calculate any similarity values for the *Dutch Governance* data set with the approach of [12]. Since there are no real log traces available, referring to [4], all possible traces were calculated. Depending on the size and on the complexity of the input models, this approach produced a mass of data leading to a memory overflow on the used hardware. Thus, the similarity values of [12] could not be calculated for this data set. All other run times for the data sets could be calculated and for two of them even in suitable time (*Student exercises* and *Birth registration*). At the same time, the usage of real execution data might improve this approach to a high degree. Nevertheless, such an analysis is not part of the work at hand.

Table 4. Run time of similarity calculations for different data sets.

Measure	Dutch Governance	Student exercises	Birth registration	University admission
[1]	3:28 min	0:00 min	0:02 min	0:02 min
[7]	8:40 min	0:01 min	0:04 min	0:05 min
[10]	n/a[a]	0:37 min	9:32 min	26:30 min
[17]	8:40 min	0:02 min	0:04 min	0:05 min
[23]	45:37 min	0:03 min	0:23 min	0:56 min
[15]	40:21 min	0:03 min	0:15 min	0:36 min
[16]	39:22 min	0:14 min	0:20 min	0:22 min
[12]	memory overflow	0:03 min	0:07 min	4:52 min

[a]For [10] the Dutch Governance processing had to be split because of a memory overflow. Since summing up the partial run times might have led to a corruption in comparison to the other calculations, it was decided to state it as not available.

6   Discussion and Limitations

Unfortunately, the availability of similarity measure implementations is quite limited. Only 22 implementations were mentioned in the publications, whereof only 8 were accessible and executable in the context of the analysis. Thereby, although the behavior-oriented measure originally works on process instances from real executions, here, all possible traces were derived from the models. This covers a slightly different case, since we implicitly considered the state space of the models instead of the observed behavior. Using real traces may lead to much lower calculation times and will most certainly lead to much lower memory consumption. On the other hand, using process logs would not measure the process model similarity but the process instance similarity. Against that background, the applied variant makes sense in the context of the work at hand.

We identified high differences in the intensity in terms of memory and time consumption. Both effects lead to trouble in the context of a practical application right up to a non-applicability. In turn, other approaches are able to calculate a similarity value within short time and with only little resources. In spite of these differences, the correlation between the similarity values is high in most cases.

The high correlation values are caused by the fact that all approaches work with an underlying node mapping, which is finally responsible for the similarity values. This is founded in the fact that the similarity measures are functions on the matchings. This leads to the result that it is necessary to separate process model similarity measurement into two components: (1) the node matching and (2) the calculation of a similarity value. As shown in the analysis at hand, this makes it possible to analyze the effects of addressing different dimensions of similarity in more detail. In fact, it might be meaningful to repeat the analysis using consistent matchings. However, the underlying scientific papers propose particular matching approaches, which is the reason for the design at hand. Moreover, the proposed procedure would lead to new challenges since the cardinality of node matchings influences the applicability of a measure. Most of them need a 1:1 matching (as, e.g., in the context of behavior based similarity measures) while complex matchings cannot easily be interpreted. At the same time, 1:1 matchings would implicitly formulate the requirement that the models need to be on the same level of detail, which is generally not given in a realistic scenario.

7 Conclusion

Based on the practical empirical evaluation, it can be stated that the computational behavior of similarity measures is diverging in concrete contexts. First of all, the conceptualization of the measures has a high impact on the execution time, which ranges from 3 to 45 min up to non-computability for the similarity measurement of a set of 80 models. Besides, it was shown that the values of most measures highly correlate to each other. However, there were also two measures showing differences to the others. Thus, there are different types of similarity measures, which might reasonably be applied in different contexts.

One special scenario might be the similarity analysis of process models which are derived through process mining. Since the data basis is automatically generated, the contained information are linguistically harmonized. Hence, the analysis of node labels with NLP techniques is of minor importance, while the usage of further information like system handbooks might be meaningful. However, because of the generally high correlations, it is recommendable to apply one of the easier and faster measures like [1] in order to get a first impression of similarity between particular process models. Only if one is interested in details, and if a reliable matching is available, it is meaningful to apply a similarity measure addressing specific dimensions. For that, the *first impression* might be seen as a preselection of relevant models.

Yet, it is still an open question whether two similarity measures measure the same pragmatic aspects as, e.g., similarity of content, of the equivalence of action, or the equivalence of objective (in contrast to the above mentioned dimensions) and how that can be determined. It is also unclear how far the automatically calculated similarity values correspond to human estimations, and thus, how valid the similarity values are. In fact, the results of the investigated similarity measures are valid with regard to their technical implementation, but how far that matches specific measurement objectives, perhaps in different application scenarios, is not analyzed so far. Amongst others, one reason is, that the requirements of different application scenarios to a similarity measure are unclear and not precisely defined. Especially concerning the underlying node mapping, it should be clarified, what a correspondence is. E.g. in case of the University admission processes (field models), some universities interview the applicants, while others prefer aptitude tests. There are good arguments for and against a match [20]. Hence, it is necessary to obtain a deeper understanding on what should be understood as a correspondence and what types of correspondences do exist. If such an understanding is reached, an application of established methods for the evaluation of process model similarity measures, e.g., in terms of validity, reliability, and objectivity, might be possible. This would considerably improve the appreciation of the capabilities of automatic similarity measurements.

References

1. Akkiraju, R., Ivan, A.: Discovering business process similarities: an empirical study with SAP best practice business processes. In: Maglio, P.P., Weske, M., Yang, J., Fantinato, M. (eds.) ICSOC 2010. LNCS, vol. 6470, pp. 515–526. Springer, Heidelberg (2010). doi:10.1007/978-3-642-17358-5_35
2. Antunes, G., et al.: The process model matching contest 2015. In: 6th International Workshop on Enterprise Modeling and Information Systems Architectures, pp. 127–155 (2015)
3. Becker, J., Delfmann, P., Dietrich, H.A., Steinhorst, M., Eggert, M.: Business process compliance checking - applying and evaluating a generic pattern matching approach for conceptual models in the financial sector. Inf. Syst. Front. **18**, 1–47 (2014)
4. Becker, M., Laue, R.: A comparative survey of business process similarity measures. Comput. Ind. **63**(2), 148–167 (2012)
5. Brockmans, S., Ehrig, M., Koschmider, A., Oberweis, A., Studer, R.: Semantic alignment of business processes. In: 8th International Conference on Enterprise Information Systems, pp. 191–196 (2006)
6. Cayoglu, U., et al.: Report: the process model matching contest 2013. In: Lohmann, N., Song, M., Wohed, P. (eds.) BPM 2013. LNBIP, vol. 171, pp. 442–463. Springer, Cham (2014). doi:10.1007/978-3-319-06257-0_35
7. Dijkman, R., Dumas, M., García-Bañuelos, L.: Graph matching algorithms for business process model similarity search. In: Dayal, U., Eder, J., Koehler, J., Reijers, H.A. (eds.) BPM 2009. LNCS, vol. 5701, pp. 48–63. Springer, Heidelberg (2009). doi:10.1007/978-3-642-03848-8_5
8. Dijkman, R.M., van Dongen, B.F., Dumas, M., García-Bañuelos, L., Kunze, M., Leopold, H., Mendling, J., Uba, R., Weidlich, M., Weske, M., Yan, Z.: A short survey on process model similarity. In: Bubenko, J., Krogstie, J., Pastor, O., Pernici, B., Rolland, C., Sølvberg, A. (eds.) Seminal Contributions to Information Systems Engineering, pp. 421–427. Springer, Heidelberg (2013)
9. Dijkman, R.M., Dumas, M., van Dongen, B.F., Käärik, R., Mendling, J.: Similarity of business process models: metrics and evaluation. Inf. Syst. **36**(2), 498–516 (2011)
10. Dongen, B., Dijkman, R., Mendling, J.: Measuring similarity between business process models. In: Bellahsène, Z., Léonard, M. (eds.) CAiSE 2008. LNCS, vol. 5074, pp. 450–464. Springer, Heidelberg (2008). doi:10.1007/978-3-540-69534-9_34
11. Fettke, P., Vella, A., Loos, P.: From measuring the quality of labels in process models to a discourse on process model quality: a case study. In: 45th Hawaii International International Conference on Systems Science, pp. 197–206 (2012)
12. Gerke, K., Cardoso, J., Claus, A.: Measuring the compliance of processes with reference models. In: Meersman, R., Dillon, T., Herrero, P. (eds.) OTM 2009. LNCS, vol. 5870, pp. 76–93. Springer, Heidelberg (2009). doi:10.1007/978-3-642-05148-7_8
13. Juntao, G., Li, Z., Wei, J.: Procuring requirements for ERP software based on semantic similarity. In: International Conference on Semantic Computing, pp. 61–70 (2007)

14. Kunze, M., Weidlich, M., Weske, M.: Behavioral similarity – a proper metric. In: Rinderle-Ma, S., Toumani, F., Wolf, K. (eds.) BPM 2011. LNCS, vol. 6896, pp. 166–181. Springer, Heidelberg (2011). doi:10.1007/978-3-642-23059-2_15

15. La Rosa, M., Dumas, M., Uba, R., Dijkman, R.M.: Merging business process models. In: On the Move to Meaningful Internet Systems, pp. 96–113 (2010)

16. La Rosa, M., Dumas, M., Uba, R., Dijkman, R.M.: Business process model merging: an approach to business process consolidation. ACM Trans. Softw. Eng. Methodol. **22**(2), 11:1–11:42 (2013)

17. Minor, M., Tartakovski, A., Bergmann, R.: Representation and structure-based similarity assessment for agile workflows. In: Weber, R.O., Richter, M.M. (eds.) ICCBR 2007. LNCS (LNAI), vol. 4626, pp. 224–238. Springer, Heidelberg (2007). doi:10.1007/978-3-540-74141-1_16

18. Niesen, T., Houy, C.: Zur Nutzung von Techniken der Natürlichen Sprachverarbeitung für die Bestimmung von Prozessmodellähnlichkeiten – Review und Konzeptentwicklung. In: 12 Internationale Tagung Wirtschaftsinformatik, pp. 1829–1843 (2015)

19. Thaler, T., Dadashnia, S., Sonntag, A., Fettke, P., Loos, P.: The IWi process model corpus. Technical report, Institute for Information Systems (IWi) at the German Research Center for Artificial Intelligence (DFKI) (2015)

20. Thaler, T., Hake, P., Fettke, P., Loos, P.: Evaluating the evaluation of process matching techniques. In: Multikonferenz Wirtschaftsinformatik, pp. 1600–1612 (2014)

21. Vogelaar, J.J.C.L., Verbeek, H.M.W., Luka, B., Aalst, W.M.P.: Comparing business processes to determine the feasibility of configurable models: a case study. In: Daniel, F., Barkaoui, K., Dustdar, S. (eds.) BPM 2011. LNBIP, vol. 100, pp. 50–61. Springer, Heidelberg (2012). doi:10.1007/978-3-642-28115-0_6

22. Webster, J., Watson, R.T.: Analyzing the past to prepare for the future: writing a literature review. MIS Q. **26**(2), xiii–xxiii (2002)

23. Yan, Z., Dijkman, R., Grefen, P.: Fast business process similarity search with feature-based similarity estimation. In: Meersman, R., Dillon, T., Herrero, P. (eds.) OTM 2010. LNCS, vol. 6426, pp. 60–77. Springer, Heidelberg (2010). doi:10.1007/978-3-642-16934-2_8

24. Zezula, P., Amato, G., Dohnal, V., Batko, M.: Similarity Search - The Metric Space Approach. Advances in Database Systems, vol. 32. Springer, New York (2006)

First International Workshop on Runtime Analysis of Process-Aware Information Systems (PRAISE2016)

The First International Workshop on Runtime Analysis of Process-Aware Information Systems (PRAISE2016)

Mustafa Hashmi[1(✉)], Guido Governatori[1], Huy Tran[2(✉)], and Uwe Zdun[2]

[1] Data61, CSIRO, 70-72 Spring Hill, Brisbane, Australia
{mustafa.hashmi,guido.governatori}@data61.csiro.au
[2] Research Group Software Architecture, University of Vienna,
1090 Vienna, Austria
{huy.tran,uwe.zdun}@univie.ac.at

Abstract. Let us introduce the first edition of *International Workshop on Runtime Analysis of Process-Aware Information Systems* (PRAISE2016) co-located with the 2016 edition of Business Process Management (BPM) conference in Rio de Janeiro, Brazil. The workshop aims to provide a high-quality and dedicate forum for researchers and practitioners to exchange research findings and ideas on (but not limited to) runtime analysis of Process-Aware Systems (PAISs) and related technologies and practice. Although we received relatively a lower number of submissions, the selected papers are of very good quality with interesting findings shared during the workshop.

Keywords: Business processes · Process-aware systems · Runtime analysis

1 Aim and Scope: Academia and Practice

Modern process–aware systems (PAISs) are increasingly large and can encompass or connect with numerous other systems ranging from traditional software systems to social networks, clouds, Internet of Things (IoT), and so on. Apart from the intrinsic complexity of integrated systems, the variability and dynamicity of the surrounding environments make it extremely challenging to ensure the correctness of such complex and large scale PAISs, especially at runtime.

Given many decades of research and development in PAISs, many existing methods and techniques have been successful in addressing software system properties at design–time. Unfortunately, less attention has been drawn to the analysis of systems at runtime. The **1st International Workshop on Runtime Analysis of Process-Aware Information Systems (PRAISE2016)** provided a dedicated forum for researchers and practitioners from different disciplines and domains interested in methods, techniques for runtime verification and compliance checking, quantitative analysis (e.g. prediction and optimization of temporal performance), their applications and validating in experimental and industrial settings. It was a forum for stimulating discussions from both theoretical and practical dimensions on investigating and addressing the problem of runtime analysis and consistencies checking of process–aware information systems, including (but not limited to):

- Runtime analysis and verification: including approaches to analyze (non-)functional properties of systems, runtime compliance checking, methods and techniques for checking and monitoring of business process consistency properties and constraints.
- Novel architectural designs for dealing with the complexity of PAISs and their surrounding runtime environments and the sheer amount of runtime data to support large scale consistency checking/monitoring. Quantitative analysis techniques/ methods for prediction and optimization of temporal performance at runtime.

Any system inconsistencies at either design–time or runtime often would lead to severe and costly consequences in system development, enactment, and maintenance. Therefore, the outcomes of this workshop can complement and strengthen the mainstream of research and raise the awareness and adoption of the resulting methods and techniques in industrial applications.

The technical program of PRAISE2016 included presentations of research papers on behavioral classification and deep learning approaches for predicting process behavior at runtime selected through a peer-review process. Each paper was reviewed by at least three members of the Program Committee for technical quality and contribution.

A very special thank is due to the excellent Program Committee for their thorough reviews and discussions of the submitted papers. Their criticisms and very useful comments and suggestions were instrumental in achieving a high-quality publication. Also, we express our gratitude to BPM Conference committees, especially to the workshops chair Marlon Dumas (University of Tartu, Estonia) and Marcelo Fantinato (University of São Paulo, Brazil) for their valuable guidance. Last but not the least, we are thankful to the authors for submitting high-quality papers, responding to the reviewers' comments, and adhering the deadlines, the presenters, and all the other workshop participants, because the workshop could not be held without their contributions and interest.

19 September 2016

Mustafa Hashmi
Huy Tran
Guido Governatori
Uwe Zdun

2 The PRAISE2016 Organizers

Mustafa Hashmi	Data61|CSIRO, Australia
Huy Tran	Vienna University, Austria
Guio Governatori	Data61|CSIRO, Australia
Uwe Zdun	Vienna University, Austria

3 Program Committee Members

Ahmed Awad	Cairo University, Egypt
Claudio Bartolini	Hewlett–Packard, USA
Aditya Ghose	University of Wollongong, Australia
Óscar González Rojas	Universidad de los Andes, Colombia
Schahram Dustdar	Vienna University of Technology, Austria
Amal Elgammal	Cairo University, Egypt
Rik Eshuis	Eindhoven University of Technology, Netherlands
Guido Governatori	Data61|CSIRO, Australia
Priotr Kulicki	The John Paul II Catholic University of Lublin, Poland
Jorge Munoz–Gama	Pontificia Univesidad Católica de Chile, Chile
Ho-Pun Lam	Data61|CSIRO, Australia
Cesare Pautasso	University of Lugano, Switzerland
Huy Tran	University of Vienna, Austria
Manfred Reichert	Ulm University, Germany
Akhil Kumar	Penn State Smeal College of Business, USA
Mustafa Hashmi	Data61|CSIRO, Australia
Stephanie Rinderle-Ma	University of Vienna, Austria
Régis Riveret	Data61|CSIRO, Australia
Markus Stumptner	University of South Australia, Australia
Jianwen Su	University of California, Santa Barbara, USA
Suriadi Suriadi	Massey University, New Zealand
Sherif Sakr	University of New South Wales, Australia
Rafael Accorsi	University of Freiburg, Germany
Schahram Dustdar	Vienna University of Technology, Austria
Uwe Zdun	University of Vienna, Austria

4 Additional Reviewers

Walid Fdhila
Hoa Khanh Dam

A Deep Learning Approach for Predicting Process Behaviour at Runtime

Joerg Evermann[1(✉)], Jana-Rebecca Rehse[2,3], and Peter Fettke[2,3]

[1] Memorial University of Newfoundland, St. John's, Canada
jevermann@mun.ca
[2] German Research Center for Artificial Intelligence, Saarbrücken, Germany
jana.rebecca.rehse@iwi.dfki.de
[3] Saarland University, Saarbrücken, Germany

Abstract. Predicting the final state of a running process, the remaining time to completion or the next activity of a running process are important aspects of runtime process management. Runtime management requires the ability to identify processes that are at risk of not meeting certain criteria in order to offer case managers decision information for timely intervention. This in turn requires accurate prediction models for process outcomes and for the next process event, based on runtime information available at the prediction and decision point. In this paper, we describe an initial application of deep learning with recurrent neural networks to the problem of predicting the next process event. This is both a novel method in process prediction, which has previously relied on explicit process models in the form of Hidden Markov Models (HMM) or annotated transition systems, and also a novel application for deep learning methods.

Keywords: Process management · Runtime support · Process prediction · Deep learning · Neural networks

1 Introduction

Managing processes at runtime has important business applications [1]. It allows customer service agents to respond to enquiries about the remaining time until a case is resolved or completed. It allows case managers to identify cases that are likely to be late or to terminate abnormally and to intervene early in order to mitigate business risk. Process prediction is an important aspect of runtime process management. In general, prediction concerns either the process outcome or the subsequent event(s) in a process. Examples of business relevant process outcomes include the final state (e.g. whether the final state is "accept client claim" or "reject client claim"), case data (e.g. whether the attribute "cost" is less than a certain amount) or LTL (linear temporal logic) compliance formulas (e.g. whether "approve claim" has occurred prior to "issue cheque" and the activities have been performed by different resources). Predictions can be made from the sequence of the activities that have occurred in the running process,

© Springer International Publishing AG 2017
M. Dumas and M. Fantinato (Eds.): BPM 2016 Workshops, LNBIP 281, pp. 327–338, 2017.
DOI: 10.1007/978-3-319-58457-7_24

the case data that has been collected, the participating resources in the case activities, the execution times of those activities and any other available case or workflow data that is available at runtime.

Previous work on process prediction at runtime has focused on predicting process outcomes and primarily the remaining time to completion, whereas there exists limited work on predicting the next process event. Most prior work is based on an explicit representation of the process, e.g. mined from event logs, and augmented with probability tables and execution time information. In contrast, our approach does not rely on an explicit representation of the underlying process model, but is based on recent work in "deep learning". While "deep learning" has only recently become a popular research topic, it is essentially an application of neural networks and thus looks back on a long history of research [2]. Recent innovations both in algorithms, allowing novel architectures of neural networks, and computing hardware, especially access to GPU processing, have led to a resurgence in interest for neural networks and popularized the term "deep learning" [3].

This work is motivated by the application of deep learning to natural language processing (NLP). In recent years, NLP research has moved from explicit representations of language models to statistical methods, specifically to recurrent neural networks (RNN) [4–6]. In this work, we apply a recurrent neural network to the problem of predicting the next event in a process from a sequence of observed events. Our idea is to treat an event trace as analogous to a natural language sentence, with the events analogous to words.

However, while there are many similarities between natural language and process traces, there are differences as well. First, the size of the vocabulary in process prediction (the number of event types) is much smaller than the size of a natural language vocabulary. Second, the length of a trace can far exceed the typical sentence length in natural language. Together, these two differences result in fewer possible prediction targets (vocabulary size or number of unique process event types) and more information to predict from (sentence or trace prefix length), suggesting that this approach may be able to achieve better results than, for example, word prediction in NLP. Third, process event sequences are determined or constrained by an internal process logic, typically determined by decision rules based on case data. However, natural language is also constrained, in the form of grammatical and morphological rules, for example the noun and verb agreement on plurals in English. Just as these linguistic rules and constraints are not explicitly captured in NLP deep-learning approaches [4–7] but are learned by the neural network when given sufficient training data, the process constraints and rules need not be explicitly represented in our approach either.

The aim of this paper is to explore the potential for applications of deep learning in business process management at runtime and to describe an initial application. We emphasize that this is an initial exploration of the feasibility of this approach, and is intended more to open this line of inquiry than to provide final answers. *The contribution therefore lies not in the performance of this particular implementation but in the demonstration of the applicability of our approach and the potential for future work using deep learning in process management.*

The remainder of the paper is structured as follows. Section 2 presents related work on process prediction, especially prediction of the next event in a process. Section 3 presents a brief introduction to deep learning. Section 4 describes our implementation of process prediction with RNNs, followed in Sect. 5 by an experimental evaluation. The paper closes with a discussion and outlook to future work in Sect. 6.

2   Related Work

Most of the prediction methods address process outcomes rather than prediction of the next event in a process, as we do here. The most frequently examined outcome is the time remaining to completion of a case. Van Dongen et al. present a first approach using boosted regression on event frequencies, event times, and case data [8]. The approach by Pandey et al. applies a Hidden Markov Model (HMM) on event sequences and execution times [9]. It is based on an annotated transition system, as is the approach by Van der Aalst et al. [10]. Folino et al. present two contributions that use clustering trees and finite state machines (FSM) to predict the remaining time of a running process case [11,12]. Schwegmann et al. report on the development of a software tool that applies Complex Event Processing (CEP) to event sequences and is trained to predict their future behavior [13]. The two approaches by Rogge-Solti et al. use stochastic petri net simulation for the same purpose [14,15]. Bevacqua et al. present a prediction technique based on clustering and regression on case data [16]. Bolt et al. employ a clustering approach on partial and completed cases [17]. Finally, Polato et al. present two approaches that are based on annotated transition systems, as well as support vector regression and naive Bayes classifiers [18,19].

Another popular objective of process prediction is the binary assessment of its outcome, i.e. whether or not a process instance will fail. This was first addressed in the early 2000s by Castellanos et al. and Grigori et al., who use decision trees on time, resource, and case data [20–22]. Decision trees are also used in the approach by Conforti et al. [23]. Kang et al. present two different approaches to predicting process failures, one using a support vector machine (SVM) [24] and another one based on clustering and local outlier detection [25]. Maggi et al. employ decision trees to predict violation of LTL (Linear Temporal Logic) restrictions [26,27]. Leontjeva et al. use random forests for the same purpose [28]. These two techniques are combined in the approach by Francescomarino et al. [29]. Both Metzger et al. and Folino et al. are concerned with the binary outcome in terms of the completion past promise; while the former employs neural networks, constraint satisfaction, and Quality-of-Service aggregation [30], the latter relies on clustering and regression [31].

Only five approaches [32–36] are concerned with predicting the next event, many of which use an explicit process model representation such as HMM and PFA (Probabilistic Finite Automatons).

The MSA approach in [32] considers each trace prefix as a state in a state-transition matrix. From the observed prefixes and their next tasks, a state transition matrix is built. When a running case has reached a state not contained in the state-transition matrix, its similarity to observed traces is computed using string edit distance. The prediction is made from the most similar observed case.

The approach described by [33,34] consists of five steps. A process model is mined from existing logs. For each XOR split in the model, a decision tree is mined from case data. These trees are then used to compute the state transition probabilities for a Markov chain specific to the running case that is to be predicted, from the case data available at that point. This HMM is then used to predict the probabilities of the following events.

The approach described in [35] uses sequence mining to identify frequent trace prefixes. For each prefix, a regression model is trained to predict remaining time to completion and a classification model (decision trees) is trained to predict the next event. The algorithm identifies the appropriate prefix of the running case to chose the regression and classification model and uses these to predict remaining completion time and next event.

RegPFA [36] is also based on explicit process models but uses a probabilistic finite automaton (PFA) instead of an HMM because it allows the future hidden state to be a function of both the previous hidden state and the previous observed event (which itself is a probabilistic consequence of the previous hidden state). RegPFA uses an EM algorithm to estimate the model parameters of the PFA, similar to the Baum-Welch algorithm used for HMM.

Our approach has the same objective, i.e. predictand, as these five previous works, but differs from the in terms of predictors, method, and process representation. Deep Learning in the form of an RNN is used to predict the next events, using event sequences and associated resources. Processes are only implicitly represented within the RNN. Overall, our method constitutes an innovative new approach to process prediction, described in the following sections.

3   Deep Learning

A neural network consists of a layer of input cells, multiple layers of "hidden" cells, and a layer of output cells. Cells in each layer are connected by weighted connections to cells in previous and following layers in various forms, allowing for different architectures (e.g. each cell connected to all others on the following layers, or other topologies). Each cell's output is a function of the weighted sum of its input. A typical network architecture is a fully connected network of cells using sigmoid activation functions:

$$a_j^l = \sigma\left(\sum_i w_i^{l-1,j} a_i^{l-1} + b_j^l\right) \qquad \text{where} \qquad \sigma(x) = \frac{1}{1 + exp(-x)}$$

Here, a_j^l is the output ("activation") of cell j in layer l, $w_j^{l,i}$ is the weight of the connection from cell i on layer $l-1$ to cell j on layer l, a_i^{l-1} is the output of cell i on layer $l-1$ and b_j^l is the "bias" of cell j on layer l.

A neural network is a supervised learning technique where the output of the neural net is compared to a target by means of a loss function. Gradients of network parameters $(w_j^{l,i}, b_i^l)$ with respect to the loss function are computed using backpropagation and parameters are then adjusted using variants of gradient descent algorithms to minimize the loss function. The type of output layer cell and the loss function are often chosen jointly for their computational properties with respect to backpropagation. A typical combination is a softmax layer with a cross-entropy loss function:

$$y_i = softmax(x)_i = \frac{exp(x_i)}{\sum_j exp(x_j)} \qquad H_{y'}(y) = -\sum_i y_i \prime log(y_i)$$

Here, $y\prime$ are the target values and y are the network outputs, computed in turn from the outputs x_i of the next to last network layer.

Recurrent Neural Networks (RNN). In a recurrent neural network, each cell also feeds back information into itself, allowing it to maintain "state" over time. In order to make this tractable within an acyclic computational graph and backpropagation, the recurrent network cells are "unrolled", that is, copies of it are produced for time $t, t-1, t-2, \ldots$. The state output of the RNN cell of time $t-1$ is state input to the cell for time t. In general, t can index any sequence, not only time. Depending on how long one wishes to maintain state, fewer or

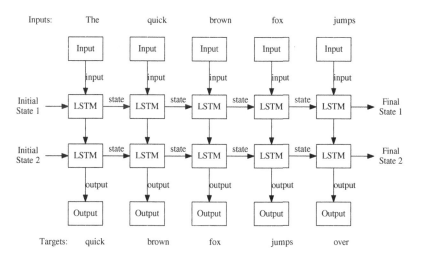

Fig. 1. RNN architecture with single hidden layer of LSTM cells, unrolled five steps

more cells are unrolled. Figure 1 shows an RNN architecture with an input layer, an output layer and two hidden layers that are unrolled five steps. Each layer (each box in Fig. 1) in turn consists of multiple input, output or hidden cells (not shown).

Long Short Term Memory (LSTM). RNN using sigmoid cells have been found to be unsatisfactory, leading to the development of long short term memory (LSTM) cells [37]. A basic LSTM cell is defined as follows, accepting C_{t-1} and h_{t-1} as state and input information from the prior unrolled cell on the same level, and accepting x_t as input from cells on the previous layers. In turn, it passes C_t and h_t as state and output information to the subsequent unrolled cell and provides h_t as output to the next layer; the various W and b are "trainable" parameters.

$$f_t = \sigma\left(W_f \cdot [h_{t-1}, x_t] + b_f\right) \qquad i_t = \sigma\left(W_i \cdot [h_{t-1}, x_t] + b_i\right)$$
$$\bar{C}_t = \tanh\left(W_C \cdot [h_{t-1}, x_t] + b_C\right) \qquad C_t = f_t \times C_{t-1} + i_t \times \bar{C}_t$$
$$o_t = \sigma\left(W_o \cdot [h_{t-1}, x_t] + b_o\right) \qquad h_t = o_t \times \tanh(C_t)$$

Intuitively, f_t represents the "forget gate", allowing the LSTM cell to selectively suppress information from C_{t-1}. Similarly, i_t represents the "input gate", allowing the LSTM cell to selectively add input to the state. Note how the new state C_t is computed by first forgetting $f_t \times C_{t-1}$ and then adding $i_t \times \bar{C}_t$ where $\bar{C}_t$ is the "candidate state information" computed from the inputs x_t and h_{t-1}. Similarly, o_t represents the output gate, allowing the LSTM cell to selectively output part of the new state to the following cells.

NLP Applications of RNN. A typical NLP application trains the RNN on sequences of input words to predict the next word, e.g. to provide word suggestions for user input. As shown in Fig. 1, the target words are simply the input words shifted by one position, so that for each input word the following word is the target to be predicted. Words are mapped into an n-dimensional "embedding" space using an "embedding matrix", which is essentially a lookup matrix of dimensions $v \times m$ where v is the size of the "vocabulary" and m is the chosen dimensionality of the embedding space and the size of each LSTM hidden layer. While the dimensionality of the space can be chosen freely, larger dimensions allow better separation of words in that space, at the cost of computational performance. The input layer in Fig. 1 is an embedding lookup function that performs the embedding lookup of each input word. The output layer in Fig. 1 is typically a softmax layer that produces a probability over the vocabulary. Training performance is usually defined in terms of the per-word perplexity, defined as $P = \exp(H/n)$, where n is the number of possible words (targets). Perplexity measures the "surprisedness" the network exhibits when encountering the next term. There are no absolute guidelines as to what an acceptable level of perplexity is, however, a network that predicts well, will show low perplexity.

4   Process Prediction Using RNN

A number of software frameworks for deep-learning, such as Caffe, Torch, Singa, and Tensorflow, have become available recently[1]. We implemented our approach using Tensorflow as it provides a suitable level of abstraction, provides RNN specific functionality, and can be used on high-performance parallel, cluster, and GPU computing platform.

The network architecture features two hidden RNN layers, unrolled to 20 steps, using basic LSTM cells. We chose $m = 500$ for the dimensionality of the embedding space and the size of the hidden layers. Thus, our network contained a total of $500 \times 20 \times 2 = 20000$ LSTM cells.

Trainable parameters are initialized using a uniform random distribution over $[-0.1, 0.1]$. Training proceeds in batches of size 20. For each batch, the backpropagation algorithm computes the mean gradients for all parameters. Training of the net is done in "epochs". Each epoch trains the net on the entire event log. Subsequent epochs maintain the weights W and biases b learned from the previous epoch but reinitialize the states for each layer and then train the net again on the entire event log. The learning rate is exponentially reduced from 1 by a factor of 0.75 after the 25th epoch. The net was trained for a maximum of 50 epochs or until maximum accuracy was reached. Because of the random initialization of parameters, we performed three runs for each dataset and report the mean results of the three runs (omitting clear outliers).

We ran our initial experiments on a single NVidia K1100M GPU. Training performance was approximately 1000 words per second. Code, data and results are available from the first author[2].

5   Experimental Results

Because only one of the related works discussed in Sect. 2 uses publicly available data, and to aid comparison of our approach to related work, we chose the same BPI Challenge 2012 and 2013 datasets that [36] used for their study. In addition to separating the BPI 2012 data set by sub-process as done in [36], we also used the combined dataset. While [36] use only activity completion events, we also tested our approach on all events (including the lifecycle transitions "start", "scheduled" and "completed"). Furthermore, we included an experimental condition where we not only extracted the event name, but combined this with the resource name (or "none" if not available). This creates a larger vocabulary which increases the prediction difficulty, but also provides more information to the training algorithm and allows prediction of next event and resource in a process. Because the number of distinct resources in the BPI 2013 Challenge dataset is very large, we combined the organizational group associated with each event, instead of the resource, with the event name. Table 1 summarizes

---

[1] http://caffe.berkelyvision.org, http://torch.ch, https://singa.incubator.apache.org, https://www.tensorflow.org.

[2] http://joerg.evermann.ca/software.html.

Table 1. Characteristics of datasets used in experimental evaluation. Event numbers for partial BPI2012 logs do not add up to that of corresponding complete log due to end-of-case marker events added to each trace.

Dataset	Number of unique event types ("vocabulary size")	Number of unique event types ("vocabulary size") (with resource or org. group)	Number of events
BPI2013.Incidents	14	3133	73087
BPI2013.Problems	8	64	11317
BPI2012 (completion events)	24	877	177593
BPI2012 (all events)	37	1349	275287
BPI2012.W (completion events)	7	264	82071
BPI2012.A (completion events)	11	302	73936
BPI2012.O (completion events)	8	313	36259
BPI2012.W (all events)	7	736	179765
BPI2012.A (all events)	11	302	73939
BPI2012.O (all events)	8	313	36259

Table 2. Results and comparison to [36]. Results are means of three runs. Precision is defined as the proportion of correct predictions among all predictions. Higher precision is better; lower perplexity is better.

Dataset	Precision in [36]	Precision	Perplexity
BPI2012.W (complete events)	.719	.623	3.128
BPI2012.A (complete events)	.801	.778	1.649
BPI2012.O (complete events)	.811	.789	1.624
BPI2012.W (complete events, resource)		.836	1.733
BPI2012.A (complete events, resource)		.941	1.226
BPI2012.O (complete events, resource)		.992	1.040
BPI2012.W (all events)		.840	1.531
BPI2012.A (all events)		.775	1.673
BPI2012.O (all events)		.793	1.591
BPI2012.W (all events, resource)		.820	1.745
BPI2012.A (all events, resource)		.941	1.225
BPI2012.O (all events, resource)		.992	1.040
BPI2012 (all events)		.852	1.462
BPI2012 (complete events)		.768	1.822
BPI2012 (all events, resource)		.724	2.368
BPI2012 (complete events, resource)		.802	1.966
BPI2013.Incidents	.714	.699	2.346
BPI2013.Problems	.690	.451	6.151
BPI2013.Incidents (with group)		.939	1.236
BPI2013.Problems (with group)		.954	1.174

the characteristics of the datasets. We transformed the published XES datasets using XSL transformations to extract traces, events, and resource information in a suitable format.

Table 2 shows our results and a comparison to the best result presented by [36], where available. While [36] report a cross-entropy H in addition to accuracy, their definition of H in their Fig. 8 appears to be the entropy, not the cross-entropy, and is therefore not comparable to the cross-entropy typically used in deep learning applications. Finally, we report the perplexity as a standard way of evaluating RNNs in the NLP context. High accuracy values close to 1 are preferable; low perplexity values close to 1 are preferable.

Comparing our initial results to those of [36] shows that they are close to the state-of-the-art on many datasets, significantly lagging only on the BPIC 2013 Problems dataset. Given that this is an initial application and evaluation, these results are encouraging. Table 2 also show many results with accuracies in excess of 90%. While we have no comparison to the state-of-the-art on these datasets available in [36], this level of precision is encouraging for practical applications. Comparing the performance of including resource or organizational groups, which dramatically increases the size of the vocabulary, shows that the predictive accuracy improves in all cases. More importantly, the improved accuracy is higher than the best results reported by [36] for the corresponding datasets without resource or group information.

6   Discussion and Conclusion

This paper presents a novel approach to predicting the behaviour of running processes. Using analogies to natural language processing, we applied deep learning, specifically recurrent neural networks with LSTM cells, to the problem of predicting the next event in a running process. Our results, close to the state-of-the-art on two real datasets and with accuracies in excess of 90% on many problems, demonstrate the feasibility of this approach and should encourage further work in this direction.

As this research is early work with the deep learning approach, we recognize the limitations of this study and the need for further work. Our immediate plans are to explore different network architectures and the parameter space. For example, more advanced LSTM cells are available [38], one can introduce additional RNN layers (currently 2), one can adjust the sequence of "unrolled" RNN cells (currently 20, which is shorter than the mean trace length for some datasets), one can adjust the dimension of the space into which terms are embedded (currently 500, which is larger than the vocabulary of all datasets), one can adjust the learning rate to be more or less "aggressive", one can adjust the clipping of gradients to allow faster, but possibly sub-optimal, convergence, and one can adjust the random initialization of network parameters. While we believe that the results we have presented were achieved using typical parameters, more work is clearly required to identify optimal architectures and sets of parameter values for the datasets considered in this work. Furthermore, additional replications and cross-validation is required.

Another area of inquiry is to add additional information into the predictors and/or the predictands. In our experiments, we have added resource information to both the predictors as well as the predictands, allowing us to predict also the next resource (as well as the next event). However, case attribute information could be added to each predictor but not the predictands, increasing the information available for prediction but not the number of possible prediction targets, and may therefore lead to better prediction accuracy.

Finally, the deep learning approach can be applied to the prediction of process outcomes. Process outcomes, such as remaining time to completion or violation of an LTL compliance expression, are continuous or categorical, but not in the form of sequences. Both are suitable for neural networks, but do not require the recurrent neural network architecture used here and more "traditional" architectures need to be explored and evaluated.

References

1. Houy, C., Fettke, P., Loos, P., Aalst, W.M.P., Krogstie, J.: BPM-in-the-large – towards a higher level of abstraction in business process management. In: Janssen, M., Lamersdorf, W., Pries-Heje, J., Rosemann, M. (eds.) EGES/GISP - 2010. IAICT, vol. 334, pp. 233–244. Springer, Heidelberg (2010). doi:10.1007/978-3-642-15346-4_19
2. Schmidhuber, J.: Deep learning in neural networks: an overview. Neural Netw. **61**, 85–117 (2015)
3. LeCun, Y., Bengio, Y., Hinton, G.: Deep learning. Nature **521**, 436–444 (2015)
4. Sutskever, I., Martens, J., Hinton, G.E.: Generating text with recurrent neural networks. In: Proceedings of the 28th International Conference on Machine Learning, ICML 2011, Bellevue, Washington, USA, June 28 - July 2, 2011, pp. 1017–1024 (2011)
5. Graves, A.: Generating sequences with recurrent neural networks. CoRR abs/1308.0850 (2013)
6. Zaremba, W., Sutskever, I., Vinyals, O.: Recurrent neural network regularization. CoRR abs/1409.2329 (2014)
7. Graves, A.: Supervised Sequence Labelling with Recurrent Neural Networks. SCI, vol. 385. Springer, New York (2012)
8. Dongen, B.F., Crooy, R.A., Aalst, W.M.P.: Cycle time prediction: When will this case finally be finished? In: Meersman, R., Tari, Z. (eds.) OTM 2008. LNCS, vol. 5331, pp. 319–336. Springer, Heidelberg (2008). doi:10.1007/978-3-540-88871-0_22
9. Pandey, S., Nepal, S., Chen, S.: A test-bed for the evaluation of business process prediction techniques. In: 7th International Conference on Collaborative Computing: Networking, Applications and Worksharing, CollaborateCom 2011, Orlando, FL, USA, 15–18 October, 2011, pp. 382–391 (2011)
10. van der Aalst, W.M.P., Schonenberg, M.H., Song, M.: Time prediction based on process mining. Inf. Syst. **36**(2), 450–475 (2011)
11. Folino, F., Guarascio, M., Pontieri, L.: Context-aware predictions on business processes: an ensemble-based solution. In: Appice, A., Ceci, M., Loglisci, C., Manco, G., Masciari, E., Ras, Z.W. (eds.) NFMCP 2012. LNCS (LNAI), vol. 7765, pp. 215–229. Springer, Heidelberg (2013). doi:10.1007/978-3-642-37382-4_15
12. Folino, F., Guarascio, M., Pontieri, L.: Discovering context-aware models for predicting business process performances. In: Meersman, R., et al. (eds.) OTM 2012. LNCS, vol. 7565, pp. 287–304. Springer, Heidelberg (2012). doi:10.1007/978-3-642-33606-5_18

13. Schwegmann, B., Matzner, M., Janiesch, C.: preCEP: facilitating predictive event-driven process analytics. In: Brocke, J., Hekkala, R., Ram, S., Rossi, M. (eds.) DESRIST 2013. LNCS, vol. 7939, pp. 448–455. Springer, Heidelberg (2013). doi:10.1007/978-3-642-38827-9_36

14. Rogge-Solti, A., Weske, M.: Prediction of remaining service execution time using stochastic petri nets with arbitrary firing delays. In: Basu, S., Pautasso, C., Zhang, L., Fu, X. (eds.) ICSOC 2013. LNCS, vol. 8274, pp. 389–403. Springer, Heidelberg (2013). doi:10.1007/978-3-642-45005-1_27

15. Rogge-Solti, A., Weske, M.: Prediction of business process durations using non-markovian stochastic petri nets. Inf. Syst. **54**, 1–14 (2015)

16. Bevacqua, A., Carnuccio, M., Folino, F., Guarascio, M., Pontieri, L.: A data-driven prediction framework for analyzing and monitoring business process performances. In: Hammoudi, S., Cordeiro, J., Maciaszek, L.A., Filipe, J. (eds.) ICEIS 2013. LNBIP, vol. 190, pp. 100–117. Springer, Cham (2014). doi:10.1007/978-3-319-09492-2_7

17. Bolt, A., Sepúlveda, M.: Process remaining time prediction using query catalogs. In: Lohmann, N., Song, M., Wohed, P. (eds.) BPM 2013. LNBIP, vol. 171, pp. 54–65. Springer, Cham (2014). doi:10.1007/978-3-319-06257-0_5

18. Polato, M., Sperduti, A., Burattin, A., de Leoni, M.: Data-aware remaining time prediction of business process instances. In: 2014 International Joint Conference on Neural Networks, IJCNN 2014, Beijing, China, July 6–11, 2014, pp. 816–823 (2014)

19. Polato, M., Sperduti, A., Burattin, A., de Leoni, M.: Time and activity sequence prediction of business process instances. CoRR abs/1602.07566 (2016)

20. Castellanos, M., Salazar, N., Casati, F., Dayal, U., Shan, M.-C.: Predictive business operations management. In: Bhalla, S. (ed.) DNIS 2005. LNCS, vol. 3433, pp. 1–14. Springer, Heidelberg (2005). doi:10.1007/978-3-540-31970-2_1

21. Grigori, D., Casati, F., Castellanos, M., Dayal, U., Sayal, M., Shan, M.: Business process intelligence. Comput. Ind. **53**(3), 321–343 (2004)

22. Grigori, D., Casati, F., Dayal, U., Shan, M.: Improving business process quality through exception understanding, prediction, and prevention. In: VLDB 2001, Proceedings of 27th International Conference on Very Large Data Bases, September 11–14, 2001, Roma, Italy, pp. 159–168 (2001)

23. Conforti, R., Leoni, M., Rosa, M., Aalst, W.M.P.: Supporting risk-informed decisions during business process execution. In: Salinesi, C., Norrie, M.C., Pastor, Ó. (eds.) CAiSE 2013. LNCS, vol. 7908, pp. 116–132. Springer, Heidelberg (2013). doi:10.1007/978-3-642-38709-8_8

24. Kang, B., Kim, D., Kang, S.: Periodic performance prediction for real-time business process monitoring. Ind. Manage. Data Syst. **112**(1), 4–23 (2011)

25. Kang, B., Kim, D., Kang, S.: Real-time business process monitoring method for prediction of abnormal termination using knni-based LOF prediction. Expert Syst. Appl. **39**(5), 6061–6068 (2012)

26. Maggi, F.M., Francescomarino, C.D., Dumas, M., Ghidini, C.: Predictive monitoring of business processes. CoRR abs/1312.4874 (2013)

27. Maggi, F.M., Francescomarino, C., Dumas, M., Ghidini, C.: Predictive monitoring of business processes. In: Jarke, M., Mylopoulos, J., Quix, C., Rolland, C., Manolopoulos, Y., Mouratidis, H., Horkoff, J. (eds.) CAiSE 2014. LNCS, vol. 8484, pp. 457–472. Springer, Cham (2014). doi:10.1007/978-3-319-07881-6_31

28. Leontjeva, A., Conforti, R., Francescomarino, C., Dumas, M., Maggi, F.M.: Complex symbolic sequence encodings for predictive monitoring of business processes. In: Motahari-Nezhad, H.R., Recker, J., Weidlich, M. (eds.) BPM 2015. LNCS, vol. 9253, pp. 297–313. Springer, Cham (2015). doi:10.1007/978-3-319-23063-4_21

29. Francescomarino, C., Dumas, M., Federici, M., Ghidini, C., Maggi, F.M., Rizzi, W.: Predictive business process monitoring framework with hyperparameter optimization. In: Nurcan, S., Soffer, P., Bajec, M., Eder, J. (eds.) CAiSE 2016. LNCS, vol. 9694, pp. 361–376. Springer, Cham (2016). doi:10.1007/978-3-319-39696-5_22

30. Metzger, A., Leitner, P., Ivanovic, D., Schmieders, E., Franklin, R., Carro, M., Dustdar, S., Pohl, K.: Comparing and combining predictive business process monitoring techniques. IEEE Trans. Syst. Man Cybern. Syst. **45**(2), 276–290 (2015)

31. Folino, F., Guarascio, M., Pontieri, L.: A prediction framework for proactively monitoring aggregate process-performance indicators. In: 19th IEEE International Enterprise Distributed Object Computing Conference, EDOC 2015, Adelaide, Australia, September 21–25, 2015, pp. 128–133 (2015)

32. Le, M., Gabrys, B., Nauck, D.: A hybrid model for business process event prediction. In: Proceedings of AI-2012, The Thirty-second SGAI International Conference on Innovative Techniques and Applications of Artificial Intelligence, Cambridge, England, UK, December 11–13, 2012, pp. 179–192 (2012)

33. Lakshmanan, G.T., Shamsi, D., Doganata, Y.N., Unuvar, M., Khalaf, R.: A markov prediction model for data-driven semi-structured business processes. Knowl. Inf. Syst. **42**(1), 97–126 (2015)

34. Unuvar, M., Lakshmanan, G.T., Doganata, Y.N.: Leveraging path information to generate predictions for parallel business processes. Knowl. Inf. Syst. **47**(2), 433–461 (2016)

35. Ceci, M., Lanotte, P.F., Fumarola, F., Cavallo, D.P., Malerba, D.: Completion time and next activity prediction of processes using sequential pattern mining. In: Džeroski, S., Panov, P., Kocev, D., Todorovski, L. (eds.) DS 2014. LNCS (LNAI), vol. 8777, pp. 49–61. Springer, Cham (2014). doi:10.1007/978-3-319-11812-3_5

36. Breuker, D., Matzner, M., Delfmann, P., Becker, J.: Comprehensible predictive models for business processes. MIS Q. **40**, 1009–1034 (2016)

37. Hochreiter, S., Schmidhuber, J.: Long short-term memory. Neural Comput. **9**(8), 1735–1780 (1997)

38. Sak, H., Senior, A.W., Beaufays, F.: Long short-term memory recurrent neural network architectures for large scale acoustic modeling. In: INTERSPEECH 2014, 15th Annual Conference of the International Speech Communication Association, Singapore, September 14–18, 2014, pp. 338–342 (2014)

Behavioral Classification of Business Process Executions at Runtime

Nick R.T.P. van Beest[1][(✉)] and Ingo Weber[2,3]

[1] Data61, CSIRO, Spring Hill, Australia
`nick.vanbeest@data61.csiro.au`
[2] Data61, CSIRO, Sydney, Australia
`ingo.weber@data61.csiro.au`
[3] University of New South Wales, Sydney, Australia

Abstract. Current automated methods to identify erroneous or malicious executions of a business process from logs, metrics, or other observable effects are based on detecting deviations from the normal behavior of the process. This requires a "single model of normative behavior": the current execution either conforms to that model, or not. In this paper, we propose a method to *automatically distinguish different behaviors during the execution of a process*, so that a timely reaction can be triggered, e.g., to mitigate the risk of an ongoing attack. The behavioral classes are learned from event logs of a process, including branching probabilities and event frequencies. Using this method, harmful or problematic behavior can be identified during or even prior to its occurrence, raising alarms as early as undesired behavior is observable. The proposed method has been implemented and evaluated on a set of artificial logs capturing different types of exceptional behavior. Pushing the method to its edge in this evaluation, we provide a first assessment of where the method can clearly discriminate between classes of behavior, and where the differences are too small to make a clear determination.

1 Introduction

Traditionally, finding erroneous or malicious executions of a business process requires expert knowledge and deep insights into the information system supporting that process. Current automated methods to identify such executions from logs, metrics, or other observable effects, such as [1], are based on detecting deviations from the normal behavior of the process. This requires a "single model of normative behavior" and the current execution either conforms to that model, or not. If the current execution deviates from the model, alarms or warnings will be triggered for each exception. The issue with such approaches is that the determination is binary: the current behavior is either fine or not; there is no distinction between less important delays and serious deviations, e.g., when an attack on the system is observed.

In this paper, we propose a method to automatically distinguish different behaviors and act appropriately on the type of erroneous or normal behavior.

© Springer International Publishing AG 2017
M. Dumas and M. Fantinato (Eds.): BPM 2016 Workshops, LNBIP 281, pp. 339–353, 2017.
DOI: 10.1007/978-3-319-58457-7_25

As such, the method presented in this paper allows for a finer and more precise distinction between different behaviors, in effect allowing to trigger alarms for crucial errors or malicious activities, but only warnings for exceptions. The behaviors are learned only from externally observable logs of a process, including event frequencies and branching probabilities.

In addition, in our method the process is not analyzed post-hoc to identify and categorize behavior, but potentially harmful or problematic behavior will be identified during or even *prior* to its occurrence (through prediction) on-the-fly, raising alarms as early as problematic behavior is observable. This enables the early launch of possible automated intervention actions as well, such as tightened firewall rules to minimize attack vectors.

The proposed method has been implemented and evaluated on a set of artificial logs capturing different types of normal and exceptional behavior and random manipulations thereof. The evaluation tests the boundaries of the proposed approach by testing edge cases: to which point can behavior be categorized clearly vs. when are the differences too miniscule to make a clear call?

The remainder of this paper is structured as follows. Section 2 introduces the formalisms we use, i.e., event structures with branching frequencies and event logs. Section 3 presents the approach, from the construction of event structures on the basis of logs to their runtime use for differentiating behavioral classes. The implementation and evaluation are discussed in Sect. 4, before Sect. 5 discusses related work and Sect. 6 concludes.

2   Preliminaries

2.1   Event Structures

A Prime Event Structure (PES) [2] is a graph of events, where each event e represents the occurrence of a task or activity in the business process. As such, multiple occurrences of the same activity are represented by different events. Similarly, an activity occurring in multiple exclusive branches will have different events representing that same activity for each branch. Events can have the following binary relations: (i) *Causality* ($e < e'$) indicates that event e is a prerequisite for e'; (ii) *Conflict* ($e\#e'$) implies that e and e' cannot occur in the same run; (iii) *Concurrency* ($e \parallel e'$) indicates that no order can be established between e and e'. The above is formalized as follows.

Definition 1 (Labeled Prime Event Structure [2]). *A Labeled Prime Event Structure over the set of event labels $\mathcal{L}$ is the tuple $\mathcal{E} = \langle E, \leq, \#, \lambda \rangle$ where*

- *E is a set of events,*
- *$\lambda : E \to \mathcal{L}$ is a labeling function.*
- *$\leq \subseteq E \times E$ is a partial order, referred to as causality,*
- *$\# \subseteq E \times E$ is an irreflexive, symmetric conflict relation,*
- *$\parallel : E^2 \setminus (< \cup <^{-1} \cup \#)$ is the concurrency relation, where $<$ denotes the irreflexive causality relation.*

The *conflict relation* satisfies the principle of conflict heredity, i.e. $e \# e' \wedge e' \leq e'' \Rightarrow e \# e''$ for $e, e', e'' \in E$. The formal details of the construction of a PES from an event log are discussed in [3].

Definition 2 (Frequency-enhanced Prime Event Structure (FPES) [3]). *A frequency-enhanced prime event structure is a tuple $\mathcal{F}(L) = \langle \mathcal{E}(L), \mathcal{O}, \mathcal{P} \rangle$ where*

- $\mathcal{E}(L) : \langle E, \leq, \#, \lambda \rangle$ *is a prime event structure of log L.*
- $\mathcal{O} : E \to \mathbb{N}$ *is a function that associates an event $[e]$ with the number of times its event label occurs in the event log, and corresponds to the cardinality of the equivalence class, i.e. $\mathcal{O}([e]) = |[e]|$.*
- $\mathcal{P} : E \times E \to [0, 1]$ *is a function that associates a pair of events $[e_i]$ and $[e_j]$ with the probability of occurrence of $[e_j]$ given that event $[e_i]$ has occurred. This function is defined as:*

$$\mathcal{P}([e_i], [e_j]) = \begin{cases} \mathcal{O}([e_j])/\mathcal{O}([e_i]) & \text{if } [e_i] \leq [e_j] \\ 0 & \text{Otherwise} \end{cases}$$

2.2 Event Logs

An event log consists of a set of traces, each containing the sequence of events produced by one execution of the process. An event in a trace refers to a task in the process and registers the start, end, abortion or other relevant state change of that task. Events in an event log are related via a total order induced by their timestamps. An event log can defined formally as follows:

Definition 3 (Event log, Trace). *Let L be an event log over the set of labels $\mathcal{L}$. Let E be a set of event occurrences and $\lambda : E \to \mathcal{L}$ a labeling function. An event trace $\sigma \in L$ is defined in terms of an order $i \in [0, n-1]$ and a set of events $E_\sigma \subseteq E$ with $|E_\sigma| = n$ such that $\sigma = \langle \lambda(e_0), \lambda(e_1), \ldots, \lambda(e_{n-1}) \rangle$. Moreover, given two events $e_i, e_{i+1} \in E_\sigma$, we write $e_i \prec e_{i+1}$ or say that event e_{i+1} immediately follows e_i.*

3 Method

To accomplish behavioral classification of process executions, any observable aspects of the process may be taken into account. However, in the current version of this work, we focus on three dimensions: *control-flow* (i.e. order of activities), *event frequencies* (i.e. occurrence frequencies of events in one behavior vs. in other behaviors), and *branching probabilities* (i.e. frequencies for branching to certain mutual exclusive execution paths). As such, each execution can be assessed based on these criteria and subsequently be classified as: (i) one of the known behavioral categories, (ii) more than one category, or (iii) no known category. For the purpose of illustration, consider the following five categories of common behaviors that process owners would want to distinguish in many cases:

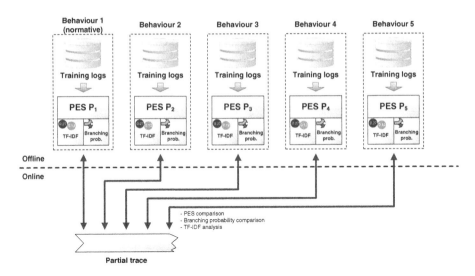

Fig. 1. Conceptual model of the proposed method

1. Normal process execution.
2. Normal process execution, but with some delays between or during activities or some minor differences in control-flow.
3. Exceptional process execution, for a number of identifiable exceptions.
4. Faulty process execution, leading to (system) errors/faults/failures.
5. Malicious process execution, either through an intruder or an internal player with malicious intent (fraud, extortion, etc.).

For each of the behavioral categories, a set of training event logs is required to learn the characteristics of each behavioral category. During run-time, current execu-

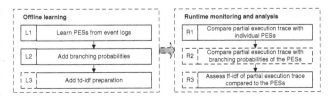

Fig. 2. Overview of offline and online steps.

tions will be analyzed and matched to the closest behavior. In Fig. 1, a graphical representation is provided of the different components of our method and their relations. In order to be able the runtime comparison and analysis, a number of steps is required to obtain the data sets representing the full behavioral characteristics of each event log. In Fig. 2, an overview is provided of the offline learning steps and the runtime analysis steps. Each of these steps will be discussed in detail below.

3.1   Offline Learning

Learning PESs from Event Logs. Our input consists of a set of labeled log traces for each behavioral category, i.e., one set of log traces for Behavior 1, one for Behavior 2, etc. From this input, we build separate models for the different behaviors in the form of *event structures*, as introduced in Sect. 2. Event structures allow us to capture all behavioral relations between events and summarize all possible behavior as represented in the underlying event log and provide a suitable mechanism for runtime comparison of the learned behaviors with a current execution of the process. In addition to the control-flow information, PESs are augmented with branching frequencies, in order to be able to more precisely distinguish the currently observed behavior.

Augmenting the PESs

Branching Probabilities. For each behavior category the branching probabilities are captured for each pair of events, using Definition 2. For instance, say event e_i is followed by event e_j in 70% of the cases, whereas it is followed by event e_k in 30% of the cases. When events in the current execution match the events as they occur in the learned event log (and are represented as such in the PES accordingly), the likelihood *within* that particular behavior of such a sequence of events occurring is highly relevant, as it allows to distinguish and categorize current behavior to learned behavior even when the control-flow occurs in multiple learned behaviors.

Term Frequency. Finally, we use *term frequency - inverse document frequency* (TF/IDF) [4] to obtain the relative "importance" of events, in order to be able to assign a weight to matches and mismatches between the current execution and the augmented event structure. If event e occurs in only one event structure, a match of e in the current execution with that event structure is much more distinguishing than a match of an event that occurs in every event structure.

As such, we compute the *term frequency (TF)* of every event in each event structure. The term frequency of e equals the number of traces that contain at least one occurrence of event e divided by the total number of traces in the log underlying the event structure. The *inverse document frequency (IDF)* of event e is computed as the logarithm of the inverse fraction of the event structures that contain e, or 0 if e does not occur in any event structure. Subsequently, *TF/IDF* equals the term frequency multiplied by the inverse document frequency.

3.2   Runtime Monitoring and Analysis

At runtime at each state, the partial trace of the currently executed business process is evaluated against each learned behavioral category for each dimension as enumerated in Sect. 3.1.

Comparing a Partial Execution Trace with Behaviors

Prime Event Structures. The current partial execution trace σ is transformed into a PES P_{cur}. As σ is a sequence of completed events $\sigma = \langle \lambda(e_0), \ldots, \lambda(e_{n-1}) \rangle$, P_{cur} will consist of a set of events $e_0, \ldots, e_{n-1}$, where $e_i < e_{i+1}$ with $0 \le i < (n-1)$.

As P_{cur} only represents a partial execution, the simulation should be performed up to the last event in P_{cur}. That is, subsequent events in the learned PESs should not be taken into account, as they did not yet occur in the current execution and can, therefore, never be matched. In addition, mismatches concerning events parallel to the last event in P_{cur} are not taken into account as well, as they are potentially not yet executed in P_{cur} due to interleaving. These events are to be removed from the learned PES, in order to prevent irrelevant events to be included in the comparison.

Let us illustrate this procedure with an example comprising two behaviors: normative behavior 1 and deviant behavior 2. Figure 3 presents the normative behavior as a BPMN model and the corresponding PES P_1. Each node in P_1 represents the event (e.g. f_0) and the corresponding activity (e.g. A). Figure 4 presents deviant behavior 2 as BPMN along with its corresponding PES P_2.

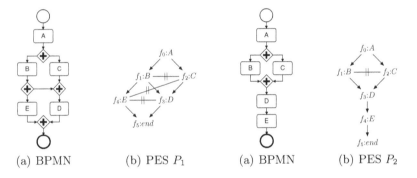

(a) BPMN   (b) PES P_1 (a) BPMN   (b) PES P_2

Fig. 3. Example of Behavior 1 (normative).

Fig. 4. Example of Behavior 2 (deviant).

Consider a current partial execution trace $\sigma_{cur} = \langle \text{A, C, E} \rangle$. This partial execution can straightforwardly be converted into a PES P_{cur}, as shown graphically in Fig. 5(a). The last event in P_{cur} that also occurs in P_1 (i.e. the last common event) is E. Consequently, event **end** can be removed from P_1. Event D does not occur in P_{cur}, but as it is concurrent with E in P_1, it can be removed from P_1 as well. Although B is also missing in P_{cur}, it should have occurred prior to E and can, therefore, not be removed from P_1.

This can be formulated formally as follows. The last *common* event $e_{L_{cur}} \in P_{cur}$ is the event for which there exists an event $e_{L_P} \in P_i$, where $\lambda(e_{L_{cur}}) = \lambda(e_{L_P})$ and $\nexists e_i \in P_{cur}$ such that $e_{L_{cur}} \le e_i$ and $\exists e_j \in P$ where $\lambda(e_i) = \lambda(e_j)$. Let E_{cur} denote the set of events in P_{cur} and E_P denote the set of events in the learned PES P_i. Any event $e \in E_P \setminus E_{cur}$ with $e_{L_P} \leq e$ or $e_{L_P} \parallel e$ is removed from P_i. For P_1 in our example, the resulting reduced PES P_1^{red} is depicted graphically in Fig. 5(b).

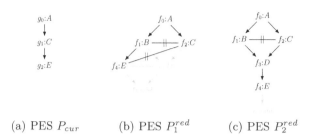

(a) PES P_{cur} (b) PES P_1^{red} (c) PES P_2^{red}

Fig. 5. Trimming PES P_1 and P_2 to length of the current execution.

Similarly, PES P_2 can be reduced by removing **end**. However, D cannot be removed here, as D occurs *before* E in P_2 instead of concurrently in P_1. As such, P_2^{red} can be depicted graphically as shown in Fig. 5(c).

After reducing the learned PESs to the relevant parts, we compare P_{cur} with P_1^{red}, then with P_2^{red}, etc. This pair-wise comparison of P_{cur} and P_i^{red} is based on a synchronized simulation of the two PESs, resulting in a Partial Synchronized Product (PSP) [3]. The synchronized simulation starts at the initial state of both PESs (i.e., no events have occurred so far). At each step, the events are matched that can occur given the current state in each of the two PESs. If the events match, they will be added to the current state of each PES and the simulation continues. If an event in the current state of one PES does not match with an enabled event in the current state in the other PES, a mismatch is declared. Consequently, the unmatched event is "hidden" in the respective PES and the simulation jumps to the next matching state. As such, the PSP comprises the following operations: (i) *match (m)*, (ii) *left hide (lh))* and (iii) *right hide (rh)*.

For each comparison, we are looking for the sequence of PSP operations that maximizes the number of event matches, which we will refer to as the optimal event matching M_O. Combinations of matches and mismatches indicate *what* is different between the learned behavior and the current execution. The following situations are distinguished in the PSP:

- *Match* e: event e occurs in both P_i and P_{cur}.
- *Right hide* e: event e does not occur in P_i. This indicates an illegal execution of e in P_{cur}.
- *Left hide* e: event e does not occur in P_{cur}. This indicates a missing execution of e in P_{cur}.
- *Match* e and *right hide* e: event e occurs in both P_i and P_{cur}, but occurs again in P_{cur} and not in P_i. This indicates an illegal repetition of e in P_{cur}.
- *Left hide* e and *right hide* e: event e occurs in both P_i and P_{cur}, but requires to be hidden in both PESs to maintain a matching state. This indicates that e occurs at the wrong time in P_{cur}.

Continuing the above example, Fig. 6(a) depicts the PSP of P_1^{red} and P_{cur}, whereas Fig. 6(b) depicts the PSP of P_2^{red} and P_{cur}. The set m records the events that have been matched in both PESs, while the sets lh and rh record the events that are hidden in P_1^{red} and P_{cur} respectively. PSP$_1$ starts with an empty state, i.e. m, lh and rh are empty. In both PESs, A can be executed and is, therefore, matched and added to m accordingly. Subsequently, C is matched. Event B occurs in P_1^{red}, but not in P_{cur} and results, therefore, in a *left hide* of B. Finally, E can be matched. Note that PSP$_2$ is very similar, but requires an additional *left hide* of D. As such, P_1^{red} is closer to P_{cur} than P_2^{red}.

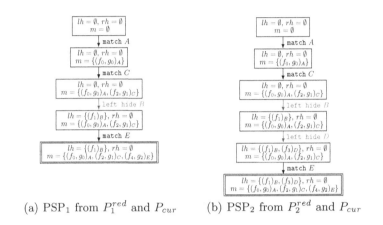

(a) PSP$_1$ from P_1^{red} and P_{cur} (b) PSP$_2$ from P_2^{red} and P_{cur}

Fig. 6. Excerpt of the resulting PSPs from comparing P_1^{red} and P_2^{red} with P_{cur}.

Branching Probabilities. When comparing event structures, there are a number of cases where the PSPs are inconclusive. That is, based on the identified difference, one cannot clearly categorize the current execution as one specific behavior. More specifically, P_{cur} may have a similar or even identical match with two or more PESs. Branching probabilities identify the probability of occurrence of e_j given that event e_i has occurred, and will be used to distinguish those similar matches.

As such, we obtain total probability of the closest matching trace in each PES: $TotalProb_{M_O} = \prod \mathcal{P}([e_i], [e_j]) \; \forall e_i, e_j \in m \cup lh$ where $e_i < e_j$ with $M_O = \langle m, lh, rh \rangle$. The PES with highest total probability of the trace would be the one that is most likely describing the behavioral category of P_{cur}. Subsequently, this score is normalized against all total probabilities of traces of length M_O in the respective PES:

$$NormProb_{M_O} = \frac{TotalProb_{M_O} - TotalProb_{max}}{TotalProb_{max} - TotalProb_{min}}$$

$TotalProb_{max}$ is the highest probability of a trace and $TotalProb_{min}$ is the lowest probability of a trace. If there exists only one trace in the PES with the length of M_O, then $NormProb_{M_O} = 1$. If all traces in the PES with the length of M_O have the same probability (i.e. $TotalProb_{max} = TotalProb_{min}$), then $NormProb_{M_O} = 0.5$.

Term Frequency. For each event e_i in the closest matching trace of the PSP, the term frequency (TF) is computed, by taking the amount of traces in the log that contain the event e_i, divided by the total amount of traces. The term frequency is subsequently multiplied by IDF-logarithm of that event, which is computed as the logarithm of the total number of event structures (i.e. number of different behaviors) divided by the total number of event structures that contain event e_i. Next, the sum of the *TF/IDF* of all mismatched events is subtracted from the sum of the *TF/IDF* of all matched events, which results in a weighted score of the respective matching trace of that behavior.

Combining the Results and Deciding on the Outcome. Above, we described how we calculate the degree of match for the various dimensions of matching. These individual values need to be combined into a single number per known behavior, such that we can make an assessment as to which behavior we are currently observing. The simplest method to achieve this is to calculate a weighted linear combination from the different factors (for behavior b, with weights ω_i and the different degrees of match as d_i):

$$d(b) = \frac{\omega_1 \times d_1 + \omega_2 \times d_2 + \cdots + \omega_n \times d_n}{\sum_{i=1\ldots n} \omega_i}$$

There are numerous more complex techniques to achieve such a combination, and our method does not preclude us from using any one of those. The above combination gives us an overall matching score $d(b)$ per behavior b. When comparing these matching scores for the different behaviors, there can be three possible outcomes:

- No matching score is high enough, i.e., $\forall_{b \in Beh} : d(b) < tr_d$ for some predefined threshold tr_d with $0 < tr_d < 1$, say 0.25, and Beh as the set of behaviors. In this case, we conclude that we're observing a previously unknown behavior.
- One matching score is a lot higher than the others, i.e., $\forall_{b \in Beh \setminus \{b'\}} : d(b) < f_m \times d(b')$ where f_m is a matching factor with $f_m > 1$, say 2. In this case, we conclude that we are reasonably certain we are observing a trace of behavior b'.
- If neither of the above is the case, then there is a set of well matching behaviors, but which one of these we are currently observing is not (yet) clear. The set of behaviors we consider "well matching" is determined as $(b \in Beh \mid d(b) \geq f_d \times max_{b' \in Beh}(d(b')) \wedge d(b) \geq tr_d)$, where f_d is a factor with $0 < f_d < 1$, say 0.8.

Each of these thresholds can be obtained by experimentation or (machine) learning. The resulting outcome depends on the current incomplete trace , and can thus change with each new event that is captured and processed.

4 Evaluation

We implemented the proposed method in a prototype, in order to demonstrate the ability of the method to correctly classify behavior. Using this implementation, we trained the method with five synthetic datasets, each representing a behavioral category. For this purpose, we started from a model of normative behavior of a customer support process in the telco sector in Australia (Fig. 7), and created four variants from it.

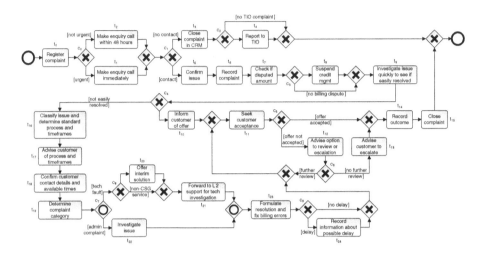

Fig. 7. Normative behavior (Behavior 1) of the customer complaint process.

4.1   Creating Log Files for Each Behavioral Category

First, we created four deviating variants from the normative model, resulting in five models in total. The variants have been modeled to represent the example behavioral categories listed in the beginning of Sect. 3 as follows:

- Behavior 1 (normal): normative behavior.
- Behavior 2 (slow): similar to Behavior 1, but with the branching probabilities on the "slowest" path set very high (as opposed to Behavior 1), to simulate slow cases and omit "shortcuts" in the process as much as possible.
- Behavior 3 (exceptional): based on Behavior 1, with similar branching frequencies but control flow differences. In this model, t_1, t_2, t_5 and t_{10} have been removed, t_4 leads directly to t_6, t_{21} is no longer optional and t_9 as well as t_{24} directly lead to t_{15}.
- Behavior 4 (faulty): based on Behavior 3, with different control flow and branching frequencies. In this model, t_3, t_6, t_{21} and t_{22} are removed, t_4 is replaced by t_5 and t_9 leads to only t_{16} and no longer to t_{15}.
- Behavior 5 (malicious): vaguely based on Behavior 1, but with many differences in control flow and branching frequencies. In this model, t_0 directly leads to t_5, followed by t_6. Subsequently, t6 is followed by either t_8 or t_{23}. Next, t_{14} can be executed or a new activity *"Remove complaint from CRM"*, which we will refer to as t_{25}. Finally, t_{15} is executed followed by *end*.

From these five models, we generated five distinct logs as input for our tool. We used the ProM plugin "Generate Event Log from Petri Net" [5], which generates a distinct log trace for each path in the model. Subsequently, the branching probabilities were applied to the generated logs to duplicate the traces according to their branching frequencies. This resulted in five event logs, containing 1664 traces (Behavior 1), 1358 traces (Behavior 2), 1480 traces (Behavior 3), 1000 traces (Behavior 4) and 1002 traces (Behavior 5) respectively, which we used for learning five different PESs.

For testing behavioral classification, we used up to 10 distinct traces for each behavior, i.e., 10 traces for Behavior 1 and 2 (drawn randomly), and the maxima of 4 traces for Behavior 4 and 6 traces for Behavior 3 and 5. Since the classification algorithm is deterministic and considers only a single trace, there is no point in testing the a single trace more than once. From these original traces, we created randomly modified traces with the following manipulation operators: (i) add an event, (ii) remove an event, or (iii) switch the order of two events. We performed this procedure with 1, 2, 3, and 4 random changes per original trace, resulting in a set of increasingly non-conforming traces for each behavior (i.e., for each original trace we test 5 variants, with 0 to 4 changes).

4.2   Results

For each of the randomized traces described above, we generated a set of partial executions having a length varying from 2 activities to the entire trace. For each of these partial executions, we performed a classification using our tool, with parameters set as follows: score threshold $tr_d = 0.25$; well-matching factor f_d = 0.8; weight 0.25 for trace frequency (normalized); and weight 0.75 for total TF/IDF respectively. For each trace, we evaluated the classification at $25\%, 50\%,$ 75% and 100% completion of the trace. To this end, we analyzed whether the partial execution was *conclusively classified* as its original behavioral category, or whether the original behavioral category was *included* in the classification (to evaluate cases where multiple behaviors were identified).

In Table 1, an overview of the results is provided. Across the board, random changes make it harder to identify the behavior clearly. The normative behavior (PES P_1) and slow normative behavior (PES P_2) were well identified, even at early stages of the process. P_1 and P_2 traces are clearly separable due to the branching frequency differences in our dataset. As the behavior becomes more deviant (P_3) or even faulty or malicious (P_4 and P_5), it is still fast to identify, but when the trace is increasingly different from the behavior in the log, it takes longer (up to 50%) to conclusively classify the behavior.

The results show that even for noisy traces and overlapping behavior, our method is in many cases able to accurately classify behavior very early in the process and in most cases well before the end of the process. We performed the tests on a laptop with an Intel Core i7-4710HQ 2.5GHz (4 cores), 16GB RAM, running JVM 8. Each full classification between a partial trace and the learned behaviors took between 0.46 and 1.81 s, which shows that the method is viable for runtime application.

Table 1. Experimental results, showing the percentage of traces where the original PES was conclusively identified (c) and/or included (i). Note that (c) is a subset of (i).

Trace		Trace completion							
Trace origin	Changes	25%		50%		75%		100%	
PES P_1	0	c: 100%	i: 100%	c: 100%	i: 100%	c: 100%	i: 100%	c: 100%	i: 100%
	1	c: 100%	i: 100%	c: 100%	i: 100%	c: 100%	i: 100%	c: 100%	i: 100%
	2	c: 100%	i: 100%	c: 100%	i: 100%	c: 100%	i: 100%	c: 100%	i: 100%
	3	c: 90%	i: 90%	c: 90%	i: 90%	c: 90%	i: 90%	c: 90%	i: 90%
	4	c: 90%	i: 90%	c: 90%	i: 90%	c: 90%	i: 90%	c: 90%	i: 90%
PES P_2	0	c: 100%	i: 100%	c: 100%	i: 100%	c: 100%	i: 100%	c: 100%	i: 100%
	1	c: 80%	i: 80%	c: 100%	i: 100%	c: 100%	i: 100%	c: 100%	i: 100%
	2	c: 80%	i: 80%	c: 100%	i: 100%	c: 100%	i: 100%	c: 100%	i: 100%
	3	c: 100%	i: 100%	c: 100%	i: 100%	c: 100%	i: 100%	c: 100%	i: 100%
	4	c: 100%	i: 100%	c: 100%	i: 100%	c: 100%	i: 100%	c: 100%	i: 100%
PES P_3	0	c: 100%	i: 100%	c: 100%	i: 100%	c: 100%	i: 100%	c: 100%	i: 100%
	1	c: 100%	i: 100%	c: 100%	i: 100%	c: 100%	i: 100%	c: 100%	i: 100%
	2	c: 83%	i: 83%	c: 100%	i: 100%	c: 83%	i: 83%	c: 100%	i: 100%
	3	c: 83%	i: 83%	c: 100%	i: 100%	c: 67%	i: 67%	c: 100%	i: 100%
	4	c: 67%	i: 67%	c: 83%	i: 83%	c: 50%	i: 50%	c: 80%	i: 80%
PES P_4	0	c: 100%	i: 100%	c: 100%	i: 100%	c: 100%	i: 100%	c: 100%	i: 100%
	1	c: 75%	i: 75%	c: 100%	i: 100%	c: 100%	i: 100%	c: 100%	i: 100%
	2	c: 50%	i: 75%	c: 100%	i: 100%	c: 100%	i: 100%	c: 100%	i: 100%
	3	c: 50%	i: 75%	c: 100%	i: 100%	c: 100%	i: 100%	c: 100%	i: 100%
	4	c: 50%	i: 75%	c: 100%	i: 100%	c: 100%	i: 100%	c: 100%	i: 100%
PES P_5	0	c: 67%	i: 100%	c: 100%	i: 100%	c: 100%	i: 100%	c: 100%	i: 100%
	1	c: 67%	i: 100%	c: 100%	i: 100%	c: 100%	i: 100%	c: 100%	i: 100%
	2	c: 0%	i: 100%	c: 100%	i: 100%	c: 100%	i: 100%	c: 100%	i: 100%
	3	c: 0%	i: 67%	c: 67%	i: 67%	c: 67%	i: 67%	c: 67%	i: 67%
	4	c: 67%	i: 100%	c: 100%	i: 100%	c: 100%	i: 100%	c: 100%	i: 100%

5   Related Work

To the best of our knowledge, a runtime solution for behavioral classification has not been adequately described in the literature. The closest related work includes our prior work on PESs [3] where we compare different behaviors, but it can only distinguish differences between logs and executions that are known to be different. In addition, [3] does not include and does not include branching probability in a similar fashion. The point there is to identify if a complete log is deviant from a single type of behavior expressed as an event structure. In this paper, we distinguish several types of behavior and detect at runtime (i.e. before the entire log is available) to which the current execution likely belongs. Neither multiple behaviors nor runtime analysis are covered in the earlier work.

In [6,7], the problem of business process monitoring is addressed, where classifiers are trained on historic data and used to predict if a business constraint will be satisfied during the execution of new instances. To this end, traces are encoded as vectors whose elements represent certain positions in the traces and data mutations during process execution. In contrast to our approach, these methods do not rely on models that capture the relations between the events and must train one classifier for each possible trace position. Moreover, while generally applicable to multiple classes of behavior, the approaches are only studied with respect to a binary classification, i.e., if the constraint holds or not.

Runtime conformance checking is also closely related. The main techniques for checking conformance in general are token replay [8], constraint checking [9,10], and alignment [11]. All of these techniques have been designed for offline conformance analysis, not runtime checking. In [1,12], we adapted token replay for runtime checking, but we only checked the current incomplete log against a single behavioral model. In contrast, here we check a runtime log against multiple behavioral models, to identify which of the known behaviors is the one "closest" to the current execution.

Model synthesis has been a topic in software engineering since the mid 1990s [13], and has been used for reverse engineering software code [14], or synthesis of behavior models from other software development artifacts, e.g. [15]. We are not aware of any model synthesis work for runtime behavioral classification.

Trace clustering is a technique in process mining where a highly diverse event log is split into homogeneous subsets of similar traces [16,17]. Although a powerful technique to prevent "spaghetti models" in process discovery, trace clustering is not suitable for partial log traces, particularly when behavioral categories are very similar, as clusters are created and evaluated by maximizing intra-cluster similarity and minimizing intercluster similarity [17].

There has been a wide range of research on *abnormal human activity detection*, to detect physical intruders or emergency situations. This research is often based on machine learning, see e.g. [18], over data collected by various types of sensors, such as wearable sensors [19] or surveillance cameras [20]. Our work operates on a different level of abstraction: we utilize event logs, not sensor data. It would be interesting to study in future work if events can be abstracted from sensor data, and if our approach can be applied in the area of activity detection.

6   Conclusion

In this paper, we presented a method that can learn models for different behaviors from logs offline and use these models to classify observed behavior at runtime: either it is clearly a known behavior, or there are possible classes that match the behavior so far, or the behavior is dissimilar to any known class. For this purpose we use event structures, which we augment with branching and event frequency information. In the evaluation, we tested when a clear determination can be made and when the observed behavior is too similar to multiple models to make a clear call. This is a first evaluation on a single dataset, where we created variants from a real-world process model. We show that, in many cases present in our dataset, our approach is able to accurately classify the observed behavior very early in the process, even when various degrees of noise are introduced. In addition, the performance of the prototype is sufficiently fast for many runtime applications.

For future work, we plan to extend the models with additional dimensions, such as timing, data, or resources. We also plan a large-scale evaluation on datasets from external real-world use cases.

References

1. Xu, X., Zhu, L., Weber, I., Bass, L., Sun, W.: POD-Diagnosis: error diagnosis of sporadic operations on cloud applications. In: IEEE/IFIP DSN (2014)
2. Nielsen, M., Plotkin, G.D., Winskel, G.: Petri nets, event structures and domains, part I. Theor. Comput. Sci. **13**, 85–108 (1981)
3. van Beest, N.R.T.P., Dumas, M., García-Bañuelos, L., La Rosa, M.: Log delta analysis: interpretable differencing of business process event logs. In: Motahari-Nezhad, H.R., Recker, J., Weidlich, M. (eds.) BPM 2015. LNCS, vol. 9253, pp. 386–405. Springer, Cham (2015). doi:10.1007/978-3-319-23063-4_26
4. Manning, C.D., Raghavan, P., Schütze, H.: Introduction to Information Retrieval. Cambridge Univ. Press, Cambridge (2008)
5. Van den Broucke, S., De Weerdt, J., Vanthienen, J., Baesens, B.: An improved process event log artificial negative event generator. Faculty of Economics and Business, KU Leuven (Belgium), Technical report KBI_1216 (2012)
6. Leontjeva, A., Conforti, R., Francescomarino, C., Dumas, M., Maggi, F.M.: Complex symbolic sequence encodings for predictive monitoring of business processes. In: Motahari-Nezhad, H.R., Recker, J., Weidlich, M. (eds.) BPM 2015. LNCS, vol. 9253, pp. 297–313. Springer, Cham (2015). doi:10.1007/978-3-319-23063-4_21
7. Teinemaa, I., Dumas, M., Maggi, F.M., Francescomarino, C.: Predictive business process monitoring with structured and unstructured data. In: La Rosa, M., Loos, P., Pastor, O. (eds.) BPM 2016. LNCS, vol. 9850, pp. 401–417. Springer, Cham (2016). doi:10.1007/978-3-319-45348-4_23
8. van der Aalst, W.: Process Mining: Discovery, Conformance and Enhancement of Business Processes. Springer, New York (2011)
9. Weidlich, M., Polyvyanyy, A., Desai, N., Mendling, J., Weske, M.: Process compliance analysis based on behavioural profiles. Inf. Syst. **36**(7), 1009–1025 (2011)
10. Maggi, F.M., Montali, M., Westergaard, M., Aalst, W.M.P.: Monitoring business constraints with linear temporal logic: an approach based on colored automata. In: Rinderle-Ma, S., Toumani, F., Wolf, K. (eds.) BPM 2011. LNCS, vol. 6896, pp. 132–147. Springer, Heidelberg (2011). doi:10.1007/978-3-642-23059-2_13
11. van der Aalst, W., Adriansyah, A., van Dongen, B.: Replaying history on process models for conformance checking and performance analysis. WIREs Data Min. Knowl. Discov. **2**(2), 182–192 (2012)
12. Weber, I., Rogge-Solti, A., Li, C., Mendling, J.: CCaaS: online conformance checking as a service. In: Proceedings of BPM Demo Track, August 2015
13. Koskimies, K., Mäkinen, E.: Automatic synthesis of state machines from trace diagrams. Softw. Pract. Exper. **24**(7), 643–658 (1994)
14. Chen, X.J., Ural, H.: Automated recovery of protocol designs from execution histories. In: Proceedings of SCI 2001, pp. 103–108, July 2001

15. Uchitel, S., Brunet, G., Chechik, M.: Synthesis of partial behavior models from properties and scenarios. IEEE TSE **35**(3), 384–406 (2009)
16. Song, M., Günther, C.W., Aalst, W.M.P.: Trace clustering in process mining. In: Ardagna, D., Mecella, M., Yang, J. (eds.) BPM 2008. LNBIP, vol. 17, pp. 109–120. Springer, Heidelberg (2009). doi:10.1007/978-3-642-00328-8_11
17. De Weerdt, J., van den Broucke, S., Vanthienen, J., Baesens, B.: Active trace clustering for improved process discovery. IEEE TKDE **25**(12), 2708–2720 (2013)
18. Yin, J., Yang, Q., Pan, J.J.: Sensor-based abnormal human-activity detection. IEEE TKDE **20**(8), 1082–1090 (2008)
19. Jin, M., Zou, H., Weekly, K., Jia, R., Bayen, A.M., Spanos, C.J.: Environmental sensing by wearable device for indoor activity and location estimation. In: IEEE IECON (2014)
20. Vishwakarma, S., Agrawal, A.: A survey on activity recognition and behavior understanding in video surveillance. Vis. Comput. **29**(10), 983–1009 (2013)

First Workshop on Resource Management in Business Processes (REMA 2016)

Introduction to the First Workshop on Resource Management in Business Processes (REMA 2016)

Cristina Cabanillas[1(✉)], Manuel Resinas[2], Alex Norta[3],
and Nanjangud C. Narendra[4]

[1] Vienna University of Economics and Business, Vienna, Austria
cristina.cabanillas@wu.ac.at
[2] University of Seville, Seville, Spain
resinas@us.es
[3] Tallinn University of Technology, Tallinn, Estonia
alexander.norta@ttu.ee
[4] Ericsson Research, Bangalore, India
ncnaren@gmail.com

Abstract. The main goal of the REMA workshop is to explore resource management in business processes from different perspectives and scenarios. In particular, contributions related to resource management in the design, modeling and analysis of processes that are executed within a single organization or distributed among several organizations, were relevant for the workshop. In this first edition, three high-quality submissions were accepted. These submissions cover both design-time and run-time aspects of resource management and they considered either single resources or whole teams.

Keywords: BPM · Resource management · Team management · Human resources

1 Aims and Scope

In business processes, the term resource jointly implies both human and non-human resources. The former are people that take part in the execution of process activities at different levels (e.g. as activity performers or people accountable for work). The latter involve all other things that are necessary to complete process activities, such as software or IT devices. Consequently, the management of both human and non-human resources is a key part of the business process lifecycle and must be supported in all of its phases (design, modeling, execution, monitoring and analysis).

The First Workshop on Resource Management in Business Processes (REMA)[1], which was held in conjunction with the BPM'16 conference in Rio de Janeiro, Brazil,

---

[1] https://ai.wu.ac.at/rema2016/.

focus on exploring how human resources are involved and can be managed in processes with intensive resource needs. The three papers that were accepted for presentation at the workshop are representative of the different challenges that are currently being addressed in the context of resource management and cover several complementary perspectives including (1) different phases of the business process lifecycle such as modeling, execution and analysis, and (2) activities performed by an individual or by a whole team. More specifically, these papers present the following contributions.

The paper "Towards Simulation- and Mining-based Translation of Resource-aware Process Models" by Lars Ackermann, Stefan Schönig and Stefan Jablonski presents a novel approach to deal with the translation between declarative and imperative resource-aware process models. Instead of relying on the definition of a set of mapping rules, the authors suggest the use of simulation and mining techniques to enable this translation and, thus, avoid the need to specify cumbersome transformation rules.

The paper "Transforming Multi-role Activities in Software Processes into Business Processes" by Juan Pulgar and María Cecilia Bastarrica also faces the problem of transforming between different models. However, in this case, the goal is to transform software processes defined using SPEM into BPMN so that they can be executed in a Business Process Management System (BPMS). In particular, the authors present two alternative approaches to deal with multi-role activities and an XSLT transformation for automatically generating each of these solutions from a software process specification.

Finally, the paper "A Multi-criteria Approach for Team Recommendation" by Michael Arias, Jorge Munoz-Gama, and Marcos Sepúlveda deals with the problem of team formation in the context of BPM. In particular, the authors present a multi-criteria framework that considers a resource request characterization, historical information, and individual and collective performance. The framework relies on the Best Position Algorithm (BPA2) to provide a recommendation.

We sincerely thank the Program Committee Members of the REMA 2016 workshop for their time and support throughout the reviewing process.

Cristina Cabanillas
Manuel Resinas
Alex Norta
Nanjangud C. Narendra
REMA 2016 Workshop Chairs

2   Workshop Chairs

Cristina Cabanillas   Vienna University of Economics and Business, Austria
                           E-mail: cristina.cabanillas@wu.ac.at
Manuel Resinas   University of Seville, Spain
                           E-mail: resinas@us.es
Alex Norta   Tallinn University of Technology, Estonia
                           E-mail: alex.norta.phd@ieee.org
Nanjangud C. Narendra   Ericsson Research, Bangalore, India
                           E-mail: ncnaren@gmail.com

3   Program Committee Members

Adela Del Río Ortega   University of Seville, Spain
Ahmed Awad   Cairo University, Egypt
Anderson Santana   SAP Labs, France
   De Oliveira
Antonio Ruiz-Cortés   University of Seville, Spain
Claudio Di Ciccio   Vienna University of Economics and Business, Austria
Daniel Schall   Siemens AG, Germany
Fabio Casati   University of Trento, Italy
Felix Garcia   University of Castilla-La Mancha, Spain
Florian Daniel   Politecnico di Milano, Italy
Joseph Davis   University of Sydney, Australia
Luis Jesús Ramón Stroppi   National Technological University of Santa Fe,
                              Argentina
Marcos Sepúlveda   Pontifical Catholic University of Chile, Chile
Mark Strembeck   Vienna University of Economics and Business, Austria
Minseok Song   POSTECH (Pohang University of Science
                              and Technology), South Korea
Schahram Dustdar   . TU Wien, Austria
Stefan Schönig   Vienna University of Economics and Business, Austria
Stefan Schulte   Vienna University of Technology, Austria
Stefanie Rinderle-Ma   University of Vienna, Austria

Towards Simulation- and Mining-based Translation of Resource-aware Process Models

Lars Ackermann$^{(\boxtimes)}$, Stefan Schönig, and Stefan Jablonski

University of Bayreuth, Bayreuth, Germany
{lars.ackermann,stefan.schoenig,stefan.jablonski}@uni-bayreuth.de

Abstract. Imperative languages like BPMN are eminently suitable for representing routine processes and are likewise cumbersome in case of flexible processes. The latter are easier to describe using declarative process modeling languages (DPMLs). However, understandability and tool support of DPMLs are comparatively poor. Additionally, there may be an affinity to a particular language caused by existing company infrastructure or individual preferences. Hence, a technique for automatically translating process models between different languages is required. Process models usually describe several aspects of a process, such as activity orderings and role assignments. Therefore, our approach focuses on translating resource-aware process models. We utilize well-established techniques for process simulation and mining to avoid the definition of cumbersome model transformation rules. Our implementation is based on a discussion of general configuration principles and a concrete configuration suggestion. The whole translation approach is discussed and evaluated at the example of BPMN and DPIL.

Keywords: Process model translation · Simulation · Process mining

1 Introduction

Business process management is a well-known discipline for improving the customer and business objective alignment using intra-company information. Processes are diverse but usually can be classified at least as either flexible or routine processes [1]. Flexible processes strongly vary, for instance, in terms of the order and participation of performed activities and involved resources. Sticking to the example of activities and resources, routine processes are stable in that terms that multiple executions show only slight variations of the order of activities and the assigned resources. Business analysts apply business process modeling techniques in order to derive a formalized representation of the actual process. Therefore, the used business process modeling languages (PMLs) need to be able to describe the different process aspects, such as the order of activities and the performing resources. Imperative modeling languages like *Business Process Model and Notation (BPMN)*[1] and *Event-driven process chain (EPC)*

---

[1] The BPMN 2.0 standard is available at http://www.omg.org/spec/BPMN/2.0/.

© Springer International Publishing AG 2017
M. Dumas and M. Fantinato (Eds.): BPM 2016 Workshops, LNBIP 281, pp. 359–371, 2017.
DOI: 10.1007/978-3-319-58457-7_26

are usually used to represent routine processes [2]. The imperative modeling style is facilitated through the stable nature of the underlying processes. In contrast, flexible processes are usually modeled using declarative languages, such as *Declare* [3], *Case Management Model and Notation (CMMN)* or *Declarative Intermediate Language (DPIL)* [4]. Though research and practical applications proved the applicability of the associated modeling paradigms there are many potential reasons where a business analyst might switch to the opposite. These are, for instance: *(i)* The understandability of a process model depends on its complexity which is heavily influenced by the chosen modeling paradigm [1], *(ii)* adopted process execution and analysis tools are usually tailored to *one* PML and most of them exclusively provide support for imperative languages – and often only for BPMN – and *(iii)* PMLs provide different support for the different process aspects. Therefore, it is necessary to be able to translate a given process model from one language into a representative of the opposite class of PMLs. However, while research extensively discussed model translations from one into another imperative language, there is no technique for the translation of resource-aware declarative process models into an imperative counterpart or vice versa, yet. Hence, we propose a corresponding translation approach, based on well-established techniques of the process simulation and mining disciplines in order to avoid the specification of cumbersome transformation rules. Though a process can be modeled according to several different perspectives, this paper focuses only on two of them: *(i)* the behavioral and *(ii)* the organizational perspective. In order to evaluate the suggested approach properly we apply the principle to two exemplary languages namely BPMN and DPIL.

The remainder of this paper is structured as follows: Sect. 2 introduces the core techniques that are involved in our approach as well as one of the involved declarative PML. The contribution (cf. Sect. 3) discusses the translation that is based on two simulation and two mining techniques. Additionally we broach preconditions regarding the involved languages and the log characteristics. Sect. 4 evaluates our approach based on five example process models whereby one describes a real-life process. Finally, we class our approach with related techniques and technologies and give an idea for future improvements.

2   Background

In this section we introduce the core building blocks of our approach, i.e., the declarative process modelling paradigm and DPIL as a language, process simulation and process mining.

2.1   Resource-Aware Declarative Process Modeling with DPIL

Research has shown that DPMLs are able to cope with a high degree of flexibility [1]. The basic idea is that, without modeling anything, everything is allowed. To restrict this maximum flexibility, DPMLs allow for formulating constraints which form a forbidden region. Independent from a specific modelling paradigm different perspectives on a process exist. The organizational perspective

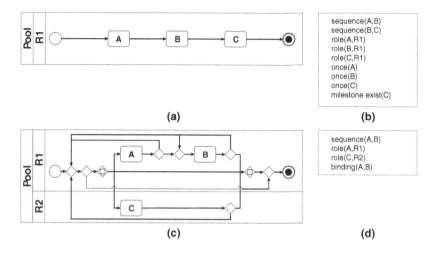

Fig. 1. Continuous example of resource-aware process models

deals with the definition and the allocation of human and non-human resources to activities. The *Declarative Process Intermediate Language (DPIL)* [4] is a declarative process modelling language that is, unlike other declarative languages multi-perspective, i.e., it allows for representing several business process perspectives, namely, control flow, data and especially resources. The expressiveness of DPIL and its suitability for business process modelling have been evaluated [4] with respect to the well-known Workflow Patterns. DPIL provides a textual notation based on the use of *macros* to define reusable rules, as it is shown in Fig. 1(b) and (d). For instance, the *sequence* macro (e.g. *sequence(A, B)*) states that the existence of an *execution event* of task b implies the previous occurrence of an execution event of task A; and the *binding* macro (*binding(A, B)*) states that an activity B is assigned to the same actor that already performed activity A. Furthermore it is possible to assign an activity directly to a certain role using the *role* macro (e.g. *role(A, R1)*). In example Fig. 1(a) each activity is executable at most once. In DPIL this can be modeled via the *once* macro (e.g. *once(A)*). In order to specify process goals the language provides the *milestone* keyword. A process instance cannot be completed until all milestone conditions are fulfilled.

2.2   Process Simulation and Resource-Aware Process Mining

Business process simulation is usually used for the investigation of correlations in and properties of business processes without resorting to an expensive and in many cases time-consuming observation of real-life process executions [5]. Instead of an exhaustive cost analysis in the paper at hand we refer to the generation of event logs only. Event logs can be generated based on the principle of *Discrete-event Simulation (DES)* [6] which assumes that all state changes of an observed system can be expressed as a discrete sequence of events. Furthermore, process mining techniques are built upon the same premise. *Process Mining* aims at discovering processes by extracting knowledge from event logs, e.g., by generating a process model reflecting the behaviour recorded in the logs [7]. There are plenty of process mining algorithms available that focus on discovering imperative process models, e.g., the simplistic *Alpha miner* [7]. Recently, tools to discover declarative process models like *MINERful* [8] have been developed. We use process mining techniques to automatically "model" the simulated behaviour in the chosen target language. In particular, we use the *DPILMiner* [9], i.e., a process mining approach to discover resource-aware, declarative process models. The tool is able to extract complex allocation rules, such as binding of duties between activities.

3   Translation of Resource-Aware Process Models

In contrast to conventional model translation approaches we propose a fully automatable two-phase technique *without* requiring the formulation of transformation rules. This decision was, on the one hand, driven by the far lower manual effort in order to develop a translation system. Assuming that we have a set of n PMLs, a translation system based on mapping-rules and with the ability to translate a model from and into any PML, has a complexity of $O\left(n^2\right)$. The specification of such rule sets can be a time-consuming task [10,11]. Secondly, there is a large gap between the contents of imperative and declarative models. In imperative models, the allowed paths through a process are specified explicitly, while in declarative models all paths which do not violate any rule are implicitly allowed. Hence, one path in an imperative model potentially conforms to a sensitive set of rules in a declarative model. It is hard to determine, which subset of information in the declarative model corresponds to which subset of information in the imperative model and we suggest to use this technique primarily for an imperative-declarative translation system. Another major advantage is that we reuse *existing* techniques for business process simulation and mining which are needed for reasons apart from our translation application. However, since there is a large variety of such techniques we provide an appropriate choice and configuration that matches the intended purpose. In summary, our core contribution is a prototypical investigation of the functional interaction of process simulation and mining techniques for the purpose of process model translation.

Our technique generates an event log using *DES* based on a source model specified in a particular PML. An event log consists of traces of valid exemplary process executions. The contained events can be ordered chronologically and provide information about the performed task as well as the associated performer. The subsequent process mining technique uses this log in order to reconstruct a process model in the target language. This indirection avoids the specification of mapping rules on meta-model level.

In [12] we already introduced some issues a translation approach for control-flow based process models has to deal with. A translation should not change the meaning of the source model but represent it in a different language. However, this is only possible if the two involved languages have the same level of expressiveness *(CP1)*. Secondly, using an event log as a transfer medium requires sufficient expressiveness, too *(CP2)*. Since both issues discuss only the general *ability* to encode the particular semantics, we additionally have to provide an appropriate component configuration that *causes* an appropriate translation *(CP3)*.

3.1   Language and Log Expressiveness

The comparison of expressiveness depends on the particular pair of modeling languages *(CP1)*. In our case we have to compare BPMN and DPIL. Though both languages in general consider both process perspectives we currently focus on, they differ widely in terms of the particular extent. Due to its imperative nature, BPMN provides explicit modeling elements for organizing the control flow. As Fig. 1 shows, there are at least three classes of predecessors for an activity, namely events, gateways or other activities. Starting from an arbitrary process activity in BPMN it is possible to backtrack the *causes* that lead to the currently selected activity. In contrast, in DPIL this is not possible since decision points and other control-flow elements cannot be modeled explicitly. In Fig. 1(d) the decision for the first activity in the process is completely free, except that A has to be executed at least once before B can be performed. This autonomy of decision is implicitly given through the *absence* of corresponding sequence rules. Conversely, in BPMN it is not possible to model milestones explicitly. However, this can be imitated using the convention that a BPMN end event must always be preceded by *something* that fulfills the milestone modeled in DPIL. Regarding Fig. 1(a) and (b) this means that no end event may occur before C has not been executed at least once. However, this strategy is valid and desired because of the completely disparate modeling paradigms BPMN and DPIL are based on.

In BPMN it is possible to assign roles to activities in order to model responsibilities. This is also possible in DPIL, as shown in Fig. 1. However, considering the contained DPIL example (d) there is a rule called $binding(A, B)$. Without the $role(A, R1)$ rule it would not be possible to find an equivalent BPMN model, because it is not possible to model dynamic organizational dependencies in BPMN. Furthermore, even the static role assignments in BPMN in the form of swimlanes are just visual model improvements but do not comprise any execution semantics. Hence, it is not possible to represent DPIL's organizational dependencies adequately. However, since a process model is also used for communicating processes, our approach was enabled to mine swimlanes, too. We do not provide a complete comparison of the expressiveness of BPMN and DPIL since this would go beyond the scope of the paper at hand. Consequently, we have chosen example process models for our evaluation which definitely have at least one equivalent in the corresponding counterpart language.

Our second precondition *(CP2)* means that a log is able to encode the behavioral and organizational information of the source model. For instance, if a model allows for repeating a particular activity arbitrarily often, this cannot be encoded explicitly in the log because of its finite length. A second issue is related to the organizational perspective. In DPIL it is possible to dynamically assign an

Fig. 2. Concept of multi-perspective process model simulation

activity to a resource that is in a certain relation to a different resource that has been assigned to a previous activity. An event log is not able to encode this relation because the comprised events just contain information about the executed task and the performer but does not encode the *reason* why the particular resource has been the performer. It is therefore impossible to discover the particular relationship between roles. There are more aspects to consider but it would go beyond the scope of the paper at hand to provide an exhaustive analysis.

3.2   Resource-Aware Process Simulation

There are two major simulation parameters [12] that are important for the whole translation approach *(CP3)*: *(i)* the *Number of simulated Traces (N)* and *(ii)* the *Maximum Trace Length (L)*. Both obviously depend on the particular source process model.

First we focus on the simulation of declarative process models. Since it is used for translation purposes we focus on generating appropriate discrete-event logs. We introduced a multi-perspective simulation approach for declarative process models [13] based on the Alloy logic [14] which is visualized in Fig. 2. A declarative process model, in our case specified in DPIL, is translated to an Alloy specification. This specification is based on a meta-model for event logs (*process event chains*) and allows for formulating all DPIL rules as an Alloy *fact* which is an invariant. Alloy ships with an analyzer that is able to *exhaustively* produce examples for a given Alloy specification and maximum example size. Hence, it is possible to configure the simulator in order to produce logs with desired characteristics like the size, the maximum trace length or the trace contents.

In contrast to the simulation of resource-aware declarative process models the simulation of the imperative counterpart has been discussed multiple times. The most important parameters are [15]: *(i)* the case-arrival time *(cat)*, *(ii)* the activity service times *(ast)* and *(iii)* the probability for choices *(pcs)*. The former is the time interval for the occurrence of start events for the whole process. The second parameter describes the time an activity usually requires to be completed. Finally, the latter parameter assigns a probability to each decision of all BPMN gateways. An appropriate setting has already been presented in [12]. However, as already discussed in Sect. 3.1 though BPMN provides a notation for modeling the involvement of resources it lacks the corresponding execution semantics. Hence, it is up to the simulation tool to reflect the assignment of resources appropriately. Consequently the organizational perspective has to be modeled twice, once in the BPMN model itself and a second time when configuring the simulation component. In order to reflect the allowed orderings of the activities in the log we use the configuration guidelines introduced in [12].

3.3   Resource-Aware Process Mining

In order to perform bidirectional process model translation both a declarative and an imperative resource-aware process mining technique as well as an appropriate configuration are required.

First, we focus on declarative mining techniques. In declarative process mining constraint templates are used for querying the provided event log. In the work at hand we make use of the *DPILMiner* [9] since the tool allows for discovering resource-aware declarative process models. First, all possible constraints need to be constructed by instantiating the given set of templates with all possible combinations of occurring process elements provided in the event log. All the resulting candidates are subsequently checked w.r.t. the log. Checking candidates provides the number of traces where a constraint holds. Constraints are separated into non-valid and valid ones by means of a support and confidence framework. Let $|\Phi|$ be the number of traces in an event log Φ. Let $|\sigma_{nv}(r)|$ be the number of traces in which a rule $r : A \rightarrow B$ is non-vacuously satisfied. This implication rule is the abstraction for many DPIL rules, e.g. the rule *sequence(A,B)* (cf. Sect. 2.1). The support $supp(r)$ and confidence $conf(r)$ values of a rule r are defined as:

$$supp(r) := \frac{|\sigma_{nv}(r)|}{|\Phi|}, \qquad conf(r) := \frac{supp(r)}{supp(A)} \tag{1}$$

Based on the confidence value a rule candidate r is classified as (i) a *valid* rule, i.e., satisfied in almost all traces or (ii) a *non-valid* rule, i.e., violated in most of the recorded traces. Constraints r with $conf(r) > minConf$ are classified as valid and all constraints r with $conf(r) < minConf$ are discarded. Since the simulated event log is completely free of noise we used $minConf = 100\%$. Furthermore, we consider every constraint as interesting that occurred in at least 5% of the available cases, i.e., $minSupp = 5\%$ (cf. Table 1, right).

Table 1. Miner configurations: MPM (l), DPILMiner (r)

Param	Value		Param	Value
Relative-to-best	0.0		Support	5.0
Dependency	49.0		Confidence	100.0
Length-one/two loop	0.0			
Long distance	100.0			
All tasks connected	true			
Long distance dep.	true			
Ignore loop dep. thresh.	false			

Furthermore, a resource-aware imperative mining approach is necessary. The control-flow-focused translation approach presented in [12] suggests a suitable mining algorithm as well as an appropriate configuration. Since the chosen *Flexible Heuristics Miner* [16] does not consider the organizational perspective either a different or a complementary mining approach is required. In [17] the authors propose a technique, for enriching a given control-flow model with swimlanes based on a *Handover of Roles (HooR)* principle. In the first step the pairs of immediately consecutive activities are analyzed in terms of potential role changes based on three rules: *(i)* pairs of immediately consecutive activities that are always executed by the same resource do not involve a HooR, *(ii)* pairs of immediately consecutive activities that are each executed by exactly the same set of resources do not involve a HooR and *(iii)* pairs of immediately consecutive activities that are, to a certain proportion w, executed by the same resources do not involve a HooR. The last rule is threshold-based whereby the corresponding threshold value is determined automatically. All rules are based upon the assumption that each resource has exactly one role. So far, the clustering-inspired algorithm generated a partition of activities *for each* HooR. However, the partitions do not correspond to roles, yet, because it is possible to switch back to a role that already has been involved in previous events. Consequently, the second and last step of the algorithm merges similar partitions in order to identify the actual roles. The merging condition is again measured by w but this time for non-immediately consecutive events and with a separate threshold. The two required thresholds are calculated automatically based on the values encountered for w. Finally, the algorithm chooses the *most suitable* final partitioning based on an entropy measure also known from the quality measurement for clustering results. The approach just summarized decorates an *existing* BPMN model, with role partitions. It is therefore necessary to discover the control flow beforehand. The configuration given in Table 1 (left) has already been suggested in [12] and can be re-used. Since the approach for discovering the role partitioning is based on pairs of directly dependent activities the result quality strongly depends on the quality of the discovered control flow.

4   Evaluation

In order to examine the efficacy of the presented approach, we provide a two-stage evaluation using the implementation discussed in the following subsection.

4.1   Implementation

Though there is no integrated implementation, yet, we identified a suitable compilation of tools for our simulation- and mining-based translation system. In order to translate BPMN models to DPIL we utilize *BIMP* [18], a web-based simulation tool. BIMP allows to specify the parameters (i)–(ii) introduced in Sect. 3.2. Furthermore, it is able to generate a process event log based on the *MXML* standard. Since the subsequent DPILMiner is based on a more recent

Table 2. Figure 1 models (a)–(d)

N	L	Fitness				Appropriateness				Org. Precision			
		(a)	(b)	(c)	(d)	(a)	(b)	(c)	(d)	(a)	(b)	(c)	(d)
10	3	1.0	1.0	0.773	0.625	1.0	1.0	0.530	0.666	1.0	1.0	1.0	1.0
100	3	1.0	1.0	0.818	0.655	1.0	1.0	0.616	0.685	1.0	1.0	1.0	1.0
1000	3	1.0	1.0	0.884	0.777	1.0	1.0	0.788	0.757	1.0	1.0	1.0	1.0
10	6	1.0	1.0	0.802	0.749	1.0	1.0	0.669	0.940	1.0	1.0	1.0	1.0
100	6	1.0	1.0	0.931	0.876	1.0	1.0	0.868	0.999	1.0	1.0	1.0	1.0
1000	6	1.0	1.0	1.0	1.0	1.0	1.0	1.0	1.0	1.0	1.0	1.0	1.0

event log standard, namely *XES* [19] we convert the generated log files using the well-established *ProM* (v 6.5.1) platform. The evaluation metrics used in the following subsection are implemented in the DPILMiner. The transformation of DPIL models to BPMN is done through the discussed Alloy-based simulation component which generates XES logs. ProM provides plugins for the desired Flexible Heuristics Miner as well as for mining the role partitioning and performing the subsequent swimlane-decoration of the BPMN model. Furthermore ProM provides plugins for conformance checking based on replaying a log on a petri net which can be easily derived from a BPMN model. Since the plugin does not consider the swimlane associations, we semi-automatically check the correctness of role assignments using a dedicated DPILMiner feature.

4.2 Transformation Result Quality

In order to measure the transformation result quality, we first apply our approach to the four simple examples shown in Fig. 1. The results are presented in Table 2. Each column (a)–(d) measures the fitness, appropriateness and precision of the role assignments for translating the particular model to the opposite language. The two measures, fitness and appropriateness are defined in [7] as follows:

(1) *Fitness*: Proportion of traces the discovered model is able to parse,
(2) *Appropriateness*: Proportion of additional behavior allowed by the model but not present in the log.

Additionally we measure the average translation quality O_p of the organizational perspective separately according to Eq. 2. Let $|\Phi|$ be the number of traces in an event log Φ whereby ϕ denotes one contained trace.

$$O_p = \frac{\sum_{\phi \in \Phi} \frac{\#correctRoleAssignments\ in\ \phi}{\#allRoleAssignments\ in\ \phi}}{|\Phi|} \tag{2}$$

The results show that the simple three-step process with one involved role (cf. Fig. 1(a)/(b)) can be translated appropriately in both directions. Since there is not a single variation point for process instances this is expected and trivial. In contrast, the remaining two simple examples perform much worse with a low number of traces and a length of three task-related events. This is expected, too, because the model contains loops that allow, for instance, that all three activities are executed twice. However, this cannot be encoded with a maximum trace length of three. Raising the number of generated traces improves the result slightly but the main influence here is the trace length.

In our second stage we apply the technique to real-life process models that have been created in order to manage university-internal business trip applications[2]. The process is strict in that terms that it does not exceed 12 process steps. In contrast, it is flexible regarding the means of conveyance as long as at least one is chosen. We confined ourselves to at most 10000 traces since there are only few variation points in the process instances. One important point is that the fitness of the generated model raises faster with increasing number of traces than the appropriateness. Table 3 show a perfect role assignment for all tasks. The reason is the limited expressiveness of BPMN which, for instance, does not allow to assign multiple roles to a single task. Since we focused on process models that can be represented using both BPMN and DPIL, we hereof omit DPIL's flexibility. This means that there is no inconstancy in terms of the role assignment. The perfect translation results are achieved due to the reason that the simulation components produce noise-free logs which allows for a very strict configuration of the mining components. Since we build our approach on existing tools that have been investigated regarding quality and performance, we forgo a second evaluation. In contrast, the computation time of our Alloy-based simulation approach has not been analyzed. A detailed performance analysis is beyond the paper at hand, because it focuses on the principle of the translation.

Table 3. Travel application: DPIL $\rightarrow$ BPMN (l), BPMN $\rightarrow$ DPIL (r)

N	L	Fitness	App.	O_p
10	12	0.856	0.696	1.0
100	12	1.0	0.772	1.0
1000	12	1.0	0.898	1.0
10000	12	1.0	1.0	1.0

N	L	Fitness	App.	O_p
10	12	0.811	0.717	1.0
100	12	0.925	0.798	1.0
1000	12	0.966	0.941	1.0
10000	12	1.0	1.0	1.0

5   Related Work

One motivation for the development of the presented technique was the lack of appropriate process model transformation approaches from and to declarative process modeling languages. In contrast, there are many approaches for the translation of imperative models to a different imperative language, e.g. BPMN to BPEL [20] or several languages to Petri Nets [21]. In [2] the authors investigated differences between imperative and declarative process modeling languages

---

[2] Available at: http://www.ai4.uni-bayreuth.de/de/research/projects/1000_Process Understanding/index.html.

regarding their understandability. The work of Prescher et al. [22] is the most relevant approach which transforms a declarative process model into an behaviorally equivalent Petri Net representation. Since the source process language, namely Declare, only models the control-flow perspective the approach is able to transform each declarative constraint into a regular expression. After transforming the latter into finite state automatons in a second step, the technique finally creates a petri net. In [12] it has been shown that we are able to establish a bidirectional translation from Declare to BPMN and vice versa. This was possible because in [23] the authors describe a simulation technique for control-flow-based declarative process models specified in Declare. However, we cannot reuse this technique for the approach discussed in the paper at hand since Declare and, consequently, the corresponding simulation technique, are *limited* to the control-flow perspective. As far as we know there is no other comparable approach. Nevertheless, our technique has been inspired by a more general approach of model transformation that has been presented in [10]. The authors prevent the user from specifying cumbersome and complex model transformation rules by interpreting examples of successful transformations and automatically deriving the rules in the background. Our technique shares the intention but is based on a different solution strategy. A successful translation depends on the equality of expressiveness of source and target language. Though this has not been analyzed concerning the language pair BPMN and DPIL, there are at least two relevant attempts to formalize a comparison of process modeling languages based on a well-established ontology that has already been applied to BPMN [24] and a semiotic quality framework [25].

6   Conclusion and Future Work

The approach presented in this paper makes use of well-established techniques in process simulation and mining in order to translate a given business process model to a particular target language. Process simulation techniques are used to generate exemplary execution traces the source process model allows. This represents an easy-to-use alternative to conventional model-to-model transformation system without backtracking from target model elements to source model elements. Since plain control-flow models neglect dependencies between process participants as well as between process participants and activities we primarily focused on models that consider the behavioral *and* the organizational perspective. The presented principle is neither limited to languages nor to the imperative-declarative setting in general. Technically the applicability of our approach depends on the availability of corresponding simulation and mining technologies as well as an appropriate configuration. Currently our evaluation exclusively focuses on the *correctness* of the translation regarding the equivalence in terms of behavior and resource allocation. We are currently conducting a study on the *gain of understandability* of the generated process models compared to particular source models. Additionally, the ability for a successful translation strongly depends on the semantic equivalence of BPMN and DPIL. Hence, we are currently analyzing both languages regarding differences in expressiveness.

References

1. Fahland, D., Lübke, D., Mendling, J., Reijers, H., Weber, B., Weidlich, M., Zugal, S.: Declarative versus imperative process modeling languages: the issue of understandability. In: Halpin, T., Krogstie, J., Nurcan, S., Proper, E., Schmidt, R., Soffer, P., Ukor, R. (eds.) BPMDS/EMMSAD -2009. LNBIP, vol. 29, pp. 353–366. Springer, Heidelberg (2009). doi:10.1007/978-3-642-01862-6_29

2. Pichler, P., Weber, B., Zugal, S., Pinggera, J., Mendling, J., Reijers, H.A.: Imperative versus declarative process modeling languages: an empirical investigation. In: Daniel, F., Barkaoui, K., Dustdar, S. (eds.) BPM 2011. LNBIP, vol. 99, pp. 383–394. Springer, Heidelberg (2012). doi:10.1007/978-3-642-28108-2_37

3. Pesic, M., Aalst, W.M.P.: A declarative approach for flexible business processes management. In: Eder, J., Dustdar, S. (eds.) BPM 2006. LNCS, vol. 4103, pp. 169–180. Springer, Heidelberg (2006). doi:10.1007/11837862_18

4. Zeising, M., Schönig, S., Jablonski, S.: Towards a common platform for the support of routine and agile business processes. In: CollaborateCom (2014)

5. Aalst, W.M.P.: Business process simulation revisited. In: Barjis, J. (ed.) EOMAS 2010. LNBIP, vol. 63, pp. 1–14. Springer, Heidelberg (2010). doi:10.1007/978-3-642-15723-3_1

6. Stewart, R.: Simulation: The Practice of Model Development and Use. Wiley, New York (2004)

7. van der Aalst, W.: Conformance and Enhancement of Business Processes, vol. 2. Springer, Heidelberg (2011)

8. Di Ciccio, C., Mecella, M.: On the discovery of declarative control flows for artful processes. TMIS **5**(4), 24 (2015)

9. Schönig, S., Cabanillas, C., Jablonski, S., Mendling, J.: Mining the organisational perspective in agile business processes. In: Gaaloul, K., Schmidt, R., Nurcan, S., Guerreiro, S., Ma, Q. (eds.) CAISE 2015. LNBIP, vol. 214, pp. 37–52. Springer, Cham (2015). doi:10.1007/978-3-319-19237-6_3

10. Wimmer, M., Strommer, M., Kargl, H., Kramler, G.: Towards model transformation generation by-example. In: HICSS, pp. 285–294 (2007)

11. Sun, Y., White, J., Gray, J.: Model transformation by demonstration. In: Schürr, A., Selic, B. (eds.) MODELS 2009. LNCS, vol. 5795, pp. 712–726. Springer, Heidelberg (2009). doi:10.1007/978-3-642-04425-0_58

12. Ackermann, L., Schönig, S., Jablonski, S.: Towards simulation- and mining-based translation of process models. In: Pergl, R., Molhanec, M., Babkin, E., Fosso Wamba, S. (eds.) EOMAS 2016. LNBIP, vol. 272, pp. 3–21. Springer, Cham (2016). doi:10.1007/978-3-319-49454-8_1

13. Ackermann, L., Schönig, S., Jablonski, S.: Simulation of multi-perspective declarative process models. In: Dumas, M., Fantinato, M. (eds.) BPM 2016 Workshops. LNBIP, vol. 281, pp. 59–71. Springer, Cham (2016)

14. Jackson, D.: Software Abstractions: Logic, Language, and Analysis. MIT press, Cambridge (2012)

15. Aalst, W.M.P.: Handbook on Business Process Management: Introduction, Methods, and Information Systems. Springer, Heidelberg (2015)

16. Weijters, A., Ribeiro, J.: Flexible heuristics miner (FHM). In: CIDM, pp. 310–317 (2011)

17. Burattin, A., Sperduti, A., Veluscek, M.: Business models enhancement through discovery of roles. In: 2013 IEEE Symposium on Computational Intelligence and Data Mining (CIDM), pp. 103–110. IEEE (2013)

18. Bimp - the business process simulator. http://code.google.com/p/ bimp-simulator/. Accessed 17 Mar 2016
19. Verbeek, H.M.W., Buijs, J.C.A.M., Dongen, B.F., Aalst, W.M.P.: XES, XESame, and ProM 6. In: Soffer, P., Proper, E. (eds.) CAiSE Forum 2010. LNBIP, vol. 72, pp. 60–75. Springer, Heidelberg (2011). doi:10.1007/978-3-642-17722-4_5
20. Recker, J.C., Mendling, J.: On the translation between bpmn and bpel: Conceptual mismatch between process modeling languages. In: CAISE Workshops, pp. 521–532 (2006)
21. Raedts, I., Petkovic, M., Usenko, Y.S., van der Werf, J.M.E., Groote, J.F., Somers, L.J.: Transformation of BPMN models for behaviour analysis. In: MSVVEIS, pp. 126–137 (2007)
22. Prescher, J., Di Ciccio, C., Mendling, J.: From declarative processes to imperative models. In: SIMPDA, pp. 162–173 (2014)
23. Ciccio, C., Bernardi, M.L., Cimitile, M., Maggi, F.M.: Generating event logs through the simulation of declare models. In: Barjis, J., Pergl, R., Babkin, E. (eds.) EOMAS 2015. LNBIP, vol. 231, pp. 20–36. Springer, Cham (2015). doi:10. 1007/978-3-319-24626-0_2
24. Johannsen, F., Leist, S., Tausch, R.: Wand and Weber's good decomposition conditions for BPMN: an interpretation and differences to event-driven process chains. Bus. Process Manage. J. **20**(5), 693–729 (2014)
25. Lindland, O.I., Sindre, G., Solvberg, A.: Understanding quality in conceptual modeling. Softw. IEEE **11**(2), 42–49 (1994)

Transforming Multi-role Activities in Software Processes into Business Processes

Juan Pulgar and María Cecilia Bastarrica[✉]

CS Department, Universidad de Chile, Santiago, Chile
jpulgar1978@gmail.com, cecilia@dcc.uchile.cl

Abstract. Software processes usually include activities involving several people playing different roles. SPEM provides primitives for defining all the roles involved in each activity. Software process specification notations are not executable and thus supporting tools cannot provide this functionality. Therefore, even having a formal software process specification we cannot achieve all the potential benefits: people have difficulties in following their responsibilities, resulting in a low productivity. The business process domain provides notations that can be executed on a BPMS. There have been attempts to transform SPEM specifications into BPMN. However, there is no natural way to model multi-role tasks in BPMN, and therefore none of these proposals has solved this issue. In this paper we discuss two promising alternatives for modeling multi-role software activities in BPMN: defining compound roles and modeling multi-role tasks as independent processes. We provide an XSLT transformation for automatically generating each of these solutions from a software process specification. We use a real world running example to illustrate the approach.

Keywords: Software process · Multi-role tasks · Model transformation

1 Introduction

Software processes are business processes in the software development domain, so most characteristics of the latter also hold for the former. They allow companies to better organize software development, and more specifically support management of project schedule, scope, resources, risks and expected software quality increasing the probability of success. Process definition is also a requirement if the company intends to achieve an ISO certification or a CMMI evaluation.

Software development is mainly an intellectual activity, so the most relevant resources to be managed in these processes are human resources. Tasks are the simplest work units in a software process, and they may be performed by one or more roles within a process.

From an enterprise point of view, a business process is a collection of activities that, given one or a set of inputs, creates a valuable output for the customer [15]. An executable process is one that may be run in a business process management system (BPMS). Specifying processes in an executable notation enables several other benefits as well:

© Springer International Publishing AG 2017
M. Dumas and M. Fantinato (Eds.): BPM 2016 Workshops, LNBIP 281, pp. 372–383, 2017.
DOI: 10.1007/978-3-319-58457-7_27

– continuous process improvement based on empirical data
– reduced time invested in changes derived from business logic, and therefore more agility and flexibility
– reduced risks due to the possibility of simulating processes before executing them

Software Process Engineering Metamodel (SPEM) [9] and Business Process Model and Notation (BPMN) [14] are the standard notations proposed by the OMG for specifying software processes and business processes, respectively. On the one hand, SPEM provides rich primitives for specifying elements that are relevant for software processes, e.g. input and output work products associated with tasks, or roles assigned to each task. SPEM organizes these elements in two parts: the method content where each process element is defined, and the process where defined process elements are combined in order to configure a process. In particular, SPEM allows the definition of tasks with several roles assigned, but SPEM is not executable. On the other hand, BPMN 2.0 is potentially executable, but it does not have inherently a clear way for specifying tasks where several roles participate, and this is a serious limitation when using BPMN for software processes [16].

Once the company has invested in specifying its software process, it is appealing to transform it into BPMN so that it can be executed. There are several proposals for automatically transforming software processes into executable notations but, to the best of our knowledge, none of them addresses the specification of multi-role tasks.

In this paper we analyze two ways of specifying multi-role tasks in BPMN: assigning the task to a compound role formed by the set of roles involved, and specifying the multi-role task as a *meeting* that is fully specified separately. We provide two XSLT transformations that take the software process specification and automatically generate each of the analyzed alternatives. We illustrate the whole procedure with a real world bookstore software process and we discuss the pros and cons of each approach. Even though the resulting business processes are not directly executable as they are, it is a significant step towards this goal.

The rest of the paper is structured as follows. Section 2 presents some related work concerning software and business process notations, and discusses former attempts to transform SPEM processes into executable business process notations. Section 3 presents existing approaches for modeling multi-role activities. Section 4 presents the proposed approach using the running example for illustrating each alternative. Section 5 concludes.

2   Related Work

In this section we discuss approaches for modeling software and business processes, as well as attempts to transform from the former to the latter.

2.1   Software Process Specification

The OMG[1] has recommended SPEM [9] as the standard for software process specification. SPEM has been designed for describing processes and their components, following an object oriented modeling approach based on UML by extending its mechanisms for software process modeling.

Formally specifying a software process is a complex activity, but it allows for using a supporting tool that eases analysis and evolution. There are several tools available for this purpose. Some of them are: Eclipse Process Framework Composer (EPFC), Objecteering Modeler, MagicDraw UML (SPEM plug-in), Rational Method Composer and Enterprise Architect. Among these we can highlight EPFC [8] provided that it is free and therefore it is an appealing option. It internally implements UMA (Unified Method Architecture), an extension of SPEM that includes an UML profile for specifying process behavior, a feature not covered by SPEM [2].

2.2   Business Process Specification

There are several notations for specifying business processes such as: Petri Nets, UML activity diagrams, business process modeling notation (BPMN) [14], XML process definition language (XPDL) [5], integration definition (IDEF) [11], business process execution language (BPEL) [13], and event-driven process chain (EPC) [12]. BPMN 2.0 is widely used in industry mainly because [17]:

– is the OMG standard and thus there are several supporting tools
– provides a graphical notation easy to understand by stakeholders
– reduces the gap between the process design and its implementation
– it is potentially executable

A business process management system (BPMS) is a platform that allows companies to coordinate the realization of business processes based on process representations as models.

2.3   Transforming Software Processes into Business Process

There are several proposals that take SPEM 2.0 software process specifications and automatically generate executable processes in notations such as XPDL or BPEL. In [6] they propose a transformation from SPEM to BPMN using RSL as a transformation language, while [18] uses QVT. None of these proposals addresses multi-role tasks. MOSKitt4ME [4] takes processes specified in EPFC as we do, and transforms them into BPMN 2.0, but it can only transform a few kinds of process elements. In [1,10] they present a formal transformation from UML AD to BPMN 1.0 using MOLA and QVT, respectively, but these transformations do not address role assignment. In [7] we have proposed an XSLT transformation from software processes specified in EPFC into BPMN.

---

[1] Object Management Group - http://www.omg.org/.

However, that work did not address multi-role tasks either. In this work we address this issue and also upgrade the transformations into BPMN 2.0 so that they will eventually be executed in a BPMS; for this purpose we use Bonita[2].

3   Multi-role Tasks in Software Process Specifications

3.1   Running Example

We have defined the software process followed in a medium size bookstore in Chile. Figure 1 depicts the incidences process for this company specified using EPFC. In order not to overload the figure, EPFC includes neither roles nor work products in the activity diagram, but internally these relationships can be specified and checked. For example, the *Developer* is responsible for *Incidence Analysis*, *Solution Design* and *Solution Programming*, while *Incidence Reproduction* and *Testing* are the responsibility of *Developer* together with the *Business Analyst*, as shown in Table 1

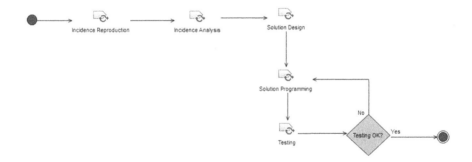

Fig. 1. Incidence process for the bookstore.

Table 1. Roles assigned to each task.

Task	Developer	Business analyst
Incidence reproduction	X	X
Incidence analysis	X	
Solution design	X	
Solution programming	X	
Testing	X	X

---

[2] http://www.bonitasoft.com/products.

3.2   Modeling Multi-role Tasks in BPMN

A pool is a modeling element in BPMN that sets the boundaries of a business process. It may be divided into lanes for organizing activities. It is a common practice to use lanes for assigning activities to roles, but there is no natural way to represent collaborative activities performed by different roles. Shapiro [16] has addressed this issue with different techniques; we describe them in what follows.

Tasks Duplicated in Each Lane. A first option is to duplicate the task in the lane of each role involved as shown in Fig. 2. A clear problem with this solution is that tasks specified in this way are not multiple copies of the same task, but different tasks with different identificators, although sharing the same label. Therefore, the diagram shows multiple tasks performing in parallel. There may also be coherence problems when properties among replicated tasks do not match, or due to coevolution inconsistencies. This solution presents problems from a graphical point of view as well. When the task is performed by two roles appearing in consecutive lanes, the meaning may be clear, but when the diagram involves a lot of lanes, the diagram becomes messy and difficult to understand. A parallel gateway should also be added for coordination.

Different Tasks Assigned to Roles. In order to address some of the problems of the previous solution, we can divide the multi-role task into smaller coordinated tasks, assigning each of them to the participating roles as shown in Fig. 3. Modeling this solution may be quite complex because it may be required

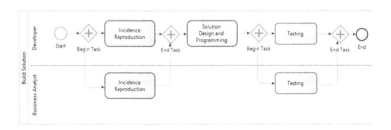

Fig. 2. Replicated multi-role task.

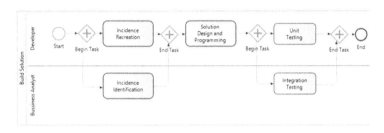

Fig. 3. Disaggregate multi-role task.

to invent several tasks that do not actually reflect the intended action, e.g., all actors that attend a meeting should perform a similar but different task instead of having just one unique activity representing the meeting in itself. Moreover, collaborative activities are not apparent just looking at the diagram.

Separate Lanes Representing Roles as Groups. Yet another proposal is to create a new lane withing the pool that represents the set of roles involved in the collaborative task. This avoids task repetition and reduces the number of activities. In this case, the resulting model is more complex since it has more lanes and this situation is even worse when there are several multi-role activities with different sets of roles involved.

Shared Activities Modeled in Separate Processes. This approach uses call activities calling a different process where the multi-role activity is modeled in a pool with a single lane. The activity is assigned to the lane of the role responsible for leading the meeting, and then the meeting is detailed in separate processes, one for each participating role. The main drawback of this approach is that each multi-role activity will have a separate process.

4   Transforming UMA Specifications into BPMN

In this paper we consider the last two options from the previous section because Shapiro [16] recommends them as the most promising ones, and we show how they can be automatically generated from the EPFC specification.

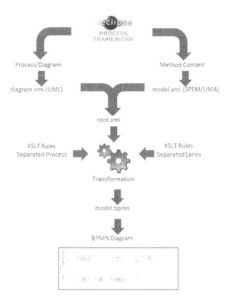

Fig. 4. Solution general structure

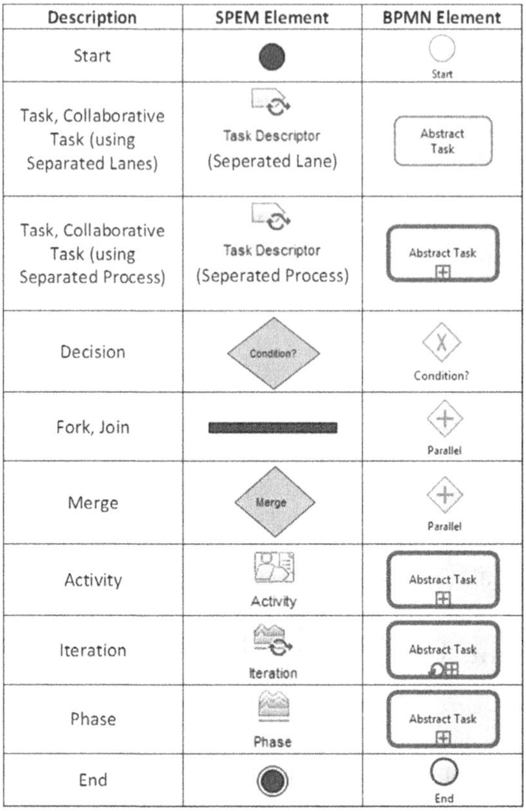

Description	SPEM Element	BPMN Element
Start	●	○ Start
Task, Collaborative Task (using Separated Lanes)	Task Descriptor (Seperated Lane)	Abstract Task
Task, Collaborative Task (using Separated Process)	Task Descriptor (Seperated Process)	Abstract Task
Decision	Condition?	X Condition?
Fork, Join	▬▬▬	+ Parallel
Merge	Merge	+ Parallel
Activity	Activity	Abstract Task
Iteration	Iteration	Abstract Task
Phase	Phase	Abstract Task
End	◉	○ End

Fig. 5. Translation of UMA elements into BPMN elements.

Figure 4 describes the structure of the proposed solution. The software process specified in EPFC includes both, the method content and the process diagram. Both are combined into a single `root.xmi` model that is taken as input for each XSLT transformation. The table in Fig. 5 indicates the way process elements are converted between notations in the transformations. In Sect. 4.1 we show the transformation of the software process into BPMN where multi-role tasks are assigned to compound roles specified as separated lanes. In Sect. 4.2 we show the transformation to automatically generate multi-role tasks represented as BPMN meetings that are then specified as independent processes.

4.1   Separate Lanes for Compound Roles

The idea is that each role has its own lane containing all the activities where he is the only responsible and, for those activities that have more than one responsible role, a separate lane will be created whose responsible is the compound role. Algorithm 1 shows the pseudocode for the XSLT transformation that takes the `root.xmi` model as input and generates this BPMN diagram.

Algorithm 1. XSLT transformation for compound roles

Require: XML file containing ⟨*diagram*⟩ and ⟨*model*⟩ nodes (root.xml)
Ensure: XML file containing ⟨*definitions*⟩ node for BONITA BPM, modeling collaborative activities in separated lanes for compound roles
1: Create a ⟨*definitions*⟩ node
2: **for all** *spemActivity* ← ⟨*uml* : *Activity*⟩ node in ⟨*diagram*⟩ **do**
3:     Create a ⟨*process*⟩ node in ⟨*definitions*⟩
4:     Set the ⟨*process*⟩ attributes (id, name) from the values of the *spemActivity* in ⟨*model*⟩
5:     Create a ⟨*laneSet*⟩ node in ⟨*process*⟩
6:     **for all** *spemTask* ← ⟨*node*[@*xmi* : *type* =′ *uml* : *ActivityParameterNode*′]⟩ node in *spemActivity* **do**
7:         *spemRole* ← *spemTask's* **role(s)** defined in ⟨*model*⟩
8:         **if** *spemRole* is empty **then**
9:             Create a ⟨*lane*⟩ node in ⟨*laneSet*⟩, setting the name attribute as *Undefined*
10:        **else**
11:            Create a ⟨*lane*⟩ node in ⟨*laneSet*⟩, setting the name attribute as *spemRole's* name
12:        **end if**
13:        Create a ⟨*flowNodeRef*⟩ node in ⟨*lane*⟩, setting its value as *spemTask's* id
14:    **end for**
15:    Group all ⟨*flowNodeRef*⟩ by ⟨*lane*⟩ node
16:    Create elements (as specified in the Node Transformation Table) appending them to the ⟨*process*⟩ node
17: **end for**
18: **return** ⟨*definitions*⟩ node

The XSLT transformation takes each activity in the EPFC `diagram.xmi` and generates a node in the BPMN *definitions* assigning the activity's attributes, as stated in lines 3–4 in Algorithm 1. Then, it creates a *laneSet* (line 5) and, for each task in `model.xmi`, creates a lane (lines 6–14). Finally, lanes assigned to the same role are combined by role in line 16. In this way, when creating lanes for multi-role activities, an independent lane will be created, and it is not grouped afterwards with any of its participating roles.

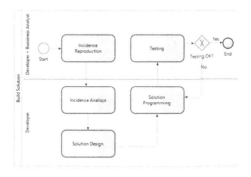

Fig. 6. Multi-role tasks assigned to compound roles

Figure 6 shows the visualization in Bonita BPMN of the result of applying this algorithm to the process in Fig. 1. We can see that *Incidence Analysis*, *Solution Design* and *Solution Programming* are assigned to the *Developer*'s lane, while *Incidence Reproduction* and *Testing* are assigned to the lane of the compound role *Developer+Business Analyst*. By default, the *Start* and *End* activities are assigned to the lane of the role that performs the first and last activities, respectively; in this case this lane is that of the compound role.

This solution is quite clear in the sense that there is no redundancy, each activity appears only once in the BPMN diagram, and the flow of control is apparent. However, for a particular role, it is not obvious that he may be involved in other activities out of those that appear in his lane. In the example, the *Developer* does not only take part in his lane's activities, but also in *Incidence Reproduction* and *Testing* that appear in another lane. This situation may be even worse since the same role may collaborate with several other roles in different activities and thus his responsibilities may be split across the whole diagram.

4.2 Independent Processes

The second solution consists of a direct translation of each activity defined in the process in EPFC into a BPMN node assigned to the lane of the first role defined in the model.xmi, as shown in lines 3–7 in Algorithm 2. Then, for each node where there are more than one role assigned, the node is marked as a *meeting* and a new lane is created for each role involved and a process describing the activities that role must perform is defined with its own *Start* and *End* nodes.

Figure 7 shows the result of applying the algorithm to the process in Fig. 1. We can see that *IncidenceReproduction* and *Testing* are assigned to the *Developer* and marked as *meeting*s. Then, in Fig. 8, the lanes corresponding to the specification of each of these activities is shown in a lane assigned to both roles *Developer+Business Analyst*.

Even though this solution seems to be clean and clear due to non duplicated activities, it may be somewhat inconvenient when multi-role tasks are defined in EPFC with its roles in different order. For example, if in the *Incidence Reproduction* tasks roles are define as *Developer* and *Business Analyst*, this activity will appear as a meeting in the *Developer*'s lane. But, if on the contrary, if the roles are assigned as *Business Analyst* and *Developer*, then it will appear in the *Business Analyst*'s lane.

4.3 Discussion

Although we do not think that any of the solutions is better than the other, we envision situations that favor one or the other. For example, if the same group of roles are usually assigned together to different tasks as in [3], then the first solution is clearer. On the other hand, if the EPFC software process specification is organized so that some of the roles are defined first in all the tasks they are assigned, then the second solution would be better.

Our initial goal was to transform SPEM processes into BPMN in order to make them executable in a BPMS. However, in order to make this possible there are certain other steps that should be performed before executablility may be fully reached. First, an application for managing the process should be built

---

Algorithm 2. XSLT transformation for separate process

Require: XML file containing ⟨*diagram*⟩ and ⟨*model*⟩ nodes (root.xml)
Ensure: XML file containing ⟨*definitions*⟩ node for BONITA BPM, modeling collaborative activities in separated processes
1: Create a ⟨*definitions*⟩ node
2: **for all** *spemActivity* ← ⟨*uml : Activity*⟩ node in ⟨*diagram*⟩ **do**
3:    Create a ⟨*process*⟩ node in ⟨*definitions*⟩
4:    Set the ⟨*process*⟩ attributes (id, name) from the values of the *spemActivity* in ⟨*model*⟩
5:    Create a ⟨*laneSet*⟩ node in ⟨*process*⟩
6:    **for all** *spemTask* ← ⟨*node*[@*xmi : type* =' *uml : ActivityParameterNode'*]⟩ node in *spemActivity* **do**
7:       *spemRole* ← *spemTask's* **first role** defined in ⟨*model*⟩
8:       **if** *spemRole* is empty **then**
9:          Create a ⟨*lane*⟩ node in ⟨*laneSet*⟩, setting the name attribute as *Undefined*
10:         **else**
11:            Create a ⟨*lane*⟩ node in ⟨*laneSet*⟩, setting the name attribute as *spemRole's* name
12:         **end if**
13:         Create a ⟨*flowNodeRef*⟩ node in ⟨*lane*⟩, setting its value as *spemTask's* id
14:      **end for**
15:      Group all ⟨*flowNodeRef*⟩ by ⟨*lane*⟩ node
16:      Create elements (as specified in the Node Transformation Table) appending them to the ⟨*process*⟩ node
17: **end for**
18: **for all** *spemTask* ← ⟨*node*[@*xmi : type* =' *uml : ActivityParameterNode'*]⟩ node in ⟨*diagram*⟩ **do**
19:    *spemRole* ← *spemTask's* **roles(s)** defined in ⟨*model*⟩
20:    **if** *spemRole* has more than one element **then**
21:       Create a ⟨*process*⟩ node in ⟨*definitions*⟩
22:       Set the ⟨*process*⟩ attributes (new id; name as the concatenation of 'Meeting' and *spemTask's* names)
23:       Create a ⟨*laneSet*⟩ node in ⟨*process*⟩
24:       Create a ⟨*lane*⟩ node in ⟨*laneSet*⟩, setting the name attribute as *spemRole's* names
25:       Create a task, a start node and an end node
26:       Create three ⟨*flowNodeRef*⟩ nodes in ⟨*lane*⟩, setting the value as the new task, start node and end node id's
27:    **end if**
28: **end for**
29: **return** ⟨*definitions*⟩ node

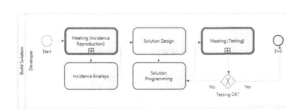

Fig. 7. Multi-role tasks assigned to the first role and marked as a meeting

together with designing the business data. Then, the process itself is built. In this stage, the first step is to create the BPMN process, and this is the step we have automated with our proposal. Then, still other information should be provided such as the business variables, contracts, and actors that will carry out the process steps.

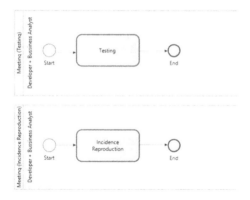

Fig. 8. Meeting tasks assigned to each role taking part

5   Conclusion

In this paper we proposed a strategy for transforming software processes specified in EPFC -conforming to the UMA metamodel- into BPMN processes. We were able to automatically generate the two most promising solutions for multi-role activity specification suggested by Shapiro et al. [16]. We provided two XSLT transformations, one for obtaining each solution, where multi-role tasks are modeled as activities in lanes assigned to compound roles or meetings that are specified in detail as independent processes. Being able to clearly identify responsibilities allows people involved in software project execution to optimize their time, and therefore the whole project productivity. Moreover, being able to automatically transform the software process already specified into a business process, reduces the effort by reusing the specification already available. Even though the resulting diagrams are not directly executable, they constitute an important step towards executing a software process that is consistent with that specified in EPFC.

We were able to transform the software process of the running example. However, and even though it is a real world software process, it is still necessary to apply the approach to larger processes in order to identify some limitations that may be hidden because of the small size of the example. So far, we were also able to transform a more complex process of a much larger organization, that also includes process patterns and iterations. These results are encouraging. However, we will be able to obtain the most out of our transformations when we build a tool that helps choosing the most appropriate approach in each case. It is also necessary to count on objective measurements of each solution quality such as understandability, complexity and completeness, among others.

Acknowledgments. This work is partly funded by Project Fondef IT13I20010, Conicyt, Chile.

References

1. Argaaraz, M., Funes, A.M., Arístides, J.: An MDA approach to business process model transformations. Electron. J. SADIO (EJS) **9**, 24–48 (2010)
2. Bendraou, R., Combemale, B., Crégut, X., Gervais, M.: Definition of an executable SPEM 2.0. In: 14th Asia-Pacific Software Engineering Conference (APSEC 2007), 5–7, Nagoya, Japan, pp. 390–397. IEEE Computer Society, December 2007
3. Cabanillas, C., Resinas, M., Mendling, J., Cortés, A.R.: Automated team selection and compliance checking in business processes. In: Proceedings of the 2015 International Conference on Software and System Process, ICSSP 2015, Tallinn, Estonia, August 24–26, 2015, pp. 42–51. ACM (2015)
4. Cervera, M., Albert, M., Torres, V., Pelechano, V.: The MOSKitt4ME approach: providing process support in a method engineering context. In: Atzeni, P., Cheung, D., Ram, S. (eds.) ER 2012. LNCS, vol. 7532, pp. 228–241. Springer, Heidelberg (2012). doi:10.1007/978-3-642-34002-4_18
5. W. M. Coalition. XML Process Definition Language (XPDL) (2012). http://www.xpdl.org/
6. Cota, M.P., Riesco, D., Lee, I., Debnath, N.C., Montejano, G.: Transformations from SPEM work sequences to BPMN sequence flows for the automation of software development process. J. Comput. Meth. Sci. Eng. **10**(3–6), 61–72 (2010)
7. Cruz, D.E., Bastarrica, M.C., Duarte-Amaya, H.: De procesos SPEM a procesos BPMN. Un enfoque basado en MDE. In: CIbSE, pp. 41–52 (2014)
8. E. Foundation. Eclipse Process Framework Project (EPF) (2015)
9. O. M. Group. Software & Systems Process Engineering Meta-Model Specification (2008). http://www.omg.org/spec/SPEM/2.0/
10. Kalnins, A., Vitolins, V.: Use of UML and model transformations for workflow process definitions. CoRR, abs/cs/0607044 (2006)
11. I. K. Knowledge Based Systems. IDEF. Integrated DEFinition Methods (2016). http://www.idef.com/
12. Korherr, B.: Business Process Modelling - Languages, Goals, and Variabilities. Ph.D. thesis, Vienna University of Technology, January 2008
13. OASIS. Web Services Business Process Execution Language Version 2.0 (2007). http://docs.oasis-open.org/wsbpel/2.0/OS/wsbpel-v2.0-OS.html
14. OMG. Business Process Model and Notation, Version 2.0 (2011). http://www.omg.org/spec/BPMN/2.0/
15. Sánchez-González, L., García, F., Ruiz, F., Velthuis, M.P.: Measurement in business processes: a systematic review. Bus. Process Manage. J. **16**(1), 114–134 (2010)
16. Shapiro, R., White, S.A., Bock, C., Palmer, N., zur Muehlen, M., Brambilla, M., Gagné, D. (eds.): BPMN 2.0 Handbook. Workflow Management Coalition, Lighthouse Point (2012)
17. Wohed, P., Aalst, W.M.P., Dumas, M., Hofstede, A.H.M., Russell, N.: On the suitability of BPMN for business process modelling. In: Dustdar, S., Fiadeiro, J.L., Sheth, A.P. (eds.) BPM 2006. LNCS, vol. 4102, pp. 161–176. Springer, Heidelberg (2006). doi:10.1007/11841760_12
18. Zorzan, F.A., Riesco, D.: Transformation in QVT of software development process based on SPEM to workflows. IEEE Latin Am. Trans. **6**(7), 655–660 (2008)

A Multi-criteria Approach for Team Recommendation

Michael Arias[✉], Jorge Munoz-Gama, and Marcos Sepúlveda

Department of Computer Science, School of Engineering,
Pontificia Universidad Católica de Chile, Santiago, Chile
m.arias@uc.cl, {jmun,marcos}@ing.puc.cl

Abstract. Team recommendation is a key and little-explored aspect within the area of business process management. The efficiency with which the team is conformed may influence the success of the process execution. The formation of work teams is often done manually, without a comparative analysis based on multiple criteria between the individual performance of the resources and their collective performance in different teams. In this article, we present a multi-criteria framework to allocate work teams dynamically. The framework considers four elements: (i) a resource request characterization, (ii) historical information on the process execution and expertise information, (iii) different metrics which calculate the suitability of the work teams taking into account both individual performance as well as collective performance of the resources, and (iv) a recommender system based on the Best Position Algorithm (BPA2) to obtain a ranking for the recommended work teams. A software development process was used to test the usefulness of our approach.

Keywords: Team recommendation · Resource allocation · Process mining · Business processes · Recommender systems · Organizational perspective

1 Introduction

The task of allocating resources to activities within a business process is a key aspect in the area of Business Process Management (BPM) [12,13,27]. The efficiency with which this task is performed has a high relevance in terms of resources productivity, quality, performance and cost of the process. The use of information about the resources that take part of the process activities is recommended to improve the task of allocating resources, for example, to support the discovery of patterns related to resource assignment [23]. In the context of processes that require a collaborative environment among their activities, the correct selection of work teams represents a challenge for those in charge, as it requires the selection of those resources that working together provide the best performance to execute the activities that conform a process. Despite its relevance, the formation of work teams is often done manually, without a comparative analysis based on multiple criteria between the individual performance of the resources

© Springer International Publishing AG 2017
M. Dumas and M. Fantinato (Eds.): BPM 2016 Workshops, LNBIP 281, pp. 384–396, 2017.
DOI: 10.1007/978-3-319-58457-7_28

and their collective performance in different teams. Seeking for a resolution to the problem of resource allocation, different methods have been proposed, including: association rules [13], decision trees [14], and Markov models [12]. Despite this, there is a lack of approaches that bring adequate automatic support to compose and allocate activities to work teams within the area of BPM [8]. From the literature, it is possible to identify different criteria which are used to evaluate resources in the context of work teams, such as: resource availability [16], abilities required to perform a task [11,24], task complexity [24], effort estimation [19], and sociometric techniques [4]. In [24], a genetic algorithm is proposed to solve the problem of team formation, including practical considerations: precedence among tasks, role, and competence level of the resource. In [5], an optimization-based approach to support staffing in a software project is introduced, where information related to the activities, the available human resources, and a set of restrictions established by the organization (e.g., schedule deadline, project budget, maximum allocation) are considered. Kumar et al. [15] present a model that measures compatibility among resources when assigning work items to collaborative groups. In [6], they focus on team formation within the context of agile software development methodologies, specifying rules to choose the best developers for a given project. Recently, in [17] a method to improve resource allocation in workflow management systems by computing the social relation among two resources was proposed. This approach only considers the time perspective. Cabanillas et al. [8] present an approach to select teams proposing a language named RAL-Team to describe the work teams, which is extended to establish selection conditions verified at the time of determining the resources that are assigned to the different activities of the business process. This approach does not take into account the historical information about past process executions, neither support the team composition at run-time. From the literature revised, we detected some limitations. First, there is a lack of methods that support the formation of teams at run-time. Second, there is a lack of mechanisms that support decision making through the recommendation of suitable teams, in which the historic performance of the resources, both individually and collectively, is analyzed. Third, the resources are always allocated considering one allocation unit: the activity level. Fourth, there is a need to count with multi-criteria approaches that are scalable and user-oriented, where the criteria used to determine the most suitable allocation can be extended in a faster and easier manner for the decision maker.

In this work, we extend the Framework for Recommending Resource Allocation based on Process Mining presented in [3]. We approach resource allocation considering work team recommendation. A role is a group of resources that are interchangeable in the sense that any member of the group can perform a given set of activities. It is also called a resource class [10]. For simplicity, a usual assumption is that each activity can be performed by only a single role. On the other hand, a resource can have multiple roles, e.g., a software architect can also be the project leader. Usually, in a given process, several roles can be identified. A work team (hereinafter and indistinctly, a team) is a group of resources, each playing a specific role, that together can perform a process instance. This work presents a novel approach that recommends the most suitable teams to execute a request defined at run-time. We consider sub-processes as the target allocation unit; however, it can also be used to allocate resources at the activity level or at the process level, as a whole. Inspired by workflow resource patters [22], three allocation types are considered: capability-based allocation, history-based allocation and role-based allocation. We proposed a multi-criteria approach in which various metrics are evaluated to incorporate the evaluation of the individual performance of the resources and their group performance through different criteria: fitting between resources capabilities and the expertise required to perform an activity, past performance (frequency, duration, quality and cost), and the current workload of each resource. The metrics are combined to generate a final recommendation ranking using BPA2 [2]. We can formulate the process of generating the ranking as a problem of obtaining the k most relevant items in a multi-dimensional dataset. To accomplish this goal, we propose the use of BPA2 to answering top-k queries using lists of data items sorted by their local scores. Also, we report experimental obtained results.

As a running example we use a traditional software development process: the Waterfall life cycle [21]. It consists of five phases run sequentially: (i) Requirements analysis (role: analyst), (ii) System design (role: designer), (iii) Implementation or coding (role: developer), (iv) Testing (role: tester), and (v) Operation and maintenance (role: support agent). The Waterfall is used within the context of software engineering [25] for systems development, where each phase is conducted by a resource that performs a specific role. Periodically, companies that develop software require the selection of resources to form the team that will execute each phase, in order to generate a specific software solution.

This article is organized in the following manner: Sect. 2 gives the preliminaries that are necessary for the rest of the article. Section 3 describes the method used to evaluate and recommend work teams. Section 4 focuses on the experiments performed and discusses the results obtained when validating the proposed approach. Finally, Sect. 5 presents the conclusions and future work.

2   Preliminaries

In this section we present the elements that form part of the framework to perform the recommendation of the most suitable work teams.

Resource Request Characterization

As we established before, a team is a group of resources, each playing a specific role, that together can perform a process instance. To recommend resources to form a team, we must first define the properties each member of the work team must comply with to fulfil its role in a process instance.

Definition 1 (*Resource Request Characterization*). *A resource request characterization $c = (f_1, \ldots, f_n)$ is a multi-factor representation of the desired request properties, where f_i is an element of a finite set.*

We used at least one of the factors to express the need for a specific role. The two-factor characterization proposed is a factor-tuple $c = (role, type)$, where $role$ defines the organizational role for which the resource is being requested, and $type$ is the typology of the process execution instance that requests the resource. For example, in the Waterfall life cycle, a different $role$ is required for each phase (e.g., an analyst or a tester), and $type$ corresponds to the project type for which the resource is requested (e.g., create a website or a mobile app). To determine the phases of the process that could require a different role, we can consider the semantics of the process itself and then perform a decomposition manually. Alternatively, this decomposition could also be done through automated approaches, such as Passages [1], o Single-Entry Single-Exit (SESE) [18].

For each resource request characterization c, we use resource allocation information that includes: historical information of previous process executions, contextual information (e.g., expertise information available in corporate information systems), and weights describing the importance of each of the allocation metrics. Our framework returns a ranking of the most suitable resources to be allocated to each request.

Resource Allocation Metrics

For our approach, we adopt the use of the six dimensions introduced in [3], where historical and contextual information is combined to evaluate the resources through different metrics in accordance with the specified criteria (dimensions). The proposed dimensions are:

- **Frequency:** measures the rate of occurrence that a resource has completed the requested characterization.
- **Performance:** measures the execution time that a resource has achieved performing the requested characterization.
- **Quality:** measures the customer evaluation of the execution of the requested characterization performed by a resource.
- **Cost:** measures the execution cost of the requested characterization performed by a resource
- **Workload:** measures the actual idle level of a resource considering the characterizations executed at the time
- **Expertise:** measures the ability level at which a resource is able to execute a characterization

We introduce a resource process cube $\mathbb{Q}$ to manage the historical information, and the expertise matrices $\mathbb{E}_r$ and $\mathbb{E}_c$ representing the expertise information. By means of the cube and the matrices, the required knowledge is generated, with the purpose of determine the suitability of the resources that take part of the recommended team. The *cube* $\mathbb{Q}$ is a semantic representation of the historical information to be analyzed. Basic OLAP operations such as slice and dice [9], allow the extraction of specific information for each request characterization.

Definition 2 (*Resource Process Cube*). *Let r, c and d, be a resource, a resource request characterization, and a dimension, respectively. A resource* *process cube $\mathbb{Q}[r][c][d]$ represents all the historical information about the resource r and the characterization c necessary to analyze the dimension d. Similarly, for a given dimension d, $\mathbb{Q}[\ ][c][d]$ represents the historical information about all resources for the execution of the characterization c, and $\mathbb{Q}[r][\ ][d]$ represents the information for all the characterizations performed by the resource r.*

Due the nature of our approach, there is a separation between the conceptual level and how data are stored, in fact, the data could be stored in different ways, e.g., in a relational database, and then computing the metrics on demand, or by first precomputing some intermediate data, and then computing the metrics.

To illustrate the definition of the metrics for each dimension, we use the *performance* dimension. Let $\mathbb{Q}[r][c][p].avg$ be an operation that returns the average duration, considering only cases in which the resource r has taken part in executing the characterization c. Let $\mathbb{Q}[\ ][c][p].min$ and $\mathbb{Q}[\ ][c][p].max$ be the minimum and maximum duration for executing the characterization c by any resource. The performance metric is defined as follows in (1):

$$Performance\_Metric(r,c) = \frac{\mathbb{Q}[\ ][c][p].max - \mathbb{Q}[r][c][p].avg}{\mathbb{Q}[\ ][c][p].max - \mathbb{Q}[\ ][c][p].min} \qquad (1)$$

Following the software development process example, if we have a characterization $c_1 = (Analyst, website)$ and a resource $r_2 = Bob$, $\mathbb{Q}[r_2][c_1][p]$ allows us to access information related to the performance dimension of resource r_2 when executing the characterization c_1. $\mathbb{Q}[r_2][c_1][p].min$ and $\mathbb{Q}[r_2][c_1][p].max$ are the minimum and maximum time *Bob* needed to execute c_1. Meanwhile, $\mathbb{Q}[r_2][c_1][p].avg$ is the average time required by *Bob* to execute c_1. In the same way, $\mathbb{Q}[\ ][c_1][p].min$ and $\mathbb{Q}[\ ][c_1][p].max$ are the minimum and maximum time required considering all resources. For example, be 54 and 120 the minimum and maximum time needed to perform the characterization c_1, and 59 the average duration of *Bob*, the *Performance_Metric(r,c)* = 0,92.

At the same time, the *expertise matrices* $\mathbb{E}_r[r][e]$ and $\mathbb{E}_c[c][e]$ allow us to represent the expertise e of a resource r and the desired level of expertise required to execute a characterization c, respectively. We represent the expertise information considering the Human Resource Meta-Model (HRMM) [20], where the expertise can be organized by competencies, skills, and knowledge. To evaluate how a resource fits with the expertise required to perform a given characterization, we compare the value of each level of expertise $\mathbb{E}_r$ with the corresponding value in $\mathbb{E}_c$. This comparison allows determining how qualified a resource r is, according to an under-qualification and an over-qualification metric. For example, given the characterization $c_1 = (Analysis, website)$ the desired expertise to perform this characterization is a mid-high level on problem solving, business analysis, and communication skills, represented as [2, 2, 2], meanwhile, the resource $r1$ has a low level in the above aspects, represented as [0, 1, 0]. If the expertise of a resource $r1$ entirely match with the required expertise, the value for both expertise metrics will be 1. Similarly, we established metrics for the other dimensions. For further detail about the definition of all the different metrics used, reader can refer to the previous publication [3]. From now on, we will refer to the metrics in a generic way as $metric_j(\mathbb{Q}(L), r, c)$, being L the event log used to compute the cube.

Best Position Algorithm Problem

The challenge of recommending a group of resources to each request can be formulated as a problem of finding the best k resources in a group of ordered lists, where each list corresponds to one of the metrics considered. With the intention of giving a recommendation ranking, we were inspired in the use of portfolio-based algorithm selection [26] as a strategy to obtain the k most relevant items in a given group of data (top-k technique) considering multiple criteria. Akbarinia et al. [2] proposed two algorithms (BPA and BPA2) to process the top-k queries starting from the ordered lists of data in accordance to each criterion considered. BPA2 is a suitable algorithm to accomplish the goal of obtaining the k most relevant items in a multi-dimensional dataset, because it requires a reduced number of data accesses, a lower query response time and execution costs, with respect to other top-k query processing algorithms, such as BPA and the Threshold Algorithm (TA), due its threshold management and early stop condition.

3   Team Recommendation

To develop a team recommendation, we consider the elements presented in Sect. 2. Figure 1 shows the general framework proposed. This framework helps to specify a request at run-time, as well as weights that describe the level of importance of each criterion established by the user. The recommendation takes account of contextual and historical information. The defined metrics enable the team resources to be evaluated in accordance with the specified criteria. Notice that in real-life scenarios, the best resources, evaluated individually, are not always those who compose the best teams collectively. This may occur in the composition of teams in software development projects or consulting projects, emergency medical teams, among others. Therefore, we created a new method of evaluation to propose suitable work teams, considering both individual and group behavior, which we will present in this section.

To obtain the recommendation, we evaluate each team based on their history, expertise, and availability, considering: (a) the individual behavior of each resource that forms the team, independent of the teams it had belong to, (b) the

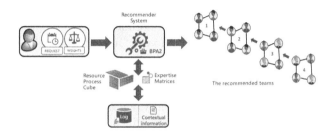

Fig. 1. General framework for team resource recommendation

behavior of each resource when it has previously formed part of the team that is being evaluated. Later, the BPA2 algorithm is used to recommend the most suitable teams. In the following, we will explain how each step is performed.

Team Filtering: A team is a group of resources that together can perform a process instance, each taking care of a specific resource request characterization (an organizational role for a specific type of process instance). To consider the history of a particular team, we extract from the event log those equivalent process instances that they have performed together in the past.

Definition 3 (Team Filtering). *Let L, tc, tr, be an event log, a tuple of resource request characterizations, and a tuple of resources, respectively. Team Filtering is a function $team\_filter(L, tc, tr) = L_{team}$ that returns an event log that contain all events of the log L that belong to those traces that contain only the characterizations tc, and where each resource request characterization in tc has been performed by the corresponding resource in tr.*

We will refer to $\mathbb{Q}(L)$ as the resource process cube created based on the historical information contained in an event log L. Moreover, $\mathbb{Q}(team\_filter(L, tc, tr))$ symbolizes the resource process cube that represents the historical information where the team tr has worked together performing different occurrences of the resource request characterizations tc associated to the process registered in L.

Individual Assessment: To determine the individual evaluation of a resource, independent of the teams it belonged, we used the *cube* $\mathbb{Q}(L)$, which represents all the historical information contained in L. Based on $\mathbb{Q}(L)$ and the expertise matrices, the different metrics are calculated; notice that all metrics are normalized between 0 and 1. For each metric, an ordered list which evaluates the behavior of each resource of the team in that dimension is created. Each metric is weighted by previously specified weights, e.g., more importance could be given to the dimensions of frequency and quality, above expertise and cost. For example, *Bob*, *John*, and *Ana* are software analyst. After performing the individual assessment, their evaluations are $Bob = 0.549$, $John = 0.525$, and $Ana = 0.514$.

Group Assessment: To determine the group evaluation of the resources that belong to a determined team, we consider the same metrics that are used in the individual evaluation, but restricting the information to those process instances in which the same team participated previously. To that end, we use the resource process cube $\mathbb{Q}(team\_filter(L, tc, tr))$, which only contains information related to process instances of similar characteristics in which the team worked together in order to execute them. Similar to the individual case, a group of lists for each metric is created, each list containing the value of each resource in the team, and weighting each metric by previously specified weights. For example, in the software development process, *Bob(analyst)*, *Teo(designer)*, *Dave(developer)*, and *Alice(tester)* are the best resources evaluated individually in each life cycle phase, respectively. However, if we consider the collective performance when the resources have worked together as a team in the past, the best evaluated resources would be: *John*, *Pete*, *Sean*, and *Sara*.

Team Recommendation: We propose a recommender system that uses a method based on BPA2 to find the most relevant k-items of a multi-dimensional dataset, i.e., to obtain a ranking of the most suitable teams to be recommended. BPA2 implicitly calculates the overall score for each data item, maintaining in a temporal set the k-data items with the highest overall score. The algorithm allows an iterative approach to access and evaluate the resources based on their local score and their position in each list. If at the same point the temporal set contains k-data items whose overall scores are higher than or equal to a generated threshold, then there is no need to continue scanning the rest of the lists. With BPA2, no position in a list is accessed by the algorithm more than once, which prevents re-accessing data items considering sorted or random access. The output of the algorithm is an ordered list, where the final score for each resource is stored. The first value represents the tuple of resources with the highest overall score and therefore the best recommendation. More details about BPA2 can be found in [2].

Definition 4 (*Team Resource Recommendation*). *Let **tc** be a tuple of n resource request characterizations that must be performed and the Cartesian product $T^{TR} = R^1 \times \cdots \times R^n$ be the set of all n-tuples of resources that can be allocated to the requests in **tc**. A team resource recommendation will propose a $top - k$ set of tuples of resources $TR' \subseteq T^{TR}$, such that $|TR'| = k$ and $\forall tr' \in TR'$ and $\forall tr'' \in T^{TR} \setminus TR'$ score(**tr'**, **tc**) $>=$ score(**tr''**, **tc**). Each tuple **tr** $\in TR'$ represents the allocation of a resource $tr_i \in$ **tr** to a characterization $tc_i \in$ **tc**, $\forall i = 1, \ldots, n$.*

A method based on BPA2 is used to solve this problem. Here, the elements of the lists are defined for each tuple **tr** $\in T^{TR}$. The lists considered are:

- Lists that measure the individual behavior of the resources: $IL_{1,1}, \ldots, IL_{1,m}$, $\ldots, IL_{i,j}, \ldots, IL_{n,m}$, being $i = 1, \ldots, n$ the different resource characteristics considered, and $j = 1, \ldots, m$ the different metrics considered. $IL_{i,j}$ contains $|R^1| \cdot \cdots \cdot |R^n|$ pairs of the form $(\mathbf{tr}, metric_j(\mathbb{Q}(L), tr_i, tc_i) \cdot w_j \cdot ind\_w_j)$ where **tr** $\in T^{TR}$, and w_j is the weight given to the metric j. $IL_{i,j}$ is sorted in descending order. The value of the metrics are computed based on the resource process cube $\mathbb{Q}(L)$.
- Lists that measure the behavior of the resources on a specific team: $GL_{1,1}$, $\ldots, GL_{1,m}, \ldots, GL_{i,j}, \ldots, GL_{n,m}$, being $i = 1, \ldots, n$ the different resource characteristics considered, and $j = 1, \ldots, m$ the different metrics considered. $GL_{i,j}$ contains $|R^1| \cdot \cdots \cdot |R^n|$ pairs of the form:
  $(\mathbf{tr}, metric_j(\mathbb{Q}(team\_filter(L, \mathbf{tr}, \mathbf{tc})), tr_i, tc_i) \cdot w_j \cdot grp\_w_j)$, where **tr** $\in T^{TR}$, and w_j is the weight given to the metric j. $GL_{i,j}$ is sorted in descending order. The value of the metrics are computed based on the resource process

cube $\mathbb{Q}(team\_filter(L, \mathbf{tr}, \mathbf{tc}))$, i.e., the resource process cube that represents the information about those cases where the team $\mathbf{tr}$ has worked together performing the different occurrences of the resource request characterization $\mathbf{tc}$ associated to the process registered in the event log L.

Notice that the length of all lists is $|R^1| \cdot \cdots \cdot |R^n|$. This is a requirement for using BPA2. The specified weight for each metric w_j is decomposed so that part of the weight is related to the respective $IL_{*,j}$ list and the other part is related to the corresponding $GL_{*,j}$ list, such that $ind\_w_j + grp\_w_j = 100\%$. The overall score $score(\mathbf{tr}, \mathbf{tc})$ is the weighted value of the different m-metrics obtained by the n-resources tuple $\mathbf{tr}$ when assigned to the characterization $\mathbf{tc}$ (2):

$$
\begin{aligned}
score(\mathbf{tr}, \mathbf{tc}) = {} & \textstyle\sum_{i=1}^{n} \sum_{j=1}^{m} metric_j(\mathbb{Q}(L), tr_i, tc_i) \cdot w_j \cdot ind\_w_j \\
& + \textstyle\sum_{i=1}^{n} \sum_{j=1}^{m} metric_j(\mathbb{Q}(team\_filter(L, \mathbf{tc}, \mathbf{tr})), tr_i, tc_i) \cdot w_j \cdot grp\_w_j
\end{aligned}
\tag{2}
$$

4   Empirical Evaluation

We adapted our evaluation considering the Waterfall life cycle as a team allocation scenario. Below, we present the dataset and the configuration of the experiments, and the discussion of the obtained results.

Datasets and Experimental Setup: For our experiments, we considered a synthetic event log with 1,000 cases. We also considered the following attributes related to the software development process: Case ID, Role, Typology, Resource, Overall quality, Incurred cost by phase, Phase creation date, Phase closing date. We created the matrices that describe the expertise required to execute each phase and the level of expertise of each resource, and an array that describes the effective availability of each resource at the moment of making the recommendation. An extension to the OpenXES library is used to standardize and represent the historical and contextual information [1]. We used 20 resources, who perform the role of analysts, designers, developers, and testers. We reproduced three different scenarios in which it was necessary to perform a team recommendation for projects developed according to the Waterfall model. For the resource request characterization, we considered as allocation unit the first four phases of the life cycle. We focused on a single project typology: mobile application development. Our evaluation involves the following experiments:

(a) Calculate the top-3 teams. Weights were specified so as to give more importance to the individual behavior of the resources through the different phases.

(b) We used a similar scenario to the one described in experiment (a), but giving more importance to the group behavior of the resources when they work together through the project phases.

---

[1] https://svn.win.tue.nl/trac/prom/browser/Packages/ResourceRecommendation/ Trunk/src/org/processmining/resourcerecommendation/utils/resrecxes.

(c) Personalized scenario were created, where weights were specified in accordance to the project being developed.

Discussion: To perform the recommendation, we evaluate the resources based on the defined metrics. We specified different weights w_j in accordance to the level of importance given to each $metric_j$. The weight is decomposed, so that a part of the weight is related to each individual assessment list $IL_{*,j}$ and the other part to each group assessment list $GL_{*,j}$. These weights are applied to the metric score for each list of each metric. We used the BPA2 algorithm to find the resource tuples that have the highest overall score. BPA2 produces an ordered list, wherein the final score for each tuple of resources is stored. The top-k tuples represent the most suitable teams to be recommended. For the three experiments, we specified the following weights for the metrics: performance: 20 - quality: 50 - cost: 20 - other dimensions: 10. Particularly, an individual weight ind_w_j of 90% and a group weight grp_w_j of 10% were specified in the experiment (a); 10% individual and 90% group weights for the experiment (b); and 80% individual and 20% group weights for the experiment (c). When creating the event log, we simulated the existence of one resource whose individual evaluation was better when compared to the other resources that participated in the same phase, i.e., there is a resource that is individually better executing each one of the phases. In a similar manner, the existence of teams that are better working together in comparison to other teams was simulated.

Table 1. Experiments results

Exp.	Recommended Team	Phase 1 Analysis	Phase 2 Design	Phase 3 Coding	Phase 4 Testing	Overall score
(a)	Top-**1**:(R1,R5,R9,R12)	0.549	0.539	0.522	0.541	**0.538**
	Top-**2**:(R1,R4,R9,R12)	0.549	0.535	0.522	0.541	**0.537**
	Top-**3**:(R1,R5,R9,R10)	0.549	0.539	0.522	0.536	**0.536**
(b)	Top-**1**:(R2,R4,R8,R12)	0.525	0.535	0.512	0.541	**0.610**
	Top-**2**:(R1,R4,R9,R10)	0.549	0.535	0.522	0.536	**0.590**
	Top-**3**:(R1,R6,R7,R10)	0.549	0.509	0.511	0.536	**0.573**
(c)	Top-**1**:(R1,R4,R9,R10)	0.525	0.535	0.522	0.536	**0.547**
	Top-**2**:(R1,R5,R9,R12)	0.549	0.539	0.522	0.541	**0.544**
	Top-**3**:(R1,R5,R8,R10)	0.549	0.539	0.512	0.536	**0.538**

(*) Phase scores correspond to the individual score of the resources allocated to each phase.

Table 1 shows the results obtained in the experiments. Experiment (a) shows that the resources defined as the best individually are effectively the recommended team (top-1) according to the ranking. In this particular case, the most suitable resources are: the analyst R1, the designer R5, the developer R9, and the tester R12. For experiment (b), we focused on forming the teams considering how well the resources had worked collectively in the past. The results are the expected ones, according to the construction of the experiments. After performing the experiment, the results recommended as the most suitable team the resources: R2, R4, R8 and R12, which represents the group of resources with the best performance working together in comparison to the rest of the teams. In experiment (c) we adjusted the given weights in accordance to specific priorities. The results obtained are different to those generated in experiment (a) and (b), which proves that our approach produces the recommendation based on the characterization of each request, generating diverse results that adapt to specific business contexts. R1, R4, R9 and R10 form the team that best adjusts to the specified priorities. Note that in each experiment a ranking is recommended with the resource tuples to be allocated in accordance to the top-k queries defined. Thus, the person in charge of making the allocation has prioritized team alternatives to allocate, with the possibility of including, if necessary, other subjective criteria not considered as part of the framework. The applicability of our approach can be supported through today corporate information systems (e.g., ERP SAP), which allow the storage and extraction of historical contextual information as well. For usual scenarios where the overall process can be decomposed into disjoint subsets (e.g., activities, sub-process, or phases) our approach is able to generate a ranking of feasible work teams. For more adaptive and complicated scenarios (e.g., agile software development), which might involve a overlapping, incremental, or iterative phases, our framework needs to be adapted.

5   Conclusions and Future Work

We extended our Framework for Recommending Resource Allocation based on Process Mining. The main contributions of this work are as follows. First, we extended our previous framework from recommending single resource rankings to recommending teams that work better collaboratively. Second, our framework is focused on improving decision making, helping the person in charge of forming teams to optimize the use of available human resources. Third, we based the team recommendation on the evaluation of multiple criteria that measure the past behavior of the resources, their expertise and current workload. Moreover, our strategy to evaluate teams combines the evaluation of the individual behavior of the resources and their behavior when they collaborate working together as a team. Fourth, due to the flexibility of this approach, the team recommendation can be executed considering any allocation unit, e.g., at the activity level, at the sub-process level, or at the whole process level, allowing it to be adapted and used in different abstraction levels. For usual team recommendation scenarios, the proposed approach generates the recommendation of the most suitable work teams based on the definition of a resource allocation request at run-time, and the use of contextual and historical information. However, for more complex scenarios, our framework needs to be adapted in order to allocate teams dynamically. Future work consider the creation of heuristics that enable determining faster those teams more likely to be recommended in accordance to the given characterizations, in order to optimize the process for computing the results. Also, we will examine other criteria to measure collaborative work which could be considered as part of our approach, as well as integrate information about the responsibility degree for each activity [7]. We have evaluated our approach over a proof-of-concept implementation. We aim to evaluate the effectiveness and efficiency of our technique through case studies using real-life event logs.

Acknowledgments. This project was partially funded by the Ph.D. Scholarship Program of CONICYT Chile (Doctorado Nacional/2014-63140181), Universidad de Costa Rica and by Fondecyt (Chile) Project No.1150365.

References

1. van der Aalst, W.M.P., Verbeek, H.M.W.: Process discovery and conformance checking using passages. Fundam. Inform. **131**(1), 103–138 (2014)
2. Akbarinia, R., Pacitti, E., Valduriez, P.: Best position algorithms for efficient top-k query processing. Inf. Syst. **36**(6), 973–989 (2011)
3. Arias, M., Rojas, E., Munoz-Gama, J., Sepúlveda, M.: A framework for recommending resource allocation based on process mining. In: Reichert, M., Reijers, H.A. (eds.) BPM 2015. LNBIP, vol. 256, pp. 458–470. Springer, Cham (2016). doi:10.1007/978-3-319-42887-1_37
4. Ballesteros-Pérez, P., González-Cruz, M.C., Fernández-Diego, M.: Human resource allocation management in multiple projects using sociometric techniques. Intl. J. Project Manage. **30**(8), 901–913 (2012)
5. Barreto, A., de Oliveira Barros, M., Werner, C.M.L.: Staffing a software project a constraint satisfaction and optimization-based approach. Comput. OR **35**(10), 3073–3089 (2008)
6. Britto, R., de Alcântara dos Santos Neto, P., Rabelo, R.A.L., Ayala, W., Soares, T.: A hybrid approach to solve the agile team allocation problem. In: Proceedings of the IEEE Congress on Evolutionary Computation, CEC, pp. 1–8 (2012)
7. Cabanillas, C., Resinas, M., Ruiz-Cortés, A.: Automated resource assignment in BPMN models using RACI matrices. In: Meersman, R., et al. (eds.) OTM 2012. LNCS, vol. 7565, pp. 56–73. Springer, Heidelberg (2012). doi:10.1007/978-3-642-33606-5_5
8. Cabanillas, C., Resinas, M., Mendling, J., Cortés, A.R.: Automated team selection and compliance checking in business processes. In: Proceedings of the 2015 International Conference on Software and System Process, ICSSP, pp. 42–51 (2015)
9. Chaudhuri, S., Dayal, U.: An overview of data warehousing and olap technology. ACM Sigmod Rec. **26**(1), 65–74 (1997)
10. Dumas, M., La Rosa, M., Mendling, J., Reijers, H.A.: Fundamentals of Business Process Management. Springer, Heidelberg (2013)
11. Gerogiannis, V.C., Rapti, E., Karageorgos, A., Fitsilis, P.: Human resource assessment in software development projects using fuzzy linguistic 2-tuples. In: Artificial Intelligence, Modelling and Simulation (AIMS), pp. 217–222. IEEE (2014)
12. Huang, Z., van der Aalst, W.M.P., Lu, X., Duan, H.: Reinforcement learning based resource allocation in business process management. DKE **70**(1), 127–145 (2011)
13. Huang, Z., Lu, X., Duan, H.: Mining association rules to support resource allocation in business process management. Expert Syst. Appl. **38**(8), 9483–9490 (2011)

14. Kim, A., Obregon, J., Jung, J.-Y.: Constructing decision trees from process logs for performer recommendation. In: Lohmann, N., Song, M., Wohed, P. (eds.) BPM 2013. LNBIP, vol. 171, pp. 224–236. Springer, Cham (2014). doi:10.1007/978-3-319-06257-0_18

15. Kumar, A., Dijkman, R., Song, M.: Optimal resource assignment in workflows for maximizing cooperation. In: Daniel, F., Wang, J., Weber, B. (eds.) BPM 2013. LNCS, vol. 8094, pp. 235–250. Springer, Heidelberg (2013). doi:10.1007/978-3-642-40176-3_20

16. Li, C., Akker, J.M., Brinkkemper, S., Diepen, G.: Integrated requirement selection and scheduling for the release planning of a software product. In: Sawyer, P., Paech, B., Heymans, P. (eds.) REFSQ 2007. LNCS, vol. 4542, pp. 93–108. Springer, Heidelberg (2007). doi:10.1007/978-3-540-73031-6_7

17. Liu, X., Chen, J., Ji, Y., Yu, Y.: Q-learning algorithm for task allocation based on social relation. In: Cao, J., Wen, L., Liu, X. (eds.) PAS 2014. CCIS, vol. 495, pp. 49–58. Springer, Heidelberg (2015). doi:10.1007/978-3-662-46170-9_5

18. Munoz-Gama, J., Carmona, J., van der Aalst, W.M.P.: Single-entry single-exit decomposed conformance checking. Inf. Syst. **46**, 102–122 (2014)

19. Narendra, N.C., Ponnalagu, K., Zhou, N., Gifford, W.M.: Towards a formal model for optimal task-site allocation and effort estimation in global software development. In: 2012 Annual SRII Global Conference, pp. 470–477 (2012)

20. Oberweis, A., Schuster, T.: A meta-model based approach to the description of resources and skills. In: AMCIS, p. 383 (2010)

21. Royce, W.W.: Managing the development of large software systems. In: proceedings of IEEE WESCON, vol. 26, pp. 1–9 (1970)

22. Russell, N., Aalst, W.M.P., Hofstede, A.H.M., Edmond, D.: Workflow resource patterns: identification, representation and tool support. In: Pastor, O., Falcão e Cunha, J. (eds.) CAiSE 2005. LNCS, vol. 3520, pp. 216–232. Springer, Heidelberg (2005). doi:10.1007/11431855_16

23. Schönig, S., Cabanillas, C., Jablonski, S., Mendling, J.: A framework for efficiently mining the organisational perspective of business processes. DSSs **89**, 87–97 (2016)

24. e Silva, L., Costa, A.P.: Decision model for allocating human resources in information system projects. Intl. J. Proj. Manage. **31**(1), 100–108 (2013)

25. Sommerville, I.: Software Engineering. Pearson, London (2015)

26. Xu, L., Hutter, F., Hoos, H.H., Leyton-Brown, K.: Satzilla: Portfolio-based algorithm selection for SAT. CoRR abs/1111.2249 (2011)

27. Zhao, W., Zhao, X.: Process mining from the organizational perspective. In: Wen, Z., Li, T. (eds.) Foundations of Intelligent Systems. AISC, vol. 277, pp. 701–708. Springer, Heidelberg (2014). doi:10.1007/978-3-642-54924-3_66

First Workshop Sustainability-Aware Business Process Management (SA-BPM'2016)

Introduction to the Workshop Sustainability-Aware Business Process Management (SA-BPM'2016)

Stefanie Betz[1(✉)], Leticia Duboc[2],
and Andréa Magalhães Magdaleno[3,4]

[1] Karlsruhe Institute of Technology, Karlsruhe, Germany
stefanie.betz@kit.edu
[2] Instituto de Matemática e Estatística,
Universidade do Estado do Rio de Janeiro (UERJ),
Rio de Janeiro, Brazil
leticia@ime.uerj.br
[3] UFF – Fluminense Federal University, IC – Computing Institute,
24210-346 Niterói, RJ, Brazil
andrea@ic.uff.br
[4] EY Brazil Consulting, Rio de Janeiro, Brazil
http://www.ey.com/br/pt/home

Abstract. The goal of this workshop is to provide a forum for researchers and professionals interested in sustainability and business process management to discuss and exchange ideas. During its first edition, the workshop achieved good results, as three research and one position paper have been selected. Moreover, during the workshop participants were divided in groups to discuss the challenges raised by the selected position paper and propose solutions.

Keywords: Sustainability · Sustainable business processes · Green business processes · Green BPM · Sustainable BPM

1 Introduction

Sustainability is, fundamentally, the capacity to endure [1], and it often refers to a set of dimensions, being the basic ones the human, environmental, societal and economic [2]. As a society, our capacity to endure is increasingly becoming a major concern that has lead to a myriads of efforts for achieving the sustainable development; these vary from simple individual behavioural changes, to national and international policies, to technological solutions, among others. This change in mentality is no different in the business world. CEOs are increasingly aware of the importance of their role in the sustainable development [3] and advantage that adopting sustainability practices can bring to their businesses, including new business opportunities and higher profits [4].

While technology can enable the offer of a variety of sustainable products and services, sustainability in businesses cannot be achieved with technological solutions alone. Truly sustainable businesses must adopt a holistic approach, which covers from their values to the implementation of their day-to-day operations. Business Process

Management (BPM) can support business in developing these embracing sustainable solutions, since it uses technology to design, implement, execute and monitor business process. This realisation has lead to an increasing interest on sustainability within the BPM practice and research [5]. A remarkable example is the growing field of Green BPM, which pays special attention to the impact of businesses in the environment [6], but often ignores the other dimensions of sustainability [5]. Yet, even Green BPM is still in early stages as a research field [7].

This workshop provides a forum for researchers and professionals interested in sustainability and BPM to discuss and exchange for incorporating sustainability into business process management. Four papers have been selected to base this discussion. Larsch et al. present an approach composed of model for to integrating sustainability into business process management and a catalogue that helps to identify potential improvements in the existing businesses operations. Ahlers and Krogstie present the results of a case study for compiling and reporting greenhouse gas (GHG) data from the Carbon Trace and Track (CCT) project. In particular, the authors describe a number of open challenges in the GHG domain that can potentially be supported by BPM. Lübbecke, Fettke and Loos propose a set of 26 Ecological Process Patterns for Green BPM, based on a qualitative study of 24 processes models for the registration, notification of changes and deregistration of German businesses. Finally, Magdaleno, Duboc and Betz, propose a research agenda for extending the modelling phase of the BPM lifecycle to support all the dimensions of sustainability. Based on this last paper, workshop participants will jointly develop a more comprehensive research agenda for Sustainability-Aware Business Process Management, which we hope to inspire other BPM researchers to embrace the sustainability cause.

2 Workshop Results

Based on the seminal position paper, the initial proposed challenges regarding Sustainability and Business Processes were:

- CH1: Definition of a conceptual model of the relations between sustainability and business processes;
- CH2: Development of a sustainable BPM Language (BPML);
- CH3: Production of a catalogue of metrics;
- CH4: Process visualization.

During group work, some alternative paths or solutions were pointed.

Challenges CH2 and CH3 should consider the existence of a proposal for an environmental extension of Business Process Model & Notation (BPMN) 2.0 together with a methodology for its application. It was also implemented on ARIS and Bizagi tools. This work is based on four environmental dimensions (Energy, Waste, Water and Emissions to Air). Each dimension is composed by specific metrics (and have their own symbols). It is important to work with a small number of metrics. These metrics can be aligned with the sustainability dimensions. They are captured through a top-down approach as part of modelling effort. Metrics can also be represented in process model.

Regarding process visualization (CH4), it is a real concern because of the overload of new information in process models. Probably it can be solved through the use of a tool with different visualization filters or layers, where one can choose which kind of elements to see in the model or not. Therefore, a process analyst can view only the process model without any annotation from sustainability and a sustainability specialist can have the complete view of process, for instance.

In addition to the proposed challenges, some others raised during the discussions:

- CH5: Need of empirical research. Green BPM has been discussed for several years as an emergent concept. It is now time to move beyond mere conceptualization.
- CH6: Need of a systematic review to study the relevant literature in depth and discuss the alternative approaches and solutions.
- CH7: How to convince the companies to invest in sustainability? It remains being a main difficulty.

3 Workshop Co-organizers

Stefanie Betz	Karlsruher Institute for Technology
Leticia Duboc	Universidade do Estado do Rio de Janeiro (UERJ)
Andréa Magalhães Magdaleno	UFF – Fluminense Federal University and EY Consulting

4 Programm Committee

Agnes Koschmider	KIT Karlsruhe, Germany
Andreas Oberweis	KIT, Karlsruhe, Germany
Carina Alves	UFPE, Brazil
Christoph Becker	University of Toronto, Canada
Colin Venters	University of Hudderfield, UK
Juliana Jansen	IBM Research, Brazil
Leonardo Murta	UFF, Brazil
Markus Nüttgens	University of Hamburg, Germany
Martina Huber	University of Zürich, Switzerland
Michael Fellmann	University of Osnbrueck, Germany
Norbert Seyff	University of Zürich, Switzerland
Peter Loos	University of Saarland, Germany
Priscila Engiel	PUC-RIO, Brazil
Sedef Kocak	Ryerson University, Canada
Stefan Seidel	University of Liechtenstein, Liechtenstein
Victor Almeida	UFF and Petrobras, Brazil

Acknowledgments. This work was partially funded by EY Consulting and by the European Social Fund, and by the Ministry of Science, Research and the Arts Baden-Wuerttemberg.

References

1. Becker, C., Chitchyan, R., Duboc, L., Easterbrook, S., Penzenstadler, B., Seyff, N., Venters, C. C.: Sustainability design and software: the Karlskrona Manifesto. In: 2015 IEEE/ACM 37th IEEE International Conference on Software Engineering (ICSE), vol. 2, issue no. 16–24, pp. 467–476 (2015)
2. Goodland, R.: Sustainability: Human, Social, Economic and Environmental (2002)
3. Accenture and United Nations Global Compact: The UN Global Compact-Accenture Strategy CEO Study 2016 (2016). https://www.accenture.com/us-en/insight-un-global-compact-ceo-study. Accessed 25 Aug 2016
4. Saul, J.: Social Innovation, Inc.: 5 Strategies for Driving Business Growth through Social Change. Jossey-Bass, San Francisco (2010)
5. Stolze, C., et al.: Sustainability in business process management research – a literature review. In: Americas Conference on Information Systems (AMCIS), Seattle, EUA (2012)
6. Seidel, S., et al.: Green business process management. In: vom Brocke, J., et al. (eds.) Green Business Process Management: Towards the Sustainable Enterprise, pp. 3–13. Springer, Berlin (2012)
7. Houy, C., et al.: Advancing business process technology for humanity: opportunities and challenges of green BPM for sustainable business activities. In: vom Brocke, J., et al. (eds.) Green Business Process Management: Towards the Sustainable Enterprise, pp. 75–92. Springer, Berlin (2012)

Integrating Sustainability Aspects in Business Process Management

Selim Larsch[1], Stefanie Betz[1(✉)], Leticia Duboc[2,3], Andréa Magalhães Magdaleno[4], and Camilla Bomfim[2]

[1] Karlsruhe Institute of Technology, Karlsruhe, Germany
selim.larsch@gmail.com, stefanie.betz@kit.edu
[2] Instituto de Matemática e Estatística, Universidade do Estado do Rio de Janeiro (UERJ), Rio de Janeiro, Brazil
leticia@ime.uerj.br, camillajbomfim@gmail.com
[3] IC – Computer Institute, UFF – Fluminense Federal University, Niterói, RJ 24210-346, Brazil
[4] EY Brazil Consulting, São Paulo, Brazil
andrea@ic.uff.br
http://www.ey.com/br/pt/home

Abstract. Business process management is an approach to improve business processes continuously. While factors like cost, quality and time are usually considered, sustainability considerations often fade into the background. With this study, we try to support the improvement of business processes with regard to sustainability. We present a process model that describes an approach of how to integrate sustainability aspects into business process management. We also present a catalogue to support the identification of improvement potential. We execute the process model and the catalogue using an existing business process. Finally, we present a summary and outline shortly the limitations and opportunities for future work.

Keywords: Sustainability · Sustainable business processes · Green business processes · Business process improvement · Business process management

1 Introduction

While efficiency factors like cost, quality and time are often considered in Business Process Management (BPM), sustainability considerations often fade into the background. Also, a lot of organizations have problems to integrate sustainability in their daily business [1, 2]. To address this problem, there exist several different frameworks and approaches regarding sustainability (e.g. Greenhouse Gas Protocol (GHG) [3], The SIGMA Project (http://www.projectsigma.co.uk), The Global Reporting Initiative (https://www.globalreporting.org)). But, those approaches have one thing in common: they give an abstract view on this topic, they provide no specific guidance of how to integrate sustainability aspects into business process design to achieve a sustainable process. It is apparent, that the process of integrating sustainability aspects into business processes is a complex procedure. On the

© Springer International Publishing AG 2017
M. Dumas and M. Fantinato (Eds.): BPM 2016 Workshops, LNBIP 281, pp. 403–415, 2017.
DOI: 10.1007/978-3-319-58457-7_29

one side, there are multiple sustainability dimensions that have to be considered – environmental, social and economical – which again influence each other and may necessitate trade-offs to achieve the desired outcome. On the other side, it is really difficult to measure the sustainability of different activities and determine their impact on different dimensions over time. To address this problem, stakeholders and modelers need concrete information and examples of how to integrate sustainability aspects into process design. This paper deals with the question of how stakeholders/modelers can integrate sustainability aspects into BPM. We are presenting a detailed process model to integrate sustainability aspects into BPM that extends the existing business process lifecycle model proposed by Weske [11]. We are providing step-by-step guidance for the stakeholders how to identify, assess, and improve possible actions (e.g. activities) that increase sustainability of a given business process. Additionally, we are providing an initial set of possible example activities to increase sustainability of business processes by improving or replacing existing activities. These activities are gathered in a catalogue, which can be accessed online[1]. This paper is structured as follows: Sect. 2 gives background information and shows a brief overview of related work. Section 3 describes the proposed solution – the process model and the catalogue. The application of the process model in an example case and the catalogue are presented in Sect. 4. Finally, the paper ends with a conclusion and an outlook.

2   Background

Different definitions of the term sustainability can be found in literature. In this paper, we use one of the most cited definitions of sustainability by Brundtland, that can be found in the Report of the World Commission on Environment and Development: Our Common Future [14]. This definition refers to sustainability as sustainable development and defines it as follows: *"Sustainable development is development that meets the needs of the present without compromising the ability of future generations to meet their own needs"* [14]. According to the triple bottom line [15] sustainability can be divided in three dimensions: economic, social and environmental:

- Environmental Sustainability: The goal of environmental sustainability is to improve human welfare for current and future generations. The focus is the protection of natural resources like water, air and land by lowering the resource consumption and improving the resource recycling [15].
- Social Sustainability: Maintaining and improving health, security, equal opportunities and justice and preserving social communities.
- Economic Sustainability: The focus of the economic sustainability is to improve competitiveness without generating negative impact on the other dimensions [15].

In order to better understand and enable the integration of sustainability aspects into business processes we developed a conceptual model of sustainability (see Fig. 1). The model is based on [16]. The presented model serves as basis for our approach.

---

[1] http://zingtree.com/host.php?tree_id=293763198&style=buttons&show_breadcrumbs=1& persist_names=Restart&persist_node_ids=119&nopermalink=1.

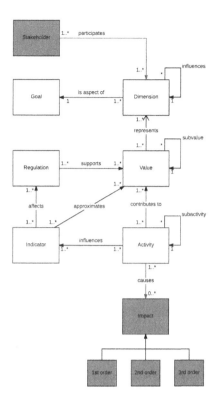

Fig. 1. Sustainability and business process meta-model based on [15]

The meta-model consists of eight elements: Goal, Dimension, Value, Indicator, Regulation, Activity, Stakeholder and Impact. The goal is the aim to be achieved regarding sustainability (development). A dimension describes an aspect of sustainability and is formed by values. Values are morals and support the strategic orientation of the company. Indicators can be qualitative or quantitative and control the compliance with values. Regulation supports values by setting limits for the indicator (e.g. emission regulations). Activities are measures taken to affect a value (e.g. use electronic receipt instead of paper receipt). We extended the original meta-model by two elements: Stakeholders and Impact. Stakeholders represent different dimensions with interest in maximizing the sustainability (development) level of their dimension. Activities can have impacts on the different sustainability dimensions. According to [11] these impacts are categorized in three levels:

- 1st order: 1st order impacts are direct impacts. They are the logical consequences of activities (e.g. energy usage).
- 2nd order: 2nd order impacts are indirect impacts. They are the consequence of the 1st order impacts and refer to changes on a micro- level (e.g. activities that are enabled through prior activities).
- 3rd order impacts: 3rd order impacts are socio-economic impacts. They refer to changes on a macro level and describe in-depth and structural changes.

Due to the nature of sustainability: its different dimensions and effects over time, sustainability "as such" can not be achieved, but several aspects can be (e.g. economic sustainability). Therefore, we are arguing that sustainability aspects need to integrated into business processes and not "sustainability".

The related work can be roughly subdivided in two parts: The first group uses sustainability indicators to measure the impact of specific activities (e.g. using CO2e) on sustainability (e.g. [5, 7]) and is working on the integration of sustainability aspects into business processes on a conceptual level (e.g. [6, 7]). The authors are focusing on modelling and monitoring of sustainability aspects in business processes. For example in [4] the authors conduct evaluation research applying sustainability performance measurement to a case study and discuss the outcome. In [5] the authors present a theoretical approach for making a process more sustainable using measurement process improvements derived from measurement of individual QoS. We think that these approaches a very interesting and metrics are a possible way to integrate sustainability aspects in business processes but, it is not possible to find metrics for all sustainability aspects and to measure their effects directly. [7] is presenting a slightly different approach to the mentioned above by providing an extended BPMN model to measure the carbon footprint of an individual process. We think it is important integrate sustainability aspects into existing business process modelling languages by extending them. But, the existing approach is limited to direct environmental effects. To improve business process sustainability it has been suggested to use process patterns [8] or process snippets [9]. In [8] the idea is to find patterns that are replaceable, to ensure more sustainable processes. This is really interesting work but, these patterns are very generic and do not take the different sustainability effects into account. Nevertheless, for the future we think it would be very interesting to combine these patterns with our approach. In [9] the authors propose a set of best practice process snippets to support environmental sustainability. This work is not providing any best practice process snippets but is rather focusing on the technical aspect of the redesign. Next, to the work on improvement, also frameworks for business process reengineering with focus on sustainability have been suggested (e.g. [12, 13]). The mentioned approaches for business process improvement and reengineering lack an holistic approach by focusing mainly on the environmental dimension and by not taking the different effects (1st, 2nd, and 3rd) into account. To summarize, especially the part of identifying company-specific solutions to replace existing activities with more sustainable ones taking all the different sustainability dimensions and all the different effects over time into account is missing.

3   The Process Model

The process model was built to facilitate the integration of sustainability aspects in business processes and supports the user from the idea generation to the model draft. The process model consists of seven steps and guides the user from the idea generation, over the identification of the relevant activity to a result table, which can be used as a modeling basis. With help of this table, filled with the gathered information, the user can improve the business process regarding sustainability. Table 1 shows an example of the result table. Below the seven steps of the process model are shown:

1. Brainstorming to identify improvement potential. This step is supports by the catalogue, which provides possible example activity examples for sustainability development. The outcomes of this step are **ideas** and their sustainability **dimensions**.
2. Identification of compatible **activities** to implement the ideas.
3. Labeling of the identified activities according to their feasibility (**status**):
   – Already implemented
   – Implement immediately
   – Implement over longer period of time
   – Don't implement
4. Gathering of additional information:
   – **Stakeholders:** Who is affected?
   – **Goals:** Why do we want to implement this idea?
   – **Activities:** How can we implement this idea?
   – **Business rules:** What rules have to receive attention?
   – **Indicators:** How can we measure the successful implementation?
   – **Impacts:** What impact does the idea have on the sustainability dimensions?
   – **Complications:** Which problems are expected to occur?
5. After every step, enter the gathered information in the "result table". The result table is the modeling basis.
6. Modify the process: Add, delete or change activities according to the result Table
7. Highlight the changed elements in the business process.

Table 1. Output of the executed process steps

Areas	Output information
Idea	Send coupons to customers
Process	Delivery process
Status	Implement immediately
Dimension	Economic
Goal	Increase rebuy rate
Stakeholder	Customers, marketing
Business rule	Respect economic viability
Indicator	Number of rebuys
Impact	Increased turnover, stable customer base
Complication	None

3.1 Integration of the Process Model into the Business Process Lifecycle

The process model can be integrated into the business process lifecycle presented by [10] as shown in Fig. 2. The business process lifecycle depicts the interaction of business process management activities. The activities are grouped into in four iteratively executed phases. In the design and analysis phase, business process are identified, renewed, validated and represented. Our process model is integrated into the design and analysis phase. After running through the seven steps of our process model, the configuration phase follows. In this phase, the improved business processes can be integrated into the process landscape. After finishing the configuration phase, we enter the enactment phase. Here, the processes are executed and monitored. Last phase of the business process lifecycle is the evaluation phase, in which the gathered data of the monitoring will be analyzed. After this step, the cycle can be ended or restarted (Continuous Process Improvement).

3.2 The Catalogue

The goal of the catalogue is to support the first step of the process model by providing (generalized) ideas to enhance the sustainability of business processes by improving or replacing activities. This idea arose during a discussion of the authors' on how to improve existing business processes concerning sustainability. We took two delivery companies and with the aim to make their business processes more sustainable, we performed several brainstorming sessions (according to step 1 of the process model). The brainstorming sessions (four all in all) have been executed with at least two authors' present. Additionally, two of the authors are or have been actually direct stakeholders of the business processes. The brainstorming was structured along the three dimensions

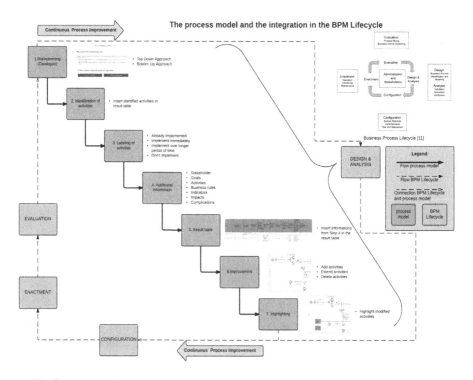

Fig. 2. Integrated process model for sustainability aware business process management

of sustainability. We ended up with a lot of ideas, which may enable more sustainable business and identified different possibilities to enhance the process sustainability. However, our results were confusing and a lot of the ideas were redundant. The redundancy happened due to the fact that the dimensions are interconnected and influence each other, thus some of the generated ideas have an impact on more than one sustainability dimension. This led us to the idea, to generate a catalogue that supports the process of brainstorming. The catalogue is structured as a decision tree. The idea is to remove the redundancy using a decision tree. Ideas/activities (modeled as vertices) that contribute to multiple dimensions can be connected to the associated dimensions through edges. When using the catalogue, the user can choose between two approaches, top-down or bottom up. Using the top-down approach, the user determines the sustainability dimension he wants to improve (e.g. social sustainability). After selecting the preferred dimension user has to confine his selection. When selecting the bottom-up approach, the user starts with a process type (e.g. delivery process) instead of a dimension. After selecting a process type, the user has to choose his preferred activity domain (e.g. customer privacy), in which he wants to improve the process. After confining the selection according to the preferred goal, both approaches (top-down or bottom-up) result in an overview with activities, which may be suitable in the specific case, as well as other relevant information that needs to be considered to improve the process (Fig. 3).

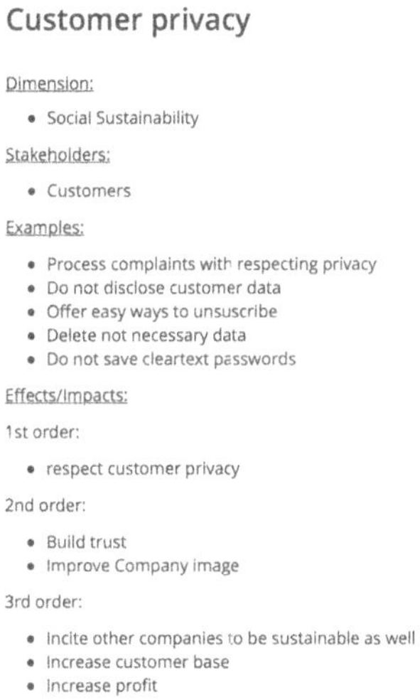

Customer privacy

Dimension:
- Social Sustainability

Stakeholders:
- Customers

Examples:
- Process complaints with respecting privacy
- Do not disclose customer data
- Offer easy ways to unsuscribe
- Delete not necessary data
- Do not save cleartext passwords

Effects/Impacts:

1st order:
- respect customer privacy

2nd order:
- Build trust
- Improve Company Image

3rd order:
- Incite other companies to be sustainable as well
- Increase customer base
- Increase profit

Fig. 3. Catalogue – result

The result page shows which dimension is affected, lists the stakeholders concerned, examples of activities, which help to achieve the chosen goal and impacts that may occur implementing those ideas. This result represents the basis for the following steps. With information provided by the catalogue, the user can start to improve existing business processes, get into direct contact with the relevant stakeholders and fill in the result table (step 5 of the process model). It should be noted that this catalogue is not comprehensive and the integrated ideas are limited. If the results provided by the catalogue don't satisfy the users requirements, the presented ideas can be used in brainstorming to generate more suitable solutions. For the future, we are planning to extend the catalogue further.

4 Example Application

In this section, we apply the process model and the catalogue presented in chapter 3 in an example process (Fig. 4) in order to check the usability and functionality. The process shown below depicts the delivery process of a bread delivery company in Rio de Janeiro (http://www.paozinhoemcasa.com.br).

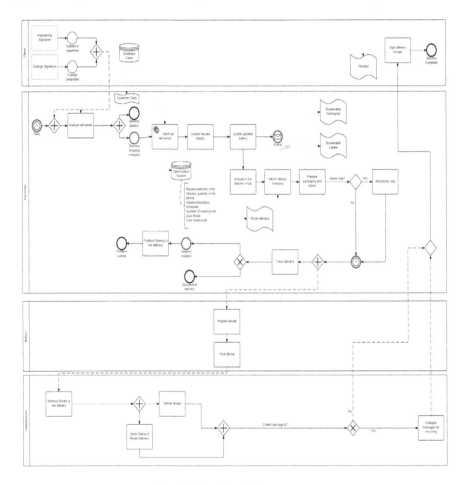

Fig. 4. Example - delivery process

We apply the process model on the business process shown in Fig. 4. This process portrays the delivery of bread ordered by a customer. On a daily basis, the company analyses and optimizes the deliveries for the day. After updating the required amount of bread needed, bakeries will be informed and start to bake the bread. The company provides the labels and the packaging. A delivery company is assigned to deliver the bread to the customers, picks off the ready bread at the bakery and delivers it to the customers. The customer signs the receipt and takes his bread. The first step of the process model is the idea generation phase. To support this step, we use the catalogue. As stated before, the user has two different paths to utilize the catalogue (top-down and bottom-up). In the following, the bottom-up approach will be executed.

Bottom-up approach: The bottom-up approach starts with the selection of the process type that shall be improved (e.g. delivery process, order process: see Fig. 5).

Sustainability Tree

Bottom up Approach

Please choose one process type from the list below to beginn

- Delivery process
- Order process
- Registration Process
- Expanding Process
- Advertisement/Marketing Process

Fig. 5. Bottom-up approach – process type selection

After selecting the type of the process, in this example delivery process, the user has to select an activity domain (e.g. Reduce air emissions) (Fig. 6).

Delivery process

Possible activity domains to support sustainability in Delivery Processes

- Assign carbon offset company
- Offer training
- Reduce air emissions
- Reduce land pollution
- Reduce procurement cost
- Reduce water pollution
- Reduce workplace accidents

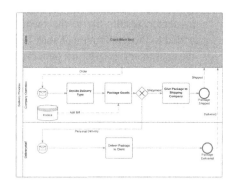

Fig. 6. Bottom-up approach – activity domain selection

The user is pursuing a reduction of the air emissions produced by the process and therefore selects the appropriate activity domain "Reduce air emissions". This leads to the result overview shown in Fig. 7.

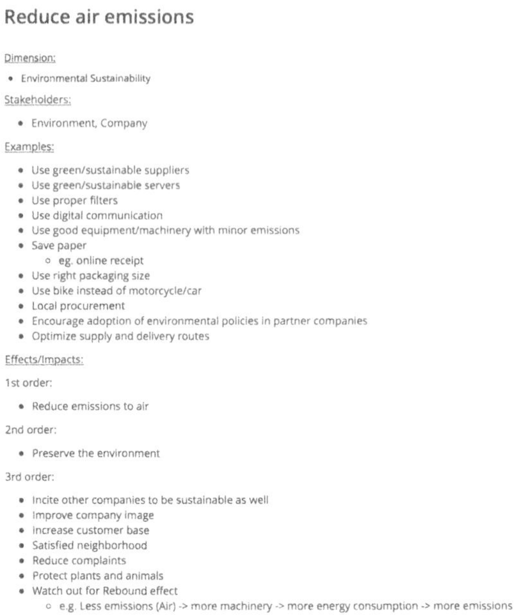

Reduce air emissions

Dimension:

- Environmental Sustainability

Stakeholders:

- Environment, Company

Examples:

- Use green/sustainable suppliers
- Use green/sustainable servers
- Use proper filters
- Use digital communication
- Use good equipment/machinery with minor emissions
- Save paper
  - eg. online receipt
- Use right packaging size
- Use bike instead of motorcycle/car
- Local procurement
- Encourage adoption of environmental policies in partner companies
- Optimize supply and delivery routes

Effects/Impacts:

1st order:

- Reduce emissions to air

2nd order:

- Preserve the environment

3rd order:

- Incite other companies to be sustainable as well
- Improve company image
- Increase customer base
- Satisfied neighborhood
- Reduce complaints
- Protect plants and animals
- Watch out for Rebound effect
  - e.g. Less emissions (Air) -> more machinery -> more energy consumption -> more emissions

Fig. 7. Environmental sustainability – reduce – reduce air pollution result overview

Now, the information gathered in the result overview can be used to identify possible improvements in the present delivery process. If additional activities are favoured a brainstorming can be executed.

Step 1: Is now executed and one can proceed to the next step

Step 2: Now, the user can select activities from the result page and check which of the presented results fits to the business process. The selected activities will be added to the result table (Table 2). For example, in the process shown in Fig. 4, the customer has to sign a paper receipt (*Sign delivery receipt*). One possibility to reduce air emissions listed in the result overview is to save paper by using electronic signature. This is technically feasible and thus an opportunity to reduce the air emission and increase the sustainability of the given business process. Therefore we add this activity to the result table

Step 3: In the next step, the labeling of the selected activities will be performed. For example "will be implemented immediately"

Step 4: In this step, more information will be gathered. To support this step, the user can answer the following questions as mentioned above (some of the points may already be covered by the catalogue):

Since every company has its own rules, it seems appropriate that the user utilizes this questionnaire to get specific answers and uses the catalogue as support. The information gathered in this step will be also added to the result table

Table 2. Supporting questions and example answers

Needed information	Question	Example
Stakeholders	Who is affected?	Suppliers, customers
Goals	Why do we want to implement this idea?	Save paper and therefore reduce emissions
Activities	How can we implement this idea?	Electronic signature instead of paper receipt
Business rules	What rules have to be respected?	Secure data transfer
Indicators	How can we measure the successful implementation?	Saved paper
Impacts	What impact does the idea have on the sustainability dimension?	Save paper, thus protect the environment
Complications	Which problems are expected to occur?	Data loss during transfer

Step 5: The result table shows, which points have to be considered to improve business processes sustainability

Step 6: Next, the user can modify the process (add, modify or delete activities) according to the information listed in the result table

Step 7: Highlight the changed process elements. To visualize the generated idea, the identified activity should be marked in the business process

Finally, according to our example a new activity (*electronic receipt*), was added to the business process. Through the change from paper to electronic receipt, this process consumes less paper and so produces less emission. To summarize, the application of the process model and the catalogue enabled the identification of different opportunities to improve the sustainability of the process. Through the guided assistance, the model and the catalogue facilitate the identification and therefore the integration of sustainability aspects. In some cases the catalogue might not be able to identify appropriate solutions. However, the results generated by the catalogue can be integrated in following brainstorming sessions to simplify the identification of satisfying solutions.

5   Summary and Outlook

In this paper we presented a process model that can help stakeholders to improve their business processes in terms of sustainability. The main focus was on the identification of activities with potential for improvement, independent from sustainability dimensions. For this purpose, we presented a catalogue. The catalogue (which can be accessed online) supports the first step of the process model – the idea generation phase, by providing suggestions to enhance the sustainability of business processes by improving or replacing activities. It also provides additional information that might be needed to model or integrate the new activities in the business process. Although, we carefully designed and applied our process model our approach has several limitations: First of all, we are aware that we have not created a catalogue of improvement patterns but rather a first step in this direction. We are planning to extend our catalogue using case studies and expert opinions. Next, due to the complexity of the term "sustainability", different parties might have different understanding of the topic. Therefore, it its crucial that a firm boundary, appropriate to the specific case, is determined. Second, due to the platform used to build the catalogue, in its current state only the administrator can extend it with new information. The extendibility by stakeholders is an essential part and has to be integrated. In addition, a general problem is to find the sweet spot in granularity. On the one side, if the ideas are too general the support might be inaccurate. On the other side, too specific ideas could have a negative impact on the versatility of the catalogue. Finally, due to the interdependence of the different dimensions, trade-offs might be necessary. The process model supports the idea generation-, identification- and modeling steps, but doesn't assist the user in deciding on trade-offs. This has to be done manually. Thus, one issue for future work is, how to treat conflicting goals and the required trade-offs in a general way. However, with our presented process model we provide useful support to integrate sustainability into business processes.

References

1. Recker, J., vom Brocke, J., Seidel, S.: Green Business Process Management. Springer, Heidelberg (2012)
2. Betz, S.: Sustainability aware business process management using XML-Nets. In: Proceedings of the 28th Conference on Environmental Informatics - Informatics for Environmental Protection, Sustainable Development and Risk Management, Oldenburg (2014)
3. GHG Protocol: GHG protocol product life cycle accounting and reporting standard ICT, sector guidance (2016)
4. Alemayehu, W., Brocke, J.: Sustainability performance measurement – the case of ethiopian airlines. In: Muehlen, M., Su, J. (eds.) BPM 2010. LNBIP, vol. 66, pp. 467–478. Springer, Heidelberg (2011). doi:10.1007/978-3-642-20511-8_43
5. Hoesch-Klohe, K., Ghose, A.: Carbon-aware business process design in Abnoba. In: Maglio, P.P., Weske, M., Yang, J., Fantinato, M. (eds.) ICSOC 2010. LNCS, vol. 6470, pp. 551–556. Springer, Heidelberg (2010). doi:10.1007/978-3-642-17358-5_38
6. Houy, C., Reiter, M., Fettke, P., Loos, P., Hoesch-Klohe, K.; Ghose, A.: Advancing business process technology for humanity: opportunities and challenges of green BPM for sustainable business activities. In: vom Brocke, J., Seidel, S., Recker, J. (eds.) Green Business Process Management – Towards the Sustainable Enterprise, pp. 75–90. Springer, Heidelberg (2012). doi:10.1007/978-3-642-27488-6_5
7. Recker, J., Rosemann, M., Hjalmarsson, A., Lind, M.: Modeling and analyzing the carbon footprint of business processes. In: vom Brocke, J., Seidel, S., Recker, J. (eds.) Green Business Process Management – Towards the Sustainable Enterprise, pp. 59–74. Springer, Heidelberg (2012). doi:10.1007/978-3-642-27488-6_6
8. Nowak, A., Leymann, F., Schleicher, D., Schumm, D., Wagner, S.: Green business process patterns. In: Proceedings of the 18th Conference on Pattern Languages of Programs (PLoP 2011), pp. 1–10. ACM (2011)
9. Hoesch-Klohe, K., Ghose, A.: Business process improvement in Abnoba. In: Maximilien, E.M., Rossi, G., Yuan, S.-T., Ludwig, H., Fantinato, M. (eds.) ICSOC 2010. LNCS, vol. 6568, pp. 193–202. Springer, Heidelberg (2011). doi:10.1007/978-3-642-19394-1_21

10. Weske, M.: Business Process Management – Concepts, Languages, Architectures, 2nd edn. Springer, Heidelberg (2012)
11. Berkhout, F., Hertin, J.: Impacts of information and communication technologies on environmental sustainability: speculations and evidence, Report to the OECD (2001)
12. Leymann, F., Nowak, A., Schumm, D., Wetzstein, B.: An architecture and methodology for a four-phased approach to green business process reengineering. In: Information and Communication on Technology for the Fight against Global Warming, pp. 150–164 (2011)
13. Mietzner, R., Nowak, A., Leymann, F.: Towards green business process reengineering. In: Service-Oriented Computing, pp. 187–192 (2011)
14. Brundtland, G.H.: World Commission on Environment and Development: Our Common Future. Oxford University Press, United Nations (1987)
15. Piotrowicz, W.: Monitoring performance. In: Cetinkaya, B., Cuthbertson, R., Ewer, G., Klaas-Wissing, T., Piotrowicz, W., Tyssen, C. (eds.) Sustainable Supply Chain Management – Practical Ideas for Moving Towards Best Practise, pp. 57–80. Springer, Heidelberg (2011). doi:10.1007/978-3-642-12023-7_3
16. Penzenstadler, B., Femmer, H.: A generic model for sustainability with process- and product-specific instances. In: GIBSE 2013. ACM (2013)

Supporting Municipal Greenhouse Gas (GHG) Emission Inventories Using Business Process Modeling: A Case Study of Trondheim Municipality

Dirk Ahlers[1], John Krogstie[1(✉)], Patrick Driscoll[1], Hans-Einar Lundli[2],
Simon-James Loveland[2], Carsten Rothballer[3], and Annemie Wyckmans[1]

[1] NTNU – Norwegian University of Science and Technology, Trondheim, Norway
{dirk.ahlers,john.krogstie}@idi.ntnu.no,
{patrick.arthur.driscoll,annemie.wyckmans}@ntnu.no
[2] Trondheim Kommune, Trondheim, Norway
[3] ICLEI Europe, Freiburg, Germany

Abstract. Business process modeling and business process management has been used to capture, support and improve a large variety of processes and practices in the private and public sector. Traditionally what is regarded as a good business process is strongly related to economic dimensions. With the increasing importance of assuring sustainable development, BPM techniques should to an increasing degree be able to be used to support the goal of sustainability of the supported or automated solution. This paper provides results from a case study in the Carbon Track and Trace (CTT) project on supporting the compilation and reporting of data on greenhouse gas (GHG) emissions on the city level in the form of GHG inventories. Although basic BPM-techniques are applicable on this levels, we have identified a number of challenges and potential improvements to represent the relevant aspects in such cases to support automated and semi-automated solutions.

Keywords: Greenhouse gas inventories · City services · Sustainable business process modeling · Green IT · Workflow analysis

1 Introduction

The concept of sustainability includes three dimensions: the ecology/ environment, the economy, and social equity, which should all be enhanced in a balanced manner over the long run in a sustainable society. Our focus in this paper is along the environmental axis, although having in mind that measures should also be economically feasible and beneficial to human conditions in a holistic approach. Today, sustainability and climate change go hand in hand. Since the early 1990s, cities around the world have been pursuing ambitious climate mitigation targets. However, setting a course towards a low-carbon city with

© Springer International Publishing AG 2017
M. Dumas and M. Fantinato (Eds.): BPM 2016 Workshops, LNBIP 281, pp. 416–427, 2017.
DOI: 10.1007/978-3-319-58457-7_30

significant reductions in greenhouse gas emissions demands a precise overview of current emissions to identify the priority areas for interventions and to track their success over time. The way to formally achieve this overview is to use methods of greenhouse gas accounting to build a yearly inventory of emissions on a city level. An emission inventory allows a city to compare its emissions over time and to other cities.

In line with Norwegian climate plans [5], the city of Trondheim, Norway has ambitious reduction goals of 70%–90% of Greenhouse Gases (GHG) by 2030 compared to a 1991 baseline [1]. Of these, transportation emissions are among the largest components of the overall GHG emissions in the city [2]. In order to provide a more sound empirical basis to support Trondheim's climate goals, the municipality is building an emission inventory. It includes direct emissions within the city with the two prioritized areas of transport and energy consumption in residential and commercial buildings.

The aim of this paper is to analyse the current workflow of data sourcing for emission inventories, compare it to recommended practices, and identify bottlenecks and issues. We use Business Process Modeling as a tool towards sustainability in the context of Green BPM [15,21]. The development process included a literature study, initial discussions at project meetings, informal interviews, and collaborative development and documentation of the used process steps towards developing the models. Through interviews and project meetings with different stakeholders from the municipality and supporting institutes, we developed an understanding of the emission inventory building process. A main issue is a large gap between mandated data input into inventories and the rather complex data sourcing. We address this gap by a close reading of the relevant GPC standard [3] and of common best practices from ICLEI [6] together with the practices and requirements of the municipality. We especially examine steps in the data gathering and compilation through interviews and iterative model development that result in a number of general models in BPMN [8] together with a list of identified challenges.

2 GPC Emission Inventory Standard

The setup of emission inventories to track Trondheim's emissions follows an emergent standard for GHG emission inventories called the Global Protocol for Community-Scale Greenhouse Gas Emission Inventories (GPC) [3], co-developed by ICLEI. GPC defines a standardized and comparable way for cities to calculate and report their GHG emissions. The protocol defines the categories of emissions to be reported as well as adaptation and scaling of data – if available data does not align with the geographical boundary of the city or the time period of the assessment – and calculations for final reporting.

The GPC reporting framework classifies emissions by their source using 6 sectors with up to 2 levels of sub-sectors, which results in a total number of 43 emission types. Sectors are Stationary Energy; Transportation; Waste, Industrial Processes and Product Use (IPPU); Agriculture, Forestry, and Other Land Use

(AFOLU); and Other Scope. Emissions are further orthogonally organized into scopes, defining whether emissions occur inside or outside the city or come from grid-supplied energy. The scopes roughly follow a complexity of emission tracking and allow cities to select a coverage of emissions that is appropriate for them.

The input to GPC consists of activity data for each defined category along with emission factors. GHG emissions are reported for 7 different greenhouse gases, which are calculated into CO_2 equivalents (CO_2e). Emissions are separated into two values per emission type. Activity data is the actual resource or energy use, or the actual emissions; the emission factor acts as an adaptation factor taking local characteristics (such as GHC emitted for electricity production) into account when transforming resource use into CO_2 equivalent emissions. Additionally, if city-level emission values are not available, scaling factors [14] have to be derived to scale down national values or to aggregate and disaggregate regional values that may not match the city boundaries. It is important to note that the GPC is designed to cover a period of one year. Shorter update cycles are not mandated but for example specific emission types or sources that are of particular interest could be tracked separately at a higher frequency, for example through better data reporting or on-the-ground emission measuring [10].

GPC accounting methods are based upon those found in financial accounting, with similar principles of relevance, completeness, consistency, transparency, accuracy, and measurability [3,7]. It defines what input of multiple different areas is needed in which formats, defined internal dependencies and consistency conditions, and contains a calculation engine to generate reports. Data quality is another aspect to consider in the case of GPC because data can originate from multiple sources that may have widely varying quality levels [17]. We focus particularly on aspects of data quality concerning fitness for use in terms of accuracy, correctness, and completeness, with a lower interest in timeliness, currency, and provenance. However, while GPC sets up requirements and data quality management, it only indirectly relates to how activity data is gathered in the first place. Yet for practical application, the process of data collection and preparation is a major aspect and furthermore can be highly varied between municipalities, which is why this work takes the form of a case study. For the actual GPC emission calculation, the ClearPath tool[1] compiles and calculates GPC-compliant inventories. Additional specialized tools are available [23] and also tool collections of smaller calculation tools for certain emissions or emission subcategories[2]. However, to the best of our knowledge, there is currently no overall workflow tool available, prompting our study.

3   Case Description: The CTT Project

The Carbon Track & Trace (CTT) project[3] [9] is intended to provide the City of Trondheim with a sound empirical basis for the development of more advanced

---

[1] ICLEI USA: http://icleiusa.org/clearpath/, http://clearpath.global/.

[2] http://www.ghgprotocol.org/calculation-tools/all-tools.

[3] http://carbontrackandtrace.com/.

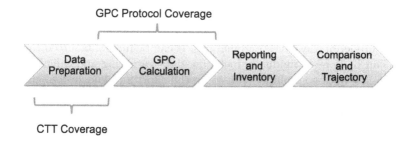

Fig. 1. Coverage of GPC and CTT in an idealized inventory workflow

Fig. 2. Detail view of the data preparation stage

greenhouse gas emissions inventory methods, including the eventual deployment of sensor systems to reduce the cost and complexity of collecting data for GHG inventories. An additional goal of the project is to help develop better methods of decision and planning support for municipal mitigation planning through integration into strategic planning instruments, cost-benefit assessments (CBA) and geo-spatial databases. This workflow analysis is the result of a research and innovation collaboration between the City of Trondheim, the Norwegian University of Science and Technology (NTNU), and ICLEI – Local Governments for Sustainability.

A simplified GPC chain is shown in Fig. 1 where we illustrate GPC coverage and the coverage of our CTT approach. GPC covers calculation and reporting and supports subsequent comparison and trajectory of reports. In our approach we take a stronger focus on input data flows and examine those in detail (Fig. 2). We assume a GPC-compliant calculation process and complementarily investigate the data acquisition and preparation phase. This part is briefly discussed, but not strictly defined in the standard. That leaves the details up to individual municipalities, who may or may not have measures in place for them. To the best of our knowledge there are no previous processes defined.

3.1 Challenges in GHG Inventory Building

Before the workflow analysis, we briefly report on a previous general gap analysis for the emission inventory development of Trondheim [16]. The main challenges found are data uncertainty, reliability, and quality, the gap between top-down (downscaled national level statistical data) and bottom-up (local or even real-time data) data in the case of calculated instead of measured data that does not reflect the actual situation, and a strong data sourcing issues. The latter consists of non-standardised workflows, very time-intensive compilation and preparation of yearly inventories, and unclear benefits and costs of inventories. The state-of-the-art on municipal emissions inventories methods is highly variable. It can be done in a number of ways and combinations: It can be fully outsourced to a third party; it can be done individually including data gathering and calculations by a city; regional data compilation can share the effort and knowledge; and there can be central national data provision (and calculation) at city granularities.

Currently, the City of Trondheim has not yet fully completed a GHG emissions inventory using the GPC, but has embedded the methodology in their new climate action plan to be adopted by the end of 2016. Therefore, this case study is based on ongoing work on setting up the GPC processes. Previously, Trondheim commissioned external studies for the municipality's carbon footprint [4]. The current practice of data acquisition and calculation is mostly top-down with municipality-developed calculations and estimations based on national data and some available local or regional sources. The availability of detailed municipal statistics from Statistics Norway has been deferred, but it is expected that the data availability situation will improve. However, it is not yet clear how this data will be derived, whether only through scaling or through more tailored methodology taking local distributions into account. For the municipal view, it is very important to recognize that purely calculated top-down data is not suitable for monitoring local actions and allowing fast feedback loops on policy decisions. Additionally, national statistical data usually is made available with a delay of one or more years, further slowing down fast impact assessments.

A main issue in building inventories is that data, both for emission factors and activity data, are usually not directly available from a structured database or similar system. This is part of a more complex issue of multiple factors that can occur alone or in combination. General challenges include:

- Direct measurements are not available
- Available statistics are scaled down from national levels and may not correspond to actual city-level emissions
- Verification of scaled statistics is difficult
- Without exact measurements, data needs to be gathered in different ways and from multiple sources.
- No sufficient tool support for data sourcing and management
- Non-standardized handling of complexity for multiple data sources and complexity of emission composition per type

The relevant scientific literature offers hints towards solutions while acknowledging that in general city-level emissions inventories are extremely complex. For example, [22] gives a good general discussion of uncertainties in emission inventories and how different inventories address them. [20] provides a survey of guidance support frameworks and discusses the issues of scaled-down data. [19] studies an emission inventory at an urban scale. Other work [18] acknowledges that emissions accounting is challenging and proposes a measure based on carbon footprints to capture indirect emissions for the scenario of Norwegian municipalities and can serve as a complementary to this study.

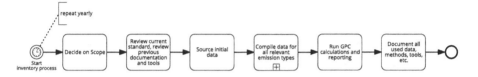

Fig. 3. General workflow for a yearly GPC inventory iteration

4 Workflow Analysis

This section details the data gathering, input, and handling necessary to build a GPC-compliant GHG inventory. As noted above, access to and readiness of data varies widely. Input values can be direct, aggregate, partitioned, or combined metrics resulting from a multi-stage process, making a streamlined approach from discovered data sources into a GPC-compliant calculation tool rather complex. We therefore work towards requirements for the data collection process and to understand how this workflow operates in general. We interviewed key players at the municipality and at ICLEI to develop an understanding of this process by analyzing current and planned workflows at the municipality and derive insights from different parts of the GPC recommendations as far as they concern the data workflow. We incorporate [7], which gives some general advice on data collection, especially accounting and reporting principles as well as considerations towards a bottom-up approach. As an additional factor to consider, the "GPC specifies the principles and rules for compiling a city-wide GHG emissions inventory; it does not require specific methodologies to be used to produce emissions data" [3]. While compilation methods are not given, the standard contains some guidance for sourcing activity data to be followed within CTT.

The overview of a complete general workflow to build a yearly GPC inventory is shown in Fig. 3. Following the GPC structure, the workflow starts with a decision on the GPC scope of emissions to be tracked. The following steps are to get an updated overview of the current standard and the documentation from previous years that can help in planning and organizing the process and the data collection and compilation. Then GPC calculations and reporting are performed. As a last wrap-up step, all used processes, data, methods, tools, etc. need to be properly documented. This will enable a learning process with a set of guidelines and best practices that make the process more easily repeatable. Following these steps, an increased automatisation of selected steps can become possible.

Compiling data for the relevant emissions is one of the major and most time-consuming steps. For each emission subcategory, it is split up into compiling the actual activity data that describes the emissions, and the emission factor which serves as an adaptation factor to transform and partition emission data into the relevant greenhouse gases or equivalents. Figure 4 shows this simple iterative process. The details of the emission factor determination are shown in Fig. 5 with the possible sources of factors listed. Determining emission factors is encouraged to be conducted at the most specific levels. This usually means

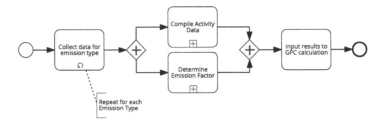

Fig. 4. Workflow for gathering data for all emission categories

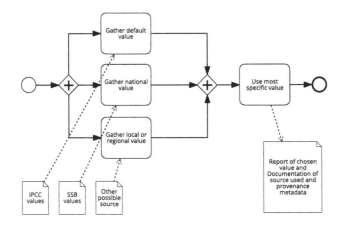

Fig. 5. Workflow detail: determining emission factors

going down from IPCC reference factors to national factors (SSB for Norway) and possible to more specific local factors, if available.

The process of compilation of activity data is much more complex. Figure 6 shows the outline of the steps involved. It makes the basic distinction between emissions where directly reported numbers can be put into GPC directly without any calculation or modelling. In this case, the only processing may be an adaptation of spatial or temporal scales. In the second case, emission activity data is not directly available and has to be estimated based on indirect data or complex measures. Figure 6 gives the overview of steps to collect data and develop models to calculate or estimate activity data and to compare modelling results on multiple levels and to external models in an iterative process. This workflow for gathering data detailed in Fig. 4 and especially for determining activity data in Fig. 6 describe a generic workflow that needs to be adapted and extended for the individual emission subcategories.

Based on the problem statement and the workflow analyses, we develop a set of recommendations to enable and improve the workflow for easier repeatability over time. Due to the complex nature of the task, it is not feasible at present to develop a fully automated workflow system, but rather to select critical steps in the process to offer tailored support either through general contributions or

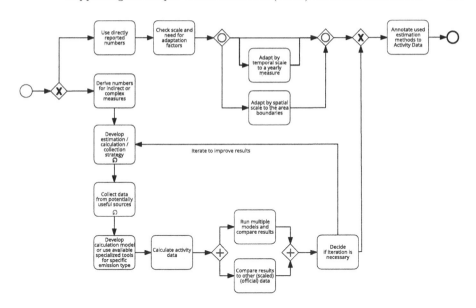

Fig. 6. Workflow detail: compiling activity data

specific focused tools. Furthermore, if previous years' constituent data sources, tools, and methods are maintained, it will enable easier reuse and discovery of artifacts, thereby speeding up the process in subsequent years and assuring an increased consistency, validity, and quality. The following general recommendations are adapted from the data collection principles mentioned in the GPC itself [3]:

- Establish collection processes that lead to continuous improvement of the data sets used in the inventory
- Prioritize improvements on the collection of data needed to improve estimates of key categories which are the largest, have the greatest potential to change, or have the greatest uncertainty
- Review data collection activities and methodological needs on a regular basis to guide progressive, and efficient, inventory improvement
- Work with data suppliers to improve the quality of the data, to better understand uncertainties, or demand better and/or more local statistics

We adapt the relevant aspects as follows. The overarching recommendation is the establishment of a management plan for the inventory process, including selection, application, and updating of inventory methodologies. The next aspect is the documentation of data, methods, assumptions, estimates, and systems. These steps then also help in maintaining quality, transparency, maintainability, repeatability and replication in following years. Furthermore, documentation should include direct data documentation, bibliographical data references, an archive of cited references, and criteria for the selection of boundaries, base years, methods, activity data, emission factors, and other parameters for the emission data. Additionally, changes in data or methodology should be documented and version control is needed.

A set of metadata for input values is defined in GPC: Definition and description of the dataset; time, period, frequency of publication; source information; information on how to access it. To these we further recommend: information on how to extract and process it; and its relation to other sources. The problem, restated, is that because of compound measures as described above, a direct annotation to the input factors is often not feasible. For example, for the transportation data discussed above, there is a complex model with multiple inputs that generates the necessary input data for GPC. A major aspect in the inventory process that is not yet covered is the need for better documentation of data sources, steps taken, and the actual process as it differs substantially between different emission categories. This implies better support for building and archiving documentation about the way information retrieval was conducted, especially which persons, organizational entities, and data sources were used and how they were accessed and converted. Data storage and processing is currently rather ad-hoc and would benefit from more attention and support as well.

On an organizational level, the comparability and replicability of inventories over the years can be enhanced by reusing the same methodology for intermediate emission data. Due to the current nature of the existing tools, documentation for the most part will happen outside of the GPC accounting tools. Obviously, a better integration is a necessary challenge for future development.

5   Future Work

Potential future work focuses on supporting, streamlining, and automating elements of GHG emissions inventory methodologies. Within this paper, it is not possible to summarize the large variation in city-level emissions inventory practices, but we aim to work with more cities to further develop an operational typology of different cities' data collection methods and data gaps and then prioritize the areas where automatization will bring the most value. A better utilization and integration of available tools to support certain emissions or emission subcategories will also need to be developed in parallel.

6   Conclusion

To summarize, the domain is categorized by a number of challenges:

- Highly complex data compilation, collection, selection, preparation, and processing.
- Little automatic data transmission and integration is possible. There would need to be large changes in current data collection, selection, pre- and post-processing, and cleaning procedures in order to automate significant parts of the GPC (or other) reporting framework.

- Emissions inventories are difficult to complete and require upwards of 3–6 person months, depending on data availability.
- Data quality aspects including accuracy, precision, and uncertainty parameters are currently poorly defined or non-existent.
- Workflow support may be possible with improved data and knowledge management procedures.

These are findings from our detailed analysis of processes in Trondheim as well as wider experiences from ICLEI. Still, situations for other municipalities may vary. From a business process modeling point of view, we note through the experience that current modeling approaches like BPMN are useful for representing and communicating the main workflow steps.

However, the main issues associated with the automation of municipal GHG emission inventory methods identified in this report are not related to inventory or workflow processes, but lie in the peculiarities of the domain as found in data location, acquisition, manipulation, and generation. A possible adaptation of BPMN to address such issues easier would be promising. As part of our ongoing work with mobile technology and services and lately IoT-solutions in Wireless Trondheim [13] we are currently working with such extensions. Frequently direct access to reliable data sources is not possible, requiring municipalities to find workarounds that are often expensive, time-consuming, and often of dubious quality. Thus also for this domain, sustainability cannot be reached by technology solutions alone, but needs a broader view incorporating organisational and other dimensions to fuel all three dimensions of sustainability. Obviously, building a GPC inventory is not an end in itself, as it helps a city develop a detailed overview about its emissions. This serves as a decision aid in prioritizing GHG reduction initiatives. In line with this larger goal, inventory building can serve as a step in organizational learning in that it aids policy learning and identification of effective practices developed by other cities. By understanding and supporting the workflow as a reflection and learning process, it can be done in such a way that the process improves over time by taking gained knowledge and lessons learned into account while still staying flexible to react to changed circumstances, e.g. data availability or quality. Then, stronger automatization can be implemented, with a special focus on high-priority areas.

Acknowledgements. This paper contains details from a project report on the municipality's workflow [11] but focuses stronger on the BPMN and sustainability perspective. It is an extended version of a SABPM workshop contribution [12]. Funding for the project was provided by the Climate KIC Low Carbon City Lab (LoCaL) flagship project along with additional in-kind contributions from the City of Trondheim, NTNU, Numascale, and Wireless Trondheim.

References

1. Energi- og klimahandlingsplan for Trondheim kommune - Mål og tiltak for perioden 2010–2020 [Energy and climate action plan for Trondheim - Targets and measures for the period 2010–2020]. Technical report, Trondheim Kommune (2010)
2. Covenant of mayors signatories: Trondheim. Technical report, Covenant of Mayors (2014). http://www.covenantofmayors.eu/about/signatories_en.html?city_id=7044
3. Global Protocol for Community-Scale Greenhouse Gas Emission Inventories - An Accounting and Reporting Standard for Cities (GPC). Technical report, World Resources Institute, GHC Protocol Initiative (2014)
4. Klimaeffektive anskaffelser i Trondheim - Komplett klimafotavtrykk og skisse til mulig styringssystem [Climate-efficient procurement in Trondheim - Complete climate footprint and outline of possible management system]. Technical report, MiSA/Asplan Viak (2014)
5. Summary of proposed action plan for Norwegian emissions of short-lived climate forcers. Technical report M135/2014, Norwegian Environment Agency (Miljødirektoratet) (2014)
6. Carbonn climate registry 5 year overview report. Technical report, ICLEI (2015)
7. A framework and guidance booklet on production of community data at the municipality level (based on national data). Technical report D2.8, Measure and share data with utilities for the Covenant of Mayors (MESHARTILITY) (2015)
8. Aagesen, G., Krogstie, J.: BPMN 2.0 for modeling business processes. In: Brocke, J., Rosemann, M. (eds.) Handbook on Business Process Management 1. IHIS, pp. 219–250. Springer, Heidelberg (2015). doi:10.1007/978-3-642-45100-3_10
9. Ahlers, D., Driscoll, P.: Understanding challenges in municipal greenhouse gas emissions inventories. In: ICE ITMC Conference (2016)
10. Ahlers, D., Driscoll, P., Kraemer, F.A., Anthonisen, F., Krogstie, J.: A measurement-driven approach to understand urban greenhouse gas emissions in Nordic cities. In: NIK2016 - Norwegian Informatics Conference (2016). To appear
11. Ahlers, D., Krogstie, J., Driscoll, P., Lundli, H.E., Loveland, S.J., Rothballer, C., Wyckmans, A.: Workflow Analysis of Greenhouse Gas (GHG) Emission Inventory Methods For Trondheim Municipality. CTT Deliverable. Technical report, NTNU (2015)
12. Ahlers, D., Krogstie, J., Driscoll, P., Lundli, H.E., Loveland, S.J., Rothballer, C., Wyckmans, A.: Supporting municipal greenhouse gas emission inventories using business process modeling: a case study of trondheim municipality. In: Workshop on Sustainability-Aware Business Process Management @ BPM 2016 (2016)
13. Andresen, S., Krogstie, J., Jelle, T.: Lab and research activities at wireless Trondheim. In: Proceedings 4th International Symposium on Wireless Communication Systems (2007)
14. Bertoldi, P., Cayuela, D.B., Monni, S., de Raveschoot, R.P.: Existing Methodologies and Tools for SEAP Development and Implementation - Report II: Methodologies and Tools for CO_2 inventories. Technical report, Covenant of Mayors (2010)
15. Betz, S.: Sustainability aware business process management using XML-Nets. In: Proceedings of the 28th Conference on Environmental Informatics - Informatics for Environmental Protection, Sustainable Development and Risk Management (2014)
16. Driscoll, P., Ahlers, D., Rothballer, C., Lundli, H.E., Loveland, S.J., Wyckmans, A.: Gap Analysis of Greenhouse Gas (GHG) Emission Inventory Methods For Trondheim Municipality. CTT Deliverable. Technical report, NTNU (2015)

17. Krogstie, J.: Evaluating data quality for integration of data sources. In: Grabis, J., Kirikova, M., Zdravkovic, J., Stirna, J. (eds.) PoEM 2013. LNBIP, vol. 165, pp. 39–53. Springer, Heidelberg (2013). doi:10.1007/978-3-642-41641-5_4

18. Larsen, H.N., Hertwich, E.G.: Implementing carbon-footprint-based calculation tools in municipal greenhouse gas inventories - the case of Norway. J. Ind. Ecol. **14**(6), 965–977 (2010)

19. Sanna, L., Ferrara, R., Zara, P., Duce, P.: GHG emissions inventory at urban scale: the Sassari case study. Energy Procedia **59**, 344–350 (2014)

20. Santovito, R., Abiko, A., Bienert, S.: Discrepancies on community-level GHG emissions inventories. In: ERES (2015)

21. Seidel, S., Recker, J., vom Brocke, J.: Green business process management. In: vom Brocke, J., Seidel, S., Recker, J. (eds.) Green Business Process Management: Towards the Sustainable Enterprise, pp. 3–13. Springer, Heidelberg (2012). doi:10.1007/978-3-642-27488-6_1

22. Wattenbach, M., Redweik, R., Lüdtke, S., Kuster, B., Ross, L., Barker, A., Nagel, C.: Uncertainties in city greenhouse gas inventories. Energy Procedia **76**, 388–397 (2015)

23. Zoellner, S.: Introduction to climate mitigation and GHG inventories. In: Urban Management Tools for Climate Change (2015)

Sustainability Patterns for the Improvement of IT-Related Business Processes with Regard to Ecological Goals

Patrick Lübbecke[(⊠)], Peter Fettke, and Peter Loos

German Research Center for Artificial Intelligence, Saarland University,
Saarbrücken, Germany
{patrick.luebbecke,peter.fettke,
peter.loos}@iwi.dfki.de

Abstract. The transformation of proven methods and knowledge to reusable artifacts in the form of patterns or anti-patterns is very common in many areas of IS research, for instance in Business Process Management (BPM). An area that lacks in support on the basis of patterns is the area of Green Business Process Management. With this paper, we introduce the concept of sustainability patterns in BPM that can be used for the improvement of existing processes or for the design of new processes in due consideration of ecological goals such as the reduction of resource consumption during the executing of these processes. We further present the results of a qualitative analysis that we conducted to extract a total number of 26 Ecological Process Patterns from real-world processes and a catalog with generic process weakness patterns. The identified patterns indicate a latent potential for process enhancement once the patterns are applied to real-world processes.

Keywords: Process patterns · Weakness patterns · Anti-patterns · Green business process management

1 Introduction

Patterns play a pivotal role in many areas related to Information Systems. The supposedly most-famous type of patterns is the design patterns proposed by (Gamma et al. 1995). In Business Process Management (BPM), the concept of patterns is prevalent as well (Lohrmann and Reichert 2014; Winter et al. 2009) and is, for example, used to propose enhancement to business processes (Lohrmann and Reichert 2014). On the other hand, weakness or anti-patterns are used in BPM for compliance checking or to find flaws in existing processes (Becker et al. 2011; Delfmann and Hübers 2015).

Many contributions addressing the application of patterns in BPM have been proposed in the last few years. These articles range from theoretical concepts (Smirnov et al. 2012) to ready-to-use pattern catalogs (Delfmann and Hübers 2015). A field in BPM that still lacks in employing the concept of patterns is the field of Green Business Process Management (Green BPM) which is devoted to the optimization of business processes in terms of ecological aspects (Fig. 1).

© Springer International Publishing AG 2017
M. Dumas and M. Fantinato (Eds.): BPM 2016 Workshops, LNBIP 281, pp. 428–439, 2017.
DOI: 10.1007/978-3-319-58457-7_31

At the moment, is a number of approaches for ecological process analysis and redesign such as reference modeling (Krcmar 2002), simulation (Lübbecke et al. 2015), or by means of key environmental indicators (KEIs) (Nowak et al. 2011). The latter two approaches, however, require a stringent collection of data for building realistic simulation models or for calculating KEIs. Since data collection does not constitute an established process in many enterprises, especially small and medium-sized enterprises often cannot provide the necessary data. Reference modeling, on the other hand, provides support at a rather superficial level and does not provide companies with the necessary guidance for improving existing processes but rather emphasizes the adoption of processes or tasks proposed by the reference model. Despite the positive impact that patterns can unfold on process performance, research has yet to come up with pattern-based approaches to enhance processes from an ecological standpoint. To the best of our knowledge, there is currently only one approach addressing "Green BPM patterns" for ecological process optimization (Nowak et al. 2013). The authors approach the topic from a strategic standpoint and propose only general optimization strategies without any relationship to real-world processes, which constitutes a major difference to our research (Table 1).

After proposing the concept of ecological process patterns (Lübbecke et al. 2016), we want to advance the topic by providing a catalog of patterns from the field of public administration (PA) in Germany. We chose this area because PA is part of every municipality and processes in the German PA are quite standardized up to a certain point. To compose the catalog, we used two sources: on the one hand, we analyzed processes for registration (1), notification of change (2) and deregistration (3) of new and existing businesses as used in eight county seats of a small federal state in Germany (24 process models). On the other hand, we analyzed a catalog of existing general process weakness patterns from the field of PA in terms of their impact on the ecological perspective. From these two sources, ecological process patterns (EPP) were derived by modeling experts familiar with ecological sustainability in BPM.

After providing a short summary on the state of the art on the use of patterns in BPM and Green BPM in Sect. 2, we outline our understanding of ecological process patterns in Sect. 3. In Sect. 4, after the concept is made clear, we will present early results of a qualitative analysis where we extract instances of ecological process patterns from existing process models. The paper concludes with a summary as well as a short overview on future research efforts that will be necessary to foster the creation and application of such patterns in the domain of Green BPM.

2 Theoretical Background

2.1 Patterns in Business Process Management

The concept of patterns was initially introduced by (Alexander 1977) in the domain of architecture. Alexander came up with the idea of people designing buildings for themselves using predefined design blocks ("patterns"). Patterns are the abstraction from a concrete form, which keeps recurring in specific, non-arbitrary contexts (Riehle and Züllighoven 1996). In Software Engineering, for example, developers use best practice in the form of design patterns to solve recurring design problems (Ambler 1998; Coplien 1995; Gamma et al. 1995; Mangalaraj et al. 2014).

(Fowler 1996) introduced patterns to the area of BPM. He proposed the analysis patterns, a group of concepts that represents a common construction in business modeling. Today, patterns in BPM are used as a template to enhance the quality of business processes by avoiding the modelling of unfavorable constructs (van der Aalst et al. 2000, 2003; van Dijk 2003), to optimize the workflow of processes (Forster 2006; Lohrmann and Reichert 2014) or to enable process analysis techniques such as process mining (Smirnov et al. 2012). One area of BPM in which patterns are applied is compliance management. In this domain, process elements in the form of patterns that violate legal requirements are searched for by using process mining techniques, for example (Becker et al. 2011; Rozinat and van der Aalst 2008).

The inversion of the paradigm of a pattern is the so-called anti-pattern, sometimes also referred to as weakness pattern. These weakness patterns depict a situation that generates decidedly negative consequences and should therefore be avoided.

2.2   Green Business Process Management

Green Information Systems (Green IS) denotes a discipline of scholarship that contributes to addressing modern day environmental issues by promoting ecologically sustainable information systems (vom Brocke et al. 2013). Green BPM is a subset of Green IS with a focus set on business processes. Green BPM can be described as defining, implementing, executing and improving business processes in corporations with the aim to support environmental objectives (Seidel and Recker 2012). Contrary to Green IT, Green BPM integrates the ideas related to Green IT with widely-accepted management principles rather than focusing on the environmental lifecycle of technological devices (ICT) (Mithas et al. 2010). Green BPM addresses, for example, the extension of conventional process models with emission annotations (Hoesch-Klohe et al. 2010) or the calculation of carbon dioxide emission of business processes (Recker et al. 2011). To create a holistic Green BPM approach, methods must be provided that support the entire lifecycle of a business in terms of ecological objectives (e.g. energy consumption or CO_2 emission) (Houy et al. 2011).

In the last couple of years the field of Green IS in general and Green BPM in particular have addressed a number of environmental issues, leading to an increasing interest and ultimately to the emergence of tracks at high-profile conferences and special issues of high-profile journals dedicated to ecological issues in Business Process Management (vom Brocke et al. 2013).

3   Introduction to the Concept of Ecological Process Patterns

Traditional approaches of process optimization aim at the reduction of costs, cycle time, or enhancement of process quality. The optimization in terms of ecological goals, however, should exploit the processes' inherent capabilities towards the reduction of consumed resources. At its simplest, this could be electrical energy spent during the execution of process steps or tasks (ICT, machines). More complex resource consumptions such as raw materials in various forms (solids, liquids, gas) can also be subject to process optimization. In both cases, the single task or process blocks have to be classified regarding their resource consumption (Houy et al. 2012) in order to extract meaningful ecological process patterns.

When considering process models at design level, there are several areas that can be addressed to reduce the resource consumption. These are represented by the three perspectives *control-flow*, *operational* and *data* (Lübbecke et al. 2016). Following the control-flow patterns of (van der Aalst et al. 2003), the configuration of control-flow elements can have an impact on the resource consumption. For example, if a process flow is split by an exclusive choice element at a certain point and the following tasks differ in terms of resource consumption, a routing over the more advantageous path should be encouraged by taking appropriate measures. The recurring traverse of process cycles (for instance due to quality problems of a manufactured item) could lead to an unnecessary consumption of resources. The variety of factors that influence the route taken by the process flow makes it hard to condense patterns from this level.

Resource consumption that results from creating, storing, processing or manipulating data objects can be optimized at the *data perspective*. Here, many weakness patterns such as the mishandling of information (media disruption) or the creation of information objects not used in later processes can be extracted.

The *operational perspective* addresses the information systems that are involved in the execution of processes. This can be any type of software (e.g. ERP/CRM, browser, office suite) or hardware (e.g. printer, scanner, fax, desktop pc or server). A simple textbook-style ecological pattern for instance could be to use e-mail instead of a "print and mail" process to save the energy for the fusing process in laser printers.

The representation of the extracted patterns depends on what the patterns are supposed to be used for. For a process designer, a textual representation in the POSA or "Gang of Four" style could be sufficient (Buschmann et al. 1998; Gamma et al. 1995). For the detection of known weakness patterns in large process model corpora, however, the representation in a formal way would be necessary (Becker et al. 2011) (Fig. 1).

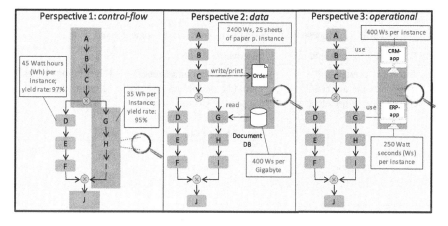

Fig. 1. Three perspectives on ecological weakness patterns

4 Extraction of Ecological Process Patterns from Processes

The creation and application of reliable Ecological Process Patterns should be carried out in certain steps. In the first step, recurring weaknesses have to be identified in process models and more ecological substitutions have to be suggested. The initial identification can be performed by a process designer. In the second step, however, the set of patterns should be further evaluated by a domain expert to make sure that the weaknesses can be substituted by the alternatives. In public administration, for example, the paper or written form of documents is oftentimes required by law which prohibits its substitution by e-mail or a digital representation in general. In the third step, the valid patterns must be formalized in either a textual, graphical or formal way, depending on the purpose of the patterns and their further use. In a fourth step, the formalized patterns can be used for process optimization. The patterns could, for example, be implemented to process modeling suites to support sustainable BPM at design-time. The patterns can also be used to search process corpora for weaknesses that can then be subject to improving measures.

Research design: We performed a search for ecological weakness patterns in the field of PA. As input data, we used 24 process models that refer to the registration (1), notification of change (2) and deregistration (3) of new and existing businesses as used by eight county seats of a small federal state in Germany. These 24 processes were collected through interviews, transcribed and later transformed to process models by a modeling expert. To expand the input data, we used a catalog with existing general modeling weakness patterns from the field of PA.[1] A modeling expert with detailed knowledge in the field of ecological sustainability analyzed the 280 presented patterns in terms of potential impact on the resource consumption. Where the expert could provide a solution for a more resource-efficient execution, we considered the as-is situation a weakness pattern and the more advantageous solution an EPP for that specific task. In the following table, we list the identified weakness patterns along with the capability for improvement and the source in which we found the pattern. For patterns found in the pattern catalog of Delfmann et al., we added the pattern number as used in the original catalog for reasons of traceability (Table 1).

Table 1. Process weakness patterns from an ecological standpoint

ID	Description of the pattern	Improvement potential	Source
1	A form is printed and afterwards the application is checked for completeness.	The check for completeness should be performed prior to the print of documents to avoid unnecessary prints in case the application is	Registration, Cities of Quierschied, Merzig

(continued)

---

[1] The pattern catalog is currently only available as internal working paper but can be requested from Patrick Delfmann of Münster University (patrick.delfmann@ercis.uni-muenster.de).

Table 1. (*continued*)

ID	Description of the pattern	Improvement potential	Source
		missing some other documents.	
2	Documents are printed and sent via mail, which results in consumption of a considerable amount of energy for the print process.	Documents should be sent via e-mail or provided through download instead to avoid unnecessary energy consumption.	Registration, deregistration, change, at several cities.
3	Payment slip for registration fee is handed to the customer personally and mailed to his address as well.	The mailing of the slip should be skipped if the customer already received it personally.	Deregistration, Sulzbach
4	Application received via third-party Information System (IS) is printed and then entered to the in-house IS.	Integration of the two Information Systems could spare the printing of the application form.	Registration, Quierschied
5	Deregistration form is printed but never used afterwards.	The form should not be printed if not used afterwards.	Deregistration, St. Ingbert
6	Document data is entered into IT and subsequently examined or formally reviewed.	The sequence of both processes should be altered to make sure the data input happens only once, thus avoiding unnecessary energy consumption for the input process.	Delfmann, Sect. 2.4.6
7	Information is sent via email in multiple steps instead of using multiple recipients, CC or BCC.	Every process of sending only a single e-mail causes unnecessary energy consumption. E-Mails should be pooled.	Delfmann, 2.8.1.2
8	A document is forwarded twice. Once as e-mail and once as print without the print version being further processed in the future.	The creation and forwarding of the printed document is unnecessary and can be avoided to save energy.	Delfmann, 3.5.4a
9	A text document is scanned. Later, the information of the document is entered manually to an Information	The scan process is either not necessary or the information should be transferred to the IS after the	Delfmann, 3.4.1

(*continued*)

Table 1. (*continued*)

ID	Description of the pattern	Improvement potential	Source
	System (e.g. word processor).	scan automatically to save energy.	
10	Redundant document storage in different media.	If a document is stored in different media (paper, HDD, SSD), the creation (print) and possibly the storage consumes energy (server). Redundant document storage should be avoided if possible (e.g. for unimportant data).	Delfmann, 3.4.2
11	A document is forwarded via mail. After the receipt, the receiver scans the document.	Document should be forwarded by e-mail to avoid energy consumption during transport, print and scan.	Delfmann, 3.5.3
12	The same document is sent through different communication channels to various recipients at various times instead of collectively sending the document by e-mail.	The forwarding through different communication channels (mail, in-house mail) consumes energy for transport and prints of the document and should be replaced by a single e-mail process.	Delfmann, 3.5.4b
13	A created document is never used again after its creation in any process.	The process of document creation could be avoided. This leads to the saving of electric energy.	Delfmann, 3.6.1
14	The process contains numerous print tasks that exceed a given threshold.	Document printing with laser or photocopier consumes a high amount of energy due to the thermal processing.	Delfmann, 3.6.4
15	A media disruption takes place from a physical to a digital representation of information or vice versa.	Energy for the transformation of the information's representation could be saved by keeping a single representation (print, scan).	Delfmann, 3.6.5
16	Document is being archived as a physical and digital representation.	Energy either for the print or the storage devices could be saved.	Delfmann, 4.1.1
17	Document is printed and scanned later on.	Energy for the transformation of the	Delfmann, 4.2.1

(*continued*)

Table 1. (*continued*)

ID	Description of the pattern	Improvement potential	Source
		information's representation could be saved (print, scan).	
18	A clerk prints a document to review it. The changes are then made in the digital representation in the Information System.	Energy for the unnecessary print.	Delfmann, 4.2.2
19	Data was burned on a CD. The CD is forwarded to a receiver who prints the documents on the CD.	The energy intensive process of burning a CD is superfluous and can be replaced by e-mail.	Delfmann, 4.2.3
20	Digitally available information is exchanged personally by using a print process.	Digitally available information should be exchanged in a digital representation to avoid energy consumption for the print of a document.	Delfmann, 5.1.3
21	Information is exchanged via FAX instead of using e-mail.	As with a photocopier, a FAX consumes a huge amount of energy even during standby. E-mail should be used to send documents instead.	Delfmann, 5.1.1
22	Information between software applications is exchanged via e-mail.	The software applications should comprise proper interfaces to support a seamless data exchange to avoid the energy consumption resulting from e-mail processes.	Delfmann, 5.1.4
23	Data is requested and exchanged via in-house mail instead of e-mail.	Document transport via in-house mail includes a thermal print of the document, which again leads to high amounts of electrical energy that is consumed.	Delfmann, 5.1.6
24	Data manually entered into an Information System.	The manual input of information to an Information System causes unnecessary energy consumption for the time the input process takes. Again, proper interfaces between	Delfmann, 5.2.3

(*continued*)

Table 1. (*continued*)

ID	Description of the pattern	Improvement potential	Source
		the applications could solve this issue.	
25	Data manually entered into multiple Information Systems.	Same as 23 but multiple energy consumption.	Delfmann, 5.2.6
26	The exchange of information or documents exceeds a reasonable threshold.	Excessive information exchange always comes along with a huge consumption for the resource that carries the information from Point A to B (energy, gas, etc.). Therefore, the number of information exchanges should be limited to a reasonable number.	Delfmann, 7.2.3

Although some of the weakness patterns seem to be very similar at first glance, they still differ from each other in details (for instance: 1 and 7). These nuances need to be considered to facilitate the transformation of the patterns to a formal representation for a (semi-)automatic discovery of occurrences in process models.

5   Conclusion

The goal of this paper was to provide an introduction to our ongoing work in the field of sustainable BPM in general and sustainability patterns in particular. Based on the ideas of patterns in software-engineering, we presented a concept for ecological process patterns in the domain of BPM. The concept provides the foundation for the evaluation of IT-related processes from the perspectives *control-flow*, *operation* and *data* from a sustainability point of view. To apply the concept for the improvement of existing processes or a sustainability-oriented design of processes from scratch, a large number of patterns have to be provided along with the concept. We could extract a total number of 26 EPPs by examining real-world processes from the PA and from a catalog of generic weakness patterns.

The limitation of our work comes from the focus on the PA domain, where saving energy is largely limited to resources for the transport of physical information (letter, document file), paper for the print of information and of course electrical energy. Furthermore, when transferring the pattern approach to alternative domains such as manufacturing, domain knowledge about alternative ways of executing tasks in scope will be necessary. Therefore, domain experts must be included in the process of identification of weakness patterns and proposal of more sustainable alternatives.

The next steps in this ongoing research will be to prepare the patterns for an evaluation together with domain experts from the field of PA to abandon patterns that cannot be used to optimize processes due to legal or commensurability reasons. The remaining patterns then need to be transformed into a formal description (rule set) and implemented in BPM software such as ARIS, a step from which process designers benefit when creating new processes or during the revision of existing ones. When modeling new processes in the field of PA, the process constructs can be validated against the rule set and problematic constructs can be highlighted or more ecological constructs can be suggested. The rule set could also be used to test existing processes and their process constructs for a violation of the rules automatically and improve the respective constructs. Thus, the collection of patterns contributes to sustainable BPM by supporting the substitution of regular processes with more ecological alternatives. The opportunities of reducing resource consumption that arise from the optimization of processes have already been discussed by (Reiter et al. 2014).

More complex patterns shall be extracted from the domain of industrial production in the near future. Production-related data from PPS systems (e.g. schedules) or process instance data from workflow management systems or from machine data logging can be combined and annotated to process models. This will allow the comparison of fungible production processes and reasonable best-practice in the form of patterns that can be proposed for production tasks. In the future, these patterns could also be used in an Industry 4.0 scenario for the instant routing of workpieces through the production line in a way that causes a minimum of resource consumption.

Acknowledgement. Parts of the research described in this paper was supported by a grant from the German Ministry of Education and Research (BMBF), project name GreenFlow, support code 01IS12050.

References

Alexander, C.: The Timeless Way of Building. Oxford University Press, New York (1977)

Ambler, S.W.: Process Patterns: Building Large-Scale Systems Using Object Technology. SIGS Books/Cambridge University Press, Cambridge (1998)

Becker, J., Bergener, P., Delfmann, P., Weiss, B.: Modeling and Checking business process compliance rules in the financial sector. In: International Conference on Information Systems (ICIS 2011), Shanghai, China, (2011)

Buschmann, F., Meunier, R., Rohnert, H., Sommerlad, P., Stal, M.: Pattern-Oriented Software Architecture. A System of Patterns. Wiley, Hoboken (1998)

Coplien, J.: A generative development-process pattern language: techniques, strategies, and applications. In: Coplien, J.O., Schmidt, D.C. (eds.) Pattern Languages of Program Design, pp. 183–237. Addison-Wesley, New York (1995)

Delfmann, P., Hübers, M.: Towards supporting business process compliance checking with compliance pattern catalogues. Enterp. Model. Inf. Syst. Architect. **10**(1), 67–88 (2015)

Forster, F.: The idea behind business process improvement: toward a business process improvement pattern framework. In: BP Trends, pp. 1–14, April 2006

Fowler, M.: Analysis Patterns: Resusable Object Models. Addison-Wesley, Reading (1996)

Gamma, E., Helm, R., Johnson, R., Vlissides, J.: Design Patterns – Elements of Reusable Object-Oriented Software. Addison-Wesley, Reading (1995)

Hoesch-Klohe, K., Ghose, A., Lê, L.-S.: Towards green business process management. In: IEEE International Conference on Services Computing (SCC) (2010)

Houy, C., Reiter, M., Fettke, P., Loos, P.: Towards green BPM – sustainability and resource efficiency through business process management. In: Muehlen, M., Su, J. (eds.) BPM 2010. LNBIP, vol. 66, pp. 501–510. Springer, Heidelberg (2011). doi:10.1007/978-3-642-205 11-8_46

Houy, C., Reiter, M., Fettke, P., Loos, P., Hoesch-Klohe, K., Ghose, A.: Advancing business process technology for humanity – opportunities and challenges of green BPM for sustainable business activities. In: vom Brocke, J., Seidel, J., Recker, J. (eds.) Green Business Process Management-Towards the Sustainable Enterprise, Progress in IS, pp. 75–92. Springer, Heidelberg (2012)

Krcmar, H.: Referenzmodelle für Informationssysteme im Umweltbereich - von der Modellierung in EcoIntegral zur Umsetzung in EcoRapid. In: Becker, J., Knackstedt, R. (eds.) Tagungsband zur 6. Fachtagung Referenzmodellierung 2002 im Rahmen der MKWI 2002, Nürnberg (2002)

Lohrmann, M., Reichert, M.: Effective application of process improvement patterns to business processes. Softw. Syst. Model. **15**, 1–23 (2014)

Lübbecke, P., Fettke, P., Loos, P.: Towards ecological workflow patterns as an instrument to optimize business processes with respect to ecological goals. In: 49th Hawaii International Conference on System Sciences (HICSS 49), Kauai, USA (2016)

Lübbecke, P., Reiter, M., Fettke, P., Loos, P.: Simulation-based decision support for the reduction of the energy consumption of complex business processes. In: 48th Hawaii International Conference on System Sciences (HICSS 48), Kauai, USA (2015)

Mangalaraj, G., Nerur, S.P., Mahapatra, R.K., Price, K.H.: Distributed cognition in software design: an experimental investigation of the role of design patterns and collaboration. MIS Q. **38**(1), 249–274 (2014)

Mithas, S., Khuntia, J., Roy, P.K.: Green information technology, energy efficiency, and profits: evidence from an emerging economy. In: 31st International Conference on Information Systems (ICIS), Saint Louis, USA (2010)

Nowak, A., Leymann, F., Schleicher, D., Schumm, D., Wagner, S.: Green business process patterns. In: 18th Conference on Pattern Languages of Programs (PLoP 2011). ACM, New York (2013)

Nowak, A., Leymann, F., Schumm, D., Wetzstein, B.: An architecture and methodology for a four-phased approach to green business process reengineering. In: Kranzlmüller, D., Toja, A.M. (eds.) ICT-GLOW 2011. LNCS, vol. 6868, pp. 150–164. Springer, Heidelberg (2011). doi:10.1007/978-3-642-23447-7_14

Recker, J., Rosemann, M., Gohar, E.R.: Measuring the carbon footprint of business processes. In: Muehlen, M., Su, J. (eds.) BPM 2010. LNBIP, vol. 66, pp. 511–520. Springer, Heidelberg (2011). doi:10.1007/978-3-642-20511-8_47

Reiter, M., Fettke, P., Loos, P.: Towards green business process management: concept and implementaion of an artifact to reduce the energy consumption of business processes. In: 47th Hawaii International Conference on System Science (HICSS 47), Waikoloa, USA (2014)

Riehle, D., Züllighoven, H.: Understanding and using patterns in software development. Theor. Pract. Object Syst. **2**(1), 3–13 (1996)

Rozinat, A., van der Aalst, W.M.: Conformance checking of processes based on monitoring real behavior. Inf. Syst. **33**(1), 64–95 (2008)

Seidel, S., Recker, J.: Implementing Green business process: the importance of functional affordances of information systems. In: Lamp, J. (ed.) 23rd Australasian Conference on Information Systems. Geelong, Australia (2012)

Smirnov, S., Weidlich, M., Mendling, J., Weske, M.: Action patterns in business process model repositories. Comput. Ind. **63**(2), 98–111 (2012)

Aalst, W.M.P., Barros, A.P., Hofstede, A.H.M., Kiepuszewski, B.: Advanced workflow patterns. In: Scheuermann, P., Etzion, O. (eds.) CoopIS 2000. LNCS, vol. 1901, pp. 18–29. Springer, Heidelberg (2000). doi:10.1007/10722620_2

van der Aalst, W.M., Barros, A.P., ter Hofstede, A.H.M., Kiepuszewski, B.: Workflow Patterns. Distrib. Parallel Databases **14**, 5–51 (2003)

Dijk, A.: Contracting workflows and protocol patterns. In: Aalst, Wil M.P., Weske, M. (eds.) BPM 2003. LNCS, vol. 2678, pp. 152–167. Springer, Heidelberg (2003). doi:10.1007/3-540-44895-0_11

vom Brocke, J., Watson, R.T., Dwyer, C., Elliot, S., Melville, N.: Green IS: directives for the IS discipline. Commun. AIS **33**(30), 508–519 (2013)

Winter, R., vom Brocke, J., Fettke, P., Loos, P., Junginger, S., Moser, C., Keller, W., Matthes, F., Ernst, A.: Patterns in business and information systems engineering. Bus. Inf. Syst. Eng. **1**(6), 468–474 (2009)

How to Incorporate Sustainability into Business Process Management Lifecycle?

Andréa Magalhães Magdaleno[1,2], Leticia Duboc[3], and Stefanie Betz[4(✉)]

[1] IC – Computing Institute, UFF – Fluminense Federal University,
Niterói 24210-346, Brazil
andrea@ic.uff.br
[2] EY Brazil Consulting, Rio de Janeiro, Brazil
[3] Instituto de Matemática e Estatística, Universidade do Estado do Rio de Janeiro (UERJ),
Rio de Janeiro, Brazil
leticia@ime.uerj.br
[4] Karlsruhe Institute of Technology, Karlsruhe, Germany
stefanie.betz@kit.edu
http://www.ey.com/br/pt/home

Abstract. The impact of human activities has become one of the greatest concerns of our society, leading to global efforts to achieve sustainable development. Business organizations play an important role on this endeavor. Business Process Management (BPM) offers a comprehensive approach for designing, implementing, executing and monitoring business processes. This position paper aims to raise the discussion about how BPM can be extended to consider the direct and indirect effects of the business processes in the environmental, economic and social dimensions of sustainability.

Keywords: Sustainability · Sustainable business processes · Green business processes · Green BPM · Sustainable BPM

1 Introduction

Sustainable development is defined [15] as "the development that meets the needs of the present without compromising the ability of future generations to meet their own needs". This ability includes the satisfaction of three interdependent dimensions: environmental, social and economic [5]. It requires integrating social justice, quality of life, health and prosperity, while respecting the Earth's capacity to support life.

The impact of sustainability has been classified in three orders [4]: First order (direct) - impacts resulted from the creation and running of technology. Second order (indirect) - impacts caused by the continuous use of technology and application of its surrounding processes. Finally, third order (systemic) - impacts resulted from the use of technology and its process by many people over medium and long term.

The objective of this work is to develop concepts, methods and tools that can be applied throughout the BPM lifecycle in order to incorporate sustainability practices into business processes. It focus on the first phase of the business process lifecycle: the modeling. Process modeling is key to describe the activities, tasks, roles, artifacts, IT systems, and

M. Dumas and M. Fantinato (Eds.): BPM 2016 Workshops, LNBIP 281, pp. 440–443, 2017.
DOI: 10.1007/978-3-319-58457-7_32

business rules of an organizational business process. Therefore, it was considered an appropriate starting point in the design of sustainable processes.

2   Why Use BPM to Promote Sustainability?

Sustainability has become a driving force in our society. Consumers' preferences for products and services with a sustainable character have led organizations to value an image of "green business". Others are obliged to comply with regulations to have permission to operate, such as companies that exploit natural resources. Finally, some have realized that sustainability practices can actually bring financial benefits [11].

In designing, implementing, executing and overall managing their business processes, organizations have traditionally focused in time, costs, efficiency, and quality drivers only [3]. With the emergence of sustainability, these classical drivers are increasingly subjected to critical scrutiny [13].

Existing approaches in the area of sustainability and BPM are focusing primarily on the environmental dimension, often overlooking social and economic dimensions [14]. They are normally concerned with "the understanding, documenting, modeling, analyzing, simulating, executing, and continuously changing of business process with dedicated consideration paid to environmental consequences of the business processes" [2] and are subsumed under the umbrella of Green BPM. Moreover, discussions on Green BPM are still in early stages and so far few approaches exist [6, 9].

In order to truly incorporate sustainability into BPM, one must execute each phase of its lifecycle with explicit consideration for the direct, indirect and systemic effects of the business processes in the three dimensions of sustainability. The BPM lifecycle consists of four different cyclic phases: process modeling, simulation and optimization, implementation and execution, monitoring and improvement. Such an holistic approach requires not only considering sustainability in every phase of the business process lifecycle, but also using concepts, methods, techniques, and tools to support the stakeholders. By doing so, stakeholders can model and simulate alternative sustainability practices into the business processes and optimize them, while considering the different degrees of impact on the environment, economy and society.

3   Research Agenda: A Roadmap to Community Discussion

Our research agenda on Sustainability and BPM topic is initially focused on modeling phase of BPM lifecycle and comprises four specific goals (SG), which we would like to discuss with the community during the workshop:

SG1: Definition of a Conceptual Model of the Relations Between Sustainability and Business Processes
From the existing definitions of Green BPM and Sustainab (le/ility) BPM, the research area have now some basic concepts used in these areas. We aim to establish a common and shared understanding of the concepts to avoid the proliferation of multiple definitions and different applications of the concepts within the community. There is already

a preliminary work of one of the authors that can serve as the starting point for this conceptual model or taxonomy [1]. We intend to invite other researchers to contribute and validate the model. The main idea is to promote the collaborative and open construction of the model to increase its soundness, dissemination and acceptance.

SG2: Development of a Sustainable BPM Language (BPML)

Study typical BPM notations - BPMN [8] and EPC (Event Process Chain) [12] - to identify if they already meet all the representation requirements of the sustainability-related concepts or if they need to be adapted to obtain a sustainable BPML. In a first literature review, we identified two possible approaches: models annotations [6] or notations extensions [10]. Models annotations include in every activity of a business process, relevant values concerning, such as the consumption of resources or the production of waste materials. On the other hand, notations extensions propose new constructs or stereotypes, to characterize the process map (and its elements). For instance, one can have a special symbol for activities related to fuel or paper consumption, or highlight gas emissions indicators. Summarizing, both approaches have the same proposal: to capture relevant information for business processes regarding the sustainable dimension, but with different levels of structure. The use of a more structured approach, as in the case of notation extensions, facilitates data collection and analysis to process simulation, monitoring and measuring.

SG3: Production of a Catalog of Metrics

We are aware that general literature on sustainability metrics already exist [7]. However, it is necessary to verify which of the metrics proposed can be applied in business processes context. In addition, the current literature on sustainability metrics for BPM seems still limited and mostly targeted at the environmental dimension. Our aim is to consolidate these works, producing a catalog of metrics that encompasses the three dimensions of sustainability and that are applicable to the context of business processes. If this reuse of metrics proves to be effective, our main contribution should be to develop a measuring method that indicate how to fit the particular business process structure, i.e., traditional process elements or new sustainable process elements (defined in the conceptual model and represented in the notation).

SG4: Process Visualization

Explore the process visualization approaches and tools that already exist to verify if they can be used to amplify the sustainability lens of business processes. To achieve process improvements regarding sustainability, it is fundamental that process specialists can understand and analyze the business process model to make decisions. The process visualization allows the observation of facts and knowledge extraction. If necessary, an initial prototype can suggest some possible visualizations that would be helpful in the sustainability context. In addition, it is also possible to offer awareness and interaction mechanisms with the process model to potentialize the results.

References

1. Betz, S.: Sustainability aware business process management using XML-Nets. In: 28th Conference on Environmental Informatics - Informatics for Environmental Protection, Sustainable Development and Risk Management, Oldenburg (2014)
2. vom Brocke, J., et al.: Green Business Process Management: Towards the Sustainable Enterprise. Springer, Heidelberg (2012)
3. Dumas, M., et al.: Fundamentals of Business Process Management. Springer, New York (2013)
4. Forum for the future: The impact of ICT on sustainable development. EITO - European Information Technology Observator, Frankfurt (2002)
5. Goodland, R.: Sustainability: Human, Social, Economic and Environmental (2002)
6. Houy, C., et al.: Advancing business process technology for humanity opportunities and challenges of green BPM for sustainable business activities. In: vom Brocke, J., et al. (eds.) Green Business Process Management: Towards the Sustainable Enterprise, pp. 75–92. Springer, Heidelberg (2012)
7. Jamous, N., Müller, K.: Environmental performance indicators. In: Dada, A., Stanoevska, K., Gómez, J.M. (eds.) Organizations' Environmental Performance Indicators Measuring, Monitoring, and Management. Springer, Heidelberg (2013)
8. OMG: Business Process Management Notation (BPMN) Version 1.2. http://www.bpmn.org/
9. Opitz, N., et al.: Environmentally sustainable business process management – developing a green BPM readiness model. In: Pacific Asia Conference on Information Systems (PACIS), Chengdu, China (2014)
10. Recker, J., et al.: Modeling and analyzing the carbon footprint of business processes. In: vom Brocke, J., et al. (eds.) Green Business Process Management: Towards the Sustainable Enterprise, pp. 93–109. Springer, Heidelberg (2012)
11. Saul, J.: Social Innovation Inc: 5 Strategies for Driving Business Growth through Social Change. Jossey-Bass, San Francisco (2010)
12. Scheer, A.-W.: ARIS - Business Process Modeling. Springer, Heidelberg (2000)
13. Seidel, S., et al.: Green business process management. In: vom Brocke, J., et al. (eds.) Green Business Process Management: Towards the Sustainable Enterprise, pp. 3–13. Springer, Heidelberg (2012)
14. Stolze, C., et al.: Sustainability in business process management research – a literature review. In: Americas Conference on Information Systems (AMCIS), Seattle, EUA (2012)
15. UN World Commission on Environment and Development: Report of the World Commission on Environment and Development: Our Common Future. Oxford University Press, Oxford, New York (1987)

5th International Workshop on Theory and Application of Visualizations and Human-centric Aspects in Processes

Introduction to the 5th International Workshop on Theory and Application of Visualizations and Human-centric Aspects in Processes

Ross Brown[1(✉)], Simone Kriglstein[2], and Stefanie Rinderle-Ma[3]

[1] Science and Engineering Faculty,
Queensland University of Technology, Brisbane, Australia
r.brown@qut.edu.au
[2] Faculty of Informatics, Vienna University of Technology, Vienna, Austria
kriglstein@cvast.tuwien.ac.at
[3] Faculty of Computer Science, University of Vienna, Vienna, Austria
stefanie.rinderle-ma@univie.ac.at

1   Introduction

This is the fifth TAProViz workshop being run at the 14th International Conference on Business Process Management (BPM). The intention this year is to consolidate on the results of the previous successful workshops by further developing this important topic, identifying the key research topics of interest to the BPM visualization community. Towards this goal, the workshop topics were extended to human computer interaction and related domains.

Submitted papers were evaluated by at least three program committee members, in a double blind manner, on the basis of significance, originality, technical quality and exposition. Three full papers were accepted for presentation at the workshop from five submissions.

The papers address a number of topics in the area of process model visualization, in particular:

- Visualization of Layout Metrics for Business Process Models
- Eye Tracking and Process of Process Modeling
- Visually Comparing Process Dynamics

In their full paper, *Visualization of the Evolution of Layout Metrics for Business Process Models*, Cornelia Haisjackl, Andrea Burattin, Pnina Soffer, and Barbara Weber, extend existing research by opening the black box of modeller behavior, introducing an enhanced technique enabling the visual analysis of the modelers layouting behavior. They demonstrate examples showing that their approach provides valuable insights to better understand and support the creation of process models.

Andrea Burattin, Michael Kaiser, Manuel Neurauter and Barbara Weber, present their full paper *Eye Tracking Meets the Process of Process Modeling: a Visual Analytic Approach*. In this paper they introduce an additional stream of data (i.e., eye tracking) to improve the analysis of the Process of Process Modelling (PPM). They show that, by exploiting this additional source of information, they can refine the detection of comprehension phases as well as provide more exploratory PPM visualizations that are both static and dynamic.

In the final full paper, *Visually Comparing Process Dynamics with Rhythm-Eye Views*, Jens Gulden presents an alternative to the common timeline projection of process event data, which allows the projection of a series of time-related events and regularities onto a circular structure. This visualization comes with advantages over timeline projections and provides more flexibility in configuration, especially for comparing process rhythms in multiple sets of event data.

2 Organizers

Ross Brown	Queensland University of Technology, Australia
Simone Kriglstein	Vienna University of Technology, Austria
Stefanie Rinderle-Ma	University of Vienna, Austria

3 Program Committee

Ralph Bobrik	Switzerland
Massimiliano De Leoni	Eindhoven University of Technology, Netherlands
Remco Dijkman	Eindhoven University of Technology, Netherlands
Phillip Effinger	Eberhard Karls Universität Tübingen, Germany
Kathrin Figl	Vienna University of Economics and Business, Austria
Hans-Georg Fill	University of Vienna, Austria
Jens Kolb	Ulm University, Germany
Agnes Koschmider	Karlsruhe Institute of Technology, Germany
Maya Lincoln	Technion, Israel Institute of Technology, Israel
Jürgen Mangler	University of Vienna, Austria
Cristiano Maciel	Universidade Federal de Mato Grosso, Brazil
Silvia Miksch	Vienna University of Technology, Austria
Margit Pohl	Vienna University of Technology, Austria
Rune Rasmussen	Queensland University of Technology, Australia
Luciana Salgado	Universidade Federal Fluminense, Brazil
Flavia Santoro	NP2Tec/UNIRIO, Brazil
Pnina Soffer	University of Haifa, Israel
Irene Vanderfeesten	Eindhoven University of Technology, Netherlands
Eric Verbeek	Eindhoven University of Technology, Netherlands
Günter Wallner	University of Applied Arts Vienna, Austria

Visualization of the Evolution of Layout Metrics for Business Process Models

Cornelia Haisjackl[1]([✉]), Andrea Burattin[1], Pnina Soffer[2], and Barbara Weber[1,3]

[1] University of Innsbruck, Innsbruck, Austria
cornelia.haisjackl@uibk.ac.at
[2] University of Haifa, Haifa, Israel
[3] Technical University of Denmark, Kongens Lyngby, Denmark

Abstract. Considerable progress regarding impact factors of process model understandability has been achieved. For example, it has been shown that layout features of process models have an effect on model understandability. Even so, it appears that our knowledge about the modeler's behavior regarding the layout of a model is very limited. In particular, research focuses on the end product or the outcome of the process modeling act rather than the act itself. This paper extends existing research by opening this black box and introducing an enhanced technique enabling the visual analysis of the modeler's behavior towards layout. We demonstrate examples showing that our approach provides valuable insights to better understand and support the creation of process models. Additionally, we sketch challenges impeding this support for future research.

Keywords: Process of process modeling · Human–centered support · Process model quality · Understandability · Layout properties · Visualization

1 Introduction

In recent years, business process models have gained significant relevance due to their critical role for the management of business processes [1]. Advantages of using business process models are, for example, obtaining a common understanding of a company's business processes [2], or enabling the discovery of improvement opportunities [3]. Literature emphasizes that a good understanding of a process model has positive influence on the success of a modeling initiative [4]. Research on impact factors of process model understandability focuses on the product or outcome of the process modeling act [5–7], identifying features that make models easier to understand [8]. In particular, features that characterize the layout of these models, which are part of their secondary notation, have been shown to have an effect on model understandability [9,10].

To promote the creation of understandable models and to overcome quality problems right from the start, we need to support humans during the creation of

M. Dumas and M. Fantinato (Eds.): BPM 2016 Workshops, LNBIP 281, pp. 449–460, 2017.
DOI: 10.1007/978-3-319-58457-7_33

business process models [11]. However, our knowledge about the modeling behavior regarding the model's layout is very limited. Investigations of the process of process modeling [12] identified distinct modeling styles that differ from each other in the amount of editing operations devoted to reconciliation of the model while creating it. Yet, we do not know what layout features (e.g., crossing edges) are addressed at what phase of the modeling process nor the relation between the modeling process and the final features of the resulting model. We believe that to efficiently support the creation of understandable models with good quality, we need to strengthen our understanding of the layouting behavior of modelers while creating process models.

This paper extends existing research on the process of process modeling, i.e., the act of formalizing a process into a process model. Pinggera et al. proposed a visualization of the process of process modeling called modeling phase diagrams [13]. The modeling phase diagram is a graphical analysis technique visualizing the low level interactions during the creation of a process model by mapping these onto phases. In this paper we introduce the advanced modeling phase diagram that overlays the modeling phase diagram with layout metrics focusing on layout properties presented in [5], allowing us to visually analyze the evolution of these metrics. We present examples showing how our analysis technique enables us to investigate the modeler's behavior toward selected layout metrics. Hence, this technique contributes insights to understand and support the modeler during the process of process modeling, fostering the creation of understandable models. In addition, we outline challenges regarding the interpretation of advanced modeling phase diagrams for further research.

The remainder of the paper is structured as follows. Section 2 gives required information about the process of process modeling (Sect. 2.1) and selected layout metrics (Sect. 2.2). Section 3 describes the advanced modeling phase diagram (Sect. 3.1) and demonstrates them along several examples (Sect. 3.2). In addition, it deals with challenges the advanced modeling phase diagram meets (Sect. 3.3) and limitations (Sect. 3.4). Finally, Sect. 5 concludes the paper.

2   Background

Before introducing the advanced modeling phase diagram, this section gives insight to required background knowledge. In particular, Sect. 2.1 introduces process of process modeling features and their operationalization through the modeling phase diagram. Section 2.2 gives an overview of selected layout metrics, which are used to demonstrate the use of the advanced modeling phase diagram.

2.1   Process of Process Modeling

The lifecycle of process model development can be characterized as an iterative and collaborative process, involving an elicitation and formalization phase [14]. In the elicitation phase information is extracted by the domain experts from the domain. This process is described in literature as a negotiation process [15]. The extracted information is then used in the formalization phase by the process modeler for creating the formal process model and validating it. This process is also denoted as process of process modeling. [13] states that the modeler's interaction with the tool, i.e., the process of process modeling, can be seen as a cycle of three successive phases, namely comprehension, modeling, and reconciliation. During the comprehension phase the modeler tries to assess the requirements to be modeled. This knowledge is used during the modeling phase to create the process model. Afterwards, the modeler enhances the understandability of the process model in the reconciliation phase (e.g., he moves modeling elements to a new position in the process model).

To analyze the process of process modeling in a systematic manner, the Cheetah Experimental Platform (CEP) [16] is used. CEP offers a basic process modeling editor and logs any user interaction together with the corresponding time stamp. These event logs describe the creation of the process model step by step, and their replay allows analyzing the process of process modeling at any point in time. [13] offers an overview of all possible interactions and their classification into comprehension, modeling, and reconciliation phases. This classification is also used to visualize the process of process modeling by mapping the model interactions onto the modeling phases (i.e., comprehension, modeling, and reconciliation), resulting into modeling phase diagrams [13].

2.2 Layout Metrics

In order to investigate the modeler's behavior regarding the model's layout, we focus on layout metrics presented in [5]. All of these metrics have been implemented [17]. For this paper, we selected two of these layout metrics. In particular, we introduce the crossing edge metric and the orthogonal segments metric. These metrics were selected as examples of clearly visible properties which are independent of one another.

Crossing Edge Metric. Literature suggests that models should not contain any crossing edges, as models without them are more comprehensible [6,18]. The crossing edge metric for a process model is described by the number of edge crossings in the model divided by the total number of the model's edges [5]. Therefore, in periods without crossing edges the value of the metric is 0. Note that not all process models can reach a value of 0 upon their completion, as some models cannot be laid out without any edge crossings, i.e., not all process models are planar (cf. [19]). In addition, a trade-off between the organization of modeling elements to patterns or structures and the number of crossing edges needs to be considered [8].

Orthogonal Segments Metric. [19] proposes that edges should make use of a Manhattan layout to establish a readable layout, i.e., edges should be aligned according an orthogonal layout that consists of horizontal and vertical lines. An edge consists of either one segment or more segments divided by bendpoints. A segment is considered orthogonal if it is parallel to either the horizontal axis or

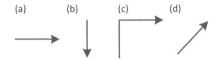

Fig. 1. Examples of orthogonal an non orthogonal segments

the vertical axis, with a threshold of up to 7 pixels. Figure 1 shows three orthogonal segments and one segment that is not orthogonal. Figure 1(a) depicts a segment that is an orthogonal horizontal line, Fig. 1(b) shows a segment that is an orthogonal vertical line, and Fig. 1(c) is an edge consisting of two orthogonal segments divided by one bendpoint. Figure 1(d) depicts a segment which is neither a horizontal nor a vertical orthogonal line. The orthogonal segments metric is calculated by the number of orthogonal segments in a process model divided by the number of all segments in the model [5]. Hence, if a process model only contains orthogonal segments, the value of the metric is 1. However, following the Manhattan layout too strictly can lead to a high number of bendpoints, hampering the readability of the model [19].

3   Investigating the Evolution of Layout Metrics

This section presents our approach of investigating the evolution of layout metrics. While Sect. 3.1 introduces the advanced modeling phase diagram, Sect. 3.2 demonstrates several examples. Section 3.3 summarizes challenges that appear when using the advanced modeling phase diagram. Finally, Sect. 3.4 lists several generalization limitations.

3.1   Advanced Modeling Phase Diagram

The advanced modeling phase diagram is an enhancement of the modeling phase diagram described in Sect. 2.1. It enables to overlay the traditional modeling phase diagram with selected metrics, allowing us to visually analyze the evolution of these metrics. In particular, each interaction with CEP is used to plot the corresponding metric values, i.e., after each interaction with the process model the new model is computed, triggering the computation of all implemented layout metrics.

For example, Fig. 2 shows an advanced modeling phase diagram with the orthogonal segments metric. Like for the modeling phase diagram, the horizontal axis depicts time while the left vertical axis represents the model's number of elements. The participant's interactions with CEP drawn in black are divided into comprehension, modeling, and reconciliation phases. In addition, the evolution of the orthogonal segments metric is depicted in blue. The value of the metric is shown on the right vertical axis.

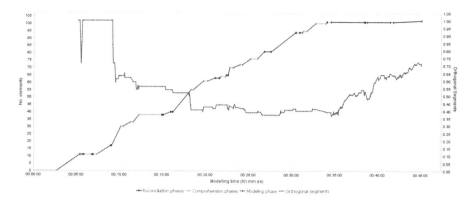

Fig. 2. An advanced modeling phase diagram with the orthogonal segments metric (Color figure online)

3.2 Demonstration

This section focuses on the evolution of selected layout metrics described in Sect. 2.2.

Methodology. We analyzed the advanced modeling phase diagrams for a dataset gained from an modeling session described in [20]. The modeling session was conducted in December 2012 with students from Eindhoven University of Technology with CEP. After an initial phase consisting of a demographic survey and a CEP tutorial, the 120 participants were asked to model a process about getting a mortgage with BPMN. CEP logged any user interaction together with the corresponding time stamp. By using the event logs we created advanced modeling phase diagrams for each process model. Afterwards, the dataset was analyzed manually, i.e., we investigated the advanced modeling phase diagrams for selected layout metrics (cf. Sect. 2.2)[1]. These metrics were selected due to their diversity, e.g., while the crossing edge metric usually is only altered from time to time (i.e., when a crossing edge is introduced), the orthogonal segments metric is affected by almost every interaction with the process model that deals with edges (e.g., add edge, delete edge, move edge, etc.). For a more qualified interpretation of the diagrams we also looked at the replays of the event logs of the user interactions (cf. [16]). In the following, we demonstrate some examples showing reoccurring behaviors that we will discuss afterwards.

Crossing Edge Metric. Figure 3 shows different examples we could repeatedly observe in our dataset showing different reoccurring behavior of how participants dealt with crossing edges. For readability considerations, we removed the modeling phase plot and left only the one related to the crossing edges metric. In

---

[1] A high resolution of all diagrams in this paper including the modeling phases as well as the corresponding process models can be downloaded from: http://bpm.q-e.at/EvolutionOfLayoutMetrics.

general, the calculation of the crossing edge metric starts with the first edge added to the model. Therefore, the initial value of the metric is typically 0 since there are usually no crossing edges in a model containing only one edge. Most participants introduced crossings edges at some point during process model creation, leading to (temporary) increases of the crossing edge metric. Overall, we observed two reasons why a crossing edge appeared: either the participant introduced it by moving a modeling element that was connected to at least one other modeling element or the tool added it as the participant connected two modeling elements. A tool induced crossing edge can happen, e.g., due to connecting two modeling elements that are on opposite sides of already connected modeling elements. In this case, the tool adds the new edge above the existing ones, without trying to find a way around the existing edges. While crossing edges occurred for almost all participants (for 119 from 120), the reactions of the participants to a crossing edge were quite different.

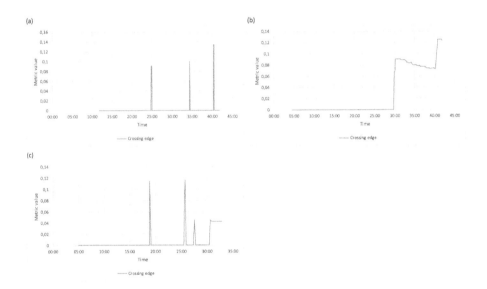

Fig. 3. Examples for the evolution of the crossing edge metric

Figure 3(a) shows an example of a participant who immediately resolved crossing edges that appeared throughout the modeling process. This is reflected by the 3 steep spikes (crossing edges appeared but were resolved within a very short timeframe). The value of the crossing edge metric was 0 throughout the whole session except for the 3 spikes. Figure 3(b), in turn, displays the opposite behavior. Edge crossings that were introduced were never resolved, leading to a continuous increase of the metric. The drop between minutes 30 and 40 is not because of an effort the participant undertook to remove the crossing edges, but can be explained by the fact that the modeler added additional modeling elements, causing the metric to drop slightly till the next crossing edge appeared. Figure 3(c) depicts another typically reoccurring behavior, showing a participant that initially resolved edge crossings immediately, but who left them unattended towards the end of the modeling process. We found three alternative explanations for this situation when examining the modeling replays. First, due to an error the process model was non-planar and therefore it was not possible to resolve the crossing edge. Second, it was too complex to resolve it, i.e., it would have taken many move operations to resolve one crossing edge. Third, the participant could have easily resolved the crossing edge with only a few move operations, but instead decided to finish the modeling session.

Orthogonal Segments Metric.   Figure 4 depicts typical reoccurring behaviors of the evolution of the orthogonal segments metric we identified. With the introduction of the first edge, which is typically orthogonal, the initial value of the metric is 1 since all existing edges (i.e., all segments of this one edge) are orthogonal. As CEP does not provide any tool support for achieving orthogonality, a model with completely orthogonal segments was not achieved by any of the 120 participants of the modeling session. Again, different examples of reoccurring behavior are discussed subsequently.

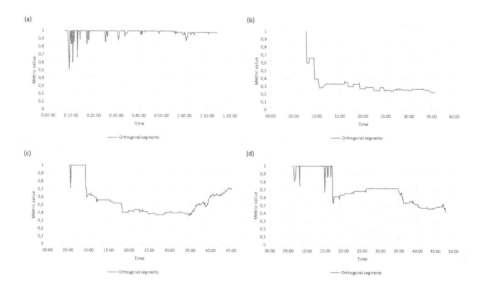

Fig. 4. Examples for the evolution of the orthogonal segments metric

For instance, Fig. 4(a) shows an evolution of the orthogonal segments metric where the participant clearly cared about this metric. The values for the orthogonal segments metric are constantly high (close to 1) except for a few downward spikes. The downward spikes of the metric were due to adding new edges that were not orthogonally aligned by the tool. However, the participant immediately reestablished the orthogonality of the segments. Overall, it can be noted that the spikes in the evolution get smaller throughout the modeling process. If the model already includes many edges, one segment which is not orthogonal does not affect the metric as much as when there are only few edges in the model (i.e., at the beginning of the modeling session). Figure 4(b) highlights another reoccurring behavior where no noticeable efforts can be observed to improve the orthogonality of the model. After modeling the first modeling elements with high segment orthogonality, the value of the metric was mainly decreasing. Figure 4(c), in turn, depicts an example of a participant who initially left the segment's orthogonality unattended (i.e., values of the metric have a downward tendency), but at the end of the modeling process made substantial efforts to improve the orthogonality (i.e., the participant applied several move operations to increase orthogonality of the segments). This is also in accordance with the modeling phase diagram of Fig. 4(c) which can be seen in Fig. 2. The modeling phase diagram shows that the increase of the metric happened during reconciliation phases. It was not always the case that only one specific behavior could be noticed, but we observed several modeling sessions with mixed behavior. For example, Fig. 4(d) shows a mix of different reactions toward the orthogonal segments metric. First the participant immediately reestablished the orthogonality of the segments. Then the value decreases, substantially at minute 15, and from then onwards no efforts could be observed to improve the metric. The slight increase after minute 15 can be explained due to adding new edges that only consisted of orthogonal segments, rather than the result of any reconciliation effort.

3.3   Interpretation Challenges

The examples shown in Sect. 3.2 demonstrate different reoccurring behaviors which contribute towards a better understanding how modelers layout their models. However, this does not come without several difficulties. This section summarizes these challenges that appeared when using the advanced modeling phase diagram for the interpretation of the evolution of layout metrics.

Challenge C1. There is a vast diversity of layout metrics which all need to be interpreted separately [5]. For example, while a low value for the crossing edge metric indicates a good quality, the opposite is the case for the orthogonal segments metric (cf. Sect. 2.2). Therefore, an automatic interpretation needs to consider each layout metric individually.

Challenge C2. Even within one metric, an increasing or decreasing value does not always indicate the same. For instance, for the orthogonal segments metric an increase usually indicates that a participant changed non orthogonal segments to be orthogonal. However, in Fig. 4(d) the increase of the metric after minute 15 happened due to adding new edges, rather than an effort made by the participant. Therefore, the modeling phase diagram is crucial for interpretation. It shows if a change of the metric happened during a modeling or a reconciliation phase, restricting possible interpretations of the metric (cf. Fig. 2).

Challenge C3. Looking at only one metric at a time does not consider any trade-offs between different layout metrics the modeler has to make. For example, as already mentioned in Sect. 2.2, a modeler might not adhere to a strict Manhattan layout in order to have a low number of bendpoints (cf. broken edges metrics [5]). The implementation of the advanced modeling phase diagram is already able to depict more than one metric at a time, allowing us to detect potential trade-offs. However, analyzing trade-offs automatically is still challenging.

Challenge C4. Some metrics require information about the context in order to be able to interpret the evolution properly. For instance, layout properties relating to the shape or area of the model (cf. [5]) are typically not intentionally handled by the modeler and simply increase. However, as the model grows bigger the modeler might switch to consciously manipulate these metrics in order to fit the process model to the screen size and to avoid scrolling. Therefore, knowing the screen size available to the modeler (i.e., the context) is important for understanding the evolution of metrics relating to the shape or area of a process model.

Challenge C5. The stability of layout metrics might change during the evolution of the metrics due to the way how they are calculated. As already mentioned in Sect. 3.2, this is the case for the orthogonal segments metric, i.e., at the start of the modeling session it is very sensitive to changes, while it gets more stable at the end (cf. Fig. 4a). Therefore, the interpretation of metrics in initial phases needs to be taken with care.

Challenge C6. Based on the data which is automatically collected by CEP and without additional verbal statements (e.g., think-aloud protocols or interviews), the modeler's intentions cannot always be inferred with certainty. For example, for the crossing edge metric we could identify three possible explanations why one last remaining crossing edge was not resolved (cf. Fig. 3c). We are able to detect if the model changed to a non-planar one due to an error. However, if that was not the case, it remains unclear why the participant did not resolve the last crossing edge (e.g., if he did not care about it or if he overlooked it).

3.4   Limitations

Our interpretation of the evolution of layout metrics for process models has to be viewed in the light of several generalization limitations. First, in the sense of notational support, we focused on the basics of BPMN, i.e., a notation that is typically taught to a large group of stakeholders [21]. The evolution of layout metrics for the complete set of BPMN or different notations was not considered. Second, all participants were students. However, the participants reported an average of 2.90 years of modeling experience (cf. [20]). Third, our investigation was focused at only one process, i.e., a process about getting a mortgage.

Nonetheless we analyzed 120 implementations of this process. Fourth, in this paper we only consider two layouting quality metrics. Fifth, we also do not consider any trade-offs decisions the participants may have taken regarding the layout, influencing the chosen layout metrics. Sixth, participants had no automatic layout support, as layouting algorithms would influence the layout metrics. It should be noted that these decision was made deliberately (cf. Sect. 2.2), highlighting future research directions.

4   Related Work

The work presented in this paper is focused on the model phase diagram described in [13,20] (cf. Sect. 2.1). While the modeling phase diagram abstracts the interactions with the process model to comprehension, modeling, and reconciliation phases, [22] introduces PPMCharts (Process of Process Modeling Charts) showing all information available of the recorded operations. These charts are based on the Dotted Chart Analysis plug-in of the process mining framework ProM, providing fine grained details of the construction of the model as well as an overview of the entire modeling process. Apart from this research, [23] investigates the process of modeling focusing on graphical modeling, showing how visualization can aid the development and assessment of computational models for large data sets.

5   Summary and Outlook

While current research focuses on the product or outcome of business process modeling, this paper motivates the need to focus on the act of formalizing a process into a process model itself. In particular, we are interested in the modeler's behavior regarding the layout of a process model. By visualizing the evolution of layout metrics, the advanced modeling phase diagram introduced in this paper contributes toward an in-depth understanding of the process of process modeling, specifically how modelers layout their process models. In particular, our examples show the evolution of two different layout metrics and present how they can be analyzed. Moreover, we list different challenges that need to be considered at future work.

In short term, our follow up research will be concerned about these challenges. For example, we consider investigating how layout metrics influence each other. The implementation of the advanced modeling phase diagram already allows to show more than one metric at a time, enabling us to detect potential trade-offs. In addition, we are planning on providing an automatic interpretation of different evolutions of layout metrics. This allows us, e.g., to quickly detect if certain layout metrics are more important to our participants than others. In the long term, our interest is how we can support the modeler during the creation of a process model. In particular, we plan a recommendation system providing suggestions regarding the layout as the user is modeling. This system is expected to point out specific layout issues the user should improve. Not only should this enhance the layout of the model, but also the overall quality of the end product by influencing other quality dimensions as well. We believe that an efficient support of modelers during the creation of process models fosters understandability and quality of business process models.

Acknowledgements. This research is supported by Austrian Science Fund (FWF): P26140.

References

1. Becker, J., Rosemann, M., von Uthmann, C.: Guidelines of business process modeling. In: Aalst, W., Desel, J., Oberweis, A. (eds.) Business Process Management. LNCS, vol. 1806, pp. 30–49. Springer, Heidelberg (2000). doi:10.1007/3-540-45594-9_3
2. Rittgen, P.: Quality and perceived usefulness of process models. In: Proceedings of the SAC 2010, pp. 65–72 (2010)
3. Scheer, A.W.: ARIS—Business Process Modeling, 3rd edn. Springer, Heidelberg (2000)
4. Kock, N., Verville, J., Danesh-Pajou, A., DeLuca, D.: Communication flow orientation in business process modeling and its effect on redesign success: results from a field study. Decis. Support Syst. **46**(2), 562–575 (2009)
5. Bernstein, V., Soffer, P.: Identifying and quantifying visual layout features of business process models. In: Gaaloul, K., Schmidt, R., Nurcan, S., Guerreiro, S., Ma, Q. (eds.) CAISE 2015. LNBIP, vol. 214, pp. 200–213. Springer, Cham (2015). doi:10.1007/978-3-319-19237-6_13
6. Purchase, H.: Which aesthetic has the greatest effect on human understanding? In: DiBattista, G. (ed.) GD 1997. LNCS, vol. 1353, pp. 248–261. Springer, Heidelberg (1997). doi:10.1007/3-540-63938-1_67
7. Leopold, H., Mendling, J., Günther, O.: Learning from quality issues of BPMN models from industry. IEEE Softw. **33**(4), 26–33 (2016)
8. Moody, D.L.: The "physics" of notations: toward a scientific basis for constructing visual notations in software engineering. IEEE Trans. Softw. Eng. **35**(6), 756–779 (2009)
9. Rosa, M.L., ter Hofstede, A., Wohed, P., Reijers, H., Mendling, J., van der Aalst, W.P.: Managing process model complexity via concrete syntax modifications. IEEE Trans. Ind. Inf. **7**(2), 255–265 (2011)
10. Schrepfer, M., Wolf, J., Mendling, J., Reijers, H.A.: The impact of secondary notation on process model understanding. In: Persson, A., Stirna, J. (eds.) PoEM 2009. LNBIP, vol. 39, pp. 161–175. Springer, Heidelberg (2009). doi:10.1007/978-3-642-05352-8_13
11. Mendling, J.: Metrics for Process Models: Empirical Foundations of Verification, Error Prediction, and Guidelines for Correctness. LNBIP, vol. 6. Springer, Heidelberg (2008). doi:10.1007/978-3-540-89224-3_4
12. Pinggera, J., Soffer, P., Fahland, D., Weidlich, M., Zugal, S., Weber, B., Reijers, H.A., Mendling, J.: Styles in business process modeling: an exploration and a model. Softw. Syst. Model **14**(3), 1055–1080 (2015). doi:10.1007/s10270-013-0349-1. ISSN:1619-1374

13. Pinggera, J., Zugal, S., Weidlich, M., Fahland, D., Weber, B., Mendling, J., Reijers, H.A.: Tracing the process of process modeling with modeling phase diagrams. In: Daniel, F., Barkaoui, K., Dustdar, S. (eds.) BPM 2011. LNBIP, vol. 99, pp. 370–382. Springer, Heidelberg (2012). doi:10.1007/978-3-642-28108-2_36

14. Hoppenbrouwers, S.J.B.A., Proper, H.A.E., Weide, T.P.: A fundamental view on the process of conceptual modeling. In: Delcambre, L., Kop, C., Mayr, H.C., Mylopoulos, J., Pastor, O. (eds.) ER 2005. LNCS, vol. 3716, pp. 128–143. Springer, Heidelberg (2005). doi:10.1007/11568322_9

15. Rittgen, P.: Negotiating models. In: Krogstie, J., Opdahl, A., Sindre, G. (eds.) CAiSE 2007. LNCS, vol. 4495, pp. 561–573. Springer, Heidelberg (2007). doi:10.1007/978-3-540-72988-4_39

16. Pinggera, J., Zugal, S., Weber, B.: Investigating the process of process modeling with cheetah experimental platform. In: Proceedings of the ER-POIS 2010, pp. 13–18 (2010)

17. Burattin, A., Bernstein, V., Neurauter, M., Soffer, P., Weber, B.: Detection and quantification of flow consistency in business process models. CoRR abs/1602.02992 (2016)

18. Petre, M.: Why looking isn't always seeing: readership skills and graphical programming. Commun. ACM **38**, 33–44 (1995)

19. Gschwind, T., Pinggera, J., Zugal, S., Reijers, H., Weber, B.: A linear time layout algorithm for business process models. Technical report RZ3830, IBM Research (2012)

20. Pinggera, J.: The process of process modeling. Ph.D. thesis, University of Innsbruck, Department of Computer Science (2014)

21. Recker, J.: Opportunities and constraints: the current struggle with bpmn. Bus. Process Manage. J. **16**(1), 181–201 (2010)

22. Claes, J., Vanderfeesten, I., Pinggera, J., Reijers, H.A., Weber, B., Poels, G.: Visualizing the process of process modeling with PPMCharts. In: Rosa, M., Soffer, P. (eds.) BPM 2012. LNBIP, vol. 132, pp. 744–755. Springer, Heidelberg (2013). doi:10.1007/978-3-642-36285-9_75

23. Crapo, A.W., Waisel, L.B., Wallace, W.A., Willemain, T.R.: Visualization and the process of modeling: a cognitive-theoretic view. In: Proceedings of the KDD 2000, pp. 218–226 (2000)

Eye Tracking Meets the Process of Process Modeling: A Visual Analytic Approach

Andrea Burattin[1(✉)], Michael Kaiser[1], Manuel Neurauter[1], and Barbara Weber[1,2]

[1] University of Innsbruck, Innsbruck, Austria
`andrea.burattin@uibk.ac.at`
[2] Technical University of Denmark, Kongens Lyngby, Denmark

Abstract. Research on the process of process modeling (PPM) studies how process models are created. It typically uses the logs of the interactions with the modeling tool to assess the modeler's behavior. In this paper we suggest to introduce an additional stream of data (i.e., eye tracking) to improve the analysis of the PPM. We show that, by exploiting this additional source of information, we can refine the detection of comprehension phases (introducing activities such as "semantic validation" or "problem understanding") as well as provide more exploratory visualizations (e.g., combined modeling phase diagram, heat maps, fixations distributions) both static and dynamic (i.e., movies with the evolution of the model and eye tracking data on top).

Keywords: Process of process modeling · Eye tracking · Modeling phase diagram

1 Introduction

The adoption of business process models is gaining significant relevance due to their importance in managing business processes [1]. Therefore, it is relevant to analyze the factors leading towards high quality models [2]. Significant research has been done to better understand the factors influencing the quality of process models [2–4]. These works, however, typically focus solely on the final outcome of the act of modeling, without considering the process of creating the process model. Instead, a more recent research field is investigating the *process of process modeling* (PPM): in this case the emphasis is on the process underlying the creation of process models [5–10]. The creation of a process model, starting from a domain description, subsumes the execution of several activities. In particular, the modeler first has to construct a mental model of the process and the expected behaviors, then she has to externalize and map the mental model into a proper representation, interacting with a modeling tool [8,9]. In general, modeling is an iterative and highly flexible process comprising three phases that are repeatedly executed in various orders, i.e., *comprehension*, *modeling*, and *reconciliation* [5]. Different techniques have been suggested to both analyze and

© Springer International Publishing AG 2017
M. Dumas and M. Fantinato (Eds.): BPM 2016 Workshops, LNBIP 281, pp. 461–473, 2017.
DOI: 10.1007/978-3-319-58457-7_34

visualize the PPM [7,11,12]. While these techniques have been demonstrated to be useful as tool for data exploration and hypothesis generation [5], their focus is on model interactions only. This might limit the analyses, since the interactions are not enough to fully reflect the activities performed by the modeler. In particular, respective techniques fall short analyzing comprehension phases, i.e., phases with no interactions with the modeling tool.

Eye tracking technologies, in turn, provide a complementary view on human behavior and allow capturing eye movements, e.g., *fixations* and *saccades*. This provides information on how visual attention is distributed and evolves within a given context. The visualization of eye tracking data serves as qualitative way of analyzing the gathered data, with the aim of finding hypotheses [13]. Besides scan path visualizations (showing sequences of fixations) and attention maps (showing fixations and their durations) a variety of visualizations for eye tracking data is used in various fields of research. However, existing eye tracking technologies are primarily tailored towards reading tasks where the content of the screen is static or only changes in response to a stimulus. The creation of process models, in contrast, constitutes a dynamic setting with constantly evolving screen content depending on user interactions. Therefore, an inter-subject comparison and hypotheses testing is not straightforward and possibilities for synchronizing and quantifying the data have to be found.

In this paper we propose to integrate both data sources (i.e., model interactions reflecting the evolution of the artifact and eye movement data highlighting the areas of attention) with the goal to provide more meaningful visualizations of the respective data. The visualizations described in this paper are suitable for depicting interactive eye tracking data gathered during process model creation as well as other human computer interaction scenarios. Moreover, the provided visualizations are potentially useful for improving existing approaches for phase detection [11] including the refinement of comprehension phases into problem understanding, method finding, as well as syntactic and semantic validation [14]. Such integrated view additionally has the potential to replay both model interactions and eye movements in an integrated manner and to build analysis techniques that allow the automatic calculation of different metrics considering the different areas of interest and the artifact being modeled (e.g., average fixation durations for modeling phases) [15]. Additional benefits of the visualizations described include validity checks on the collected data, such as biases.

Section 2 of the paper describes background information; Sect. 3 reports the new data streams and corresponding analyses; Sect. 4 describes few demonstrations; and Sect. 5 provides conclusions and implications for future research.

2   Background

2.1   Visualizing the Process of Process Modeling

The process of creating a process model is a highly flexible and iterative process that involves three phases:

- *Comprehension*: when creating a process model, modelers first need to understand the problem (i.e., the requirements and the process model created so far). During these phases modelers build an internal representation (i.e., a mental model) of the process parts to be modeled within their working memory [9] independently of any modeling notation, which then needs to be mapped to the constructs provided by the modeling notation.
- *Modeling*: during modeling phases the modeler interacts with the modeling environment to externalize the internal representation of the problem stored in working memory. For instance, modelers might insert nodes like activities and gateways into the model and connect those using edges.
- *Reconciliation*: after a modeling phase is over, modelers might work on improving the understandability of the process model through changes to the labeling of modeling elements [3], but also layouting [16].

To enable the systematic investigation of the modeling process the research tool Cheetah Experimental Platform (CEP) has been developed [17][1]. It includes a simple graphical process editor that logs all interactions of the user with the process editor (e.g., creation, deletion, renaming and moving of nodes, edges, conditions, bend points, but also scrolling) together with a timestamp. The graphical user interface of CEP is shown in Fig. 1. The textual description contains the natural language representation of the process to be modeled. The modeling canvas is the area where the modeler places all the modeling elements (i.e., activities, events and edges). The toolbox contains the items the modeler can use to translate the textual description into the graphical representation.

To automatically detect PPM phases from the logged interactions existing algorithms map the model interactions recorded by CEP onto PPM phases (i.e., comprehension, modeling, and reconciliation) [5,11]. In general, create and delete interactions are classified as modeling phase; renaming and moving interactions characterize reconciliation phases; and comprehension phases are phases with no model interactions. The detected phases can then be plotted on the so called Modeling Phase Diagram (MPD). An example of this diagram is reported in Fig. 2. In particular, x axis reports the time, and y axis indicates the number of items currently shown on the modeling canvas.

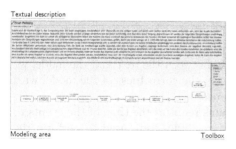
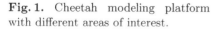
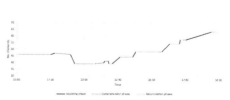

Fig. 1. Cheetah modeling platform with different areas of interest.

Fig. 2. Modeling phase diagram. In this case only 15 min of modeling are reported.

---

[1] Cheetah Experimental Platform is available at http://www.cheetahplatform.org.

2.2   Eye Tracking

Using eye tracking techniques, researchers can gain knowledge about the information processed by subjects at any moment in time [18]. Eye tracking manufacturer's software gathers the collected raw data to *fixations* and *saccades* in a first abstraction step. Fixations are time frames in which the gaze stops on a stimulus, lasting for about 200 ms up to several seconds. Therefore, fixations are representing the amount of time a specific area caught the subject's attention [18]. Saccades are fast eye movements (30 to 80 ms), with the purpose of shifting the attention to different sources of information. During saccades, no information encoding takes place [19]. To analyze specific areas of the screen, areas of interest (AOI) can be defined as an additional abstraction step. Thus, deeper insights into the attention focus of the participants is possible, such as saccades from one AOI towards another AOI (called transitions) and different AOIs can be compared regarding fixation durations and fixations counts.

This type of instruments has been fruitfully used, e.g., to analyze websites' usability [20]. By conducting eye tracking while process modeling, the actual attention focus is recorded allowing to gain knowledge about cognitive processes in place. For analyzing the data in this paper we split the screen into three AOIs (cf. Fig. 1): the textual description, the modeling canvas and the toolbox.

3   Analysis with Additional Data Streams

In this paper we propose to integrate the model interactions and eye movement data highlighting the areas of attention, in order to provide more meaningful visualizations of the respective data.

The MPD depicts the different high-level phases a modeler performed. However, as described, these phases are generated solely based on the list of interactions with the modeling tool [11]. While modeling and reconciliation phases can be easily derived from interaction logs, comprehension phases remain black boxes. To tackle this problem we add eye movement data as an additional event stream. This way, it becomes possible to dig into time fractions with no tool interactions and gain knowledge regarding user's actions. For example, if during a period with no interaction, the user is just looking at the text, this might suggest that she is reading; if the modeler, in turn, focuses her attention at the model already represented, we might conclude she is validating the syntactic correctness; if the modeler's look is jumping between the text and the model we might infer that she is semantically validating the correctness of the model with respect to the textual description of the process [14].

Note that combining the two data streams requires the time of the events to be synchronized. In our scenario, different machines were used to perform the recording of the model interactions and the recording of the eye movement data. We used the Microsoft utility `net time` to achieve synchronisation.

3.1   Data Visualizations

This section describes the visualizations combining model interactions and eye movement data.

Fixations Distribution. The first representation we would like to focus on shows the distribution of fixations on the screen (cf. Fig. 3). In this visualization, for each fixation, one black point is plotted onto the screen with the modeling environment. This visualization is useful to obtain an initial overview of the distribution of fixations. For example, Fig. 3 highlights that fixations

Fig. 3. Fixations distribution over CEP.

almost entirely cover the textual description, indicating that the user focused on the entire AOI. Fixations are also observed on the toolbox and on the central area of the modeling canvas. Some fixations on the canvas scrollbar can be detected as well.

Combining fixations with the stream of model interactions enables a "smart filtering" that can be applied to the visualization of fixations. For example, it is possible to filter only the fixations occurring during comprehension phases. This visualization can also be useful to identify (systematic) biases of the eye tracker which might be a problem for an automatic phase detection (see also Sect. 4).

Fixations Heat Map. While the first visualization focuses on the distribution of fixations, the second visualization focuses on the time spent looking at specific parts of the screen (i.e., fixations durations) using heat maps [21]. These visualizations map values to a color scheme. In our case, the color is associated to the sum of fixations durations on each point[2] of an area of interest.

The collection of model interactions and their synchronization with the fixations enable us to compute the heat map for each AOI independently and not just for the modeling AOI that is visible to the user, but for the whole modeling canvas. The idea is that the modeling AOI can be seen as a *viewport*, showing a portion of the whole model (scrollbars are employed to host large models). Therefore, by considering the "offset" given by the scrollbars positions, we map each fixation to the actual model (which might be larger that the modeling AOI). Fixations are thereby plotted onto the final process model, i.e., the model version at the end of the modeling session. Please note that if during the creation of the model objects are moved around or deleted, the final process model is not adequate. The solution to this problem is described in Sect. 3.2.

The picture on top of Fig. 4 reports the heat map for the textual description, the one below contains the heat map of the the entire process model.

---

[2] Actually, an approximation is applied by diving the AOIs with a "grid system".

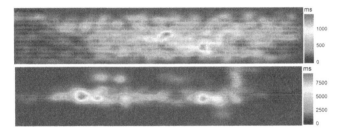

Fig. 4. Heatmap representations of the textual description and the process model.

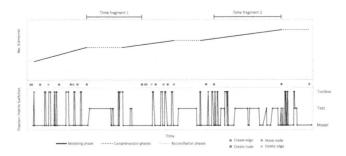

Fig. 5. Combined MPD with logged events and transitions between AOIs.

Combined Modeling Phase Diagram. Integration of data sources allows to enhance modeling phase diagrams with additional information regarding fixations. Specifically, the idea is to introduce the list of model interaction as well as the list of fixations on each AOI using the same diagram. These different elements share the time dimension and can thus be synchronized via their timestamps. An example of such chart is depicted in Fig. 5 which reports a fragment of a complete modeling session. Thereby, the x axis refers to the time dimension. On the upper part of the screen the figure shows a standard representation of a MPD with the y axis indicating the number of elements on the canvas and the line type depicting the different phase types (i.e., modeling, comprehension or reconciliation). Below the MPD, the logged events are reported. Specifically, the position of an event indicates the time of its occurrence and its color depicts the type of interaction. The bottom part of the figure, in turn, shows the AOIs the user was focusing on at a particular point in time, i.e., resulting into a depiction of the transition between AOIs over time.

The combined MPD is especially suited to obtain a better understanding of comprehension phases, i.e., phases where no tool interactions occur. The example depicted in Fig. 5 highlights two such time frames. In the first case, the user initially spent time reading the text. Then, after a quick look at the model, she concentrated on the toolbox and finally on the model again. This behavior might indicate that the user needed to understand the problem (i.e., focus on text and model) and then to find a way how to externalize it (i.e., focus on toolbox and model). In the second case, the user flipped her gaze between text and model many times, which might indicate semantic validation.

Please note that these patterns of transitions between AOIs are used for the improvement and refinement of the automatic phases detection. Such automatic phase detection might be useful in experiments, to calculate specific metrics for phases, and also as a building block of neuro-adaptive IS [14].

3.2 From Static Visualizations to Movies

The first two visualizations described do not consider the time dimension: in the first case all points are plotted, one on top of the other; in the second case, if the total time spent looking at two areas is the same, it is not distinguished whether such distribution occurred at the beginning or at the end of the modeling session. To cope with this problem, we not only provide static images but *movies* reflecting the actual distribution of fixations and the distribution of the time spent at each point. In order to generate these animations, the basic idea is to concatenate frames each of them providing a snapshot of the evolution of the modeling process (in terms of model interactions and fixations). In particular, we build time frames following two different approaches: sliding window and incremental. Following a *sliding window* approach, statistics (and, therefore, the frames) are computed just for the events observed in the latest sliding window. The *incremental* approach, instead, computes the statistics and the frames with all events from the beginning until the current point.

A fundamental characterization of these movies is the fact that we can depict not only the evolution of information obtained from the eye tracker but the evolution of the underlying model as well. Therefore, the first frames provide the fixations on top of an almost empty canvas, whereas the final frames report the fixations on top of the finished model. Figures 6 and 7 report few frames from movies created using the approach just mentioned. Frames reported in Fig. 6 have been created with sliding window approach (window length: 10s). Instead, frames in Fig. 7 have been created adopting the incremental policy.[3]

Fig. 6. Frames sampled from a video: evolution of the fixations distribution.

Finally, please note that there is no need for such representation for the last visualization (the Combined MDP): in this case the time dimension is explicitly reported on the x-axis of each chart.

---

[3] Animations are available at http://bpm.q-e.at/eye_tracking_ppm.

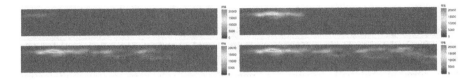

Fig. 7. Frames sampled from a video: evolution of the heat map over time.

4   Demonstration

This section demonstrates the application of the introduced visualizations in the context of an experiment performed in 2015 were model interactions and eye tracking data of 116 subjects (all novices) were assessed, amongst other data, while creating a business process model. Details on the design, research questions and hypotheses can be found in [22]. The examples shown in this section rely on visualisations and logs of six out of the 116 subjects.

By visualizing the existing MPDs, the user interactions, and the fixation data in an integrated manner we could gain additional information about the comprehension phases and refine the fixation patterns initially proposed in [14]. For this, the fixation patterns were assessed manually for six subjects by two persons using the combined modeling phase diagram followed by a cross validation. *Problem understanding* as a pattern is characterized by exclusive fixations on the textual description (see Fig. 8): subjects were reading the textual description and building a mental model of the process. Usually, after the problem understanding pattern, subjects are focusing on the toolbox, followed by the modeling canvas before they start modeling (see Fig. 8). The participants seem to select the proper tool to externalize their mental model within that time frames. We called this pattern *method finding*. Additionally, as depicted in Fig. 9, fixation behavior alternately focusing on the textual description and the modeling canvas seems to be typical for checking if the current model corresponds to the textual description. We named that pattern *semantic validation*. The second validation pattern, also in Fig. 9, is characterized by exclusive fixations on the model. This type of validation appears to be used to check if the immanent modeling language rules are obeyed, thus, we called the described pattern *syntactic validation* as in [14].

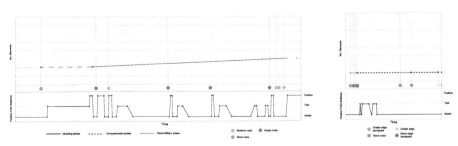

Fig. 8. Problem understanding and method finding.   **Fig. 9.** Validation.

Exploring our visualizations not only provides deeper insights into comprehension phases (that were treated as black boxes in MPDs so far), but also into modeling and reconciliation phases. For example, modeling and comprehension phases alternate more often than previously thought, as depicted in Fig. 8: in this case the combined MPD shows several phases of modeling and comprehension, which where aggregated to a long modeling phase by the original classifier. This misclassification happened due to the data used by the previous classifier which did not considered eye tracking. Phases of inactivity below a threshold value were aggregated with the time frames before and after. We found similar evidence regarding reconciliation phases (Fig. 10).

Here, the reconciliation phase identified by the original phase detection is interrupted by a phase of semantic validation, which was not correctly detected because the duration of this phase was below the threshold. These examples demonstrate the benefit of the combined MDP and give advice to improve phase detection algorithm to distinguish different patterns with higher accuracy. At the moment we are implementing a refined version of the

Fig. 10. Comprehension interrupting reconciliation.

automatic detection algorithm and plan to use the combined MPDs to assess the reliability of the classifier again.

By means of heat maps, the specific parts within the process which received most attention can be identified for every subject. Figure 11 depicts heatmap plots of the textual description for (i) the complete modeling session and additionally for the timeframes of (ii) comprehension, (iii) modeling, and (iv) reconciliation. Figure 12, in turn, is an example for the heatmap plots of the model

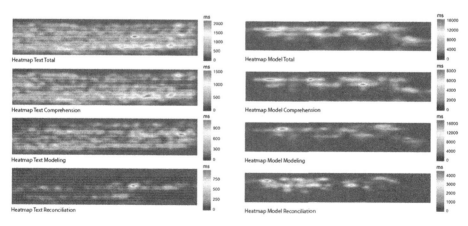

Fig. 11. Heat maps with text.   **Fig. 12.** Heat maps with model.

itself for the same timeframes as used in Fig. 11. The shown heat maps are a representative example of a single subject within the data set.[4] In line with [18], we expect that parts of a task which are causing difficulties will receive more attention and should thus be identifiable on a subjective level. The combined MPDs can not directly visualize the amount of attention given to a certain area of the screen while working on the task. Heat maps, in turn, are suitable to identify the most difficult parts by coloring areas depending on the total fixation durations and thus provide a very intuitive way of interpreting the given attention. Comparing the parts of highest attention on the text and the model for the total time spent on the task revealed that the difficult parts within the textual description do not necessarily correspond to the most difficult parts within the model. The parts of highest attention within the textual description deal with the entering point of a long back loop and a part describing an XOR-split and the corresponding conditions and alternative branches (Fig. 11). One of those alternative branches is the origin of the long back loop. Within the model, the XOR-splits are generally receiving more attention than the rest of the elements. But here the fact that conditions for the splits need to be defined has to be taken into account. Still, most difficult parts within the modeling canvas seem to be a loop at the beginning of the process; the entering point of the long back loop; and one ending point of the process (Fig. 12). Several subjects had problems modeling the loop at the beginning of the process, but since the subjects are novices we do not think this is representative for process modeling in general. In addition to the separation into textual description and modeling canvas we provide visualizations for the respective areas corresponding to comprehension, modeling, and reconciliation phases. For instance, as depicted in Fig. 11 "Heatmap Text Comprehension", the attention on the text clearly focuses on the XOR-split described above, and therefore on the origin of the long back loop. The identification of the correct entering point within the text for the respective loop seems to be identified during reconciliation phases as shown in "Heatmap Text Reconciliation". When examining the heatmap with the modeling canvas in modeling phases (Fig. 12, "Heatmap Model Modeling"), the most difficult part seems to be the small loop at the beginning of the process, pointing towards a lack of modeling knowledge once again. During phases of comprehension and reconciliation (Fig. 12, "Heatmap Model Comprehension" and "Heatmap Model Comprehension"), the origin of the long back loop received a high amount of attention, along with one ending point of the process. In sum, since the long back loop within the process occurred as difficult part within the textual description and within the model itself, we expect this part to be the most difficult overall. This fits into the statistical data: the back loop is the part with the highest error rate, with about 50% correct solutions out of 116 subjects.

To draw reliable conclusions and to use an automated PPM classifier eye tracking data of high quality is necessary. As mentioned in Sect. 3, we used fixations distribution plots to identify biases (cf. Fig. 3). In the upper corners of the screen no fixations were recorded by the eye tracker even though it is very likely

---

[4] Higher resolution versions can be found at http://bpm.q-e.at/eye_tracking_ppm.

that the subject actually fixated on those spots, since reading the whole text without skipping parts of sentences is much more natural. This deviation seems to be bend on the top corners of the screen, reminiscent of *fisheye lenses*. This bending effect seems to occur unrelated to visual aids, since the depicted participant does not wear glasses or lenses, this is also true for the opposite shaping of the bending effect we also found within the dataset. In general, eye tracking accuracy tends to be poorest in the corners of the recorded screen and is strongly dependent to particular participant's characteristics [23]. Even though the most interesting part of the screen in our case is the area around the horizontal middle line, this bending causes information loss and might be disadvantageous when using a classifier. Since the classifier is based on fixations on different AOIs, fixations falsely assigned to the modeling canvas instead of assigning it to the textual description could lead to wrong classifications. As an example, see the fixation points slightly below the textual description, within the left side of the modeling canvas in Fig. 3. Manual or automated offset compensation methods could be employed during the analysis phase, but should be avoided [19]. The described bending effect is a limitation of eye tracking technology, along with reported gaze inaccuracies up to 2 degrees which represents about 2 cm at a recording distance of 70 cm [19]. Although we do not have such general tracking inaccuracies within the bending effect free data sets, this has to be considered and shows the importance of visualization tools to assess eye tracking quality.

5   Conclusions and Future Work

This paper reports an innovative contribution on the inspection of the process of graphical artifact modeling and, in particular, to the modeling of business processes. Specifically, the literature describes analysis techniques focusing on the interactions with the modeling environment. In this paper we also analyze data referring to eye tracking. By combining and synchronizing these two streams, it is possible to provide visual tools which enable a deeper understanding and a more fine grained representation of users' actions and intentions. Demonstrations show that such analysis is indeed capable of providing useful insights, such as the identification of "problem understanding", and "validations" actions.

The importance of this work is based on the innovative possibilities it enables: the automatic analysis of eye tracking data, during process modeling sessions, is the main future work. Such automatic analysis will allow the extraction of detailed actions and the refinement of the MPD. In order to be able to automatically detect modeling phases, the conducted fixation patterns have to be validated. Therefore, a qualitative study with post hoc interviews would be a suitable approach. Additionally, we are going to examine modeling and reconciliation phases to clarify if the same or similar patterns can be found here. However, before starting such analysis, a deeper understanding of the behavior is needed, and the tools described in this paper represents optimal instruments.

Acknowledgements. This work is partially funded by the Austrian Science Fund project "The Modeling Mind: Behavior Patterns in Process Modeling" (P26609).

References

1. Becker, J., Rosemann, M., Uthmann, C.: Guidelines of business process modeling. In: Aalst, W., Desel, J., Oberweis, A. (eds.) Business Process Management. LNCS, vol. 1806, pp. 30–49. Springer, Heidelberg (2000). doi:10.1007/3-540-45594-9_3
2. Mendling, J.: Metrics for Process Models: Empirical Foundations of Verification, Error Prediction, and Guidelines for Correctness. Springer, Heidelberg (2008)
3. Mendling, J., Reijers, H.A., Recker, J.: Activity labeling in process modeling: Empirical insights and recommendations. Inf. Syst. **35**(4), 467–482 (2010)
4. Reijers, H.A., Mendling, J.: A study into the factors that influence the understandability of business process models. IEEE Trans. Syst. Man Cybern. Part A Syst. Hum. **41**(3), 449–462 (2011)
5. Pinggera, J.: The Process of Process Modeling. Ph.D. thesis, Innsbruck University (2014)
6. Pinggera, J., Soffer, P., Fahland, D., Weidlich, M., Zugal, S., Weber, B., Reijers, H., Mendling, J.: Styles in business process modeling: An exploration and a model. Softw. Syst. Model. **14**(3), 1055–1080 (2015)
7. Claes, J., Vanderfeesten, I., Pinggera, J., Reijers, H., Weber, B., Poels, G.: A visual analysis of the process of process modeling. Inf. Syst. e-Bus. Manage. **13**(1), 147–190 (2015)
8. Hoppenbrouwers, S.J.B.A., Proper, H.A.E., Weide, T.P.: A fundamental view on the process of conceptual modeling. In: Delcambre, L., Kop, C., Mayr, H.C., Mylopoulos, J., Pastor, O. (eds.) ER 2005. LNCS, vol. 3716, pp. 128–143. Springer, Heidelberg (2005). doi:10.1007/11568322_9
9. Soffer, P., Kaner, M., Wand, Y.: Towards understanding the process of process modeling: Theoretical and empirical considerations. In: Daniel, F., Barkaoui, K., Dustdar, S. (eds.) BPM 2011. LNBIP, vol. 99, pp. 357–369. Springer, Heidelberg (2012). doi:10.1007/978-3-642-28108-2_35
10. Martini, M., Pinggera, J., Neurauter, M., Sachse, P., Furtner, M.R., Weber, B.: The impact of working memory and the "process of process modelling" on model quality: Investigating experienced versus inexperienced modellers. Sci Rep. 6 (2016)
11. Pinggera, J., Zugal, S., Weidlich, M., Fahland, D., Weber, B., Mendling, J., Reijers, H.A.: Tracing the process of process modeling with modeling phase diagrams. In: Daniel, F., Barkaoui, K., Dustdar, S. (eds.) BPM 2011. LNBIP, vol. 99, pp. 370–382. Springer, Heidelberg (2012). doi:10.1007/978-3-642-28108-2_36
12. Claes, J., Vanderfeesten, I., Pinggera, J., Reijers, H.A., Weber, B., Poels, G.: Visualizing the process of process modeling with PPMCharts. In: Rosa, M., Soffer, P. (eds.) BPM 2012. LNBIP, vol. 132, pp. 744–755. Springer, Heidelberg (2013). doi:10.1007/978-3-642-36285-9_75
13. Blascheck, T., Kurzhals, K., Raschke, M., Burch, M., Weiskopf, D., Ertl, T.: State-of-the-Art of visualization for eye tracking data. In: EuroVis - STARs. The Eurographics Association (2014)
14. Weber, B., Pinggera, J., Neurauter, M., Zugal, S., Martini, M., Furtner, M., Sachse, P., Schnitzer, D.: Fixation patterns during process model creation: Initial steps toward neuro-adaptive process modeling environments. In: HICSS. IEEE (2016)
15. Weber, B., Neurauter, M., Pinggera, J., Zugal, S., Furtner, M., Martini, M., Sachse, P.: Measuring cognitive load during process model creation. In: Davis, F.D., Riedl, R., vom Brocke, J., Léger, P.-M., Randolph, A.B. (eds.) Information Systems and Neuroscience. LNISO, vol. 10, pp. 129–136. Springer, Cham (2015). doi:10.1007/978-3-319-18702-0_17

16. Schrepfer, M., Wolf, J., Mendling, J., Reijers, H.A.: The impact of secondary notation on process model understanding. In: Persson, A., Stirna, J. (eds.) PoEM 2009. LNBIP, vol. 39, pp. 161–175. Springer, Heidelberg (2009). doi:10.1007/978-3-642-05352-8_13

17. Pinggera, J., Zugal, S., Weber, B.: Investigating the process of process modeling with cheetah experimental platform. In: ER-POIS, 13–18(2010)

18. Riedl, R., Léger, P.M.: Fundamentals of NeuroIS. Springer, Heidelberg (2016)

19. Holmqvist, K., Nyström, M., Andersson, R., Dewhurst, R., Jarodzka, H., Van de Weijer, J.: Eye Tracking: A Comprehensive Guide to Methods and Measures. Oxford University Press, Oxford (2011)

20. Ehmke, C., Wilson, S.: Identifying web usability problems from eye-tracking data. In: British HCI, British Computer Society, pp. 119–128 (2007)

21. Grinstein, G., Trutschl, M., Cvek, U.: High-dimensional visualizations. In: Visual Data Mining Workshop, KDD. Citeseer (2001)

22. Neurauter, M., Pinggera, J., Martini, M., Burattin, A., Furtner, M., Sachse, P., Weber, B.: The influence of cognitive abilities and cognitive load on business process models and their creation. In: Davis, F.D., Riedl, R., vom Brocke, J., Léger, P.-M., Randolph, A.B. (eds.) Information Systems and Neuroscience. LNISO, vol. 10, pp. 107–115. Springer, Cham (2015). doi:10.1007/978-3-319-18702-0_14

23. Hornof, A.J., Halverson, T.: Cleaning up systematic error in eye-tracking data by using required fixation locations. Behav. Res. Methods Instrum. Comput. **34**(4), 592–604 (2002)

Visually Comparing Process Dynamics with Rhythm-Eye Views

Jens Gulden[(✉)]

University of Duisburg-Essen, Universitätsstr. 9, 45141 Essen, Germany
jens.gulden@uni-due.de

Abstract. To visualize information about process behavior over time, typically timeline based visualizations are used in contemporary analysis tools. When an overview over a large range of process instances and possible repetitive behavior is to be displayed, however, the timeline projection comes with several limitations. In this article, an alternative to the common timeline projection of process event data is elaborated, which allows to project series of time-related events and regularities therein onto a circular structure. Especially for comparing process rhythms in multiple sets of event data, this visualization comes with advantages over timeline projections and provides more flexibility in configuration. A conceptual elaboration of the approach together with a prototypical implementation is presented in this paper.

Keywords: Process event analysis · Event visualization · Process visualization · Rhythm visualization · Business analytics · Business intelligence

1   Questions of Process Rhythmics Analysis

In the domain of Process Rhythmics analysis, knowledge about the dynamic behavior of business processes over time is derived from process run-time execution data.

A number of analyses approaches exist which focus on deriving information about the structural relationships among events, especially Business Process Mining [1] which provides means for (re-)constructing process type models from given event instance data. Analytical means for deriving knowledge from time-related information about process execution, however, are still under-represented [2,3] among knowledge discovery approaches in Business Process Management (BPM) [4,5] and Business Intelligence (BI) [6]. In [7] it is pointed out that reflecting on the rhythms in work processes constitutes a relevant activity for process design and monitoring, especially when it comes to establishing processes which incorporate human interaction. Working together in the right rhythm psychologically causes the incorporated actors to feel as belonging to a larger whole.

Analyzing time-related properties of process execution behavior allows to complement existing analytical approaches by providing means to answer, among others, the following questions:

© Springer International Publishing AG 2017
M. Dumas and M. Fantinato (Eds.): BPM 2016 Workshops, LNBIP 281, pp. 474–485, 2017.
DOI: 10.1007/978-3-319-58457-7_35

- Does the execution of specific process types or groups of process instances follow re-identifiable patterns over time?
- Do event rhythms of different process types overlap or resonate, i.e., do events from one type occur in parallel to other types' events?
- Do some event types occur with similar or different intensity than others?

While Process Rhythm analysis addresses a wide field of analytical views on time-related properties of process event data, one specific interest lies in the analysis of the time-related behavior of process instance events, which belong to a known process type and are associated with a unique process instance identifier (also called "case id" [1]). Having such data at hand, a number of additional analytical questions can be asked toward it, such as the following:

- How wide-spread do individual event data occurrences differ from the average time of occurrence in a process instance?
- How do different subsets of process instances of the same type differ with respect to the event rhythm?
- How do the rhythms of instances of different process types differ?
- How can an analytical perspective be taken in that abstracts from individual lengths of processes and allows to compare their rhythmical shapes only?

As a suggestion for an analysis approach which is capable of answering the above listed questions, the work at hand elaborates the *rhythm-eye* view as a visual analysis tool.

The article is structured as follows: First, the dominant approach for visualizing events and processes over time, which is a time-related scatter plot, or so called *piano-roll* view, is introduced as existing related work in Sect. 2. The rhythm-eye view provides a novel extension to the analytical capabilities of the piano-roll view and is elaborated in Sect. 3, including a detailed description of its configuration options. An implementation of a rhythm-eye renderer written on top of the D3 visualization library is presented in Sect. 4. In the closing Sect. 5, the results achieved with the proposed visualization approach are summarized.

2   Scatter-Plots and Piano-Roll Views

A common visual analytical tool for the temporal occurrences of event and process instances is a plot of dots or bars over a timeline [8]. Such scatter plot approaches can be applied to diverse analytical cases, e.g., Fig. 1 shows simulated event data about passengers of an airplane flight.

Scatter plots can generally be described as 4-dimensional information spaces [9] that project points in time (or any value from an interval scale), and optionally durations (any value from a ratio scale) onto the x-axis of the diagram space. Any kind of categorical or numerical value can additionally be projected onto the y-axis, and, optionally a categorical or ordered value can be displayed by the use of color.

By Instances of Process Flight

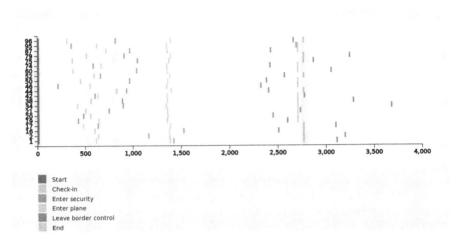

Fig. 1. Piano-roll view showing events from a simulated process type "Airplane Flight"

From this abstract perspective as a multi-dimensional information space, a piano-roll view can be described with the information feature signature as shown in Table 1. This way of expressing characteristics of information visualizations is an application of the approach described in [10].

Table 1. 4-dimensional feature signature of a piano-roll view

#	Meaning	Nominal	Ordered	Interval	Ratio	Continuous	Discrete	Projection	Optional
1	Point in time			✓		✓		x-axis position	
2	Duration			✓	✓	✓		x-axis section	✓
3	Category/ value	✓	✓	✓	✓	✓	✓	y-axis position	✓
4	Category/ value	✓	✓				✓	Color	✓

While applicable to a large class of visualization use cases in a versatile way, the piano-roll view also comes with a number of inherent disadvantages. The use of a 1-dimensional time axis can distract perception of process rhythm patterns as a whole, because the viewer of a piano-roll is forced to actively navigate along the timeline to perceive all available information. Along the linear structure, Gestalt patterns can superimpose perception, e.g., the middle area of a timeline projection can be perceived as more relevant than peripheral areas, which can possibly distract perception from gaining relevant knowledge from pattern occurrences in these areas. As a consequence, in some cases a linear timeline projection can hinder perception of process event data at a glance, and thus cannot leverage the human cognitive pattern perception apparatus in an optimal way.

To overcome this disadvantage, the following work proposes a modification to the 1-dimensional piano-roll, and rather suggests to project time-related events onto a 2-dimensional circular structure. The resulting visualization type is called *rhythm-eye*.

3   Rhythm-Eye Process Visualization

To overcome the previously discussed deficiencies of the classical piano-roll view, this section develops an alternative visual analysis instrument that can not only replace piano-roll views, but potentially offers better cognitive access to patterns of process behavior visible in event plots. It also uses space on a display device more efficiently.

3.1   Description of the Rhythm-Eye View

The rhythm-eye view uses a circular timeline to represent the running time of processes. Events are projected onto a ring, rather than onto a linear timeline, according to their relative time of occurrence during process execution. With this circular view, rhythmic patterns in processes can be made visible at a glance and better be compared across boundaries of process types.

The rhythm-eye analysis view is a "twisted" piano-roll view that forms an up to 360° circle to represent a full run of a process. It can be configured in various ways to display either events from individual or multiple process instances at a time.

By folding the original timeline to a ring, a 2-dimensional structure is created that represents the relative progress of time. According to the considerations made in Sect. 2, using such a visual configuration can address cognitive pattern matching capabilities better than a linear timeline structure. Projecting event patterns into a 2-dimensional space relieves the viewer from having to navigate explicitly along a timeline, i.e., cognitive load is reduced and perceiving the visualization can be assumed to happen more efficiently than with the use of timelines in piano-roll views.

The multi-dimensional feature signature of the rhythm-eye view is compatible to the one of the piano-roll view (Table 1). Therefore, the rhythm-eye view has the same formal expressiveness as the piano-roll view.

It should be noted that the rhythm-eye visualization, unlike the piano-roll, is not well suited for displaying absolute process lengths or to compare differences in lengths of processes. These features rather get distorted by the circular structure. For such use cases, piano-roll views are more suitable, which linearly map durations of time to horizontal positions.

Figure 2 shows an example rhythm-eye view. Events are rendered as thin lines on top of the rhythm-eye ring. Average time values of each event type are represented by semi-transparent slightly thicker circle segments, one per event type. Different event types are distinguished by colors.

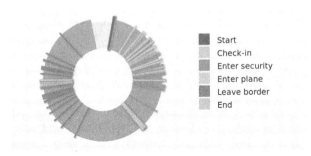

Fig. 2. Example rhythm-eye view showing events from a simulated process type "Airplane Flight" (Color figure online)

By combining multiple rhythm-eye views, rhythmical characteristics of different groups of process instances can be displayed, e.g., to compare instance characteristics of different types of processes (mixed-type analysis), or to compare subsets of process instances of the same process type (single-type analysis).

The fundamental advantage of the rhythm-eye view analysis instrument is its ability to elucidate an impression of the rhythms of processes at a single glance, using clearly visible markers running from the center of each rhythm-eye's circle to its periphery. Different coloring of each associated event type makes the event groups distinguishable at a basic level of visual cognitive processing.

The circular structure of the rhythm-eye, in contrast to a piano-roll projection, allows to abstract from the actual lengths of processes, but focus on perceiving them as solely having a start and an end, and distinguishable phases in between. This property of the analysis tool makes it suitable for comparing different types of processes as part of mixed-type analyses, in which multiple rhythm-eye views are shown together, to allow for a quick comparison of different process types or otherwise distinguished groups of process event data.

3.2 Displaying Multiple Rhythm-Eye Views

It is inherent to the conceptualization of the rhythm-eye view that its use in overall constellations with multiple rhythm-eye views is anticipated, which allows to compare rhythmic characteristics and dynamic behavior between different sets of event instances.

Therefore, two options for displaying multiple rhythm-eye views in an overall analysis scenario exist. The first one, not surprisingly, is the option to place multiple rhythm-eye views besides each other in a common setting, e.g., in a dash-board configuration. The normalized projection of time intervals provided by the rhythm-eye views allows to compare characteristics of process data as shown in Fig. 3.

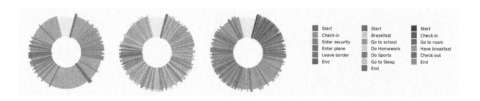

Fig. 3. Multiple rhythm-eye views placed besides each other for visual comparison

A second option for combining multiple rhythm-eye views into one overall analytical perspective is to nest rhythm-eye views inside each other. The free inner space in each rhythm-eye can efficiently be used for this. This option optimizes the use of available space, and creates visual patterns which allow to clearly distinguish where different sets of event data match with respect to their rhythm, and where not.

The nesting of multiple rhythm-eye views into each other is exemplified in Fig. 4.

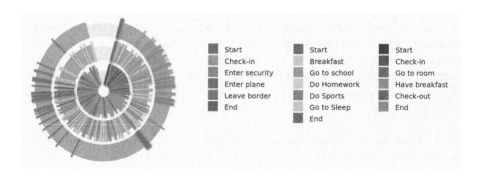

Fig. 4. Multiple rhythm-eye views nested into each other for visual comparison

3.3   Configuration Options of the Rhythm-Eye View

The rendering of the rhythm-eye view can be fine-tuned by using various parameters that shape the actual visual appearance of the display. Besides general options for the center position (parameters x and y) and control over the size of the rhythm-eye (parameter `radius`), a number of parametrization options exist which address possible variations in the use of the rhythm-eye. The elements of the visualization which are subject to this configuration are schematically sketched in Fig. 5.

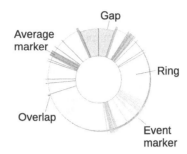

Fig. 5. Schematic sketch of a rhythm-eye showing configurable elements

Four configuration options control the possible variations in the appearance of a rhythm-eye rendering. They are introduced in the following, together with the names of the corresponding input parameters of the implemented `renderRhythmEye(...)` function (Sect. 4).

Start-end gap . (parameter `gap`) The top gap in the display ring fulfills the function of visually separating the start area of the ring (which begins to the right of the gap) and its end area (to the left of the gap).

The size of the gap between the start of the display ring and its end, which is visible as a gap in the top section of the display ring, is contingent and thus subject to configuration, independent from the input data provided to a rhythm-eye view renderer. It makes sense to consider the gap as symmetric around the middle position on the top of the display ring, so there is no need to separately configure the size of the gap to the left and to the right. This means, a single configuration option `gap`, which represents a radian fraction that has half the size of the gap, is applied equally to the left and to the right of the vertical middle line.

Width of the display ring . (parameter `ringwidth`) The width of the display ring determines the density of the appearance of a rhythm-eye view. The sum of the radius of the empty inner zone of a rhythm-eye, the ring's width, and the overlap with which event markers and average markers exceed the ring, equals the entire radius of the rhythm-eye view.

Overlap exceeding display ring . (parameter `overlap`) The small overlap, with which event and average markers exceed the base ring of the rhythm-eye to make it reach the size specified as overall radius, can be controlled via this option. It is recommended to use a comparably small value in relation to the overall radius.

3.4   Displaying Event Indicators and Mean Value Indicators

The configuration also allows to optionally choose whether to include the rendering of individual event markers, average mean value markers, or both into the

rhythm-eye display (parameters `showAverages` and `showEvents`, both default-
ing to `true`). A rendering without individual event markers, but showing the
mean occurrence times of events, can in some cases be of more value than a view
including individual event markers, and can provide a more distinctive view on
different process characteristics at a glance.

A set of rhythm-eye views which do not include the display of individual
event markers is shown in Fig. 6.

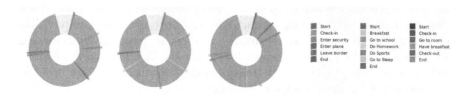

Fig. 6. Multiple rhythm-eye views showing average mean values of event types only

4 Prototype Implementation with the D3 Graphics Library

The examples shown in Figs. 2, 3, 4 and 6 have been rendered using a prototype
implementation of a rhythm-eye renderer component implemented with the D3
visualization framework [11].

The D3 graphics library has matured to one of the standard data visualization
libraries today. It provides abstractions for binding data to the structure and
content of arbitrary XML documents, and is typically used, like also in the
prototype, for creating Scalabale Vector Graphics (SVG) in a client-side web-
browser page programmed with JavaScript [12].

The rhythm-eye implementation consists of two externally callable functions.
The main one is the function `renderRhythmEye(...)` which renders one rhythm-
eye representation with diverse parameters onto an SVG 2D graphics component.

The second one is the auxiliary function `renderLegend(...)` which serves
to render explanatory legends about the color coding of event types besides the
rhythm-eye views. The implementation of this is based on the `d3-legend.js`
extension to D3, which needs to be included when using the rhythm-eye implementation.

`collectAll(...)` is a helper function which builds an array of all uniquely
occurring values in a set of data entries. It is used to get a list of all available
event types in the data. The helper functions `xCircle(...)` and `yCircle(...)`
serve to calculate radial coordinates.

Data is passed to `renderRhythmEye(...)` in form of a JavaScript Object
Notation (JSON) [13] array. This is the standard data format used by the
D3 library. Every data object is expected to have the slots `relEventTime` and
`eventType`. The first one contains the occurrence time of an event relative to

the start of the process it belongs to. The second one indicates the type of an event with a unique string identifier. All JSON arrays that contain objects with these two slots can be processed by the rhythm-eye renderer.

The implementation is provided as source code in Listing 1. It can also be downloaded from www.wi-inf.uni-due.de/FGFrank/download/Rhythm-Eye/, together with a usage example of how the renderer functionality is embedded into an HTML web page.

It should be noted that there is some potential for improvement in the implementation. While for the rendering of individual event markers the abstract D3 binding mechanism is used which allows to declaratively associate data to visible SVG elements [11] (lines 45–59), the average markers are imperatively generated for each event type in a for-loop (lines 64–81). The binding mechanism provided by D3 could be used for the latter task in future versions as well, in order to fully make use of the framework's abstractions.

Listing 1. Rhythm-Eye prototype implementation

```
1   /**
2    * Renders a Rhythm-Eye view for one set of event data.
3    * Written by Jens Gulden, jens.gulden@uni-due.de.
4    *
5    * @param data JSON encoded data to render
6    * @param startTime start of the time-range displayed by the Rhythm-Eye
7    * @param endTime end of the time-range displayed by the Rhythm-Eye
8    * @param svg SVG element onto which to render
9    * @param x horizontal coordinate where to render
10   * @param y vertical coordinate where to render
11   * @param radius outer radius of the entire rhythm eye
12   * @param ringwidth width of the display ring
13   * @param overlap length with which event lines exceed the ring towards the
14   *        outer direction (to reach the total radius)
15   * @param gap half size of the top gap in radians
16   * @param colorScale D3 color scale object to associate event types with
17   * @param showEvents switch whether to display event markers (optional, true
18   *        by default)
19   * @param showAverages switch whether to display average markers (optional,
20   *        true by default)
21   */
22  function renderRhythmEye(data, startTime, endTime, svg, x, y, radius,
23              ringwidth, overlap, gap, colorScale, showEvents = true,
24              showAverages = true) {
25     var allEventTypes = collectAll(data, function(d){return d.eventType});
26     var totalLength = endTime - startTime;
27     var meanSize = Math.PI / 128;
28     var g = svg.append("g").attr("transform", "translate("+x+", "+y+")");
29
30     var arc = d3.svg.arc()
31       .innerRadius(radius-overlap-ringwidth)
32       .outerRadius(radius-overlap * 2)
33       .startAngle( - gap )
34       .endAngle( + gap );
35     g.append("path").attr("d", arc).attr("class", "arcBg");
36
37     arc = d3.svg.arc()
38       .innerRadius(radius-overlap-ringwidth)
39       .outerRadius(radius-overlap)
40       .startAngle( gap )
41       .endAngle( Math.PI * 2 - gap );
42     g.append("path").attr("d", arc).attr("class", "arc");
43
```

```
44    if (showEvents) {
45      var sel = g.selectAll("line").data(data);
46      sel.enter().append("line");
47      sel.exit().remove("line");
48      sel
49      .attr("x1", function(d){return xCircle(radius-overlap-ringwidth,
50          totalLength, d.relEventTime, gap)})
51      .attr("y1", function(d){return yCircle(radius-overlap-ringwidth,
52          totalLength, d.relEventTime, gap)})
53      .attr("x2", function(d){return xCircle(radius, totalLength,
54          d.relEventTime, gap)})
55      .attr("y2", function(d){return yCircle(radius, totalLength,
56          d.relEventTime, gap)})
57      .attr("stroke", function(d){return colorScale(d.eventType)})
58      .attr("class", "line")
59      .append("title").text(function(d){return d.eventType+"␣"+d.eventTime});
60    }
61
62    if (showAverages) {
63      // TODO implement using entry/exit sets
64      for (var i=0; i < allEventTypes.length; i++) {
65        var type = allEventTypes[i];
66        var filteredData = data.filter(
67            function(d){ return d.eventType == type } );
68        var meanRelEventTime = d3.mean(filteredData,
69            function(d){ return d.relEventTime });
70        var a = d3.svg.arc()
71        .innerRadius(radius-overlap-ringwidth)
72        .outerRadius(radius)
73        .startAngle( ((meanRelEventTime / totalLength) *
74            (Math.PI * 2 - gap * 2) + gap ) - meanSize)
75        .endAngle( ((meanRelEventTime / totalLength) *
76            (Math.PI * 2 - gap * 2) + gap ) + meanSize);
77        g.append("path")
78        .attr("d", a)
79        .attr("fill", colorScale(type))
80        .attr("class", "avg");
81      }
82    }
83  }
84
85  /**
86   * Renders a legend which associates colors from a qualitative scale to labels.
87   * This wraps around the d3-legend.js implementation.
88   *
89   * @param colorScale D3 color scale object previously used for rendering a /
90   *        multiple Rhythm-Eye(s)
91   * @param svg SVG element onto which to render
92   * @param x horizontal coordinate where to render
93   * @param y vertical coordinate where to render
94   */
95  function renderLegend(title, colorScale, svg, x, y) {
96    var legend = d3.legend.color().scale(colorScale);
97    svg.append("text")
98    .attr("transform", "translate("+x+","+y+")")
99    .attr("class", "legend")
100   .text(title);
101   svg.append("g")
102   .attr("transform", "translate("+x+","+(y+20)+")")
103   .attr("class", "legend").call(legend);
104 }
105
106 // private helper functions
107
108 function collectAll(data, valueFunction) {
109   var all= [];
110   data.forEach(function(d)
111     {var val=valueFunction(d); if(all.indexOf(val)==-1){all.push(val)}});
```

```
112 │   return all;
113 │ }
114 │
115 │ function xCircle(radius, total, pos, gap) {
116 │   return Math.sin( (pos/total) * (Math.PI*2 - gap*2) + gap ) * radius;
117 │ }
118 │
119 │ function yCircle(radius, total, pos, gap) {
120 │   return - Math.cos( (pos/total) * (Math.PI*2 - gap*2) + gap ) * radius;
121 │ }
```

5   Conclusion and Outlook

This paper has presented the idea of an alternative visualization type to the piano-roll scatter plot approach for displaying time-related characteristics of process event data.

The formalism of a multi-dimensional feature signature comparison shows that the rhythm-eye view is compatible to the piano-roll view with regard to the information dimensions it can display. In addition, it has been argued that folding the 1-dimensional structure of the piano-roll into a circular projection allows to better leverage the capabilities of the human perceptual apparatus to perceive patterns in a 2-dimensional space. At the same time, this projection reduces the need for a human to actively navigate along the 1-dimensional information space, i.e., cognitive workload can be expected to be lower when visually analyzing rhythm patterns of process behavior.

With these theoretical considerations that are mainly based on informed arguments about the cognitive nature of the human visual perception apparatus, an approach has been suggested which requires to be further evaluated by an empirical user study. Performing such a study as part of an overarching research project is a next step to continue with the presented work.

References

1. van der Aalst, W.M.P.: Process Mining: Discovery, Conformance and Enhancement of Business Processes. Springer, Heidelberg (2011)
2. van der Aalst, W.M.P., de Leoni, M., ter Hofstede, A.H.: Process mining and visual analytics: breathing life into business process models. In: Floares, A. (ed.) Computational Intelligence. Nova Science Publishers, New York (2011)
3. Mendling, J., Recker, J.: The state of the art of business process management research. Bus. Inf. Syst. Eng. (BISE) **58**(1), 55–72 (2016)
4. van der Aalst, W.M.P., Desel, J., Oberweis, A. (eds.): Business Process Management: Models, Techniques, and Empirical Studies. LNCS, vol. 1806. Springer, Heidelberg (2000). doi:10.1007/3-540-45594-9
5. Weske, M.: Business Process Management: Concepts, Languages, Architectures. Springer, Heidelberg (2007)
6. Grossmann, W., Ma, S.R.: Fundamentals of Business Intelligence. Springer, Heidelberg (2015)
7. Herrmann, T.: Kreatives Prozessdesign. Springer, Heidelberg (2012)

8. Song, M., van der Aalst, W.M.P.: Supporting process mining by showing events at a glance. In: Chari, K., Kumar, A. (eds.) Seventeenth Annual Workshop on Information Technologies and Systems (WITS 2007), Montreal, Canada, 8–9 December 2007, Social Science Research Network, pp. 139–145 (2007)

9. Gärdenfors, P.: Conceptual Spaces. MIT Press, Cambridge (2000)

10. Gulden, J.: A description framework for data visualizations in enterprise information systems. In: Proceedings of the 19th IEEE International Enterprise Distributed Object Computing Conference (EDOC 2015), Adelaide, Australia, 22–25 September 2015. IEEE Xplore (2015)

11. Bostock, M., Ogievetsky, V., Heer, J.: D3: data-driven documents. IEEE Trans. Vis. Comput. Graph. **17**(12), 2301–2309 (2011)

12. Flanagan, D.: JavaScript: The Definitive Guide, 6th edn. O'Reilly Media, Sebastopol (2011)

13. Bassett, L.: Introduction to JavaScript Object Notation. O'Reilly Media, Sebastopol (2015)

Author Index

Printed in the United States
By Bookmasters